OPTICAL ELECTRONICS

AJOY GHATAK and K. THYAGARAJAN

Physics Department, Indian Institute of Technology, New Delhi

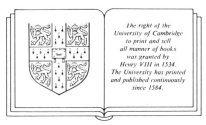

The right of the University of Cambridge to print and sell all manner of books was granted by Henry VIII in 1534. The University has printed and published continuously since 1584.

CAMBRIDGE UNIVERSITY PRESS

Cambridge

New York   New Rochelle   Melbourne   Sydney

Published by the Press Syndicate of the University of Cambridge
The Pitt Building, Trumpington Street, Cambridge CB2 1RP
32 East 57th Street, New York, NY 10022, USA
10 Stamford Road, Oakleigh Melbourne 3166, Australia

First published 1989

Printed in Great Britain at the University Press, Cambridge

British Library cataloguing in publication data

Ghatak, A.K. (Ajoy Kumar), 1939–
   Optical electronics.
   1. Electro-optics
   I. Title II. Thyagarajan, K.
   535

Library of Congress cataloguing in publication data

Ghatak, A.K. (Ajoy K.), 1939–
   Optical electronics/A.K. Ghatak and K. Thyagarajan.
      p. cm.
   Bibliography: p.
   Includes index.
   ISBN 0–521–30643–4.   ISBN 0–521–31408–9 (pbk.)
   1.Optoelectronics. I. Thyagarajan, K. II. Title.
TA1750.G48   1989
621.38'0414–dc19   88–10291 CIP

ISBN 0 521 30643 4 hard covers

ISBN 0 521 31408 9 paperback

TM

Contents

Preface

Ever since the invention of the laser in 1960, there has been a renaissance in the field of optics and the field of optical electronics encompassing generation, modulation, transmission etc. of optical radiation has gained tremendous importance. With optics and optical electronics now finding applications in almost all branches of science and engineering, study of these subjects is becoming extremely important. The present book intended for senior undergraduate and first year graduate students is an attempt at a coherent presentation of the basic physical principles involved in the understanding of some of the important optoelectronic effects and devices.

The book starts with the basic formulation of the study of propagation of electromagnetic waves, reflection and refraction and propagation through anisotropic media. This is followed by diffraction and its application in the study of spatial frequency filtering and holography. Basic physics behind laser operation is treated next with a brief discussion on different laser types. The next four chapters deal with the subject of optical waveguides including fibre and integrated optics which are already revolutionizing the field of information transmission. The next five chapters deal with three very important effects which are used in many opto-electronic devices namely the electrooptic, acoustooptic and nonlinear optical effects.

The various concepts in the book have been derived from first principles and hence it can also be used for self study. A large number of solved and unsolved problems have been scattered throughout the book. This should particularly help the reader to a better appreciation of the concepts developed and also to get a feel for the numbers involved. Some of the problems are intended to extend the range of understanding beyond what is derived in the book.

The writing of the book started in 1979 and during the past eight years portions of the book have been used in various courses, workshops and summer schools. The feedback received from the students and participants of these courses has been of immense value and has helped us in putting the book in its present form. During the preparation of the book, we have had numerous discussions with our colleagues, in particular with Professor M.S. Sodha, Professor I.C. Goyal, Dr B.P. Pal, Dr Arun Kumar, Dr Anurag Sharma, Dr Enakshi Sharma, Dr G. Umesh and Dr M.R. Shenoy; to them we are greatly indebted for many invaluable suggestions. Portions of the book have also been used by Professor M.S. Sodha, Professor I.C. Goyal, Dr B.P. Pal, Dr Arun Kumar and Dr G. Umesh at IIT Delhi and Dr Enakshi Sharma at the University of Delhi. We are very grateful to them for their constructive criticisms. One of us (AG) used a part of this book in presenting a course of lectures at University of Karlsruhe, West Germany. The many stimulating discussions with Professor G. Grau, Dr W. Freude and Dr E.G. Sauter are gratefully acknowledged. We would also like to thank Ms Swagatha Banerjee, Mr U.K. Das, Ms Supriya Diggavi, Ms Vrinda Kalia, Ms Jacintha Kompella, Mr Verghese Paulose, Mr Saeed Pilevar, Mr Vishnu Priye, Mr M.R. Ramadas, Mr R.K. Sinha and Dr R.K. Varshney for their help during the preparation of the manuscript. We are grateful to Dr R.W. Terhune, Dr H. Kogelnik, Dr R.A. Phillips, Dr W. Freude and Dr M. Papuchon for providing some of the photographs appearing in the book. We are also grateful to Professor N.M. Swani, Director, IIT Delhi and Professor M.S. Sodha, Head of our department for their encouragement and support of this work.

New Delhi
14 September 1987

Ajoy Ghatak
K. Thyagarajan

1

Maxwell's equations and propagation of electromagnetic waves

1.1 Introduction

In this chapter we will use Maxwell's equations to derive the wave equation and study its solutions in a homogeneous, isotropic and linear medium; the medium could be either absorbing or nonabsorbing. The results derived in this chapter will be used almost throughout the book – in particular, the solutions will be the starting point in the next chapter in which we will study the reflection and refraction of electromagnetic waves by a dielectric and a metal surface.

1.2 Maxwell's equations

All electromagnetic phenomena can be said to follow from Maxwell's equations. These equations are based on experimental laws and are given by

$$\mathbf{V} \cdot \mathscr{D} = \rho \tag{1.1}$$

$$\mathbf{V} \cdot \mathscr{B} = 0 \tag{1.2}$$

$$\mathbf{V} \times \mathscr{E} = -\partial \mathscr{B}/\partial t \tag{1.3}$$

$$\mathbf{V} \times \mathscr{H} = \mathbf{J} + \partial \mathscr{D}/\partial t \tag{1.4}$$

where ρ represents the charge density and $\mathbf{J}$ the current density; $\mathscr{E}, \mathscr{D}, \mathscr{B}$ and $\mathscr{H}$ represent the electric field, electric displacement, magnetic induction and magnetic field respectively. We will consistently be using the MKS system of units.

We will discuss the solution of above equations in a linear, isotropic and homogeneous medium where the following *constitutive relations* are satisfied

$$\mathscr{D} = \epsilon \mathscr{E} \tag{1.5}$$

$$\mathscr{B} = \mu \mathscr{H} \tag{1.6}$$

and

$$\mathbf{J} = \sigma \mathscr{E} \tag{1.7}$$

The parameters ϵ, μ and σ are known as the dielectric permittivity, magnetic permeability and conductivity of the medium and since the medium has been assumed to be linear and homogeneous, these parameters have a constant value.[†] Using the constitutive relations and assuming the medium to be charge free (i.e., $\rho = 0$), Eqs. (1.1)–(1.4) become

$$\mathbf{\nabla} \cdot \mathscr{E} = 0 \tag{1.8}$$

$$\mathbf{\nabla} \cdot \mathscr{H} = 0 \tag{1.9}$$

$$\mathbf{\nabla} \times \mathscr{E} = -\mu(\partial \mathscr{H}/\partial t) \tag{1.10}$$

$$\mathbf{\nabla} \times \mathscr{H} = \sigma \mathscr{E} + \epsilon(\partial \mathscr{E}/\partial t) \tag{1.11}$$

Taking the curl of Eq. (1.10) and using Eq. (1.11) we obtain

$$\mathbf{\nabla} \times (\mathbf{\nabla} \times \mathscr{E}) = -\mu \frac{\partial}{\partial t}(\mathbf{\nabla} \times \mathscr{H})$$

$$= -\mu\sigma \frac{\partial \mathscr{E}}{\partial t} - \mu\epsilon \frac{\partial^2 \mathscr{E}}{\partial t^2} \tag{1.12}$$

But[‡]

$$\nabla^2 \mathscr{E} \equiv \mathbf{\nabla}(\mathbf{\nabla} \cdot \mathscr{E}) - \mathbf{\nabla} \times (\mathbf{\nabla} \times \mathscr{E}) \tag{1.13}$$

[†] For an anisotropic medium, the parameters form a tensor so that, for example, $\mathscr{D}$ and $\mathscr{E}$ are not in the same direction; see Chapter 3.

[‡] We should mention here that (contrary to what is written in many books) Eq. (1.13) is *not* a vector identity. Eq. (1.13) *defines* the operator ∇^2 acting on a *vector*. However, simple vector manipulations show that if we take a Cartesian component of Eq. (1.13), we would obtain

$$(\nabla^2 \mathscr{E})_x = \frac{\partial^2 \mathscr{E}_x}{\partial x^2} + \frac{\partial^2 \mathscr{E}_x}{\partial y^2} + \frac{\partial^2 \mathscr{E}_x}{\partial z^2}$$

$$= \nabla^2 \mathscr{E}_x$$

where the ∇^2 operator on the RHS is now divgrad. Thus

$$(\nabla^2 \mathscr{E})_x = \mathbf{\nabla} \cdot (\mathbf{\nabla} \mathscr{E}_x)$$

On the other hand, if we take a non-Cartesian component, the above equation is no longer valid. For example working in the cylindrical coordinates, it can easily be shown that

$$(\nabla^2 \mathscr{E})_\phi = \frac{\partial^2 \mathscr{E}_\phi}{\partial r^2} + \frac{1}{r}\frac{\partial \mathscr{E}_\phi}{\partial r} + \frac{1}{r^2}\frac{\partial^2 \mathscr{E}_\phi}{\partial \phi^2} + \frac{\partial^2 \mathscr{E}_\phi}{\partial z^2} + \frac{2}{r^2}\frac{\partial \mathscr{E}_r}{\partial \phi} - \frac{1}{r^2}\mathscr{E}_\phi$$

$$= \text{divgrad } \mathscr{E}_\phi + \left[\frac{2}{r^2}\frac{\partial \mathscr{E}_r}{\partial \phi} - \frac{1}{r^2}\mathscr{E}_\phi\right]$$

which contains two extra terms in addition to divgrad $\mathscr{E}_\phi$.

Since $\nabla \cdot \mathscr{E} = 0$ (see Eq. (1.8)), Eq. (1.12) becomes

$$\nabla^2 \mathscr{E} = \mu\sigma \frac{\partial \mathscr{E}}{\partial t} + \mu\epsilon \frac{\partial^2 \mathscr{E}}{\partial t^2} \tag{1.14}$$

Similarly, taking the curl of Eq. (1.11) and using Eqs. (1.9) and (1.10) we get

$$\nabla^2 \mathscr{H} = \mu\sigma \frac{\partial \mathscr{H}}{\partial t} + \mu\epsilon \frac{\partial^2 \mathscr{H}}{\partial t^2} \tag{1.15}$$

1.3   Plane waves in a dielectric

We consider a perfect dielectric for which $\sigma = 0$. Thus Eqs. (1.14) and (1.15) simplify to

$$\nabla^2 \mathscr{E} = \mu\epsilon (\partial^2 \mathscr{E}/\partial t^2) \tag{1.16}$$

$$\nabla^2 \mathscr{H} = \mu\epsilon (\partial^2 \mathscr{H}/\partial t^2) \tag{1.17}$$

which are known as vector wave equations. If we consider a Cartesian component of either of the two equations, we would obtain the *scalar wave equation*

$$\nabla^2 \Psi = \epsilon\mu (\partial^2 \Psi/\partial t^2) \tag{1.18}$$

where Ψ may represent $\mathscr{E}_x$, $\mathscr{E}_y$ or $\mathscr{E}_z$, or $\mathscr{H}_x$, $\mathscr{H}_y$ or $\mathscr{H}_z$. The solution of the above equation represents waves (see Appendix A) and therefore, Maxwell's equations (which were used to derive the wave equation) *predict* the existence of electromagnetic waves. The speed of these waves is given by (see Appendix A)

$$v = 1/(\epsilon\mu)^{\frac{1}{2}} \tag{1.19}$$

For free space

$$\epsilon = \epsilon_0 \approx 8.854 \times 10^{-12} \text{ C}^2/\text{N m}^2$$
$$\mu = \mu_0 = 4\pi \times 10^{-7} \text{ N/A}^2 \tag{1.20}$$

The speed of electromagnetic waves in free space is denoted by the symbol c and is given by

$$c = 1/(\epsilon_0 \mu_0)^{\frac{1}{2}}$$
$$\approx 2.99794 \times 10^8 \text{ m/s} \tag{1.21}$$

It may be worthwhile mentioning that except for the term corresponding to the displacement current ($= \partial \mathscr{D}/\partial t$) in Eq. (1.4), all the experimental laws which are described in Eqs. (1.1)–(1.4) were known before Maxwell. By introducing the concept of displacement current, Maxwell (around 1860) could derive the wave equation (Eqs. (1.16) and (1.17)) and *predict* the existence of electromagnetic waves.[†] Further, using the value of ϵ_0 available to him, Maxwell found that the velocity of these electromagnetic waves should be about 3.1074×10^8 m/s. During the time of Maxwell the best known value of the speed of light was 3.14858×10^8 m/s (measured by Fizeau in 1849) and with 'faith in rationality of nature', Maxwell said that these two numbers cannot be accidentally equal and therefore light must be an electromagnetic wave. In the words of Maxwell himself, the speed of electromagnetic waves

> ···calculated from the electromagnetic measurements of Kohlrausch and Weber, agrees so exactly with the velocity of light calculated from the optical experiments of M. Fizeau, that we can scarcely avoid the inference that light consists *in the transverse undulations of the same medium which is the cause of electric and magnetic phenomena.*

Now, in a dielectric, the velocity of propagation of the electromagnetic wave can be written in the form (see Eqs. (1.19) and (1.20))

$$v = c/n \tag{1.22}$$

where n, known as the refractive index of the medium, is given by

$$n = \left(\frac{\epsilon}{\epsilon_0} \frac{\mu}{\mu_0} \right)^{\frac{1}{2}} \tag{1.23}$$

For most dielectrics μ is very close to μ_0 and we have

$$n = K^{\frac{1}{2}} \tag{1.24}$$

where

$$K = \epsilon/\epsilon_0 \tag{1.25}$$

is known as the dielectric constant of the medium.

In Appendix A we have shown that the solution of the wave equation (Eq. (1.18)) can be written in the form

$$\Psi = A \, e^{i(\omega t - \mathbf{k} \cdot \mathbf{r})} \tag{1.26}$$

where A is a constant and k_x, k_y and k_z (which represent the components of the vector $\mathbf{k}$) and ω can take arbitrary values subject to the condition that

$$k^2 = k_x^2 + k_y^2 + k_z^2 = \omega^2/v^2 = \omega^2 \epsilon \mu \tag{1.27}$$

[†] It was only in 1888 that Hertz carried out experiments which could produce and detect electromagnetic waves of frequencies much smaller than that of light.

or

$$\omega = k/(\epsilon\mu)^{\frac{1}{2}}. \tag{1.28}$$

Eq. (1.26) represents a plane wave propagating in the direction of $\mathbf{k}$ and the phase fronts are normal to $\mathbf{k}$ (see Appendix A). It should be mentioned that for a given frequency ω, the value of k is fixed (see Eq. (1.28)); however, we can have waves propagating in different directions depending on the relative values of k_x, k_y and k_z. For example, for a wave propagating along the x-direction,

$$k = k\hat{\mathbf{x}}$$

and the phase fronts are parallel to the y-z plane. On the other hand, for,

$$k = \frac{\sqrt{3}}{2} k\hat{\mathbf{x}} + \tfrac{1}{2}k\hat{\mathbf{y}}$$

we have a plane wave which is propagating in the x-y plane making 30°, 60° and 90° with the x, y and z-axes respectively (see Fig. 1.1). For all points on a plane perpendicular to $\mathbf{k}$, the quantity $\mathbf{k} \cdot \mathbf{r}$, and therefore the phase, is a constant.

Returning to Eqs. (1.16) and (1.17) we see that since each Cartesian component of $\mathcal{E}$ and $\mathcal{H}$ satisfies the scalar wave equation, the plane wave

Fig. 1.1 A propagating plane wave with its propagation vector $\mathbf{k}$ making angles 30°, 60° and 90° with the x, y and z-axes respectively. For all points on a plane perpendicular to $\mathbf{k}$, the quantity $\mathbf{k} \cdot \mathbf{r}(=|\mathbf{k}| \cdot OL)$, and hence the phase, is a constant.

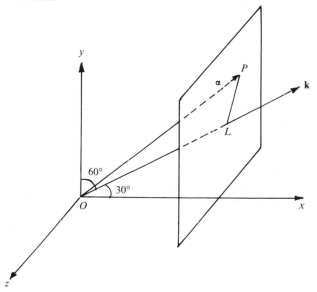

solutions of Eqs. (1.16) and (1.17) can be written in the form

$$\mathscr{E} = \mathbf{E}e^{i(\omega t - \mathbf{k}\cdot\mathbf{r})} \tag{1.29}$$

$$\mathscr{H} = \mathbf{H}e^{i(\omega t - \mathbf{k}\cdot\mathbf{r})} \tag{1.30}$$

where the vectors **E** and **H** are assumed to be independent of space and time. The various components of $\mathscr{E}$ and $\mathscr{H}$ are related to each other through Maxwell's equations. Let us consider the x-component of Eq. (1.10):

$$\partial\mathscr{E}_z/\partial y - \partial\mathscr{E}_y/\partial z = -\mu(\partial\mathscr{H}_x/\partial t) \tag{1.31}$$

If we now substitute the various components of Eqs. (1.29) and (1.30) in Eq. (1.31) we obtain

$$-ik_y\mathscr{E}_z + ik_z\mathscr{E}_y = -i\omega\mu\mathscr{H}_x$$

or

$$(\mathbf{k} \times \mathscr{E})_x = \omega\mu(\mathscr{H})_x$$

and similar equations for the y and z-components. Thus

$$\mathscr{H} = (1/\omega\mu)\mathbf{k} \times \mathscr{E} \tag{1.32}$$

Similarly, by substituting the various components of Eqs. (1.29) and (1.30) in the various components of Eq. (1.11) (with $\sigma = 0$) we would obtain

$$\mathscr{E} = -\mathbf{k} \times \mathscr{H}/\omega\epsilon \tag{1.33}$$

From Eqs. (1.32) and (1.33) it is obvious that the vectors $\mathscr{E}$ and $\mathscr{H}$ and $\mathbf{k}$ form a rectangular triad of vectors. $\mathscr{E}$ and $\mathscr{H}$ are at right angles to each other and also to $\mathbf{k}$; thus the fields associated with a plane electromagnetic wave are transverse to the direction of propagation; the transverse character of waves could have been directly obtained by substituting Eqs. (1.29) and (1.30) in Eqs. (1.8) and (1.9) respectively to obtain

$$\mathbf{k}\cdot\mathscr{E} = 0 \tag{1.34}$$

and

$$\mathbf{k}\cdot\mathscr{H} = 0 \tag{1.35}$$

However, the above two equations do not show that $\mathscr{E}$ and $\mathscr{H}$ are at right angles to each other. It may be noted that if we substitute for $\mathscr{H}$ from Eq. (1.32) in Eq. (1.33) we obtain

$$k^2 = \omega^2\epsilon\mu$$

which is consistent with Eq. (1.28).

Without any loss of generality we may assume the propagation to be along the z-axis, i.e.,

$$\mathbf{k} = \hat{z}k \tag{1.36}$$

As a special case we assume the electric vector to be along the x-axis; thus the actual electric field variation is assumed to be of the form

$$\mathscr{E} = \hat{x}E_0 \cos(\omega t - kz) \tag{1.37}$$

Using Eq. (1.32) we readily obtain the corresponding magnetic field variation

$$\mathscr{H} = \hat{y}H_0 \cos(\omega t - kz) \tag{1.38}$$

with the amplitudes of the electric and magnetic fields (E_0 and H_0) related through the following equation

$$H_0 = (k/\omega\mu)E_0 = (\epsilon/\mu)^{\frac{1}{2}} E_0 \tag{1.39}$$

Notice that the $\mathscr{E}$ and $\mathscr{H}$ fields are in phase, i.e., at a particular value of time and on a plane $z = z_0$ (z_0 being arbitrary), if the $\mathscr{E}$ field has attained its maximum value then the $\mathscr{H}$ field will also be at its maximum value etc. The wave described by Eqs. (1.37) and (1.38) is said to be linearly polarized because the $\mathscr{E}$ (or $\mathscr{H}$) fields are always along a particular direction. It is also referred to as a plane polarized wave because the electric vector always lies in the x–z plane (and the magnetic vector in y–z plane – see Fig. 1.2). The most common terminology (to describe the state of polarization corresponding to Eqs. (1.37) and (1.38)) is to refer it as an x-polarized wave.

We next consider a y-polarized wave (with an additional phase of $\frac{1}{2}\pi$) described by

$$\mathscr{E} = \hat{y}E_0 \cos(\omega t - kz + \tfrac{1}{2}\pi) = -\hat{y}E_0 \sin(\omega t - kz) \tag{1.40}$$

and

$$\mathscr{H} = -\hat{x}H_0 \cos(\omega t - kz + \tfrac{1}{2}\pi) = \hat{x}H_0 \sin(\omega t - kz) \tag{1.41}$$

Fig. 1.2 An x-polarized plane electromagnetic wave propagating along the z-direction. The arrows represent the direction and magnitude of the $\mathscr{E}$ and $\mathscr{H}$ vectors at a particular instant of time. The electric vector always lies in the x–z plane and the magnetic vector always lies in the y–z plane.

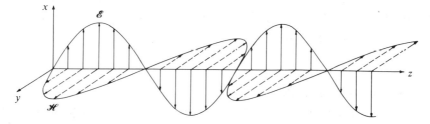

with E_0 and H_0 related through Eq. (1.39). Since Maxwell's equations are linear, a linear superposition of Eqs. (1.37) and (1.40) (and a corresponding superposition of Eqs. (1.38) and (1.41)) will also be a valid solution leading, in general, to an elliptically polarized wave. In particular, just a simple addition of the two solutions would lead to

$$\begin{aligned} \mathscr{E} &= E_0[\hat{\mathbf{x}}\cos(\omega t - kz) - \hat{\mathbf{y}}\sin(\omega t - kz)] \\ \mathscr{H} &= H_0[\hat{\mathbf{y}}\cos(\omega t - kz) + \hat{\mathbf{x}}\sin(\omega t - kz)] \end{aligned} \right\} \tag{1.42}$$

and

If we consider any plane perpendicular to the direction of propagation – in particular, if we consider the plane $z = 0$ then at *all* points on the plane, the time dependence of the fields would be given by

$$\begin{aligned} \mathscr{E}_x &= E_0 \cos\omega t \quad \text{and} \quad \mathscr{E}_y = -E_0 \sin\omega t \\ \mathscr{H}_x &= H_0 \sin\omega t \quad \text{and} \quad \mathscr{H}_y = H_0 \cos\omega t \end{aligned} \right\} \tag{1.43}$$

Thus

$$\mathscr{E}^2 = \mathscr{E}_x^2 + \mathscr{E}_y^2 = E_0^2$$

and

$$\mathscr{H}^2 = \mathscr{H}_x^2 + \mathscr{H}_y^2 = H_0^2$$

implying that the electric and magnetic vectors rotate on the circumference of a circle and if we look along the direction of propagation of the wave (i.e., along the z-axis) then the electric vector will rotate in an anticlockwise direction (see Fig. 1.3(a)). Such a wave is known as a left circularly polarized wave.

On the other hand, at a particular instant of time (say $t = 0$), we have

$$\mathscr{E}_x = E_0 \cos kz, \quad \mathscr{E}_y = E_0 \sin kz \tag{1.44}$$

and

$$\mathscr{E}^2 = \mathscr{E}_x^2 + \mathscr{E}_y^2 = E_0^2$$

Thus at a particular instant of time the tip of the electric vector (and similarly the tip of the magnetic vector) traces a right handed helix along the z-direction (see Fig. 1.3(b)).

In general, if we superpose two mutually orthogonal linearly polarized waves propagating along the same direction but with different amplitudes and phases, we would have (at a particular value of z, say $z = 0$)

$$\begin{aligned} \mathscr{E}_x &= E_1 \cos(\omega t - \theta_1) \\ \mathscr{E}_y &= E_2 \cos(\omega t - \theta_2) \end{aligned} \right\} \tag{1.45}$$

Thus

$$\mathscr{E}_x/E_1 = \cos\omega t \cos\theta_1 + \sin\omega t \sin\theta_1$$

and

$$\mathcal{E}_y/E_2 = \cos \omega t \cos \theta_2 + \sin \omega t \sin \theta_2$$

Simple manipulations give us

$$(\mathcal{E}_x/E_1) \sin \theta_2 - (\mathcal{E}_y/E_2) \sin \theta_1 = \cos \omega t \sin (\theta_2 - \theta_1)$$

Fig. 1.3 (*a*) For the electric (and magnetic) fields given by Eq. (1.44), the electric and the magnetic vectors rotate on the circumference of a circle in an anticlockwise direction. The propagation is into the plane of the page and the wave is said to be left circularly polarized. (*b*) At a particular instant of time, the tip of the electric vector traces a right handed helix along the *z*-direction.

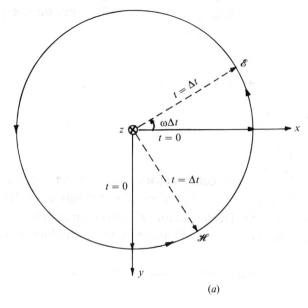

(*a*)

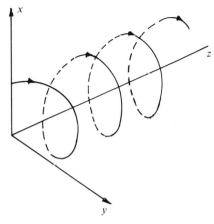

(*b*)

and

$$(\mathscr{E}_x/E_1)\cos\theta_2 - (\mathscr{E}_y/E_2)\cos\theta_1 = -\sin\omega t\sin(\theta_2 - \theta_1)$$

If we square and add we would obtain

$$(\mathscr{E}_x/E_1)^2 + (\mathscr{E}_y/E_2)^2 - 2(\mathscr{E}_x/E_1)(\mathscr{E}_y/E_2)\cos\theta = \sin^2\theta \qquad (1.46)$$

where $\theta = \theta_2 - \theta_1$ represents the phase difference between $\mathscr{E}_x$ and $\mathscr{E}_y$. Eq. (1.46) represents an ellipse (see Problem 1.1 and Fig. 1.4) and the electric

Fig. 1.4 The superposition of two mutually orthogonal linearly polarized waves leads, in general, to elliptically polarized waves. The figure below shows the states of polarization corresponding to different values of θ for $E_1 = E_2$.

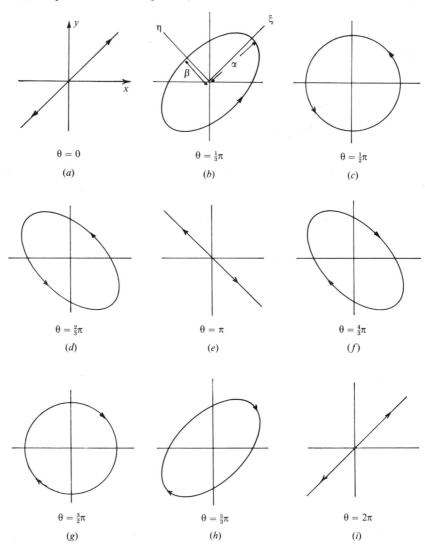

vector rotates on the circumference of the ellipse; such an electromagnetic wave is said to be elliptically polarized. For $\theta = \frac{1}{2}\pi, \frac{3}{2}\pi, \ldots$, the major and minor axes of the ellipse are along the x and y-axes and in addition if $E_1 = E_2$ the ellipse becomes a circle. In general, the axes of the ellipse are tilted with respect to the x and y-axes (see Fig. 1.4). For $\theta = 0, \pi, 2\pi, \ldots$ the ellipse degenerates into a straight line

$$\mathscr{E}_y / \mathscr{E}_x = \pm E_2 / E_1$$

which represents a linearly polarized wave. Fig. 1.4 shows the states of polarization corresponding to different values of θ for $E_1 = E_2$. The production and detection of different types of polarized light is discussed in Chapter 3.

From the above analysis and Fig. 1.4, it is obvious that any general elliptical polarization can be written as a superposition of two mutually orthogonal linearly polarized waves.

Problem 1.1: Show that the equation

$$\frac{x^2}{a^2} + \frac{y^2}{b^2} - \frac{2xy}{ab} \cos \theta = \sin^2 \theta \tag{1.47}$$

represents an ellipse. Determine the direction of the major and minor axes and their lengths.

Solution: We consider a rotated coordinate system

$$\begin{aligned} x &= \xi \cos \phi - \eta \sin \phi \\ y &= \xi \sin \phi + \eta \cos \phi \end{aligned} \tag{1.48}$$

We substitute in Eq. (1.47) and equate the coefficient of the $\xi\eta$ term to zero which gives us

$$\tan 2\phi = \frac{2ab}{(a^2 - b^2)} \cos \theta \tag{1.49}$$

In the rotated coordinate system, Eq. (1.47) becomes

$$\frac{\xi^2}{\alpha^2} + \frac{\eta^2}{\beta^2} = 1 \tag{1.50}$$

where

$$\left. \begin{aligned} \frac{1}{\alpha^2} &= \left(\frac{\cos^2 \phi}{a^2} + \frac{\sin^2 \phi}{b^2} - \frac{\cos \theta}{ab} \sin 2\phi \right) \frac{1}{\sin^2 \theta} \\ \frac{1}{\beta^2} &= \left(\frac{\sin^2 \phi}{a^2} + \frac{\cos^2 \phi}{b^2} + \frac{\cos \theta}{ab} \sin 2\phi \right) \frac{1}{\sin^2 \theta} \end{aligned} \right\} \tag{1.51}$$

Eq. (1.50) represents an ellipse with semiaxes of lengths α and β and oriented along the ξ and η-axes. For example for the ellipse corresponding to Fig. 1.4(*b*), since $a = b$, $\phi = \frac{1}{4}\pi$.

1.4   The Poynting vector

In this section we will discuss the power flow associated with an electromagnetic wave. We start with Eqs. (1.3) and (1.4)

$$\mathbf{\nabla} \times \boldsymbol{\mathcal{E}} = -\partial \boldsymbol{\mathcal{B}}/\partial t \tag{1.52}$$

and

$$\mathbf{\nabla} \times \boldsymbol{\mathcal{H}} = \mathbf{J} + \partial \boldsymbol{\mathcal{D}}/\partial t \tag{1.53}$$

We next use the vector identity[†]

$$\mathbf{\nabla} \cdot (\boldsymbol{\mathcal{E}} \times \boldsymbol{\mathcal{H}}) = -\boldsymbol{\mathcal{E}} \cdot (\mathbf{\nabla} \times \boldsymbol{\mathcal{H}}) + \boldsymbol{\mathcal{H}} \cdot (\mathbf{\nabla} \times \boldsymbol{\mathcal{E}}) \tag{1.54}$$

and substitute Eqs. (1.52) and (1.53) on the RHS to obtain

$$\mathbf{\nabla} \cdot (\boldsymbol{\mathcal{E}} \times \boldsymbol{\mathcal{H}}) = -\mathbf{J} \cdot \boldsymbol{\mathcal{E}} - (\boldsymbol{\mathcal{E}} \cdot \partial \boldsymbol{\mathcal{D}}/\partial t + \boldsymbol{\mathcal{H}} \cdot \partial \boldsymbol{\mathcal{B}}/\partial t) \tag{1.55}$$

If we integrate over an arbitrary volume and use Gauss' theorem[‡] we would obtain

$$\oint_A \mathbf{S} \cdot \hat{\mathbf{n}} \, \mathrm{d}a = -\int_V \mathbf{J} \cdot \boldsymbol{\mathcal{E}} \, \mathrm{d}V - \int_V (\boldsymbol{\mathcal{E}} \cdot \partial \boldsymbol{\mathcal{D}}/\partial t + \boldsymbol{\mathcal{H}} \cdot \partial \boldsymbol{\mathcal{B}}/\partial t) \, \mathrm{d}V \tag{1.56}$$

where

$$\mathbf{S} \equiv \boldsymbol{\mathcal{E}} \times \boldsymbol{\mathcal{H}} \tag{1.57}$$

is known as the Poynting vector and the surface integral is over the surface bounding the volume V. Eq. (1.56) is usually referred to as the *energy law* associated with the electromagnetic field. It may be pointed out that Eq. (1.56) is rigorously correct in the sense that we have used only Maxwell's equations and *not* the constitutive relations. If we do use Eqs. (1.5) and (1.6) we can write

$$\boldsymbol{\mathcal{E}} \cdot \frac{\partial \boldsymbol{\mathcal{D}}}{\partial t} = \epsilon \boldsymbol{\mathcal{E}} \cdot \frac{\partial \boldsymbol{\mathcal{E}}}{\partial t} = \frac{1}{2}\epsilon \frac{\partial \boldsymbol{\mathcal{E}}^2}{\partial t} = \frac{1}{2} \frac{\partial}{\partial t} (\boldsymbol{\mathcal{E}} \cdot \boldsymbol{\mathcal{D}}) \tag{1.58}$$

[†] The vector identity can easily be proved in Cartesian coordinates.
[‡] According to Gauss' theorem

$$\int_V \mathbf{\nabla} \cdot \mathbf{F} \, \mathrm{d}V = \int_A \mathbf{F} \cdot \hat{\mathbf{n}} \, \mathrm{d}a$$

where A represents the surface bounding the volume V and $\hat{\mathbf{n}}$ represents the unit (outward) normal to the surface.

Similarly

$$\mathscr{H} \cdot \frac{\partial \mathscr{B}}{\partial t} = \mu \mathscr{H} \cdot \frac{\partial \mathscr{H}}{\partial t} = \tfrac{1}{2} \mu \frac{\partial \mathscr{H}^2}{\partial t} = \frac{1}{2} \frac{\partial}{\partial t} (\mathscr{H} \cdot \mathscr{B}) \qquad (1.59)$$

Thus

$$\mathscr{E} \cdot \frac{\partial \mathscr{D}}{\partial t} + \mathscr{H} \cdot \frac{\partial \mathscr{B}}{\partial t} = \frac{\partial}{\partial t} (w_e + w_m) = \frac{\partial w}{\partial t} \qquad (1.60)$$

where

$$w_e = \tfrac{1}{2} \epsilon \mathscr{E}^2 = \tfrac{1}{2} \mathscr{E} \cdot \mathscr{D} \qquad (1.61)$$

and

$$w_m = \tfrac{1}{2} \mu \mathscr{H}^2 = \tfrac{1}{2} \mathscr{H} \cdot \mathscr{B} \qquad (1.62)$$

represent the energy densities associated with the electric and magnetic fields respectively.

Thus, if we assume the validity of the constitutive relations, Eqs. (1.55) and (1.56) may be written in the form

$$\mathbf{\nabla} \cdot \mathbf{S} + \partial w / \partial t = - \mathbf{J} \cdot \mathscr{E} \qquad (1.63)$$

and

$$\oint_A \mathbf{S} \cdot \hat{\mathbf{n}} \, da = - \int_V \mathbf{J} \cdot \mathscr{E} \, dV - (d/dt) \int_V w \, dV \qquad (1.64)$$

The term

$$\int_V \mathbf{J} \cdot \mathscr{E} \, dV$$

on the RHS of Eq. (1.64) represents the total dissipated power within the volume V. For example, in a conductor $\mathbf{J} = \sigma \mathscr{E}$ and $\sigma \mathscr{E}^2$ represents the ohmic power dissipated per unit volume. We should mention here that the current density $\mathbf{J}$ in Eq. (1.53) can be considered to be the sum of two parts

$$\mathbf{J} = \mathbf{J}_c + \mathbf{J}_v \qquad (1.65)$$

where $\mathbf{J}_c$ represents the conduction current density and $\mathbf{J}_v$ the convection current density due to moving charges. Thus, if a charge Q (moving with velocity $\mathbf{v}$) is acted on by an electromagnetic field then the work done *by* the field in moving it through a distance $d\mathbf{s}$ would be $\mathbf{F} \cdot d\mathbf{s}$ where

$$\mathbf{F} = Q(\mathscr{E} + \mathbf{v} \times \mathbf{B}) \qquad (1.66)$$

represents the Lorentz force. Thus the work done per unit time would be

$$\begin{aligned} \mathbf{F} \cdot d\mathbf{s}/dt &= Q(\mathscr{E} + \mathbf{v} \times \mathscr{B}) \cdot \mathbf{v} \\ &= Q \mathscr{E} \cdot \mathbf{v} \end{aligned}$$

If there are N particles per unit volume, each carrying a charge Q then the work done per unit volume per unit time would be

$$NQ\mathbf{v}\cdot\mathscr{E} = \mathbf{J}_v\cdot\mathscr{E} \tag{1.67}$$

where

$$\mathbf{J}_v = NQ\mathbf{v} \tag{1.68}$$

represents the convection current density due to moving charges. The energy given by Eq. (1.67) appears in the form of kinetic (or heat) energy of the charged particles.

Returning to Eq. (1.64), if we consider electromagnetic fields which vanish at large distances from the origin and if the surface area A is very far away (so that the surface integral vanishes), Eq. (1.64) can be written in the form

$$\mathrm{d}W/\mathrm{d}t = -\int_V \mathbf{J}\cdot\mathscr{E}\,\mathrm{d}V \tag{1.69}$$

where

$$W \equiv \int_V w\,\mathrm{d}V = \int_V (w_e + w_m)\,\mathrm{d}V \tag{1.70}$$

Since the RHS of Eq. (1.69) has been shown to represent the total dissipated power within the volume V, we may interpret W to represent the total electromagnetic energy contained within the volume V; w_e and w_m may therefore be interpreted to represent the energy per unit volume associated with the electric and magnetic fields respectively.

Thus in Eq. (1.64) (considering the integrals over an *arbitrary* volume V) we have the following physical interpretation of the various terms.

(a) The term $\int_V \mathbf{J}\cdot\mathscr{E}\,\mathrm{d}V$ represents the total (instantaneous) dissipated power in the volume V.

(b) The term $\int_V w\,\mathrm{d}V$ represents the total electromagnetic energy in the volume V.

(c) Because of the above, the term

$$\int_A \mathbf{S}\cdot\hat{\mathbf{n}}\,\mathrm{d}a = \int_A (\mathscr{E} \times \mathscr{H})\cdot\hat{\mathbf{n}}\,\mathrm{d}a$$

represents the total power flowing *out* of the volume V.

For a better understanding, we consider a few specific cases:

Case 1: For plane waves in a dielectric we can write for the electric and magnetic fields (see Eqs. (1.37) and (1.38)):

$$\mathscr{E} = \hat{\mathbf{x}}E_0 \cos{(\omega t - kz)} \tag{1.71}$$

$$\mathscr{H} = \hat{\mathbf{y}} H_0 \cos(\omega t - kz) \tag{1.72}$$

Thus

$$w_e = \tfrac{1}{2}\epsilon E_0^2 \cos^2(\omega t - kz) \tag{1.73}$$

For optical frequencies, the $\cos^2$ term fluctuates with extreme rapidity so that we should take a time average of Eq. (1.73) to obtain

$$\langle w_e \rangle = \tfrac{1}{2}\epsilon E_0^2 \langle \cos^2(\omega t - kz) \rangle$$

where $\langle \cdots \rangle$ denotes the time average of the quantity inside the angular brackets; the time average of a time dependent function is defined by the following equation

$$\langle F(t) \rangle = (1/T) \int_0^T F(t)\, dt$$

for a periodic function, one may choose T as the time period $(= 2\pi/\omega)$. Thus

$$\langle \cos^2(\omega t - kz) \rangle = (\omega/2\pi) \int_0^{2\pi/\omega} \cos^2(\omega t - kz)\, dt = \tfrac{1}{2} \tag{1.74}$$

Thus

$$\langle w_e \rangle = \tfrac{1}{4}\epsilon E_0^2$$

Similarly

$$\langle w_m \rangle = \tfrac{1}{4}\mu H_0^2$$

From Eq. (1.39)

$$\langle w_e \rangle = \langle w_m \rangle = \tfrac{1}{4}\epsilon E_0^2$$

and therefore

$$\langle w \rangle = 2\langle w_e \rangle = \tfrac{1}{2}\epsilon E_0^2 \tag{1.75}$$

The Poynting vector is given by

$$\mathbf{S} = \mathscr{E} \times \mathscr{H} = E_0 H_0 \cos^2(\omega t - kz)\hat{\mathbf{z}}$$

Taking the time average and using Eq. (1.39) we get

$$\langle \mathbf{S} \rangle = \tfrac{1}{2}(k/\omega\mu) E_0^2 \hat{\mathbf{z}} = \tfrac{1}{2}(\epsilon/\mu)^{\frac{1}{2}} E_0^2 \hat{\mathbf{z}} \tag{1.76}$$

Since the velocity of the wave, v is equal to $\omega/k(= \omega(\epsilon\mu)^{\frac{1}{2}})$, the above expression for the Poynting vector can be written as

$$\langle \mathbf{S} \rangle = \langle w \rangle v\hat{\mathbf{z}} \tag{1.77}$$

Thus for a plane wave in a dielectric we may interpret $\mathbf{S}\cdot\hat{\mathbf{n}}\, da$ as the rate of energy flow through the area da.

Case 2: The quantity $\mathbf{S}\cdot\hat{\mathbf{n}}\,da$ does not *always* represent the rate of energy flow through the area da; for example, we may have static electric and magnetic fields where $\mathscr{E} \times \mathscr{H}$ is finite but we know that there is no energy flow. However, the integral

$$\int_A \mathbf{S}\cdot\hat{\mathbf{n}}\,da$$

over a closed surface rigorously represents the net energy flowing out of the surface. As a specific example we consider a long cylindrical conductor carrying a steady current I. Assuming constant current density (along the z-direction) the electric field inside the conductor is given by

$$\mathscr{E} = \hat{\mathbf{z}}(J_z/\sigma) = \hat{\mathbf{z}}(I/\sigma A) \tag{1.78}$$

where A represents the area of cross section of the conductor. Using Ampère's law, the magnetic field at the outer surface of the conductor is given by (see, e.g., Johnk (1975) p. 272 and p. 405)

$$\mathscr{H} = \hat{\boldsymbol{\phi}}(I/2\pi\rho_0) \tag{1.79}$$

where ρ_0 represents the radius of the cylindrical conductor. We consider the integral

$$\int (\mathscr{E} \times \mathscr{H})\cdot\hat{\mathbf{n}}\,da$$

over the surface of a cylinder of length l whose curved surface coincides with the surface of the conductor. Since

$$\mathscr{E} \times \mathscr{H} = -\hat{\boldsymbol{\rho}}\frac{I^2}{(2\pi\rho_0)\sigma A}$$

(using Eqs. (1.78) and (1.79)), $\mathscr{E} \times \mathscr{H}$ is normal to the flat surface at the ends and therefore there is no contribution (to the surface integral) from the ends. From the curved surface we have

$$\int_{\phi=0}^{2\pi}\int_{z=0}^{l}(\mathscr{E} \times \mathscr{H})\cdot\hat{\mathbf{n}}\rho_0\,d\phi\,dz = -\frac{I^2}{(2\pi\rho_0)\sigma A}\rho_0\left(\int_0^{2\pi}d\phi\right)\left(\int_0^l dz\right)$$

Thus

$$\oint(\mathscr{E} \times \mathscr{H})\cdot\hat{\mathbf{n}}\,da = -I^2(l/A\sigma) = -I^2R \tag{1.80}$$

where we have used the fact that the quantity $l/A\sigma$ is the resistance R of the length l of the conductor. The quantity I^2R is the power dissipated in the length l of the conductor. Further

$$\int \mathbf{J} \cdot \mathscr{E}\, dV = \sigma \int \mathscr{E}^2\, dV = \sigma(I^2/\sigma^2 A^2)(Al) = I^2 R \tag{1.81}$$

consistent with Eq. (1.80)

Case 3: We consider an infinitesimal oscillating dipole at the origin whose dipole moment is along the z-axis and is assumed to be given by

$$\mathbf{p} = \hat{\mathbf{z}} p_0 \cos \omega t \tag{1.82}$$

The far field radiation pattern is given by (see, e.g., Panofsky and Phillips (1962) p. 258):

$$\mathscr{E} = -\frac{k^2 p_0}{4\pi\epsilon_0} \sin\theta \frac{\cos(\omega t - kr)}{r} \hat{\boldsymbol{\theta}} \tag{1.83}$$

$$\mathscr{H} = -\frac{\omega k p_0}{4\pi} \sin\theta \frac{\cos(\omega t - kr)}{r} \hat{\boldsymbol{\phi}} \tag{1.84}$$

where we have used spherical polar coordinates (r, θ, ϕ) and have assumed free space surrounding the dipole (see Fig. 1.5). Eqs. (1.83) and (1.84) represent (outgoing) spherical waves and are valid when $kr \gg 1$. Notice that

Fig. 1.5 An infinitesimal oscillating dipole at the origin whose dipole moment is along the z-axis. At large distances from the dipole, $\mathscr{E}$, $\mathscr{H}$ and S are mutually orthogonal and are along $\hat{\boldsymbol{\theta}}, \hat{\boldsymbol{\phi}}$ and $\hat{\mathbf{r}}$ respectively.

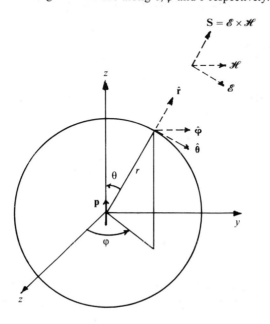

the $\mathscr{E}$ and $\mathscr{H}$ fields are in phase and the magnitudes are related through Eq. (1.39). The Poynting vector is given by

$$\langle \mathbf{S} \rangle = \langle \mathscr{E} \times \mathscr{H} \rangle$$

$$= \frac{\omega k^3 p_0^2}{32\pi^2 \epsilon_0} \frac{\sin^2 \theta}{r^2} \hat{\mathbf{r}} \tag{1.85}$$

where we have taken the time average. The energy flows radially and the intensity falls off as $1/r^2$ which is the inverse square law. The quantity $\mathbf{S} \cdot \hat{\mathbf{n}} \, da$ would represent the energy flow per unit time through the area da and if we integrate over a sphere of radius R we would obtain

$$\int_A \langle \mathbf{S} \rangle \cdot \hat{\mathbf{n}} \, da = \frac{\omega k^3 p_0^2}{32\pi^2 \epsilon_0 R^2} \int_0^\pi \int_0^{2\pi} \sin^2 \theta R^2 \sin \theta \, d\theta \, d\phi$$

$$= \frac{\omega k^3 p_0^2}{12\pi \epsilon_0} \tag{1.86}$$

which represents the total radiated power.

1.5   The complex notation

In many problems it is very convenient to use the complex notation in which the electric and magnetic fields are written in the form

$$\mathscr{E}_1 = \mathbf{E}_1 e^{i(\omega t - \mathbf{k} \cdot \mathbf{r} + \theta_1)} \tag{1.87}$$

and

$$\mathscr{H}_1 = \mathbf{H}_1 e^{i(\omega t - \mathbf{k} \cdot \mathbf{r} + \phi_1)} \tag{1.88}$$

where $\mathbf{E}_1$ and $\mathbf{H}_1$ are assumed to be real vectors. The actual electric and magnetic fields will be the real parts of the RHSs, i.e.,

$$\mathbf{E}_1 \cos(\omega t - \mathbf{k} \cdot \mathbf{r} + \theta_1) \quad \text{and} \quad \mathbf{H}_1 \cos(\omega t - \mathbf{k} \cdot \mathbf{r} + \phi_1)$$

respectively. Now, if we have another electromagnetic wave characterized by the fields $\mathscr{E}_2$ and $\mathscr{H}_2$, then the resultant electric and magnetic fields will be

$$\left. \begin{aligned} \text{Re}\,\mathscr{E}_1 + \text{Re}\,\mathscr{E}_2 &= \text{Re}(\mathscr{E}_1 + \mathscr{E}_2) \\ \text{Re}\,\mathscr{H}_1 + \text{Re}\,\mathscr{H}_2 &= \text{Re}(\mathscr{H}_1 + \mathscr{H}_2) \end{aligned} \right\} \tag{1.89}$$

Thus we may superpose the complex fields and *then* take the real part. This leads to considerable simplification in many problems. One must, however, be careful in calculating any quantity which is not linear in $\mathscr{E}$ and $\mathscr{H}$ such as the energy density w or the Poynting vector $\mathbf{S}$. For example, for the fields given by Eqs. (1.71) and (1.72), the complex fields are given by

$$\mathscr{E} = \hat{\mathbf{x}} E_0 e^{i(\omega t - kz)}$$

and

$$\mathscr{H} = \hat{\mathbf{y}} H_0 e^{i(\omega t - kz)}$$

The energy density is given by (see Eq. (1.75))

$$\langle w \rangle = 2 \langle w_e \rangle = 2 \times \tfrac{1}{2} \epsilon \langle \operatorname{Re} \mathscr{E} \cdot \operatorname{Re} \mathscr{E} \rangle \tag{1.90}$$

where $\operatorname{Re} \mathscr{E}$ represents the real part of $\mathscr{E}$. Thus

$$\langle w \rangle = \tfrac{1}{4} \epsilon \langle (\mathscr{E} + \mathscr{E}^*) \cdot (\mathscr{E} + \mathscr{E}^*) \rangle$$

where $\mathscr{E}^*$ represents the complex conjugate of $\mathscr{E}$. The time average of $\mathscr{E} \cdot \mathscr{E}$ and $\mathscr{E}^* \cdot \mathscr{E}^*$ are zero and therefore (in the complex notation)

$$\langle w \rangle = \tfrac{1}{2} \epsilon \mathscr{E} \cdot \mathscr{E}^* = \tfrac{1}{2} \epsilon E_0^2 \tag{1.91}$$

We next consider the time average of the Poynting vector

$$\langle \mathbf{S} \rangle = \langle \operatorname{Re} \mathscr{E} \times \operatorname{Re} \mathscr{H} \rangle$$
$$= \tfrac{1}{4} \langle (\mathscr{E} + \mathscr{E}^*) \times (\mathscr{H} + \mathscr{H}^*) \rangle$$

Since

$$\langle \mathscr{E} \times \mathscr{H} \rangle = 0 = \langle \mathscr{E}^* \times \mathscr{H}^* \rangle$$

we obtain

$$\langle \mathbf{S} \rangle = \tfrac{1}{2} \operatorname{Re} \langle \mathscr{E} \times \mathscr{H}^* \rangle \tag{1.92}$$

In Sec. 1.6 we will show that in an absorbing medium the electric and magnetic fields are given by

$$\mathscr{E} = \hat{\mathbf{x}} E_0 e^{-\beta z} e^{i(\omega t - \alpha z)} \tag{1.93}$$

and

$$\mathscr{H} = \hat{\mathbf{y}} \left(\frac{\epsilon}{\mu} \right)^{\frac{1}{2}} \left[1 + \left(\frac{\sigma}{\omega \epsilon} \right)^2 \right]^{\frac{1}{4}} E_0 e^{-\beta z} e^{i(\omega t - \alpha z - \phi)} \tag{1.94}$$

Substituting in Eq. (1.92) we obtain

$$\langle \mathbf{S} \rangle = \hat{\mathbf{z}} \frac{1}{2} \left(\frac{\epsilon}{\mu} \right)^{\frac{1}{2}} \left[1 + \left(\frac{\sigma}{\omega \epsilon} \right)^2 \right]^{\frac{1}{4}} E_0^2 e^{-2\beta z} \cos \phi \tag{1.95}$$

1.6   Wave propagation in an absorbing medium

Let us first consider propagation of electromagnetic waves through a conducting medium. The corresponding wave equation is given by (see

Eq. (1.14))

$$\nabla^2 \mathscr{E} = \mu\sigma(\partial\mathscr{E}/\partial t) + \mu\epsilon(\partial^2\mathscr{E}/\partial t^2) \tag{1.96}$$

If we assume a solution of the form

$$\mathscr{E} = \hat{\mathbf{x}} E_0 e^{i(\omega t - kz)} \tag{1.97}$$

we would obtain

$$-k^2 = i\omega\mu\sigma - \mu\epsilon\omega^2 \tag{1.98}$$

The above equation implies that k must be complex. We write

$$k = \alpha - i\beta \tag{1.99}$$

substitute in Eq. (1.98), and equate real and imaginary parts to obtain

$$\begin{aligned} \alpha^2 - \beta^2 &= \mu\epsilon\omega^2 \\ \alpha\beta &= \tfrac{1}{2}\omega\mu\sigma \end{aligned} \tag{1.100}$$

The above two equations can readily be solved to obtain

$$\alpha = \omega(\epsilon\mu)^{\frac{1}{2}} \left[\tfrac{1}{2} + \tfrac{1}{2}(1 + \sigma^2/\omega^2\epsilon^2)^{\frac{1}{2}} \right]^{\frac{1}{2}} \tag{1.101}$$

$$\beta = \frac{\omega\mu\sigma}{2\alpha} = \left(\frac{\mu}{\epsilon}\right)^{\frac{1}{2}} \frac{\sigma}{2} \left[\frac{1}{2} + \frac{1}{2}\left(1 + \frac{\sigma^2}{\omega^2\epsilon^2}\right)^{\frac{1}{2}} \right]^{-\frac{1}{2}} \tag{1.102}$$

Since k is now complex, Eq. (1.97) becomes

$$\mathscr{E} = \hat{\mathbf{x}} E_0 e^{-\beta z} e^{i(\omega t - \alpha z)} \tag{1.103}$$

which shows attenuation in the z-direction. The quantity

$$\delta = \frac{1}{\beta} = \frac{2}{\sigma}\left(\frac{\epsilon}{\mu}\right)^{\frac{1}{2}} \left[\frac{1}{2} + \frac{1}{2}\left(1 + \frac{\sigma^2}{\omega^2\epsilon^2}\right)^{\frac{1}{2}} \right]^{\frac{1}{2}} \tag{1.104}$$

is referred to as the 'skin depth' and represents the distance in which the amplitude falls off by a factor of e.

It may be noted that assuming harmonic solutions (i.e., time dependence of the form $e^{i\omega t}$) Eq. (1.11) takes the form

$$\begin{aligned} \nabla \times \mathscr{H} &= (\sigma + i\omega\epsilon)\mathscr{E} \\ &= i\omega(\epsilon - i\sigma/\omega)\mathscr{E} \end{aligned} \tag{1.105}$$

Thus all the analysis in the previous section remains valid except for the fact that ϵ has to be replaced by $(\epsilon - i\sigma/\omega)$. Thus instead of Eq. (1.27) we have

$$k^2 = \omega^2(\epsilon - i\sigma/\omega)\mu \tag{1.106}$$

which is identical to Eq. (1.98). Furthermore, using Eq. (1.97), Eq. (1.10) gives

$$\hat{\mathbf{y}}(\partial \mathscr{E}_x / \partial z) = -\mathrm{i}\omega\mu\mathscr{H}$$

Thus

$$\mathscr{H} = \hat{\mathbf{y}}(k/\omega\mu)E_0 \mathrm{e}^{\mathrm{i}(\omega t - kz)}$$

$$= \hat{\mathbf{y}}\left(\frac{\epsilon}{\mu}\right)^{\frac{1}{2}}\left(1 - \frac{\mathrm{i}\sigma}{\omega\epsilon}\right)^{\frac{1}{2}}E_0 \mathrm{e}^{-\beta z}\mathrm{e}^{\mathrm{i}(\omega t - \alpha z)}$$

$$= \hat{\mathbf{y}}\left(\frac{\epsilon}{\mu}\right)^{\frac{1}{2}}\left[1 + \left(\frac{\sigma}{\omega\epsilon}\right)^2\right]^{\frac{1}{4}}E_0 \mathrm{e}^{-\beta z}\mathrm{e}^{\mathrm{i}(\omega t - \alpha z - \phi)} \qquad (1.107)$$

where

$$\phi = \tfrac{1}{2}\tan^{-1}(\sigma/\omega\epsilon) \qquad (1.108)$$

Thus the electric and magnetic fields are no longer in phase. For a good conductor (at low frequencies) we may assume

$$\sigma/\omega\epsilon \gg 1 \qquad (1.109)$$

so that

$$\alpha \approx \beta \approx (\tfrac{1}{2}\omega\mu\sigma)^{\frac{1}{2}} \qquad (1.110)$$

and

$$\phi \approx \tfrac{1}{4}\pi \qquad (1.111)$$

As an example, we consider copper for which at low frequencies σ is a constant and $\approx 5.8 \times 10^7$ mhos/m, further $\mu \approx \mu_0$ so that

$$\delta = \frac{1}{\alpha} \approx \begin{cases} 6.6 \times 10^{-3}\,\mathrm{m} & \text{at } 100\,\mathrm{Hz} \\ 2.1 \times 10^{-4}\,\mathrm{m} & \text{at } 100\,\mathrm{kHz} \end{cases} \qquad (1.112)$$

implying that the skin depth decreases with increase in frequency.

We next calculate the Poynting vector and for the sake of simplicity we assume $\sigma/\omega\epsilon \gg 1$ so that $\alpha \approx \beta$ and $\phi \approx \tfrac{1}{4}\pi$. In this limit, the actual fields are given by

$$\mathscr{E} = \hat{\mathbf{x}}E_0 \mathrm{e}^{-\alpha z}\cos(\omega t - \alpha z) \qquad (1.113)$$

and

$$\mathscr{H} = \hat{\mathbf{y}}H_0 \mathrm{e}^{-\alpha z}\cos(\omega t - \alpha z - \phi) \qquad (1.114)$$

where

$$H_0 = (\epsilon/\mu)^{\frac{1}{2}}[1 + (\sigma/\omega\epsilon)^2]^{\frac{1}{4}}E_0 \qquad (1.115)$$

If we calculate $\mathscr{E} \times \mathscr{H}$ and then take the time average we obtain

$$\langle \mathbf{S} \rangle = \langle \mathscr{E} \times \mathscr{H} \rangle = \tfrac{1}{2}E_0 H_0 \mathrm{e}^{-2\alpha z}\cos\phi\,\hat{\mathbf{z}} \qquad (1.116)$$

which again shows the exponential attenuation of the power flow.

We should point out here that for a metal (such as sodium, copper, silver, etc.) σ may be assumed to be a constant only for frequencies $\lesssim 10^{13}$ Hz. At higher frequencies σ becomes both frequency dependent and complex, as does ϵ, and we may write

$$\sigma = \sigma' - i\sigma'' \tag{1.117}$$

and

$$\epsilon = \epsilon' - i\epsilon'' \tag{1.118}$$

Substituting in Eq. (1.98) we obtain

$$
\begin{aligned}
-k^2 &= i\omega\mu(\sigma' - i\sigma'') - \mu(\epsilon' - i\epsilon'')\omega^2 \\
&= i\omega\mu(\sigma' + \omega\epsilon'') - \mu(\epsilon' - \sigma''/\omega)\omega^2
\end{aligned}
\tag{1.119}
$$

Thus the entire analysis given earlier in this section remains valid provided

Fig. 1.6 The variation of η and κ with wavelength for copper (adapted from Lynch and Hunter, 1985).

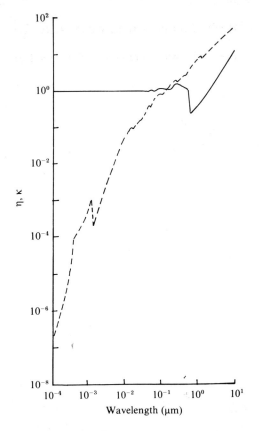

we make the following transformations

$$\left.\begin{array}{l} \sigma \to \sigma' + \omega\epsilon'' \\ \epsilon \to \epsilon' - \sigma''/\omega \end{array}\right\} \tag{1.120}$$

Often it is more convenient to write (cf. Eq. (1.99))

$$k = (\omega/c)n = (\omega/c)(\eta - i\kappa) \tag{1.121}$$

where η and κ represent the real and imaginary parts of the (complex) refractive index n. The plane wave solution is therefore written in the form

$$\begin{aligned} \mathscr{E} &= \mathbf{E}_0 \exp\{i[\omega t - (\omega/c)(\eta - i\kappa)z]\} \\ &= \mathbf{E}_0 \exp[-(\omega/c)\kappa z]\exp\{i[\omega t - (\omega/c)\eta z]\} \end{aligned} \tag{1.122}$$

The quantity κ is usually referred to as the extinction coefficient. Substituting for k from Eq. (1.121) in Eq. (1.119) we readily get

$$\eta^2 - \kappa^2 = c^2\mu(\epsilon' - \sigma''/\omega) \tag{1.123}$$

$$2\eta\kappa = (c^2\mu/\omega)(\sigma' + \omega\epsilon'') \tag{1.124}$$

Figs. 1.6 and 1.7 show the variation of η and κ with wavelength for copper

Fig. 1.7 The variation of η and κ with wavelength for gold (adapted from Lynch and Hunter, 1985).

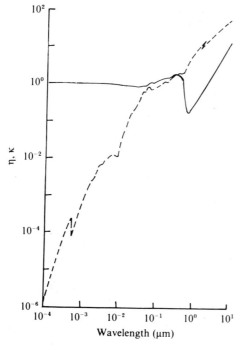

Table 1.1. *Optical constants of some metals*

Metal	$\lambda = 31\,\text{Å}$		$\lambda = 6888\,\text{Å}$	
	η	κ	η	κ
Copper	0.993	2.45×10^{-3}	0.213	4.05
Gold	0.995	6.15×10^{-3}	0.160	3.80
Indium	0.992	8.14×10^{-3}	2.64	4.81
Molybdenum	0.995	5.09×10^{-3}	3.81	3.58
Nickel	0.993	2.03×10^{-3}	2.14	4.00
Platinum	0.993	8.08×10^{-3}	2.51	4.43
Silver	0.998	8.82×10^{-3}	0.140	4.44
Tungsten	0.993	6.50×10^{-3}	3.82	2.91

Adapted from Lynch and Hunter (1985)

and gold. Notice that at low wavelenghts (i.e., high frequencies) the metal becomes almost transparent. Table 1.1 gives the values of optical constants for some metals at $\lambda = 31\,\text{Å}$ and $6888\,\text{Å}$. Detailed tables for such variations can be found in Lynch and Hunter (1985).

1.6.1 Refractive index of a metal

In a metal the conductivity is approximately given by

$$\sigma(\omega) = \frac{Nq^2}{m(v_c + i\omega)} \tag{1.125}$$

where N represents the number of free electrons per unit volume, q the electronic charge, m the electronic mass and v_c the collision frequency. Thus

$$\sigma' = \frac{Nq^2 v_c}{m(v_c^2 + \omega^2)} \tag{1.126}$$

and

$$\sigma'' = \frac{Nq^2 \omega}{m(v_c^2 + \omega^2)} \tag{1.127}$$

Assuming $\epsilon \approx \epsilon_0$ and $\mu \approx \mu_0$ (which is true for most metals) we obtain by using Eq. (1.98)

$$(\omega^2/c^2)n^2 \approx -i\omega\mu_0(\sigma' - i\sigma'') + \mu_0\epsilon_0\omega^2$$

or

$$n^2 = 1 - \frac{\omega_p^2}{\omega^2}\left(1 + \frac{v_c^2}{\omega^2}\right)^{-1}\left(1 + \frac{iv_c}{\omega}\right) \tag{1.128}$$

where

$$\omega_p = (Nq^2/m\epsilon_0)^{\frac{1}{2}} \tag{1.129}$$

Table 1.2. *Values of the wavelength corresponding to the plasma frequency for different metals*

Element	ρ (g/cm^3)	A	$N = 10^6 N_0 \rho / A$ (m^{-3})	$(\lambda_c)_{th} = 2\pi c (m\epsilon_0/q^2 N)^{\frac{1}{2}}$ (μm)	$(\lambda_p)_{expt}$ (μm)	n_{eff}
Lithium	0.534	6.94	4.634×10^{28}	1.552	2.050	0.573
Sodium	0.9712	22.99	2.544×10^{28}	2.095	2.100	0.995
Potassium	0.870	39.10	1.340×10^{28}	2.887	3.150	0.840
Rubidium	1.532	85.48	1.079×10^{28}	3.217	3.600	0.799

The values of various parameters used are $m = 9.109 \times 10^{-31}$ kg, $q = 1.602 \times 10^{-19}$ C, $\epsilon_0 = 8.854 \times 10^{-12}$ C^2/N m^2, $N_0 = 6.023 \times 10^{23}$ mole^{-1}.

is known as the plasma frequency. The above equation tells us that for high frequencies ($\omega \gg v_c$), the refractive index is essentially real with a frequency dependence of the form

$$n^2 \approx 1 - \frac{\omega_p^2}{\omega^2} \tag{1.130}$$

Thus for $\omega > \omega_p$, the refractive index is real. Indeed, in 1933 Wood discovered that alkali metals are transparent to ultraviolet light. Assuming that the refractive index is primarily due to the free electrons and that there is one free electron per atom, the calculated values of

$$\lambda_p = 2\pi c/\omega_p = 2\pi c (m\epsilon_0/Nq^2)^{\frac{1}{2}}$$

are tabulated in Table 1.2 for lithium, sodium, potassium and rubidium. The experimental values are also given and as can be seen although the calculated values are of the same order as the experimental values, the agreement is not really perfect (except for the case of sodium where the theoretical and experimental values agree very well). One of the shortcomings of the theory is the assumption that there is one free electron per atom; indeed, one may use the experimental values of λ_c to calculate the effective number of free electrons per atom from the formula

$$n_{eff} = (\lambda_c^2)_{th}/(\lambda_c^2)_{expt}$$

Additional problems

Problem 1.2: Discuss the state of polarization of a plane electromagnetic wave when the x and y-components of the electric field are given by the following equations:

(a) $E_x = \frac{1}{2}E_0 \cos(\omega t - kz)$

$E_y = \frac{\sqrt{3}}{2} E_0 \sin(\omega t - kz)$

(b) $E_x = E_0 \cos(\omega t - kz)$

   $E_y = \frac{1}{2}E_0 \cos(\omega t - kz + \pi)$

(c) $E_x = E_0 \cos(\omega t - kz)$

   $E_y = - E_0 \sin(\omega t - kz)$

(Answer: (a) Right elliptically polarized; (b) linearly polarized; (c) left circularly polarized.)

Problem 1.3: Determine the direction of propagation of the following waves:

(a) $E = E_0 \cos\left(\omega t - \frac{k}{2}x - \frac{\sqrt{3}}{2}kz\right)$

(b) $E = E_0 \cos\left(\omega t + \frac{k}{\sqrt{3}}x - \frac{k}{\sqrt{3}}y + \frac{k}{\sqrt{3}}z\right)$

Problem 1.4: Consider a displacement of the form

$$\psi = 0.2 \cos\left(t - \frac{x}{\sqrt{2}} - \frac{y}{\sqrt{2}}\right)$$

where t is measured in seconds and ψ, x and y are measured in centimetres. Determine the direction of propagation and calculate the wavelength and frequency.

Problem 1.5: A circularly polarized electromagnetic wave is propagating in the z-direction in free space and is described by the following equation

$$\mathscr{E} = \hat{x}5 \cos(\omega t - kz) + \hat{y}5 \sin(\omega t - kz) \text{ V/m}$$

The wavelength is 6×10^{-7} m. Find the corresponding magnetic field and show that the average of the Poynting vector is given by

$$\langle \mathbf{S} \rangle = 6.64 \times 10^{-2}\hat{z} \quad \text{W/m}^2$$

Problem 1.6: For a plane electromagnetic wave in a dielectric medium the ratio of the magnitude of the electric and magnetic fields is

$$(\mu/\epsilon)^{\frac{1}{2}}$$

which has the units of ohms and is referred to as the intrinsic impedance of the medium and is denoted by Z. Show that for free space

$$Z = Z_0 \approx 376.7 \, \Omega \simeq 120\pi \, \Omega$$

2

Reflection and refraction of electromagnetic waves

2.1   Introduction

In this chapter we will study the reflection and refraction of electro-magnetic waves from an interface separating two media and from a stack of films. Such studies are very important in understanding many practical optical devices such as Fabry–Perot etalons, interference filters, special optical coatings etc. Furthermore, by studying the state of polarization of a light beam reflected from a medium, one can obtain its optical character-istics; this forms the basis of the field of ellipsometry.

In deriving the reflection and transmission coefficients we will use the following continuity conditions[†] at the interface:

(a) continuity of the tangential components of the electric vector $\mathscr{E}$;
(b) continuity of the normal components of the displacement vector $\mathscr{D}$;
(c) continuity of the tangential components of the magnetic field vector $\mathscr{H}$; and
(d) continuity of the normal components of the magnetic induction vector $\mathscr{B}$.

We will find that the equations determining the reflection and transmis-sion coefficients fall into two groups: one of the groups contains only the components of $\mathscr{E}$ parallel to the plane of incidence (and $\mathscr{H}$ perpendicular to the plane) and the other group contains only the components of $\mathscr{E}$ perpendicular to the plane of incidence (and $\mathscr{H}$ parallel to the plane). Therefore the two cases (being independent of each other) will be considered separately and using them we can study the reflection (and refraction) of electromagnetic waves which have arbitrary states of polarization. We will, for example, show that a circularly polarized wave on reflection from a

[†] These boundary conditions can be derived from Maxwell's equations and since the field distributions that we would be using would satisfy Maxwell's equations, it will not be necessary to use all the boundary conditions.

dielectric surface can become elliptically polarized. Similarly, a linearly polarized wave reflected from a metal surface may be linearly polarized or circularly polarized or elliptically polarized depending on the angle of incidence and the direction of the electric vector associated with the incident wave.

2.2   Reflection and refraction at the interface of two homogeneous nonabsorbing dielectrics

We first consider the reflection and refraction of a plane electromagnetic wave incident at the interface of two dielectrics characterized by (ϵ_1, μ_1) and (ϵ_2, μ_2) (see Fig. 2.1). We assume the media to be nonabsorbing, isotropic and homogeneous. Let

$$\begin{aligned}
\mathscr{E}_1 &= \mathbf{E}_1 e^{i(\omega t - \mathbf{k}_1 \cdot \mathbf{r})} \\
\mathscr{E}_2 &= \mathbf{E}_2 e^{i(\omega_2 - \mathbf{k}_2 \cdot \mathbf{r})} \\
\mathscr{E}_3 &= \mathbf{E}_3 e^{i(\omega_3 t - \mathbf{k}_3 \cdot \mathbf{r})}
\end{aligned} \qquad (2.1)$$

and

represent the electric fields associated with the incident, refracted and reflected waves respectively; the vectors $\mathbf{E}_1, \mathbf{E}_2$ and $\mathbf{E}_3$ are independent of space and time. Let θ_1, θ_2 and θ_3 represent the (acute) angles that the vectors $\mathbf{k}_1, \mathbf{k}_2$, and $\mathbf{k}_3$ make with the normal to the interface (see Fig. 2.1). Fig. 2.1 also shows the direction of the Cartesian axes; the plane $x = 0$ represents the interface of the two dielectrics, and the direction of the y-axis

Fig. 2.1 The reflection and refraction of a plane wave incident at the interface of two dielectrics; the electric vector is assumed to be in the plane of incidence (i.e., *p*-polarized).

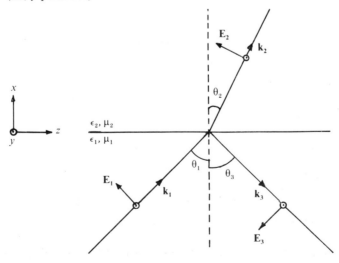

is such that

$$k_{1y} = 0 \tag{2.2}$$

i.e., the propagation vector associated with the incident wave is parallel to the x–z plane.

Since the fields satisfy the wave equation we must have

$$k_1^2 = \omega^2 \epsilon_1 \mu_1 \tag{2.3a}$$

$$k_2^2 = \omega_2^2 \epsilon_2 \mu_2 \tag{2.3b}$$

$$k_3^2 = \omega_3^2 \epsilon_1 \mu_1 \tag{2.3c}$$

We consider two special cases: one in which the electric vector lies in the plane of incidence and the other in which the electric vector lies perpendicular to the plane of incidence. By using appropriate boundary conditions it can easily be shown that if the electric vector associated with the incident plane wave lies in the plane of incidence then the electric vectors associated with the reflected and refracted waves will also lie in the plane of incidence (see Problem 2.1) and the same is true for the perpendicular case.

Case 1: $\mathscr{E}$ lying in the plane of incidence. We will first consider the case when the electric field associated with the incident wave lies in the plane of incidence (see Fig. 2.1). We resolve the electric vector $\mathscr{E}$ along the x and z-axis and since the z-component is tangential to the surface we must have $\mathscr{E}_z$ continuous across the interface; thus

$$\mathscr{E}_{1z} + \mathscr{E}_{3z} = \mathscr{E}_{2z} \tag{2.4}$$

or

$$[-E_1 e^{i(\omega t - \mathbf{k}_1 \cdot \mathbf{r})} \cos \theta_1 - E_3 e^{i(\omega_3 t - \mathbf{k}_3 \cdot \mathbf{r})} \cos \theta_3]_{x=0}$$
$$= [-E_2 e^{i(\omega_2 t - \mathbf{k}_2 \cdot \mathbf{r})} \cos \theta_2]_{x=0}$$

where all expressions are to be evaluated at the interface $x = 0$. Now

$$\mathbf{k} \cdot \mathbf{r} = k_x x + k_y y + k_z z$$
$$= k_y y + k_z z \qquad \text{(at the interface } x = 0\text{)}$$

Thus, we must have

$$-E_1 e^{i(\omega t - k_{1y}y - k_{1z}z)} \cos \theta_1 - E_3 e^{i(\omega_3 t - k_{3y}y - k_{3z}z)} \cos \theta_3$$
$$= -E_2 e^{i(\omega_2 t - k_{2y}y - k_{2z}z)} \cos \theta_2 \tag{2.5}$$

The above equation has to be valid at *all* times and for *all* values of y and

z (on the plane $x = 0$) and therefore we must have

$$\omega = \omega_2 = \omega_3 \tag{2.6}$$

$$k_{1y} = k_{2y} = k_{3y} \tag{2.7}$$

$$k_{1z} = k_{2z} = k_{3z} \tag{2.8}$$

Thus the frequencies associated with the reflected and refracted waves must be the same as that of the incident wave – which is also physically obvious. Thus

$$k_1 = \omega(\epsilon_1 \mu_1)^{\frac{1}{2}} = k_3 \tag{2.9}$$

and

$$k_2 = \omega(\epsilon_2 \mu_2)^{\frac{1}{2}} \tag{2.10}$$

Now, since $k_{1y} = 0$ (see Eq. (2.2)) we must have

$$k_{2y} = k_{3y} = 0 \tag{2.11}$$

i.e., $\mathbf{k}_1, \mathbf{k}_2$ and $\mathbf{k}_3$ will all be parallel to the x–z plane. Eq. (2.8) gives

$$k_1 \sin \theta_1 = k_2 \sin \theta_2 = k_1 \sin \theta_3 \tag{2.12}$$

where we have used Eq. (2.9). Thus

$$\theta_1 = \theta_3 \tag{2.13}$$

which says that the angle of incidence equals the angle of reflection. Further

$$\frac{\sin \theta_1}{\sin \theta_2} = \frac{k_2}{k_1} = \left(\frac{\epsilon_2 \mu_2}{\epsilon_1 \mu_1} \right)^{\frac{1}{2}} = \frac{n_2}{n_1} \tag{2.14}$$

where

$$n_1 = \frac{c}{v_1} = \left(\frac{\epsilon_1 \mu_1}{\epsilon_0 \mu_0} \right)^{\frac{1}{2}} \tag{2.15}$$

and

$$n_2 = \frac{c}{v_2} = \left(\frac{\epsilon_2 \mu_2}{\epsilon_0 \mu_0} \right)^{\frac{1}{2}} \tag{2.16}$$

represent the refractive indices of media 1 and 2 respectively and $c = (\epsilon_0 \mu_0)^{-\frac{1}{2}}$ represents the velocity of light in free space. Eq. (2.14) gives us

$$n_1 \sin \theta_1 = n_2 \sin \theta_2 \tag{2.17}$$

which is Snell's law of refraction.

We should mention here that in the derivation of the laws of reflection and refraction (*viz.* Eqs. (2.6)–(2.17)) we have only used the fact that *a* particular continuity condition should be valid at all times for all values of y and z on the plane $x = 0$. Since this argument is valid for *all* continuity conditions, Eqs (2.6)–(2.17) will be valid for any *arbitrary* state of polarization associated with the incident wave.

From now on we will assume that both media are non-magnetic, i.e.,

$$\mu_1 \approx \mu_2 \approx \mu_0 \tag{2.18}$$

which is indeed true for all dielectrics. Thus

$$n_1 = \left(\frac{\epsilon_1}{\epsilon_0}\right)^{\frac{1}{2}} = K_1^{\frac{1}{2}} \tag{2.19}$$

and

$$n_2 = \left(\frac{\epsilon_2}{\epsilon_0}\right)^{\frac{1}{2}} = K_2^{\frac{1}{2}} \tag{2.20}$$

K_1 and K_2 being the dielectric constants of media 1 and 2 respectively.

Now, using Eqs. (2.6)–(2.8), Eq. (2.5) becomes

$$- E_1 \cos \theta_1 - E_3 \cos \theta_1 = - E_2 \cos \theta_2 \tag{2.21}$$

where we are have also used Eq. (2.12). Similarly, since the normal component of $\mathscr{D}(= \mathscr{D}_x)$ is also continuous across the interface we must have

$$\mathscr{D}_{1x} + \mathscr{D}_{3x} = \mathscr{D}_{2x} \tag{2.22}$$

or

$$\epsilon_1 E_1 \sin \theta_1 - \epsilon_1 E_3 \sin \theta_3 = \epsilon_2 E_2 \sin \theta_2 \tag{2.23}$$

where we have used the relation $\mathscr{D} = \epsilon \mathscr{E}$ and Eq. (2.1). Substituting for E_2 from Eq. (2.23) in Eq. (2.21) and carrying out elementary manipulations we get

$$r_p \equiv \frac{E_3}{E_1} = \frac{\epsilon_1 \sin \theta_1 \cos \theta_2 - \epsilon_2 \sin \theta_2 \cos \theta_1}{\epsilon_2 \sin \theta_2 \cos \theta_1 + \epsilon_1 \sin \theta_1 \cos \theta_2} \tag{2.24}$$

where r_p denotes the *amplitude reflection coefficient* for parallel polarization.[†] If we now substitute for E_3 from Eq. (2.24) in Eq. (2.21) and carry out elementry simplifications, we get

$$t_p \equiv \frac{E_2}{E_1} = \frac{2\epsilon_1 \sin \theta_1 \cos \theta_1}{\epsilon_2 \cos \theta_1 \sin \theta_2 + \epsilon_1 \sin \theta_1 \cos \theta_2} \tag{2.25}$$

where t_p denotes the *amplitude transmission coefficient* for parallel polarization.

[†] We use the standard convention of using the subscripts p and s for parallel and perpendicular polarizations.

The reflectivity R and the transmittivity T are defined through the following equations

$$R = \frac{\text{Amount of energy reflected (per unit time) from an area } da \text{ of the interface}}{\text{Amount of energy incident (per unit time) on the same area } da \text{ of the interface}}$$

$$= \frac{|\langle \mathbf{S}_{\text{ref}} \cdot \mathbf{da} \rangle|}{|\langle \mathbf{S}_{\text{inc}} \cdot \mathbf{da} \rangle|} = \frac{|\langle (\mathbf{S}_{\text{ref}})_x \rangle|}{|\langle (\mathbf{S}_{\text{inc}})_x \rangle|} \tag{2.26}$$

$$T = \frac{\text{Amount of energy transmitted (per unit time) from an area } da \text{ of the interface}}{\text{Amount of energy incident (per unit time) on the same area } da \text{ of the interface}}$$

$$= \frac{|\langle \mathbf{S}_{\text{tr}} \cdot \mathbf{da} \rangle|}{|\langle \mathbf{S}_{\text{inc}} \cdot \mathbf{da} \rangle|} = \frac{|\langle (\mathbf{S}_{\text{tr}})_x \rangle|}{|\langle (\mathbf{S}_{\text{inc}})_x \rangle|} \tag{2.27}$$

where $\mathbf{S}_{\text{inc}}$, $\mathbf{S}_{\text{ref}}$ and $\mathbf{S}_{\text{tr}}$ represent the Poynting vectors (see Sec. 1.4) associated with the incident wave, reflected wave and transmitted wave respectively and $\langle \cdots \rangle$ represents the time average of the quantity inside the angular bracket. Recalling Eq. (1.76), we have

$$|\langle S_x \rangle| = |\langle (\mathscr{E} \times \mathscr{H})_x \rangle| = \tfrac{1}{2}(k/\omega\mu)|E|^2 \cos\theta$$
$$= \tfrac{1}{2}(\epsilon/\mu)^{\frac{1}{2}}|E|^2 \cos\theta \tag{2.28}$$

where θ is the (acute) angle that $\mathscr{E} \times \mathscr{H}$ (i.e., the propagation vector $\mathbf{k}$) makes with the x-axis. Thus

$$R_p = \frac{\tfrac{1}{2}(\epsilon_1/\mu_1)^{\frac{1}{2}}|E_3|^2 \cos\theta_3}{\tfrac{1}{2}(\epsilon_1/\mu_1)^{\frac{1}{2}}|E_1|^2 \cos\theta_1} = \left|\frac{E_3}{E_1}\right|^2 = |r_p|^2 \tag{2.29a}$$

or

$$R_p = |r_p|^2 = \left(\frac{\epsilon_2 \cos\theta_1 \sin\theta_2 - \epsilon_1 \sin\theta_1 \cos\theta_2}{\epsilon_2 \cos\theta_1 \sin\theta_2 + \epsilon_1 \sin\theta_1 \cos\theta_2}\right)^2 \tag{2.29b}$$

Similarly

$$T_p = \frac{\tfrac{1}{2}(\epsilon_2/\mu_2)^{\frac{1}{2}}|E_2|^2 \cos\theta_2}{\tfrac{1}{2}(\epsilon_1/\mu_1)^{\frac{1}{2}}|E_1|^2 \cos\theta_1} \tag{2.30a}$$

$$= \left(\frac{\epsilon_2}{\epsilon_1}\right)^{\frac{1}{2}} \left[\left(\frac{\epsilon_2}{\epsilon_1}\right)^{\frac{1}{2}} \frac{\sin\theta_2}{\sin\theta_1}\right]$$

$$\times \left(\frac{2\epsilon_1 \sin\theta_1 \cos\theta_1}{\epsilon_2 \cos\theta_1 \sin\theta_2 + \epsilon_1 \sin\theta_1 \cos\theta_2}\right)^2 \frac{\cos\theta_2}{\cos\theta_1}$$

where we have use Eq. (2.14) to substitute for $(\mu_1/\mu_2)^{\frac{1}{2}}$. Simplifying, we get

$$T_p = \frac{4\epsilon_1\epsilon_2 \sin\theta_1 \cos\theta_1 \sin\theta_2 \cos\theta_2}{(\epsilon_2 \cos\theta_1 \sin\theta_2 + \epsilon_1 \sin\theta_1 \cos\theta_2)^2} \qquad (2.30b)$$

It can readily be verified that

$$R_p + T_p = 1 \qquad (2.31)$$

as it indeed should be.

We return to Eq. (2.24) and use Eqs. (2.19) and (2.20) to write it in the form

$$r_p = \frac{n_1^2 \sin\theta_1 \cos\theta_2 - n_2^2 \cos\theta_1 \sin\theta_2}{n_1^2 \sin\theta_1 \cos\theta_2 + n_2^2 \cos\theta_1 \sin\theta_2} \qquad (2.32)$$

From the above equation we can make the following observations.

(a) When $n_2 = n_1$, $\theta_2 = \theta_1$ and

$$r_p = 0$$

implying that there is no reflected wave. Thus if we put a transparent substance (like glass) inside a liquid which has the same refractive index as glass then the glass will not be visible from outside!

(b) Polarization by reflection (Brewster's law): the amplitude reflection coefficient r_p vanishes when

$$n_1^2 \sin\theta_1 \cos\theta_2 = n_2^2 \cos\theta_1 \sin\theta_2$$

If we multiply both sides by $\sin\theta_2$ and use Snell's law (i.e., $n_1 \sin\theta_1 = n_2 \sin\theta_2$) we get

$$n_1^2 \sin\theta_1 \cos\theta_2 \sin\theta_2 = n_1^2 \sin^2\theta_1 \cos\theta_1$$

or

$$\sin 2\theta_1 = \sin 2\theta_2$$

Thus either $\theta_1 = \theta_2$ (which corresponds to the case (a) above) or

$$\theta_1 = \tfrac{1}{2}\pi - \theta_2.$$

Thus when

$$\theta_1 + \theta_2 = \tfrac{1}{2}\pi \qquad (2.33)$$

(i.e., when the reflected and transmitted rays are at right angles to each other) there is no reflection for the parallel component of $\mathscr{E}$. The corresponding angle of incidence can readily be found:

$$\cos\theta_1 = \sin\theta_2 = n_1 \sin\theta_1/n_2$$

where we have used Eq. (2.17). Thus when

$$\theta_1 = \theta_p \equiv \tan^{-1}(n_2/n_1) \tag{2.34}$$

$r_p = 0$ and if the incident wave is unpolarized then only the perpendicular component of $\mathcal{E}$ will be reflected and the (reflected) light will be linearly polarized (see Fig. 2.2). This is Brewster's law and the angle θ_p is usually referred to as the *polarizing angle* or the *Brewster angle*. For the air–glass interface $n_1 = 1.0$, $n_2 = 1.5$ and

$$\theta_p = \tan^{-1} 1.5 \approx 56.3° \tag{2.35}$$

(c) Reflection at normal and grazing incidence: in order to study reflection at normal incidence we use Snell's law ($n_2 \sin\theta_2 = n_1 \sin\theta_1$) to put Eq. (2.32) in the following form:

$$r_p = \frac{n_1 \cos\theta_2 - n_2 \cos\theta_1}{n_1 \cos\theta_2 + n_2 \cos\theta_1} \tag{2.36}$$

At normal incidence, $\theta_1 = \theta_2 = 0$ and we have

$$r_p = \frac{n_1 - n_2}{n_1 + n_2} \tag{2.37}$$

For the air–glass interface $n_1 = 1.0$, $n_2 = 1.5$ giving

$$r_p = -0.2 \quad \text{and} \quad R = 0.04 \tag{2.38}$$

Fig. 2.2 When an electromagnetic wave is incident at the Brewster angle $(\theta_1 = \theta_p = \tan^{-1}(n_2/n_1))$, the reflected wave is linearly polarized with its electric vector oscillating at right angles to the plane of incidence; thus an incident unpolarized wave becomes s-polarized on reflection.

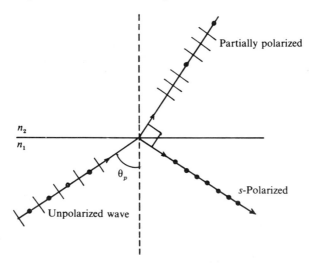

implying that only 4% of the light intensity is reflected.

On the other hand, for grazing incidence (i.e., $\theta_1 \to \frac{1}{2}\pi$) we have the following limiting formula (see Problem 2.2)

$$r_p \approx 1 - \frac{2(n_2/n_1)^2 \alpha_1}{[(n_2/n_1)^2 - 1]^{\frac{1}{2}}} \tag{2.39}$$

where $\alpha_1 \equiv \frac{1}{2}\pi - \theta_1$, and is measured in radians. (Obviously, we are assuming $n_2 > n_1$, otherwise total internal reflection will occur – see Sec. 2.3.) Eq. (2.39) tells us that at grazing incidence, $|r_p|$ tends to unity and reflection is almost complete. Thus if we hold a glass plate at the level of the eye, it will almost act as a mirror!

(*d*) Complete expressions for the electric and magnetic fields: using the definitions for r_p and t_p and Eq. (1.32) we can write down the complete expressions for the electric and magnetic fields associated with the incident, transmitted and reflected waves:

$$\boldsymbol{\mathscr{E}}_i = \boldsymbol{\mathscr{E}}_1 = (\hat{\mathbf{x}} \sin \theta_1 - \hat{\mathbf{z}} \cos \theta_1) E_1 e^{i[\omega t - k_1(x \cos \theta_1 + z \sin \theta_1)]} \tag{2.40}$$

$$\boldsymbol{\mathscr{H}}_i = \boldsymbol{\mathscr{H}}_1 = \hat{\mathbf{y}}(E_1/Z_1) e^{i[\omega t - k_1(x \cos \theta_1 + z \sin \theta_1)]} \tag{2.41}$$

$$\boldsymbol{\mathscr{E}}_t = \boldsymbol{\mathscr{E}}_2 = (\hat{\mathbf{x}} \sin \theta_2 - \hat{\mathbf{z}} \cos \theta_2) t_p E_1 e^{i[\omega t - k_2(x \cos \theta_2 + z \sin \theta_2)]} \tag{2.42}$$

$$\boldsymbol{\mathscr{H}}_t = \boldsymbol{\mathscr{H}}_2 = \hat{\mathbf{y}}(t_p E_1/Z_2) e^{i[\omega t - k_2(x \cos \theta_2 + z \sin \theta_2)]} \tag{2.43}$$

$$\boldsymbol{\mathscr{E}}_r = \boldsymbol{\mathscr{E}}_3 = (-\hat{\mathbf{x}} \sin \theta_1 - \hat{\mathbf{z}} \cos \theta_1) r_p E_1 e^{i[\omega t - k_1(-x \cos \theta_1 + z \sin \theta_1)]} \tag{2.44}$$

$$\boldsymbol{\mathscr{H}}_r = \boldsymbol{\mathscr{H}}_3 = -\hat{\mathbf{y}}(r_p E_1/Z_1) e^{i[\omega t - k_1(-x \cos \theta_1 + z \sin \theta_1)]} \tag{2.45}$$

where

$$Z_1 = (\mu_1/\epsilon_1)^{\frac{1}{2}} \quad \text{and} \quad Z_2 = (\mu_2/\epsilon_2)^{\frac{1}{2}} \tag{2.46}$$

are known as the characteristic impedances of media 1 and 2 respectively. Notice that the direction of the magnetic field is along the *y*-axis which is tangential to the surface. Thus, continuity of the tangential component of $\boldsymbol{\mathscr{H}}$ at the interface leads to

$$\frac{t_p E_1}{Z_2} = \frac{E_1}{Z_1} - \frac{r_p E_1}{Z_1}$$

or

$$t_p = \left(\frac{\mu_2 \epsilon_1}{\mu_1 \epsilon_2}\right)^{\frac{1}{2}} (1 - r_p)$$

$$= \frac{\epsilon_1 \sin \theta_1}{\epsilon_2 \sin \theta_2} (1 - r_p) \tag{2.47}$$

If we substitute for r_p from Eq. (2.24) we obtain Eq. (2.25). Thus the continuity condition for $\mathcal{H}_y$ does not give any additional result. For nonmagnetic media (see Eqs. (2.19) and (2.20)), Eq. (2.47) becomes

$$t_p = (n_1/n_2)(1 - r_p) \tag{2.48}$$

Case 2: $\mathcal{E}$ *lying perpendicular to the plane of incidence*. We next consider the case in which the electric vector associated with the incident wave is perpendicular to the plane of incidence, i.e., along the y-axis (see Fig. 2.3). The electric vectors associated with the reflected and transmitted waves will also be perpendicular to the plane of incidence (see Problem 2.1). Thus, the electric fields associated with the incident, transmitted and reflected waves will be given by (cf. Eq. (2.1))

$$\mathcal{E}_1 = \hat{\mathbf{y}} E_1 e^{i(\omega t - \mathbf{k}_1 \cdot \mathbf{r})} \tag{2.49a}$$

$$\mathcal{E}_2 = \hat{\mathbf{y}} E_2 e^{i(\omega t - \mathbf{k}_2 \cdot \mathbf{r})} \tag{2.49b}$$

and

$$\mathcal{E}_3 = \hat{\mathbf{y}} E_3 e^{i(\omega t - \mathbf{k}_3 \cdot \mathbf{r})} \tag{2.49c}$$

The magnetic field is given by

$$\mathcal{H} = \frac{\mathbf{k} \times \mathcal{E}}{\omega \mu} = \frac{1}{\omega \mu} [\hat{\mathbf{x}}(-k_z \mathcal{E}_y) + \hat{\mathbf{z}}(k_x \mathcal{E}_y)] \tag{2.50}$$

Fig. 2.3 The reflection and refraction of a plane wave incident at the interface of two dielectrics; the electric vector is assumed to be perpendicular to the plane of incidence (i.e., s-polarized).

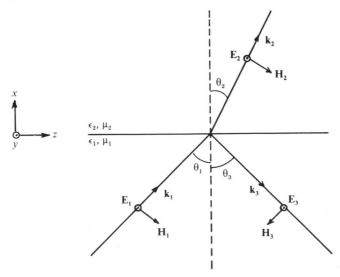

where we have used the fact that $k_y = 0$ (see Eqs. (2.2) and (2.11)). Thus if we write

$$\mathscr{H}_1 = \mathbf{H}_1 e^{i(\omega t - \mathbf{k}_i \cdot \mathbf{r})} \tag{2.51}$$

(and similar expressions for $\mathscr{H}_2$ and $\mathscr{H}_3$), then

$$\mathbf{H}_1 = \left(\frac{\epsilon_1}{\mu_1}\right)^{\frac{1}{2}} (-\hat{\mathbf{x}} \sin \theta_1 + \hat{\mathbf{z}} \cos \theta_1) E_1 \tag{2.52}$$

$$\mathbf{H}_2 = (\epsilon_2/\mu_2)^{\frac{1}{2}} [-\hat{\mathbf{x}} \sin \theta_2 + \hat{\mathbf{z}} \cos \theta_2] E_2 \tag{2.53}$$

and

$$\mathbf{H}_3 = (\epsilon_1/\mu_1)^{\frac{1}{2}} [-\hat{\mathbf{x}} \sin \theta_1 - \hat{\mathbf{z}} \cos \theta_1) E_3 \tag{2.54}$$

will represent the space and time independent part of the magnetic field associated with the incident, transmitted and reflected waves respectively. Since H_z represents a tangential component, its continuity across the interface leads to

$$(\epsilon_1/\mu_1)^{\frac{1}{2}} (\cos \theta_1) E_1 - (\epsilon_1/\mu_1)^{\frac{1}{2}} (\cos \theta_1) E_3 = (\epsilon_2/\mu_2)^{\frac{1}{2}} (\cos \theta_2) E_2 \tag{2.55}$$

Also, since $\mathscr{E}$ has only a y-component (which is tangential to the interface) we have

$$E_1 + E_3 = E_2 \tag{2.56}$$

Substituting for E_2 from Eq. (2.56) in Eq. (2.55) we get after simple manipulations

$$r_s \equiv \frac{E_3}{E_1} = \frac{(\epsilon_1/\mu_1)^{\frac{1}{2}} \cos \theta_1 - (\epsilon_2/\mu_2)^{\frac{1}{2}} \cos \theta_2}{(\epsilon_1/\mu_1)^{\frac{1}{2}} \cos \theta_1 + (\epsilon_2/\mu_2)^{\frac{1}{2}} \cos \theta_2} \tag{2.57}$$

where the subscript s refers to perpendicular polarization. For nonmagnetic media

$$r_s = \frac{n_1 \cos \theta_1 - n_2 \cos \theta_2}{n_1 \cos \theta_1 + n_2 \cos \theta_2} \tag{2.58}$$

Further

$$t_s \equiv \frac{E_2}{E_1} = 1 + r_s = \frac{2n_1 \cos \theta_1}{n_1 \cos \theta_1 + n_2 \cos \theta_2} \tag{2.59}$$

The variation of $|r_p|$ and $|r_s|$ with the angle of incidence θ_1 is plotted in Fig. 2.4 for the air–glass interface $n_1 = 1.0$, $n_2 = 1.5$. Notice that $|r_p|$ vanishes at the Brewster angle. Also $|r_p|$ and $|r_s|$ tend to the same value at normal and grazing incidence.

As for the case of parallel polarization we write the complete expressions for the electric and magnetic fields associated with the incident, transmitted and reflected waves:

$$\mathscr{E}_i = \mathscr{E}_1 = \hat{y}E_1 \exp\{i[\omega t - k_1(x\cos\theta_1 + z\sin\theta_1)]\} \tag{2.60}$$

$$\begin{aligned}\mathscr{H}_i = \mathscr{H}_1 &= (-\hat{x}\sin\theta_1 + \hat{z}\cos\theta_1)(E_1/Z_1)\\ &\times \exp\{i[\omega t - k_1(x\cos\theta_1 + z\sin\theta_1)]\}\end{aligned} \tag{2.61}$$

$$\mathscr{E}_t = \mathscr{E}_2 = \hat{y}t_s E_1 \exp\{i[\omega t - k_2(x\cos\theta_2 + z\sin\theta_2)]\} \tag{2.62}$$

$$\begin{aligned}\mathscr{H}_t = \mathscr{H}_2 &= (-\hat{x}\sin\theta_2 + \hat{z}\cos\theta_2)(t_s E_1/Z_2)\\ &\times \exp\{i[\omega t - k_2(x\cos\theta_2 + z\sin\theta_2)]\}\end{aligned} \tag{2.63}$$

$$\mathscr{E}_r = \mathscr{E}_3 = \hat{y}r_s E_1 \exp\{i[\omega t - k_1(-x\cos\theta_1 + z\sin\theta_1)]\} \tag{2.64}$$

$$\begin{aligned}\mathscr{H}_r = \mathscr{H}_3 &= (-\hat{x}\sin\theta_1 - \hat{z}\cos\theta_1)(r_s E_1/Z_1)\\ &\times \exp\{i[\omega t - k_1(-x\cos\theta_1 + z\sin\theta_1)]\}\end{aligned} \tag{2.65}$$

Problem 2.1: Consider the case in which $\mathscr{E}$ associated with the incident wave lies in the plane of incidence. Show that reflected or transmitted wave will not have any y-component of $\mathscr{E}$. Similarly for perpendicular polarization.

Fig. 2.4 Variation of the reflection coefficients $|r_p|$ and $|r_s|$ as a function of the angle of incidence for an air–glass interface.

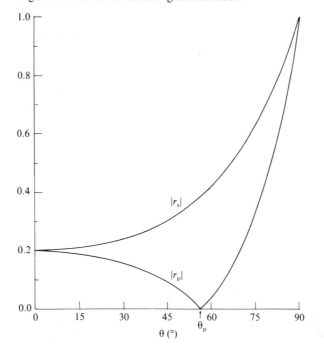

Solution: Let us suppose that we do have a y-component of $\mathscr{E}$ in the refracted and transmitted waves. Thus although the y-component of $\mathscr{E}_1$ is zero, y-components of $\mathscr{E}_2$ and $\mathscr{E}_3$ are nonvanishing; let their corresponding amplitudes be E_{2y} and E_{3y}. The continiuty of $\mathscr{E}_y$ and $\mathscr{H}_z$ across the interface gives us

$$E_{3y} = E_{2y} \tag{2.66}$$

and

$$-(\epsilon_1/\mu_1)^{\frac{1}{2}} \cos\theta_1 E_{3y} = (\epsilon_2/\mu_2)^{\frac{1}{2}} \cos\theta_2 E_{2y} \tag{2.67}$$

which represent an inconsistent set of equations and therefore

$$E_{2y} = E_{3y} = 0.$$

Problem 2.2: Derive Eq. (2.39).

Solution: We define $\alpha_1 = \frac{1}{2}\pi - \theta_1$ and $\alpha_2 = \frac{1}{2}\pi - \theta_2$. Thus Snell's law gives us

$$\cos\alpha_2 = (n_1/n_2)\cos\alpha_1 = (1/n)\cos\alpha_1 \tag{2.68}$$

where

$$n = n_2/n_1$$

Thus

$$\sin\alpha_2 = [1 - (1/n^2)\cos^2\alpha_1]^{\frac{1}{2}} \approx (1 - 1/n^2)^{\frac{1}{2}}$$

Now

$$r_p = -\frac{n\sin\alpha_1 - \sin\alpha_2}{n\sin\alpha_1 + \sin\alpha_2} \approx -\frac{n\alpha_1 - (1 - 1/n^2)^{\frac{1}{2}}}{n\alpha_1 + (1 - 1/n^2)^{\frac{1}{2}}} \tag{2.69}$$

to first order in α_1. Expanding the denominator in a power series in α_1, gives Eq. (2.39).

Problem 2.3: Using Snell's law, show that Eqs. (2.36), (2.48), (2.58) and (2.59) can be put in the following forms

$$r_p = \frac{-\sin\theta_1\cos\theta_1 + \sin\theta_2\cos\theta_2}{\sin\theta_1\cos\theta_1 + \sin\theta_2\cos\theta_2} = -\frac{\tan(\theta_1 - \theta_2)}{\tan(\theta_1 + \theta_2)} \tag{2.70}$$

$$t_p = -\frac{2\sin\theta_2\cos\theta_1}{\sin\theta_1\cos\theta_1 + \sin\theta_2\cos\theta_2} \tag{2.71}$$

$$r_s = -\frac{\sin(\theta_1 - \theta_2)}{\sin(\theta_1 + \theta_2)} \tag{2.72}$$

$$t_s = 1 + r_s = \frac{2\cos\theta_1\sin\theta_2}{\sin(\theta_1 + \theta_2)} \tag{2.73}$$

where all the above relations are valid for nonmagnetic media (see Eqs. (2.18)–(2.20)).

Problem 2.4: In continuation of the previous problem, let r_p', t_p', r_s' and t_s' denote the corresponding reflection and transmission coefficients when the media are interchanged and the angles of incidence and refraction are reversed (i.e., the incident wave is propagating in a medium of refractive index n_2 hitting the interface at an angle of incidence θ_2). Show that

$$1 + r_p r'_p = t_p t'_p \tag{2.74a}$$

$$1 + r_s r'_s = t_s t'_s \tag{2.74b}$$

$$r'_p = -r_p \tag{2.74c}$$

$$r'_s = -r_s \tag{2.74d}$$

which are known as Stoke's relations and which can also be derived by using the principle of optical reversibility (see e.g., Ghatak (1977) Sec. 11.12).

Problem 2.5: At normal incidence $\theta_1 = \theta_2 = 0$ and any orientation of linear polarization can be thought of as parallel polarization or perpendicular polarization; show that r_p and r_s tend to the same value at normal incidence.

Problem 2.6: Consider a plane electromagnetic wave described by the following equation

$$\mathcal{E}_i = (\hat{x} \sin \theta_1 - \hat{z} \cos \theta_1) E_0 e^{i(\omega t - \mathbf{k}_1 \cdot \mathbf{r})} + \hat{y} E_0 e^{i(\omega t - \mathbf{k}_1 \cdot \mathbf{r} - \frac{1}{2}\pi)} \tag{2.75}$$

where

$$\mathbf{k}_1 = \hat{x} k_1 \cos \theta_1 + \hat{z} k_1 \sin \theta_1 \tag{2.76}$$

Discuss the state of polarization of the electromagnetic wave. Assuming the above wave to be incident at the air–glass interface (at $x = 0$), determine the electric vector, associated with the reflected wave (and the corresponding state of polarization) when $\theta_1 = \tan^{-1}(1.5)$ and $\theta_1 = 45°$ ($n_1 = 1.0$ and $n_2 = 1.5$).

Solution: It is readily seen that $\mathcal{E} \cdot \mathbf{k}_1 = 0$. We choose a new axis $\hat{z}'$ along $\mathbf{k}_1$; thus

$$\left. \begin{array}{l} \hat{z}' = \hat{x} \cos \theta_1 + \hat{z} \sin \theta_1 \\ \hat{x}' = \hat{x} \sin \theta_1 - \hat{z} \cos \theta_1 \\ \hat{y}' = \hat{y} \end{array} \right\} \tag{2.77}$$

see Fig. 2.5. Thus taking the real parts

$$\mathcal{E}_{ix'} = E_0 \cos(\omega t - k_1 z')$$

$$\mathcal{E}_{iy'} = E_0 \sin(\omega t - k_1 z')$$

which represents a right circularly polarized wave (see Fig. 2.5(a)). The reflected wave will be (see Eqs. (2.44) and (2.64))

$$\mathcal{E}_r = -(\hat{x} \sin \theta_1 + \hat{z} \cos \theta_1) r_p E_0 e^{i(\omega t - \mathbf{k}_2 \cdot \mathbf{r})} + \hat{y} r_s E_0 e^{i(\omega t - \mathbf{k}_2 \cdot \mathbf{r} - \frac{1}{2}\pi)} \tag{2.78}$$

Choosing

$$\hat{x}'' = \hat{x} \sin \theta_1 + \hat{z} \cos \theta_1, \quad \hat{y}'' = \hat{y}, \quad \hat{z}'' = -\hat{x} \cos \theta_1 + \hat{z} \sin \theta_1$$

We get

$$\mathcal{E}_{rx''} = -r_p E_0 \cos(\omega t - k_2 z'')$$

and

$$\mathcal{E}_{ry''} = r_s E_0 \sin(\omega t - k_2 z'')$$

When $\theta_1 = \tan^{-1}(1.5)$, $r_p = 0$ and the reflected wave is y-polarized. When $\theta_1 = 45°$,

$$\theta_2 = \sin^{-1}\left(\frac{n_1 \sin \theta_1}{n_2}\right) = \sin^{-1}\left(\frac{1/\sqrt{2}}{1.5}\right) \approx 28.1°$$

Thus

$$r_p = -\frac{\tan(45° - 28.1°)}{\tan(45° + 28.1°)} \approx -0.092$$

$$r_s = -\frac{\sin(45° - 28.1°)}{\sin(45° + 28.1°)} \approx -0.303$$

and

$$\mathscr{E}_{rx''} = 0.092 E_0 \cos(\omega t - k_2 z'')$$

$$\mathscr{E}_{ry''} = -0.303 E_0 \sin(\omega t - k_z z'')$$

Fig. 2.5 (*a*) The reflection and refraction of a right circularly polarized light wave incident on the interface of two dielectrics. (*b*) The states of polarization corresponding to incident and reflected waves.

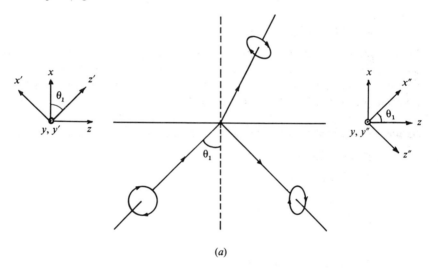

(*a*)

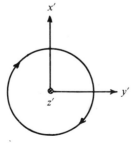

Right circular

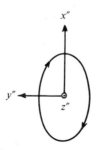

Left elliptical

(*b*)

which represents a left elliptically polarized wave (see (Fig. 2.5(*b*)). Further, for $\theta_1 = 45°$

$$t_p = (n_1/n_2)(1 - r_p) \approx 0.728$$
$$t_s = 1 + r_s \approx 0.697$$

and the transmitted beam will be right elliptically polarized as shown in Fig. 2.5(*a*)

2.3   Total internal reflection and evanescent waves
According to Snell's law

$$n_1 \sin \theta_1 = n_2 \sin \theta_2 \qquad (2.79)$$

and thus if the wave is incident on a rarer medium (i.e., if $n_2 < n_1$) then $\sin \theta_2$ will be greater than unity if

$$\theta_1 > \theta_c \qquad (2.80)$$

where

$$\theta_c = \sin^{-1}(n_2/n_1) \qquad (2.81)$$

This angle is known as the critical angle. For $\theta_1 > \theta_c$ we have the phenomenon known as total internal reflection. Since Eq. (2.79) is derived by applying the continuity conditions, it remains valid even for $\theta_1 > \theta_c$; however, for $\theta_1 > \theta_c$, we have a wave for which $\sin \theta_2$ exceeds unity (i.e., $k_{2z} = k_2 \sin \theta_2 > k_2$) and therefore $\cos \theta_2$ is a purely imaginary quantity (i.e., $k_{2x} = k_2 \cos \theta_2$ is purely imaginary). Now

$$\cos^2 \theta_2 = 1 - \sin^2 \theta_2 = -(n_1^2 \sin^2 \theta_1/n_2^2 - 1)$$

where the quantity inside the brackets is positive when $\theta_1 > \theta_c$. Thus

$$\cos \theta_2 = -i(n_1^2 \sin^2 \theta_1/n_2^2 - 1)^{\frac{1}{2}} \qquad (2.82)$$

(the reason for choosing the minus sign will soon become obvious). For definiteness we consider perpendicular polarization for which the amplitude transmission coefficient is given by Eq. (2.59). Since $\cos \theta_2$ is now an imaginary quantity, t_s is complex and we may write

$$t_s = |t_s| e^{i\alpha} \qquad (2.83)$$

Thus the transmitted electric field can be written in the form

$$\mathscr{E}_t = \hat{y} |t_s| E_1 \exp\{i[\omega t - k_2(x \cos \theta_2 + z \sin \theta_2) + \alpha]\} \qquad (2.84)$$

If now substitute for $\sin \theta_2$ and $\cos \theta_2$ from Eqs. (2.79), and (2.82) respectively and use $k_2 = (\omega/c)n_2$ we obtain[†]

$$\mathscr{E}_t = \hat{\mathbf{y}}|t_s|E_1 \exp(-\gamma x)\exp\{i[\omega t - \omega(n_1/c)z\sin\theta_1 + \alpha]\} \qquad (2.85)$$

where

$$\gamma = (\omega/c)(n_1^2\sin^2\theta_1 - n_2^2)^{\frac{1}{2}} \qquad (2.86)$$

Eq. (2.85) represents a wave which decays exponentially in the x-direction and propagates in the z-direction; such a wave is known as an *evanescent wave*. The corresponding magnetic field is given by (see Eq. (2.63)):

$$\mathscr{H}_t = \left[-\hat{\mathbf{x}}\frac{n_1\sin\theta_1}{n_2} - \hat{\mathbf{z}}i\left(\frac{n_1^2\sin^2\theta_1}{n_2^2} - 1\right)^{\frac{1}{2}} \right]$$
$$\times \frac{|t_s|E_1}{Z_2}\exp(-\gamma x)\exp\left[i\left(\omega t - \frac{\omega n_1}{c}z\sin\theta_1 + \alpha\right)\right] \qquad (2.87)$$

Since $-i = e^{-\frac{1}{2}i\pi}$, the z-component of $\mathscr{H}$ is $\frac{1}{2}\pi$ out of phase with respect to $\mathscr{H}_x$ and $\mathscr{E}_y$. The actual electric and magnetic fields (which would be the real parts of Eqs. (2.85) and (2.87)) would be given by

$$\mathscr{E}_y = |t_s|E_1 e^{-\gamma x}\cos\left(\omega t - \frac{\omega n_1}{c}z\sin\theta_1 + \alpha\right) \qquad (2.88)$$

$$\mathscr{H}_x = -\frac{n_1\sin\theta_1}{n_2}|t_s|\frac{E_1}{Z_2}e^{-\gamma x}\cos\left(\omega t - \frac{\omega n_1}{c}z\sin\theta_1 + \alpha\right) \qquad (2.89)$$

and

$$\mathscr{H}_z = \left(\frac{n_1^2\sin^2\theta_1}{n_2^2} - 1\right)^{\frac{1}{2}}\frac{|t_s|E_1}{Z_2}e^{-\gamma x}\sin\left(\omega t - \frac{\omega n_1}{c}z\sin\theta_1 + \alpha\right)$$
$$(2.90)$$

Thus

$$\langle S_x \rangle = \langle \mathscr{E}_y\mathscr{H}_z \rangle = 0$$

implying that there is *no* power flow along the x-direction and therefore

---

[†] We should mention here that if, instead of Eq. (2.82) we had taken

$$\cos\theta_2 = +i(n_1^2\sin^2\theta_1/n_2^2 - 1)^{\frac{1}{2}}$$

we would have obtained a wave whose amplitude would have *increased* exponentially with x; clearly this would be physically impossible.

the transmission coefficient (see Eq. (2.27)) is zero. However,

$$\langle S_z \rangle = \langle -\mathscr{E}_y \mathscr{H}_x \rangle = \frac{1}{2} \frac{n_1 \sin\theta_1}{n_2} |t_s|^2 \frac{E_1^2}{Z_2} e^{-2\gamma x} \tag{2.91}$$

Thus there *is* power flow along the z-axis in the second medium (see Fig. 2.6). We should point out that since $\cos\theta_2$ is purely imaginary[†].

$$|r_p| = |r_s| = 1 \tag{2.92}$$

(see Eqs. (2.36) and (2.57)). Thus although the entire energy is reflected back, there is power flowing in the second medium. Physically, we can understand this by considering the incidence of a spatially bounded beam at the interface (see Fig. 2.7). As shown in the figure, the beam undergoes

Fig. 2.6 When a plane wave undergoes total internal reflection (at the interface of two dielectrics) there is an evanescent wave in the rarer medium which propagates along the z-axis and whose amplitude decreases exponentially along the x-axis.

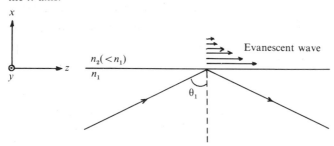

Fig. 2.7 When a spatially bounded beam is incident at an interface making an angle greater than the critical angle, then the beam undergoes a lateral shift which can be interpreted as the beam entering the rarer medium and reemerging (from the rarer medium) after reflection.

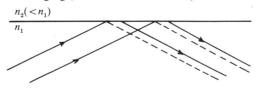

[†] Using Eqs. (2.58) and (2.82), we obtain $r_s = e^{-i\chi}$ where
$\chi = -2\tan^{-1}\{[n_1^2 \sin^2\theta_1 - n_2^2]^{\frac{1}{2}}/n_1 \cos\theta_1\}$.

a lateral shift which can be interpreted as the beam entering the rarer medium and reemerging (from the rarer medium) after reflection. This shift is known as the Goos–Hanchen shift (see, e.g., Ghatak and Thyagarajan (1978) for a detailed account of the Goos–Hanchen shift). It is now physically obvious that if, instead of a spatially bounded beam, we have an infinitely extended plane wave incident on the interface then although the reflection is complete, energy will flow along the z-axis in the rarer medium; the magnitude of this energy decays along the x-axis.

A very important application of evanescent waves is the prism–film coupler (see Fig. 2.8) which is extensively used in the coupling of a laser beam into a thin film waveguide used in integrated optics (see Chapter 14). It consists of a high refractive index prism placed in close contact with a thin film waveguide and the laser beam is incident on the base of the prism at an angle greater than the critical angle (i.e., $\theta > \sin^{-1}(n_c/n_p)$; see Fig. 2.8). In the air space between the prism and the film, an evanescent wave is generated which couples energy from the incident beam into the waveguide. Detailed calculation shows that maximum power transfer occurs when the angle of incidence satisfies the following condition: $\theta = \sin^{-1}(\beta/k_0 n_p)$ where k_0 is the free space wave number and β is the propagation constant of the waveguide mode[†]. Thus different modes can be excited independently by choosing different angles of incidence.

Fig. 2.8 The prism–film coupler arrangement: n_f, n_s and n_c represent the refractive indices of the film, substrate and cover respectively, $n_f > n_s, n_c$.

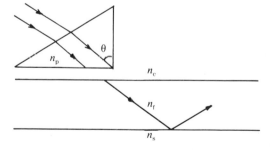

[†] In Chapters 11 and 14 we will show that a thin film waveguide can support various modes of propagation and each mode is characterized by a specific value of the propagation constant β. The phase velocity of the mode is given by ω/β which can be written as $c/(\beta/k_0)$ where we have used $\omega = ck_0$. Thus β/k_0 can be interpreted as the effective refractive index of the medium for the mode.

Problem 2.7: Consider the incidence of a plane wave on a glass–air interface (n_1 = 1.5, n_2 = 1.0). Assume the incident plane wave to be described by the following equation

$$\mathscr{E} = 5\hat{\mathbf{y}}\exp\left[i(\omega t - kx\cos\theta - kz\sin\theta)\right] V/m$$

For $\theta = 52°$, $60°$ and $80°$ calculate the transmitted field. Assume the free space wavelength of the wave to be 6000 Å. Show that the depths of penetration ($= 1/\gamma$) of the evanescent wave are 1152 Å and 878 Å for $\theta = 60°$ and $80°$ respectively.

2.4   Reflection and transmission by a film

In this section we will consider the reflection (and transmission) of an electromagnetic wave incident normally on a film of thickness d_2, having refractice index n_2 as shown in Fig. 2.9. The electric fields in the three media are given by

$$\mathscr{E} = \begin{cases} \hat{\mathbf{y}}E_1^+\,e^{-ik_1x} + \hat{\mathbf{y}}E_1^-\,e^{ik_1x}; & x < 0 \\ \hat{\mathbf{y}}E_2^+\,e^{-ik_2x} + \hat{\mathbf{y}}E_2^-\,e^{ik_2x}; & 0 < x < d_2 \\ \hat{\mathbf{y}}E_3^+\,e^{ik_3d_2}e^{-ik_3x} + \hat{\mathbf{y}}E_3^-\,e^{-ik_3d_2}e^{ik_3x}; & x > d_2 \end{cases} \tag{2.93}$$

where

$$k_i = (\omega/c)n_i = k_0 n_i; \quad i = 1, 2, 3 \tag{2.94}$$

n_1', n_2 and n_3 being the refractive indices in regions $x < 0$, $0 < x < d_2$ and $x > d_2$ respectively and where

$$k_0 = \omega/c = 2\pi/\lambda_0 \tag{2.95}$$

Fig. 2.9 Reflection and transmission of an electromagnetic wave incident on a dielectric film of thickness d_2.

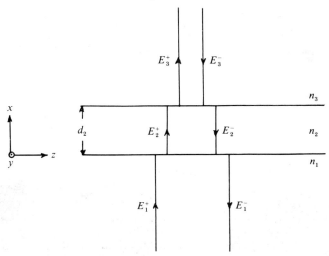

represents the free space wave number. The implicit $e^{i\omega t}$ time dependence has been suppressed and therefore the superscripts $+$ and $-$ correspond to waves propagating in the $+x$ and $-x$-directions respectively. The phase factors (for the field in the region $x > d_2$) have been chosen so that E_3^+ and E_3^- represent the fields at $x = d_2$. The corresponding magnetic fields can be calculated by using the formula

$$\mathcal{H} = \mathbf{k} \times \mathcal{E}/\omega\mu \tag{2.96}$$

and are given by

$$\mathcal{H} = \begin{cases} \hat{\mathbf{z}}\dfrac{n_1}{c\mu_0} E_1^+ e^{-ik_1 x} - \hat{\mathbf{z}}\dfrac{n_1}{c\mu_0} E_1^- e^{ik_1 x}; & x < 0 \\[2mm] \hat{\mathbf{z}}\dfrac{n_2}{c\mu_0} E_2^+ e^{-ik_2 x} - \hat{\mathbf{z}}\dfrac{n_2}{c\mu_0} E_2^- e^{ik_2 x}; & 0 < x < d_2; \\[2mm] \hat{\mathbf{z}}\dfrac{n_3}{c\mu_0} E_3^+ e^{ik_3 d_2} e^{-ik_3 x} - \hat{\mathbf{z}}\dfrac{n_3}{c\mu_0} E_3^- e^{-ik_3 d_2} e^{ik_3 x}; & x > d_2 \end{cases} \tag{2.97}$$

where the media are assumed to be nonmagnetic with $\mu = \mu_0$. Since both $\mathcal{E}$ and $\mathcal{H}$ represent tangential components, they must be continuous at $x = 0$ and at $x = d_2$. Continuity at $x = 0$ gives us

$$\begin{aligned} E_1^+ + E_1^- &= E_2^+ + E_2^- \\ n_1(E_1^+ - E_1^-) &= n_2(E_2^+ - E_2^-) \end{aligned} \tag{2.98}$$

Simple manipulations enable us to write the above equation in the following form

$$\begin{pmatrix} E_1^+ \\ E_1^- \end{pmatrix} = S_1 \begin{pmatrix} E_2^+ \\ E_2^- \end{pmatrix} = \frac{1}{t_1}\begin{pmatrix} 1 & r_1 \\ r_1 & 1 \end{pmatrix}\begin{pmatrix} E_2^+ \\ E_2^- \end{pmatrix} \tag{2.99}$$

where r_1 and t_1 are the amplitude reflection and transmission coefficients at the first interface and are given by (cf. Eqs. (2.58) and (2.59))

$$r_1 = \frac{n_1 - n_2}{n_1 + n_2} \tag{2.100}$$

$$t_1 = \frac{2n_1}{n_1 + n_2} \tag{2.101}$$

Continuity of the fields at $x = d_2$ readily gives us

$$\begin{pmatrix} E_2^+ \\ E_2^- \end{pmatrix} = S_2 \begin{pmatrix} E_3^+ \\ E_3^- \end{pmatrix} = \frac{1}{t_2}\begin{pmatrix} e^{i\delta_2} & r_2 e^{i\delta_2} \\ r_2 e^{-i\delta_2} & e^{-i\delta_2} \end{pmatrix}\begin{pmatrix} E_3^+ \\ E_3^- \end{pmatrix} \tag{2.102}$$

where r_2 and t_2 are once again the amplitude reflection and transmission coefficients at the second interface (i.e., at the interface of media n_2 and n_3) and

$$\delta_2 = k_2 d_2 = k_0 n_2 d_2 \tag{2.103}$$

We can combine Eqs. (2.99) and (2.102) to obtain

$$\begin{pmatrix} E_1^+ \\ E_1^- \end{pmatrix} = S \begin{pmatrix} E_3^+ \\ E_3^- \end{pmatrix} = \begin{pmatrix} a & b \\ c & d \end{pmatrix} \begin{pmatrix} E_3^+ \\ E_3^- \end{pmatrix} \tag{2.104}$$

where

$$
\left.
\begin{aligned}
a &= \frac{1}{t_1 t_2} (e^{i\delta_2} + r_1 r_2 e^{-i\delta_2}) \\[2mm]
b &= \frac{1}{t_1 t_2} (r_2 e^{i\delta_2} + r_1 e^{-i\delta_2}) \\[2mm]
c &= \frac{1}{t_1 t_2} (r_1 e^{i\delta_2} + r_2 e^{-i\delta_2}) \\[2mm]
d &= \frac{1}{t_1 t_2} (r_1 r_2 e^{i\delta_2} + e^{-i\delta_2})
\end{aligned}
\right\} \tag{2.105}
$$

Now, in the third medium there will not be any reflected wave and as such $E_3^- = 0$. This immediately gives us

$$
\left.
\begin{aligned}
E_1^+ &= a E_3^+ \\
E_1^- &= c E_3^+
\end{aligned}
\right\} \tag{2.106}
$$

Therefore the amplitude reflection and transmission coefficients of the film are given by

$$r = \frac{E_1^-}{E_1^+} = \frac{c}{a} = \frac{r_1 e^{i\delta_2} + r_2 e^{-i\delta_2}}{e^{i\delta_2} + r_1 r_2 e^{-i\delta_2}} \tag{2.107}$$

$$t = \frac{E_3^+}{E_1^+} = \frac{1}{a} = \frac{t_1 t_2}{e^{i\delta_2} + r_1 r_2 e^{-i\delta_2}} \tag{2.108}$$

The reflectivity and transmittivity are given by (see Eqs. (2.29a) and (2.30a)):

$$R = |r|^2 = \frac{r_1^2 + r_2^2 + 2 r_1 r_2 \cos 2\delta_2}{1 + r_1^2 r_2^2 + 2 r_1 r_2 \cos 2\delta_2} \tag{2.109}$$

$$T = \frac{\frac{1}{2}(\epsilon_3/\mu_0)^{\frac{1}{2}} |E_3^+|^2}{\frac{1}{2}(\epsilon_1/\mu_0)^{\frac{1}{2}} |E_1^+|^2} = |t|^2 \frac{n_3}{n_1} = \frac{n_3}{n_1} \frac{t_1^2 t_2^2}{1 + r_1^2 r_2^2 + 2 r_1 r_2 \cos 2\delta_2} \tag{2.110}$$

On substituting the values of r_1, r_2, t_1, t_2 we would get

$$R + T = 1 \tag{2.111}$$

Furthermore, when $\cos 2\delta_2 = -1$, i.e., when

$$\delta_2 = (m + \tfrac{1}{2})\pi; \quad m = 0, 1, 2, \ldots$$

or

$$d_2 = \lambda_0/4n_2, 3\lambda_0/4n_2, 5\lambda_0/4n_2, \ldots \tag{2.112}$$

the reflectivity of the film is a minimum (for $r_1 r_2 > 0$) and is given by

$$R = \left(\frac{r_1 - r_2}{1 - r_1 r_2} \right)^2 = \left(\frac{n_1 n_3 - n_2^2}{n_1 n_3 + n_2^2} \right)^2 \tag{2.113}$$

Thus the reflectivity would vanish if $n_2 = (n_1 n_3)^{\frac{1}{2}}$. Hence by using such a film, the reflectivity of the surface can be made zero at any chosen wavelength λ_c by choosing the film thickness to be $\lambda_c/4n_2$. This technique of reducing the reflectivity of a surface is known as blooming and the surface is said to be antireflection coated.

As an example, we consider a glass surface (of refractive index 1.62) which has to be made antireflecting around $\lambda_c = 5500\,\text{Å}$. Thus

$$n_1 = 1, \quad n_3 = 1.62$$

and therefore for the minimum reflectivity to be zero, we must choose a film whose refractive index is given by

$$n_2 = (n_1 n_3)^{\frac{1}{2}} = (1.62)^{\frac{1}{2}} \approx 1.273$$

having a thickness given by (see Eq. (2.112)):

$$d_2 = (2m + 1)\,\lambda_c/4n_2 \approx (2m + 1)1080.13\,\text{Å}$$

In Fig. 2.10 we have plotted the variation of the reflectivity with wavelength for $d_2 = \lambda_c/4n_2 = 1080\,\text{Å}$ and $d_2 = 5\lambda_c/4n_2 \approx 5400\,\text{Å}$. In plotting the figure we have neglected the dependence of the refractive indices on the wavelength which, in general, have a small effect. It can be seen that whereas the reflectivity is zero at $5500\,\text{Å}$ for both values of d_2, the variation of reflectivity over the visible range of the spectrum is much smaller when $d_2 = \lambda_c/4n_2$. Thus, as can be seen from the figure, at $\lambda_0 = 6500\,\text{Å}$, the reflectivity increases to 0.0034 and 0.0493 for $d_2 = \lambda_c/4n_2$ and $d_2 = 5\lambda_c/4n_2$ respectively. It is for this reason that one should use a film of thickness $\lambda_c/4n_2$ rather than of thickness $3\lambda_c/4n_2$ or $5\lambda_c/4n_2$.

Returning to Eq. (2.109) when $\cos 2\delta_2 = 1$ i.e., when

$$d_2 = \lambda_0/2n_2, \quad 2\lambda_0/2n_2, \quad 3\lambda_0/2n_2 \cdots \tag{2.114}$$

the reflectivity is given by

$$R = \left(\frac{n_3 - n_1}{n_3 + n_1}\right)^2 \tag{2.115}$$

which is independent of the refractive index of the intermediate film.

We should mention here that even for oblique incidence Eqs (2.109) and (2.110) for R and T remain the same, provided we redefine δ_2 by

$$\delta_2 = (2\pi/\lambda_0)n_2 d_2 \cos\theta_2 \tag{2.116}$$

where θ_2 represents the angle that the ray in medium 2 makes with the normal (see Fig. 2.11); furthermore r_1, r_2, t_1 and t_2 would then represent the amplitude reflection and transmission coefficients (corresponding to the appropriate polarization) for the first and second surfaces.

Problem 2.8: Lenses are usually coated with magnesium fluoride which has a refractive index of 1.38. Assuming the refractive index of glass to be 1.5, and the thickness of the film to be $\lambda_c/4n_2$ ($= 996.38$ Å), plot the variation of the reflectivity as a function of wavelength in the visible region of the spectrum and show that the minimum reflectivity is about 1.4% and increases to a little over 2% at the extremities of the visible spectrum.

Fig. 2.10 The variation of the reflectivity with wavelength for $d_2 = \lambda_c/4n_2 = 1080$ Å and $d_2 = 5\lambda_c/4n_2 = 5400$ Å; $\lambda_c = 5500$ Å and $n_1 = 1$, $n_2 = 1.273$ and $n_3 = 1.62$. The horizontal line corresponds to 5% reflectivity of the bare surface without any coating and corresponds to Eq. (2.115).

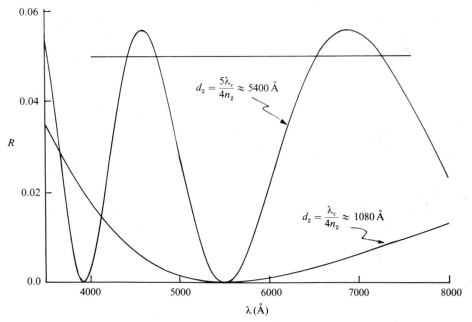

We next consider two simple examples:

Example 1: The Fabry–Perot etalon Let us consider a film of refractive index n_2 on both sides of which are media each having refractive index n_1, i.e., $n_3 = n_1$ in Fig. 2.11. For such a case for both s and p-polarizations

$$r_2 = -r_1 = r \qquad \text{(say)}$$

and Eq. (2.109) becomes

$$R = \frac{2r^2(1 - \cos 2\delta_2)}{1 + r^4 - 2r^2 \cos 2\delta_2}$$

$$= \frac{F \sin^2 \delta_2}{1 + F \sin^2 \delta_2} \tag{2.117}$$

where

$$F = 4r^2/(1 - r^2)^2 \tag{2.118}$$

is called the coefficient of finesse. The transmittivity can be calculated by using Eq. (2.110) or could be written directly:

$$T = 1 - R = \frac{1}{1 + F \sin^2 \delta_2} \tag{2.119}$$

Fig. 2.12 shows the variation of transmittivity as a function of δ_2 for different values of F. As can be seen from Eq. (2.119), the transmittivity is unity for

$$\delta_2 = (2\pi/\lambda_0)n_2 d_2 \cos \theta_2 = m\pi \tag{2.120}$$

Furthermore, as can be seen from Fig. 2.12 the width of the transmission peaks reduces as F is increased. This principle is used in the Fabry–Perot etalon which consists of a pair of highly reflecting glass plates which are held parallel to each

Fig. 2.11 Oblique incidence of a beam on a dielectric film.

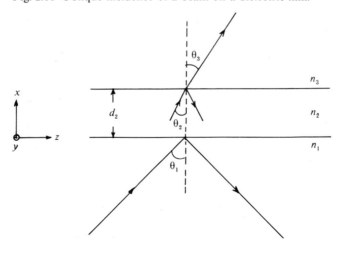

other at a fixed separation. Each glass plate is usually coated by a highly reflecting ($r \approx 0.9$) metallic film and each surface has to be flat within about $\lambda/20$–$\lambda/100$. In the schematic arrangement shown in Fig. 2.13, light from a broad source is collimated with the help of a lens and is passed through the Fabry–Perot etalon. Considering light of a specific wavelength λ_0, for angles of incidence such that

$$\cos \theta_2 = m \lambda_0 / 2 d_2 n_2 \tag{2.121}$$

Fig. 2.12 The variation of the transmittivity T as a function of δ_2 for different values of F. For large values of F, the transmittivity becomes very sharply peaked around $\delta_2 = m\pi$, $m = 1, 2, 3, \ldots$.

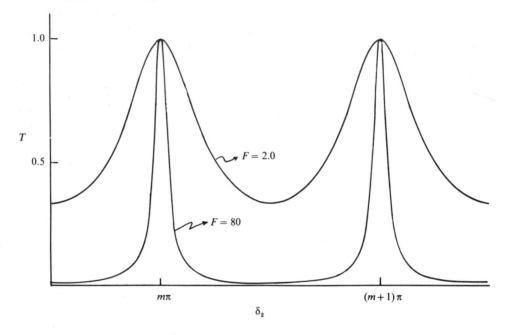

Fig. 2.13 Light waves from a broad source of light are collimated by the lens L_1 and are incident on a Fabry–Perot etalon which consists of a pair of highly reflecting glass plates which are parallel to each other. Whenever the angle of incidence satisfies Eq. (2.121) a ring pattern is observed on the focal plane of lens L_2. Different rings correspond to different values of λ_0 and m.

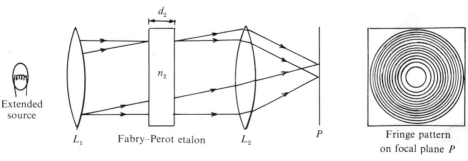

the incident light is completely transmitted. For large values of F, when θ_2 is slightly different from the values given by Eq. (2.121), the transmittivity will be very small. Hence for a given wavelength, at the focal plane P of the lens L_2, one will obtain a fringe system consisting of concentric rings corresponding to different values of m. The sharpness of the fringes will increase with the value of F.

It is obvious from Eq. (2.121) that the diameter of the rings will be different for different wavelengths and since the fringes can be made extremely sharp, two closely spaced wavelengths (such as in hyperfine splitting) can be resolved.

If one of the plates is kept fixed while the other is capable of moving to vary the separation between the two plates, the system is called a Fabry–Perot interferometer. In some cases, although the plates are fixed, the optical spacing (nd) between the plates can be varied by changing the pressure of the gas between the two plates. In the scanning method employing a Fabry–Perot interferometer, we allow a laser beam to be incident normally on the Fabry–Perot interferometer and scan the variation of the intensity at the focal spot of lens L_2 with the help of a photodetector by varying the optical spacing nd (see Fig. 2.14(a)). A typical output of the photodetector is shown in Fig. 2.14(b) in which we can see the various modes oscillating in the laser. The width of the peaks shown in Fig. 2.14(b) could be due either to the actual width of the spectral line or to the resolution limitation of the interferometer.

Fig. 2.14 (a) The schematic of a scanning Fabry–Perot interferometer in which the optical spacing (nd) between the plates is varied continuously and at the focal plane of the lens L_2 a photodetector collects the light behind a pinhole. (b) A typical output photodetector scan as a function of the optical spacing nd for a multi-longitudinal mode laser output. The peaks correspond to the various cavity modes.

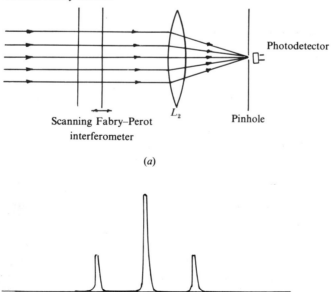

(a)

(b)

Example 2: Using the formulas for reflection and transmission coefficients developed earlier in the chapter, one can show that for a wave polarized in the plane of incidence and incident at the Brewster angle

$$\theta_1 = \tan^{-1}(n_2/n_1)$$

the reflectivity of the system is the same as that in the absence of the film. Thus by measuring this angle, one can calculate the refractive index of the film accuately. This method is known as *Abeles' method for direct measurement of refractive index of a transparent film by measuring the reflectivity at the Brewster angle.*

2.5   Extension to two films

We next consider an arrangement as shown in Fig. 2.15. The expression for the fields in the first three regions are given by Eqs. (2.93) and (2.97); in region 4, the fields are given by

$$\mathscr{E} = \hat{\mathbf{y}} E_4^+ e^{ik_4(d_2+d_3)} e^{-ik_4x} + \hat{\mathbf{y}} E_4^- e^{-ik_4(d_2+d_3)} e^{ik_4x} \tag{2.122}$$

$$\mathscr{H} = \hat{\mathbf{z}} \frac{n_4}{c\mu_0} E_4^+ e^{ik_4(d_2+d_3)} e^{-ik_4x} - \hat{\mathbf{z}} \frac{n_4}{c\mu_0} E_4^- e^{-ik_4(d_2+d_3)} e^{ik_4x} \tag{2.123}$$

Continuity of the fields at $x = d_2 + d_3$ gives us

$$\begin{pmatrix} E_3^+ \\ E_3^- \end{pmatrix} = S_3 \begin{pmatrix} E_4^+ \\ E_4^- \end{pmatrix} = \frac{1}{t_3} \begin{pmatrix} e^{i\delta_3} & r_3 e^{i\delta_3} \\ r_3 e^{-i\delta_3} & e^{-i\delta_3} \end{pmatrix} \begin{pmatrix} E_4^+ \\ E_4^- \end{pmatrix} \tag{2.124}$$

where r_3 and t_3 are the amplitude reflection and transmission coefficients between media 3 and 4 and

$$\delta_3 = k_3 d_3 = (2\pi/\lambda_0) n_3 d_3 \tag{2.125}$$

Fig. 2.15 Normal incidence of a plane wave on a two film structure.

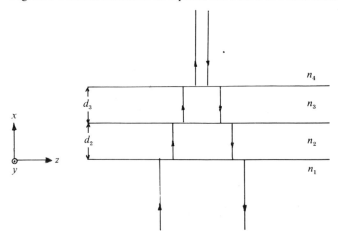

Thus

$$\begin{pmatrix} E_1^+ \\ E_1^- \end{pmatrix} = S \begin{pmatrix} E_4^+ \\ E_4^- \end{pmatrix} \tag{2.126}$$

where $S = S_1 S_2 S_3$. Proceeding in a manner similar to that used earlier one can readily obtain expressions for reflectivity and transmittivity.

Generalizing the above analysis for N films and for oblique incidence (see Fig. 2.16), we may write

$$\begin{pmatrix} E_1^+ \\ E_1^- \end{pmatrix} = S \begin{pmatrix} E_{N+2}^+ \\ E_{N+2}^- \end{pmatrix} \tag{2.127}$$

where

$$S = S_1 S_2 \cdots S_{N+1} \tag{2.128}$$

and

$$S_j = \frac{1}{t_j} \begin{pmatrix} e^{i\delta_j} & r_j e^{i\delta_j} \\ r_j e^{-i\delta_j} & e^{-i\delta_j} \end{pmatrix} \tag{2.129}$$

$$\delta_1 = 0$$

$$\delta_j = k_j d_j \cos\theta_j = (2\pi/\lambda_0) n_j d_j \cos\theta_j; \quad j = 2, 3, \ldots, (N+1) \tag{2.130}$$

Fig. 2.16 Oblique incidence of a plane wave on a layer of films.

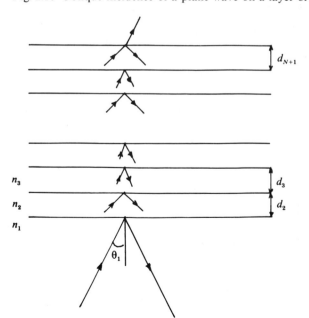

and r_j and t_j are the amplitude reflection and transmission coefficients between media of refractive indices n_j and n_{j+1}. The matrix multiplication can easily be carried out to determine the elements of S from which we can obtain the reflection and transmission coefficients (cf. Eqs. (2.106)–(2.108)). It should be mentioned that using Snell's law

$$n_1 \sin \theta_1 = n_2 \sin \theta_2 = \cdots = n_j \sin \theta_j = \cdots \tag{2.131}$$

Eq. (2.130) can be written in the form

$$\delta_j = (2\pi/\lambda_0)d_j(n_j^2 - n_1^2 \sin^2 \theta_1)^{\frac{1}{2}} \tag{2.132}$$

Fig. 2.17 (*a*) A three film antireflecting structure consisting of magnesium fluoride, zinc sulphide and aluminium oxide on a light flint glass substrate. (*b*) The solid curve shows the variation of reflectivity with wavelength for the structure shown in (*a*); the dashed curve shows the reflectivity for a single $\lambda_c/4n_2$ magnesium fluoride film. Note that using three films the reflectivity can be made very low over a large range of wavelengths.

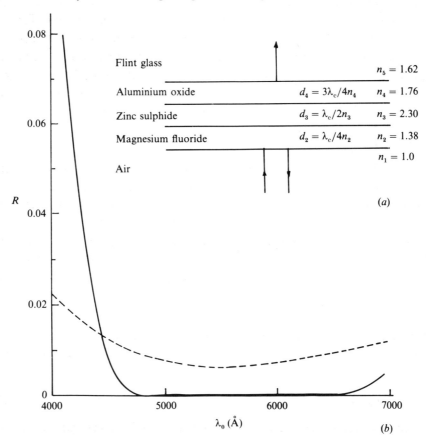

Thus knowing the angle of incidence at the first interface (θ_1), δ_j can readily be calculated.

The advantage of the above method is that it can very easily be programmed on a small computer and one can optimize the film thicknesses and corresponding refractive indices for specific requirements. For example, in some applications it may be necessary to have almost zero reflectance over a large wavelength region and indeed if we use a three layer system, as shown in Fig. 2.17(a) the corresponding reflectivity is extremely flat over a large wavelength region as shown in Fig. (2.17(b)).

2.6   Interference filters

In an interference filter we require a very high value of transmittivity over a small region of wavelength so that when white light is incident, the output will have a very small spread in wavelength. Fig. 2.18(a) shows an interference filter consisting of alternate layers of high and low refractive index on either side of a film of refractive index n_H whose thickness is $\lambda_c/2n_H$. Each film is of thickness $\lambda_c/4n$ as shown in Fig. 2.18(a). In Fig. 2.18(b) we have shown the spectral dependence of the transmittivity for a seven layer structure (shown in (a)) and a similar fifteen layer structure. As can be seen the transmittivity for the fifteen layer structure is very sharply peaked and therefore would allow only a small range of wavelengths to be transmitted. Such interference filters are extensively used in various optical devices/experiments requiring spectrally pure light.

2.7   Periodic media

We next consider the use of the matrix method in calculating the reflectivity and transmittivity of a periodic medium. In Fig. 2.19(a) we have shown a stack of layers consisting of alternate layers of equal thicknesses of media of refractive indices $(n_0 - \Delta n)$ and $(n_0 + \Delta n)$; in Fig. 2.19(b) we have plotted the corresponding variation of reflectivity with wavelength for a stack of 100 layers. As can be seen the wavelength corresponding to the maximum reflectivity corresponds to a free space wavelength given by

$$\lambda_0 = 2n_0 d \tag{2.133}$$

where d represents the spatial periodicity of the structure. It is interesting to note that this is indeed the Bragg condition according to which (see Chapter 18).

$$k_{dx} = k_{ix} - K \tag{2.134}$$

where $\mathbf{k}_d$ and $\mathbf{k}_i$ represent the wave vectors of the reflected and incident

Fig. 2.18 (*a*) A seven layer film of zinc sulphide and magnesium fluoride. The central film (zinc sulphide) is of thickness $\lambda_c/2n_H$; all other films are $\lambda_c/4n$ thick. (*b*) Wavelength variation of the transmittivity for stacks of seven and fifteen layer films.

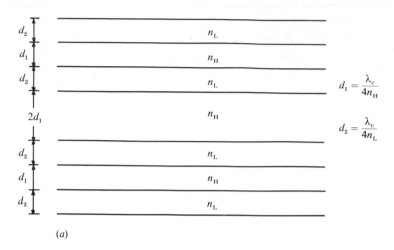

(*a*)

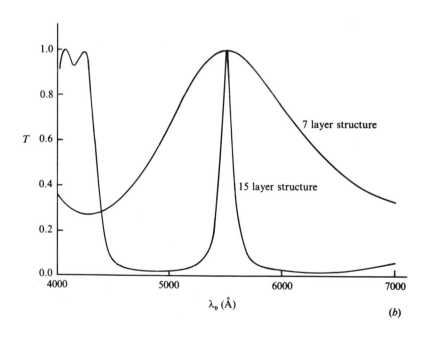

(*b*)

Fig. 2.19 (*a*) A periodic medium consisting of alternate layers of refractive indices $(n_0 + \Delta n)$ and $(n_0 - \Delta n)$ each of thickness $d/2 = \lambda_c/4n_0$. (*b*) The variation of reflectivity of a 100 layer periodic structure ($n_0 = 1.5$, $\Delta n = 0.01$, $d = \lambda_c/2n_0 = 1833.3$ Å with wavelength. Observe that the peak reflectivity appears at $\lambda_0 = 2n_0d$.

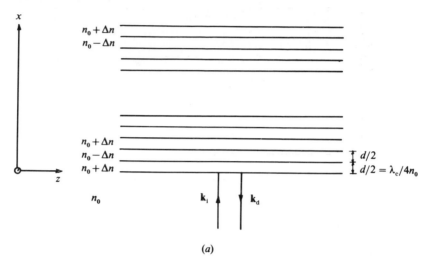

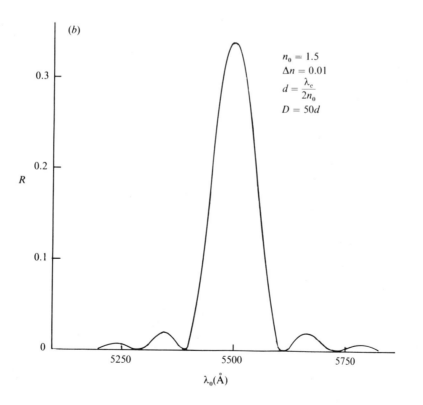

waves and

$$K = 2\pi/d \tag{2.135}$$

For normal incidence

$$k_{dx} = -(2\pi/\lambda_0)n_0, \quad k_{ix} = (2\pi/\lambda_0)n_0 \tag{2.136}$$

and Eq. (2.134) gives us Eq. (2.133).

We should mention here that such periodic media are of great importance in various optoelectronic devices such as acousto-optic modulators in which such a periodic refractive index grating is produced by a propagating acoustic wave in the medium (see Chapters 17–19); it is also used in a distributed Bragg reflector or distributed feedback lasers to provide highly wavelength selective reflection etc. (see e.g., Kressel and Butler (1977), Casey and Panish (1978)). The matrix method can also be used conveniently for a medium having a sinusoidal variation of refractive index by replacing the refractive index variation by a large number of steps.

2.8 Reflection and transmission in the presence of absorbing media

In the previous sections, we considered the reflection and refraction of a plane electromagnetic wave at the interface of two dielectrics as well as by a stack of dielectric films. In all cases we assumed the dielectrics to be nonabsorbing. In this section we will discuss how the theory has to be modified if the refractive indices become complex. This, indeed, happens in many devices using metallic films.

We first consider the reflection and refraction of a plane electromagnetic wave incident at the interface of a dielectric and an absorbing medium characterized by refractive indices n_1 and n_2; n_1 is assumed to be real and n_2, is, in general, complex. The incident, reflected and refracted waves are given by

$$\left.\begin{aligned}
\mathscr{E}_1 &= \mathbf{E}_1 e^{i(\omega t - \mathbf{k}_1 \cdot \mathbf{r})} = \mathbf{E}_1 e^{i\omega t} e^{-ik_1(x\cos\theta_1 + z\sin\theta_1)} \\
\mathscr{E}_2 &= \mathbf{E}_2 e^{i(\omega t - \mathbf{k}_2 \cdot \mathbf{r})} = \mathbf{E}_2 e^{i\omega t} e^{-ik_2(x\cos\theta_2 + z\sin\theta_1)} \\
\mathscr{E}_3 &= \mathbf{E}_3 e^{i(\omega t - \mathbf{k}_3 \cdot \mathbf{r})} = \mathbf{E}_3 e^{i\omega t} e^{-ik_1(x\cos\theta_3 - z\sin\theta_3)}
\end{aligned}\right\} \tag{2.137}$$

where we have assumed the plane of incidence to be the x–z plane (see Fig. 2.1). Here $k_1 = (\omega/c)n_1$ is a real quantity and

$$k_2 = (\omega/c)n_2 = (\omega/c)(\eta - i\kappa) \tag{2.138}$$

is a complex quantity (see Sec. 1.6). Use of the continuity conditions at

$x = 0$ gives us Snell's law

$$k_1 \sin \theta_1 = k_2 \sin \theta_2 = k_1 \sin \theta_3 \qquad (2.139)$$

Thus $\theta_1 = \theta_3$ and since n_2 is complex

$$\sin \theta_2 = \frac{n_1 \sin \theta_1}{n_2} = \frac{n_1 \sin \theta_1}{(\eta - i\kappa)} \qquad (2.140)$$

is also a complex quantity; however, $n_2 \sin \theta_2 (= n_1 \sin \theta_1)$ is real. Thus the field in the metal will be of the form

$$\mathscr{E}_2 = \mathbf{E}_2 e^{-\beta x} e^{i(\omega t - \alpha x - k_1 z \sin \theta_1)} \qquad (2.141)$$

where α and β represent the real and imaginary parts of

$$k_2 \cos \theta_2 = (\omega/c)[(\eta - i\kappa)^2 - n_1^2 \sin^2 \theta_1]^{\frac{1}{2}} = \alpha - i\beta \qquad (2.142)$$

Eq. (2.141) represents a wave which attenuates along the x-direction.

As an example we consider reflection at an air–silver interface for which (see Table 1.1)

$$n_1 = 1.0,$$
$$n_2 = \eta - i\kappa = 0.140 - i4.4 \qquad \text{(at } \lambda = 0.6888 \, \mu\text{m)}$$

For normal incidence, $\theta_1 = \theta_2 = 0$ and

$$\begin{aligned} \alpha &= (\omega/c)\eta = 1.28 \, \mu\text{m}^{-1} \\ \beta &= (\omega/c)\kappa = 40.1 \, \mu\text{m}^{-1} \end{aligned} \qquad (2.143)$$

which represents a penetration depth of $0.025 \, \mu\text{m}$.

At oblique incidence α and β have to be evaluated from Eq. (2.142). The corresponding expressions for the reflection coefficients remain the same as in Sec. 2.2, i.e.,

$$\left. \begin{aligned} r_p &= \frac{n_1 \cos \theta_2 - n_2 \cos \theta_1}{n_1 \cos \theta_2 + n_2 \cos \theta_1} \\[2mm] r_s &= \frac{n_1 \cos \theta_1 - n_2 \cos \theta_2}{n_1 \cos \theta_1 + n_2 \cos \theta_2} \end{aligned} \right\} \qquad (2.144)$$

with $\sin \theta_2$ and $\cos \theta_2$ now being complex. Thus the p and s-polarizations undergo different phase changes on reflection and therefore linearly polarized light would, in general, become elliptically polarized on reflection. By making measurements on the polarization characteristics of the reflected light, one can obtain the optical constants of the metal.

For normal incidence $\theta_1 = \theta_2 = 0$ and for an air–metal interface, the reflectivity is given by

$$R = \frac{(1 - \eta)^2 + \kappa^2}{(1 + \eta)^2 + \kappa^2} \tag{2.145}$$

Additional problems

Problem 2.9: At $\omega = 2\pi \times 10^{10}\,\mathrm{s}^{-1}$, for silver $\sigma \approx 3 \times 10^7\,\mathrm{mhos/m}$ and $\epsilon \approx \epsilon_0 = 8.854 \times 10^{-12}\,\mathrm{C^2/Nm^2}$. Show that for a wave incident normally at the above frequency, the reflectivity is about 0.9996.

Problem 2.10: For gold, at $\lambda = 0.200\,\mu\mathrm{m}$, $0.310\,\mu\mathrm{m}$, $0.400\,\mu\mathrm{m}$, $0.517\,\mu\mathrm{m}$, $0.653\,\mu\mathrm{m}$, $1.59\,\mu\mathrm{m}$, the values of (η, κ) are respectively $(1.427, 1.215)$, $(1.83, 1.916)$, $(1.658, 1.956)$, $(0.608, 2.12)$, $(0.166, 3.15)$, $(0.583, 10.1)$; calculate the reflectivity at normal incidence. (Answer: $23\%, 37\%, 39\%, 66\%, 94\%, 98\%$)

Problem 2.11: For a perfect conductor, $\sigma \to \infty$ and from Eq. (1.98) we have

$$k^2 = (\omega^2/c^2)n^2 \approx -i\omega\mu\sigma$$

and if we write

$$n = \eta - i\kappa = \eta(1 - i\delta)$$

then $\eta \to \infty$ and $\delta \to 1$. For such a perfect conductor, show that $r_s, r_p = -1$

Problem 2.12: Consider a linearly polarized beam with (see Fig. 2.1.)

$$\mathscr{E}_x = \tfrac{1}{2}E_0 \cos(\omega t - \mathbf{k} \cdot \mathbf{r})$$
$$\mathscr{E}_y = \tfrac{1}{\sqrt{2}}E_0 \cos(\omega t - \mathbf{k} \cdot \mathbf{r})$$
$$\mathscr{E}_z = -\tfrac{1}{2}E_0 \cos(\omega t - \mathbf{k} \cdot \mathbf{r})$$

where $\mathbf{k} = k/\sqrt{2}(\hat{\mathbf{x}} + \hat{\mathbf{z}})$. Show that if the above wave is incident on a perfect conductor, then the reflected wave is also linearly polarized with $\mathscr{E}$ rotated by $90°$.

3

Wave propagation in anisotropic media

3.1   Introduction

In Chapter 1 we discussed wave propagation in isotropic media in which the velocity of propagation of an electromagnetic wave is independent of the direction of propagation. In this chapter we will discuss wave propagation in anisotropic media in which the velocity of propagation, in general, depends on the propagation direction and also on the state of polarization and one observes the phenomenon of double refraction. Anisotropic media form the basis of a large number of polarization devices such as quarter wave and half wave plates, the Soleil–Babinet compensator, the Wollaston prism, etc. Their study is also very important for understanding various light modulators based on the electrooptic effect (see Chapter 15).

In Sec. 3.2 we will discuss the phenomenon of double refraction and in Sec. 3.3 we will discuss some important polarization devices based on anisotropic media. Secs. 3.4–3.7 will discuss the electromagnetics of anisotropic media. In Sec. 3.8 we will introduce the index ellipsoid and show how from the index ellipsoid one can obtain the velocities of propagation and the polarizations of the two waves which can propagate along any given direction.

3.2   Double refraction

If we place a crystal of calcite or quartz on a point marked on a piece of paper, we will, in general, observe two images of the point. This phenomenon is referred to as double refraction or birefringence. This happens because when a ray enters the crystal it splits up, in general, into two rays which propagate along different directions and which are orthogonally polarized. In these crystals one of the rays follows Snell's laws of refraction and is called the ordinary ray (o-ray) but the other, referred to as the extraordinary ray (e-ray), does not follow the laws of refraction. If we consider a point source embedded in such a crystal, we will have two different sets of wavefronts emanating from the point source, one of which is spherical corresponding to the o-ray and the other is an ellipsoid of revolution corresponding to the e-ray (see Fig. 3.1). The o-ray is characterized by the same velocity for all directions of propagation (and hence gives rise to a spherical wavefront) and the e-ray has different velocities along different propagation directions (and hence gives rise to a nonspherical wavefront). Along one particular direction, the o and e-rays travel with the same velocities. This direction is called the optic axis (see Fig. 3.1). In these crystals there is just one direction where the velocities are equal. Hence they are called uniaxial crystals. In general, a crystal can have at most two directions along which the velocities may be equal.[†] Such crystals are called biaxial. A more detailed discussion is given in Secs. 3.4–3.7.

If the o-ray velocity is greater than the e-ray velocity along all directions then the spherical wavefront would be completely outside the ellipsoidal wavefront (Fig. 3.1(*b*)) and such a crystal is called a positive uniaxial crystal. On the other hand, if the converse is true, the spherical wavefront lies completely inside the ellipsoidal wavefront and the crystal is a negative uniaxial crystal (Fig. 3.1(*a*)). Quartz is an example of a positive uniaxial

Fig. 3.1 A point source placed in a uniaxial crystal generates two wavefronts one of which is spherical and the other is an ellipsoid of revolution.
The two surfaces touch at two diametrically opposite points. The direction along which the two wavefronts travel with equal velocities is the optic axis; (*a*) a negative uniaxial crystal for which $n_o > n_e$ and (*b*) a positive uniaxial crystal for which $n_o < n_e$.

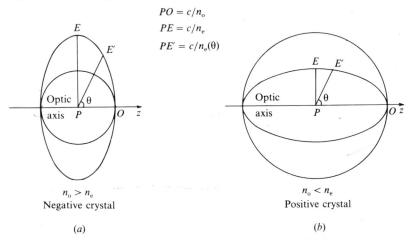

$$PO = c/n_o$$
$$PE = c/n_e$$
$$PE' = c/n_e(\theta)$$

$n_o > n_e$
Negative crystal

(*a*)

$n_o < n_e$
Positive crystal

(*b*)

[†] We should point out that in biaxial crystals, the wave velocities (and not the ray velocities) are equal along the optic axes. However, in uniaxial crystals, both the ray and wave velocities are equal along the optic axis.

Table 3.1 *Principal indices of refraction of some crystals*

	n_x	n_y	n_z	α
Biaxial crystals at sodium wavelength ($\lambda_0 = 0.589\,\mu m$)				
Mica	1.5601	1.5936	1.5977	71.0°
Sulphur	1.9500	2.0430	2.2400	37.3°
Topaz	1.6190	1.6200	1.6270	20.8°
Lead oxide, (Lithargite)	2.5120	2.6100	2.7100	46.3°
Uniaxial crystals				
		n_o	n_e	
Calcite (0.589 μm)		1.65835	1.48640	
Quartz (0.589 μm)		1.54424	1.55335	
Rutile (titanium dioxide)				
(0.5791 μm)		2.621	2.919	
Lithium niobate (0.6 μm)		2.2967	2.2082	
Lithium tantalate (0.6 μm)		2.1834	2.1878	*Negative*
KDP (0.6328 μm)		1.50737	1.46685	
ADP (0.6328 μm)		1.52166	1.47685	

Table adapted from Jenkins and White (1957), Wolfe (1978) and Hartfield and Thompson (1978).

crystal[†] and calcite is a negative uniaxial crystal. Table 3.1 lists some uniaxial crystals with the corresponding indices.

The o-ray velocity can be written as

$$v_{ro} = c/n_o \tag{3.1}$$

where n_o is called the ordinary refractive index. The e-ray velocity is

$$\frac{1}{v_{re}^2(\theta)} = \frac{n_{re}^2(\theta)}{c^2} = \frac{\cos^2\theta}{c^2/n_o^2} + \frac{\sin^2\theta}{c^2/n_e^2} \tag{3.2}$$

$\Rightarrow$

$$n_{re}^2(\theta) = n_o^2 \cos^2\theta + n_e^2 \sin^2\theta \tag{3.3}$$

where θ is the angle made by the e-ray with the optic axis and n_e is called the extraordinary refractive index. Eq. (3.2) which describes the variation of the velocity of the e-ray with the direction of propagation will be derived in Sec. 3.6. For positive uniaxial crystals $n_o < n_e$ and for negative uniaxial crystals $n_o > n_e$.

As we will show in detail in Sec. 3.6 both the o and e-rays are linearly polarized. The **E** and **D** of the o-ray are perpendicular to the direction of

---

[†] Indeed for quartz, the two surfaces do not exactly touch each other. This leads to the phenomenon of optical activity. Here we will neglect this.

propagation and to the optic axis. On the other hand, the **E** of the e-ray is perpendicular to the direction of propagation and to the **E** of the o-ray. The **D** of the e-ray is, in general, *not* parallel to the **E** of the e-ray and hence is not perpendicular to the ray propagation direction. As we will show in Sec. 3.5, **D** of the e-ray is perpendicular to the propagation vector **k** of the e-ray.

We should mention here that a beam of arbitrary polarization state will, in general, propagate in a quite complicated manner, changing its polarization state as it propagates. However, for studying the propagation one can

Fig. 3.2 The phenomenon of double refraction when a wavefront *BD* is normally incident on a uniaxial medium. The dashed line shows the direction of the optic axis. If the incident light has components both in the plane of the paper and perpendicular to it then there are two refracted rays. The o-ray (along *BO*) propagates without any deflection and has its **D** perpendicular to the plane of the paper. The e-ray (along *BE*) is deflected and has its **D** in the plane containing the direction of propagation and the optic axis. Observe that the extraordinary wavefront is still parallel to the interface. (*a*) and (*b*) show that the direction of the e-ray depends on the direction of the optic axis. Rotation of the crystal about the normal will rotate the e-ray as shown in (*c*).

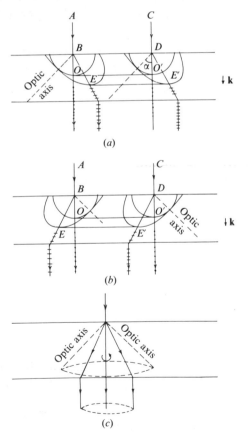

consider an arbitrary polarization as a superposition of ordinary and extraordinary polarization states which propagate with, in general, different velocities but without changing their state of polarization.

We will now study the propagation of waves in a uniaxial medium using Huygens' principle.[†] According to this principle, each point on the wavefront can be considered to be a source of secondary wavelets and the position and shape of the wavefront at a later time is obtained by drawing the envelope of the secondary wavelets.

In order to understand the origin of double refraction, we consider a plane electromagnetic wave incident normally on a uniaxial crystal (say a negative crystal) as shown in Fig. 3.2. The optic axis of the crystal is assumed to lie in the plane of incidence at some angle to the interface. We now use the Huygens' construction to obtain the shape and position of the refracted wavefronts. Let the incident wavefront be represented by *BD* (see Fig. 3.2). Using Huygens' construction, we now draw with *B* and *D* as centres, two wavefronts one corresponding to the o-ray and the other the e-ray. These, as we have already seen, correspond to a sphere and an ellipsoid of revolution respectively which touch each other along the direction of the optic axis. We then draw the common tangent to the two spheres which gives us the wavefront corresponding to the o-ray. The direction of the o-ray is obtained by joining the point *B* to the point of intersection *O* of the tangent and the sphere. The extraordinary wavefront is obtained by drawing a tangent plane to the ellipsoids of revolution. The direction of the e-ray is obtained by joining *B* to the point of intersection *E* of the ellipsoid with the tangent plane. As obvious from the figure, even for a normally incident ray, the e-ray is refracted obliquely. On the other hand the extraordinary wavefront is still parallel to the interface. The **D** of the o-ray is perpendicular to the direction of propagation and to the optic axis and hence is pointed perpendicular to the plane of the paper. On the other hand, the **D** of the e-ray is perpendicular to the direction of **k** (i.e., parallel to the wavefront) and to the **D** of the ordinary ray. These are shown as short lines in Fig. 3.2.

Since the o-ray and e-ray are propagating along different directions, they will come out of the other face of the crystal as two rays with orthogonal polarization states as shown in Fig. 3.2. These rays produce two distinct images.

Thus we see that one incident ray leads to two refracted rays. This phenomenon is known as double refraction. We should mention here that if the input rays were polarized perpendicular to the plane of the paper then

---

[†] Huygens' principle is discussed in detail in many books – see, e.g., Ghatak (1977).

only the o-ray will be observed and if the input rays are polarized in the plane of the paper then only the e-ray will be observed. If the input rays contain both these polarization states then both o-rays and e-rays will be produced. The orthogonal polarization states of the o-ray and e-ray can easily be experimentally verified with the help of a polarizer.

When the optic axis is parallel or perpendicular to the surface, the situation is simpler as shown in Fig. 3.3. For the optic axis parallel to the interface the ordinary and the extraordinary wavefronts are parallel to one another and are not laterally shifted but as can be seen from Fig. 3.3(a), they travel with different velocities. The o-ray travels at a velocity c/n_o and the e-ray at a velocity c/n_e. The o-ray is polarized perpendicular to the plane of the paper and the e-ray is in the plane of the paper. The two different velocities of the o-ray and e-ray lead to a phase difference between the two polarizations as they propagate. This phenomenon is used in the construction of various polarization devices such as quarter wave plates, half wave plates,

Fig. 3.3 Propagation of a wave in a uniaxial crystal, (a) and (c) perpendicular to the optic axis and (b) parallel to the optic axis. In all cases both o-ray and e-ray travel along the same direction; in the former case they travel with different velocities c/n_o and c/n_e respectively and in the latter case they travel with the same velocity c/n_o.

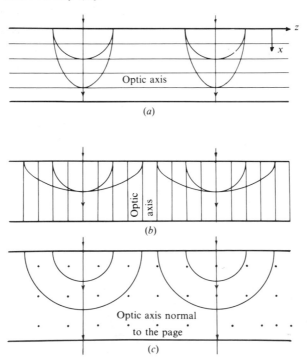

z

x

Optic axis

(a)

Optic axis

(b)

Optic axis normal
to the page

(c)

Babinet–Soleil compensator, etc. For the optic axis perpendicular to the interface, the two rays travel with the same velocity c/n_o and in the same direction.

Similar constructions can be used to analyse oblique incidence also; see, e.g., Ghatak (1977). See also Problem 3.5.

3.3 Some polarization devices

3.3.1 The Nicol prism

One technique for producing a linearly polarized beam from an unpolarized beam is to use the phenomenon of double refraction. We have just seen that the o and the e-rays are orthogonally polarized and hence if one of these components can be removed then the emergent beam must be linearly polarized. Such a technique is used in the Nicol prism, shown in Fig. 3.4. A crystal of calcite is cut into two prisms and then cemented using canada balsam or some similar optically transparent material whose refractive index lies between the ordinary and extraordinary index of calcite; for calcite $n_o = 1.66$, $n_e = 1.49$ and for canada balsam $n = 1.55$. The direction of the optic axis in the crystals is shown in Fig. 3.4 by the dashed line. A ray of unpolarized light entering the first prism is split into o and e-rays. The angles of the prims are so chosen that the o-ray (which has a higher refractive index in calcite) is total internally reflected at the calcite–cement interface while the e-ray (having a lower refractive index) is transmitted. Thus for rays entering the Nicol prism within a certain angular region, the beam coming out of the prism will be linearly polarized as shown in Fig. 3.4.

A device such as the Nicol prism is called a linear polarizer or simply a polarizer since it produces a linearly polarized beam from an incident beam of an arbitrary polarization state. One can also use the Nicol prism as an

Fig. 3.4 A Nicol prism made of calcite: the two prisms are held together with a cement such as canada balsam whose refractive index is between n_o and n_e. The angles are such that the o-ray suffers total internal refection at the interface between the first prism and the cement while the e-ray is transmitted.

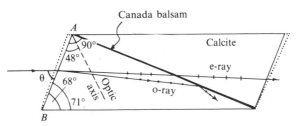

analyser for detecting a linearly polarized beam and to measure the direction of the polarization.

It must be noted that the angle of incidence of the light is limited due to two reasons: (*a*) if the angle θ of the incident beam with the face *AB* (see Fig. 3.4) is decreased then an angle will be reached at which the o-ray no longer undergoes total internal reflection and the beam at the exit is not linearly polarized; (*b*) if the angle θ increases, then beyond a certain angle the e-ray also undergoes total internal reflection and there is no exit beam.

3.3.2 The Glan-Thomson prism

This is again a polarizer consisting of two prism sections of calcite cemented together (see Fig. 3.5). The input and output faces of the device are perpendicular to the light beam. The optic axes of the prisms are perpendicular to the plane of the figure and are shown as dots in Fig. 3.5. The

Fig. 3.5 The Glan–Thomson prism in which the optic axis is normal to the plane of the page. The o-ray with its polarization in the plane of the paper suffers total internal reflection while the e-ray with its polarization perpendicular to the plane of the paper is transmitted.

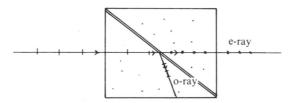

Fig. 3.6 A retarder of thickness *l* made of a uniaxial medium with its optic axis parallel to the interface. Since the propagation is perpendicular to the optic axis both o-ray and e-ray travel along the same direction with velocities c/n_o and c/n_e. This difference in velocities leads to a phase difference between the two polarizations at any point along the propagation direction. If the phase difference at the output is $\frac{1}{2}\pi$ we get a quarter wave plate and if the phase difference is π we get a half wave plate.

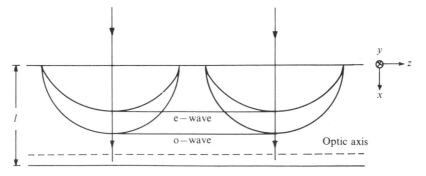

index of refraction of the cement and the angle of the cut are so chosen that the o-wave undergoes total internal reflection while the e-wave is transmitted.

3.3.3   *Retardation plates*

Consider a slice of quartz with plane parallel faces and of thickness l, cut so that the optic axis is parallel to the surfaces as shown in Fig. 3.6. A light beam incident normally on the plate will propagate normally to the optic axis. Thus if the incident beam is linearly polarized perpendicular to the optic axis and to the direction of propagation, i.e., normal to the plane of the page, it will travel as an o-wave with a velocity c/n_o. Similarly a wave polarized in the plane of the paper will propagate as an e-wave with a velocity c/n_e. The phase shift suffered by the o-wave when it emerges from the crystal will be

$$\phi_o = k_0 n_o l$$
(3.4a)

where $k_0 = \omega/c$ is the free space wave number. Similarly, the phase shift suffered by the e-wave will be

$$\phi_e = k_0 n_e l$$
(3.4b)

Hence the crystal has introduced a phase difference

$$\Delta\phi = k_0(n_e - n_o)l$$
(3.5)

between the two polarization components. Thus by choosing a particular value of l, one can introduce a predetermined phase difference between the two polarization components for a given wavelength. Thus, if linearly polarized light is incident on the crystal input face, the output light will, in general, be elliptically polarized.

If the phase difference introduced between the two polarization components is $\frac{1}{2}\pi$, then such a device is known as a quarter wave plate or $\frac{1}{4}\lambda$ plate (since a phase difference of $\frac{1}{2}\pi$ corresponds to a path difference of $\frac{1}{4}\lambda$). The required thickness is given by

$$k_0(n_e - n_o)l = \frac{1}{2}\pi$$

or

$$l = \lambda_0/4(n_e - n_o)$$
(3.6)

Observe that a plate of such a thickness introduces a phase difference of $\frac{1}{2}\pi$ only for the particular wavelength λ_0.

As an example, if we consider a $\frac{1}{4}\lambda$ plate made of quartz, the thickness required at an operating wavelength of 5893 Å (sodium wavelength) will be

$$l = \frac{0.5893 \times 10^{-4}}{4(1.55335 - 1.54424)} \approx 16.17 \, \mu m$$

It should be mentioned here that any odd integral multiple of the thickness given by Eq. (3.6) can also be used but due to their stronger dependence of phase difference on λ they are less advantageous.

Such a quarter wave plate can be used for converting a linearly polarized wave into a circularly polarized wave and vice versa. Thus if the input beam is polarized at 45° to the optic axis then the two polarization components (parallel and perpendicular to the optic axis) will be of equal magnitude and phase. On emerging from the crystal they would have a phase difference of $\frac{1}{2}\pi$ and since they have equal magnitude, the resultant beam will be circularly polarized (see Sec. 1.3 and Problem 3.10). Quarter wave plates also find application as isolators – see Problem 3.13, in electrooptic modulators (see Chapter 15), in ellipsometry, measurement of strain birefrigence etc.

If the thickness of the crystal is such that it introduces a phase difference of π (i.e., a path difference of $\frac{1}{2}\lambda$) then such a device is known as a half wave plate. The required thickness is

$$l = \lambda_0/2(n_e - n_o) \tag{3.7}$$

Such half wave plates can be used for rotating the plane of polarization of linearly polarized beams (see Problem 3.12). The directions parallel and perpendicular to the optic axis of the $\frac{1}{4}\lambda$ and $\frac{1}{2}\lambda$ plates are called the axes of the retardation plates. For positive uniaxial crystals since $n_o < n_e$, the o-wave travels faster than the e-wave and hence the directions parallel and perpendicular to the optic axis directions are called the slow and fast axes. For negative uniaxial crystals, this is reversed.

3.3.4   The Soleil–Babinet compensator

In Sec. 3.3.3 we saw that a thin slice of a properly cut anisotropic crystal can be used to generate fixed phase or path differences between two polarization components. The compensator we will now discuss is a device which can be adjusted to give a continuous range of phase difference between two polarization components. Fig. 3.7 shows the construction of a Soleil–Babinet compensator which consists of two similar wedges of quartz, the lower wedge being cemented to a plane parallel plate of quartz. The optic axes in the two wedges are parallel to the edges of the prisms and the optic axis in the plate is perpendicular to the edges (see Fig. 3.7). The upper wedge can be displaced as shown in Fig. 3.7 and thus the pair of wedges essentially represents a plate of variable thickness. Consider a beam polarized perpendicular to the plane of the paper and incident normally on the compensator as shown in Fig. 3.7. Let t_1 be the variable thickness of the two wedges taken together and t_2 the thickness of the plate. The incident beam is polarized parallel to the optic axis in the wedges and is hence an e-wave. Thus the phase shift suffered by the beam after passing through the wedges is

$$k_0 n_e t_1$$

The beam is polarized perpendicular to the optic axis in the plane parallel plate and hence is an o-wave. Thus the phase shift suffered is

$$k_0 n_o t_2$$

Thus the total phase shift is

$$\phi_1 = k_0(n_e t_1 + n_o t_2) \tag{3.8}$$

Similarly the total phase shift suffered by a beam polarized parallel to the plane of the paper (which is an o-wave in the wedges and an e-wave in the plane parallel plate) will be

$$\phi_2 = k_0(n_o t_1 + n_e t_2) \tag{3.9}$$

Any input polarization state can be considered as a superposition of the in-plane and perpendicular components and the phase difference introduced between the two polarization components is

$$\Delta\phi = \phi_1 - \phi_2 = k_0(n_e - n_o)(t_1 - t_2) \tag{3.10}$$

Fig. 3.7 A Soleil–Babinet compensator consisting of a pair of wedges with optic axes lying perpendicular to the plane of the paper and a plane parallel plate with its optic axis parallel to the surface and perpendicular to the optic axis direction in the wedges. One of the wedges can be moved relative to the other and thus the pair of wedges form a variable thickness plate. Depending on the position of the wedge, one can introduce a variable phase difference between the o and e-waves.

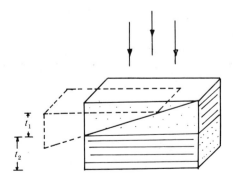

Since the front wedge is movable, t_1 which is the total thickness of both wedges taken together is variable and thus $\Delta\phi$ can be varied continuously. Observe that since $(t_1 - t_2)$ is constant over the whole area of the compensator, the phase difference introduced between the two polarization states is constant throughout the whole field (compare with the Babinet compensator discussed in Problem 3.4).

The Soleil–Babinet compensator is very useful for analyzing the state of polarization in a beam, for generating any polarization state and in various applications such as in strain induced birefringence measurements, thickness measurements of thin films using ellipsometry etc.

3.4   Plane waves in anisotropic media

In the earlier sections we discussed wave propagation through anisotropic media using Huygens' construction. In this section we will consider plane wave propagation through an anisotropic medium and show that corresponding to any direction of propagation, there are two orthogonal linearly polarized waves which propagate without changing their state of polarization and, in general, with different velocities. We restrict ourselves to electrically anisotropic and magnetically isotropic media; the magnetic permeability is assumed to be μ_0.

Referring to Eq. (1.5), we see that if the dielectric permittivity ϵ is a scalar, then the direction of **D** is parallel to the direction of **E**. This is true for all possible directions of **E** only in isotropic media. In certain crystals it is found that the directions of **D** and **E** are not parallel for all possible directions of **E**. Thus, for example in an isotropic medium, an applied electric field E_x generates a displacement **D** lying only along the x-direction, i.e., $D_y = 0 = D_z$. On the other hand, in an anistropic medium, an electric field **E** applied along the x-direction generates, in general, a **D** which has all the three components. Thus we may write

$$D_x = \epsilon_{xx}E_x, \quad D_y = \epsilon_{yx}E_x, \quad D_z = \epsilon_{zx}E_x \tag{3.11}$$

where $E_y = 0 = E_z$; $\epsilon_{xx}, \epsilon_{yx}$ and ϵ_{zx} are the corresponding permittivity components. Generalizing the above, if the applied electric field has components E_x, E_y and E_z then the components of **D** may be written in the form

$$D_x = \epsilon_{xx}E_x + \epsilon_{xy}E_y + \epsilon_{xz}E_z \tag{3.12}$$

$$D_y = \epsilon_{yx}E_x + \epsilon_{yy}E_y + \epsilon_{yz}E_z \tag{3.13}$$

$$D_z = \epsilon_{zx}E_x + \epsilon_{zy}E_y + \epsilon_{zz}E_z \tag{3.14}$$

or in the following matrix form

$$
\begin{pmatrix} D_x \\ D_y \\ D_z \end{pmatrix} = \begin{pmatrix} \epsilon_{xx} & \epsilon_{xy} & \epsilon_{xz} \\ \epsilon_{yx} & \epsilon_{yy} & \epsilon_{yz} \\ \epsilon_{zx} & \epsilon_{zy} & \epsilon_{zz} \end{pmatrix} \begin{pmatrix} E_x \\ E_y \\ E_z \end{pmatrix}
\tag{3.15}
$$

The 3×3 matrix of the nine coefficients $\epsilon_{xx}, \epsilon_{xy}, \ldots, \epsilon_{zz}$ constitute what is known as the dielectric tensor of the medium. If we denote $\mathbf{D}$ and $\mathbf{E}$ by the column vectors as in Eq. (3.15), one can also write

$$
\mathbf{D} = \bar{\epsilon} \mathbf{E}
\tag{3.16}
$$

where $\bar{\epsilon}$ is the dielectric tensor. It can be shown (see, e.g., Born and Wolf, 1975) that $\bar{\epsilon}$ is a symmetric tensor, i.e.,

$$
\epsilon_{xy} = \epsilon_{yx}; \quad \epsilon_{xz} = \epsilon_{zx}; \quad \epsilon_{yz} = \epsilon_{zy}
\tag{3.17}
$$

We can always choose a coordinate system (i.e., choose the directions of the $x, y,$ and z-axes appropriately) in which the dielectric tensor ϵ is diagonal:

$$
\bar{\epsilon} = \begin{pmatrix} \epsilon_x & 0 & 0 \\ 0 & \epsilon_y & 0 \\ 0 & 0 & \epsilon_z \end{pmatrix}
\tag{3.18}
$$

Such a coordinate system is known as the principal axis system and Eqs. (3.12)–(3.14) reduce to

$$
D_x = \epsilon_x E_x; \quad D_y = \epsilon_y E_y; \quad D_z = \epsilon_z E_z
\tag{3.19}
$$

Here ϵ_x, ϵ_y and ϵ_z are known as the principal dielectric permittivities. The principal axis directions are special directions in the crystal such that an electric field applied along any one of them generates a displacement vector parallel to it; the constant of proportionality ($\epsilon_x, \epsilon_y, \epsilon_z$) is in general different for different directions. Depending upon the relationship between ϵ_x, ϵ_y and ϵ_z, we have the following three media;

$$
\left.
\begin{array}{ll}
\epsilon_x = \epsilon_y = \epsilon_z & \text{isotropic} \\
\epsilon_x = \epsilon_y \neq \epsilon_z & \text{uniaxial} \\
\epsilon_x \neq \epsilon_y \neq \epsilon_z & \text{biaxial}
\end{array}
\right\}
\tag{3.20}
$$

Table 3.1 shows the refractive indices defined as $n_i = (\epsilon_i/\epsilon_0)^{\frac{1}{2}}, i = x, y, z$ for some uniaxial and biaxial crystals. For a uniaxial crystal

$$
n_o = (\epsilon_x/\epsilon_0)^{\frac{1}{2}} = (\epsilon_y/\epsilon_0)^{\frac{1}{2}}; \quad n_e = (\epsilon_2/\epsilon_0)^{\frac{1}{2}}
$$

We now consider a plane wave propagating through an anistropic medium. The time and space dependence of the electric and magnetic fields will be given by

$$\mathbf{E} = \mathbf{E}_0\, e^{i(\omega t - \mathbf{k}\cdot\mathbf{r})} \tag{3.21}$$

$$\mathbf{H} = \mathbf{H}_0\, e^{i(\omega t - \mathbf{k}\cdot\mathbf{r})} \tag{3.22}$$

where $\mathbf{E}_0$ and $\mathbf{H}_0$ are constants, $\mathbf{k}$ is the propagation vector of the plane wave and ω is the frequency of the electromagnetic wave. The wave refractive index n_w is defined by the equation

$$|\mathbf{k}| = k = (\omega/c)n_w \tag{3.23}$$

In the absence of any currents, Maxwell's equations (Eqs. (1.1)–(1.4)) become

$$\nabla \times \mathbf{E} = -\,\partial\mathbf{B}/\partial t = -\,i\omega\mu_0\mathbf{H} \tag{3.24}$$

$$\nabla \times \mathbf{H} = \partial\mathbf{D}/\partial t = i\omega\mathbf{D} \tag{3.25}$$

where we have used

$$\mathbf{B} = \mu_0\mathbf{H} \tag{3.26}$$

Using the expression for $\mathbf{E}$ given by Eq. (3.21) we get

$$\begin{aligned}
(\nabla \times \mathbf{E})_x &= (\partial E_z/\partial y - \partial E_y/\partial z)\\
&= (-\,ik_y E_{0z} + ik_z E_{0y})\, e^{i(\omega t - \mathbf{k}\cdot\mathbf{r})}\\
&= -\,i(\mathbf{k} \times \mathbf{E})_x
\end{aligned}$$

and similarly for the other components. Thus we obtain

$$\nabla \times \mathbf{E} = -\,i(\mathbf{k} \times \mathbf{E}) = -\,i\omega\mu_0\mathbf{H}$$

or

$$\mathbf{H} = \frac{1}{\omega\mu_0}(\mathbf{k} \times \mathbf{E}) \tag{3.27}$$

Similarly

$$\mathbf{D} = -\frac{1}{\omega}(\mathbf{k} \times \mathbf{H}) \tag{3.28}$$

Fig. 3.8 The relative directions of **E**, **H**, **D**, **k** and **S** in an anisotropic medium. The vectors **E**, **D**, **S** and **k** lie in one plane perpendicular to **H**; (**k**, **D**, **H**) and (**S**, **E**, **H**) form pairs of orthogonal sets of vectors.

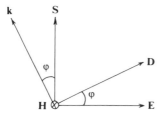

From Eqs. (3.27) and (3.28) it follows that **H** is perpendicular to **k**, **E** and **D**, and **D** is perpendicular to **k**. Thus **k**, **E** and **D** must lie in a plane normal to **H** and in this plane **D** is perpendicular to **k** (see Fig. 3.8). Since, in general, **D** and **E** are not parallel to each other, **E** is not, in general, perpendicular to **k**.

The energy propagation direction is along the Poynting vector defined by

$$\mathbf{S} = \mathbf{E} \times \mathbf{H} \tag{3.29}$$

and is perpendicular to **E** and **H**. Thus the vectors **k**, **E**, **D** and **S** lie in one plane perpendicular to **H**. It can thus be seen that in an anisotropic medium the energy propagation direction (defined by **S**) is not, in general, along the wave normal **k**. Since **k** is perpendicular to **D** and **S** is perpendicular to **E**, the angle between **E** and **D** is the same as that between **k** and **S**. Furthermore, since a ray represents the direction of propagation of energy, **S** also represents the ray propagation direction. The vectors (**k**, **D**, **H**) and (**S**, **E**, **H**) form a pair of orthogonal vectors rotated with respect to each other about **H** (see Fig. 3.8). We now obtain a relation between the phase (wave) velocity $v_w = c/n_w$ and the energy propagation velocity or the ray velocity v_r. The velocity of energy propagation will simply be the energy crossing an area perpendicular to the direction of energy flow per unit time divided by the energy density. Thus if W represents the energy density, then the ray (or energy) velocity v_r, will be

$$v_r = |\mathbf{S}|/W \tag{3.30}$$

We assume that the expression for the energy density derived for isotropic media remains valid even for an anisotropic medium and thus (see Eqs. (1.61) and (1.62))

$$W = \tfrac{1}{2}(\mathbf{E}\cdot\mathbf{D} + \mathbf{B}\cdot\mathbf{H}) = \tfrac{1}{2}(\mathbf{E}\cdot\mathbf{D} + \mu_0\mathbf{H}\cdot\mathbf{H}) \tag{3.31}$$

Substituting for **H** and **D** from Eqs. (3.27) and (3.28) we get

$$\begin{aligned} W &= \tfrac{1}{2}\left[-\frac{1}{\omega}\mathbf{E}\cdot\mathbf{k} \times \mathbf{H} + \mu_0(1/\omega\mu_0)(\mathbf{k} \times \mathbf{E})\cdot\mathbf{H} \right] \\ &= (1/\omega)\mathbf{k}\cdot(\mathbf{E} \times \mathbf{H}) \\ &= (1/\omega)\mathbf{k}\cdot\mathbf{S} \end{aligned} \tag{3.32}$$

Thus

$$v_r = \frac{\omega|\mathbf{S}|}{\mathbf{k}\cdot\mathbf{S}} = \frac{\omega}{k\cos\phi} = \frac{v_w}{\cos\phi} \tag{3.33}$$

where $k = |\mathbf{k}|$,

$$v_w = \omega/k \tag{3.34}$$

represents the wave (or phase) velocity and ϕ represents the angle between the wave normal $\mathbf{k}$ and the ray direction $\mathbf{S}$ (see Fig. 3.8). The phase velocity v_w is the projection of the ray velocity v_r along the wave normal.

3.5   Wave refractive index

We will now show that corresponding to any direction of propagation, there are two orthogonally linearly polarized plane waves which can propagate without changing their state of polarization and, in general, with two different velocities. Indeed when we try to find solutions of Maxwell's equations in the form represented by Eqs. (3.21)–(3.22), we are finding these possible polarization states and their corresponding phase velocities.

From Eqs. (3.27) and (3.28), we get

$$\mathbf{D} = -\frac{1}{\omega}\mathbf{k} \times \left(\frac{1}{\omega\mu_0}\mathbf{k} \times \mathbf{E}\right)$$

$$= -\frac{1}{\omega^2\mu_0}\mathbf{k} \times (\mathbf{k} \times \mathbf{E})$$

or

$$(\mathbf{k}\cdot\mathbf{E})\mathbf{k} - k^2\mathbf{E} = -\omega^2\mu_0\bar{\bar{\epsilon}}\mathbf{E} \tag{3.35}$$

Let $\hat{\boldsymbol{\kappa}}$ represent the unit vector along $\mathbf{k}$ i.e, along the normal to the wavefront. Thus (see Eq. (3.23))

$$\mathbf{k} = (\omega/c)n_w\hat{\boldsymbol{\kappa}} \tag{3.36}$$

We can also write

$$\bar{\bar{\mathbf{K}}} = \bar{\bar{\epsilon}}/\epsilon_0 \tag{3.37}$$

where $\bar{\bar{\mathbf{K}}}$ is the dielectric constant tensor and ϵ_0 is the free space permittivity. Using Eqs. (3.36) and (3.37), Eq. (3.35) becomes

$$(\hat{\boldsymbol{\kappa}}\cdot\mathbf{E})\hat{\boldsymbol{\kappa}}(\omega^2/c^2)n_w^2 - (\omega^2/c^2)n_w^2\mathbf{E} = -\omega^2\epsilon_0\mu_0\bar{\bar{\mathbf{K}}}\mathbf{E} = -(\omega^2/c^2)\bar{\bar{\mathbf{K}}}\mathbf{E}$$

or

$$(\hat{\boldsymbol{\kappa}}\cdot\mathbf{E})\hat{\boldsymbol{\kappa}} - \mathbf{E} = -(1/n_w^2)\bar{\bar{\mathbf{K}}}\mathbf{E} \tag{3.38}$$

If we refer to the principal axis system, $\bar{\bar{\mathbf{K}}}$ is diagonal (since $\bar{\bar{\epsilon}}$ is diagonal) and the x, y and z-components of Eq. (3.38) give us

$$[(1/n_w^2)K_x - \kappa_y^2 - \kappa_z^2]E_x + \kappa_x\kappa_yE_y + \kappa_x\kappa_zE_z = 0 \tag{3.39}$$

$$\kappa_x\kappa_yE_x + [(1/n_w^2)K_y - \kappa_x^2 - \kappa_z^2]E_y + \kappa_y\kappa_zE_z = 0 \tag{3.40}$$

$$\kappa_x\kappa_zE_x + \kappa_y\kappa_zE_y + [(1/n_w^2)K_z - \kappa_x^2 - \kappa_y^2]E_z = 0 \tag{3.41}$$

where we have used the fact that

$$\kappa_x^2 + \kappa_y^2 + \kappa_z^2 = 1 \tag{3.42}$$

For a nontrivial solution of Eqs. (3.39)–(3.41), the determinant of the coefficients of E_x, E_y and E_z must vanish giving us

$$\begin{vmatrix} K_x/n_w^2 - \kappa_y^2 - \kappa_z^2 & \kappa_x\kappa_y & \kappa_x\kappa_z \\ \kappa_y\kappa_x & K_y/n_w^2 - \kappa_x^2 - \kappa_z^2 & \kappa_y\kappa_z \\ \kappa_x\kappa_z & \kappa_y\kappa_z & K_z/n_w^2 - \kappa_x^2 - \kappa_y^2 \end{vmatrix} = 0 \tag{3.43}$$

The above determinant is an eigenvalue equation; for a given direction of propagation of the plane wave, i.e., for a given set κ_x, κ_y and κ_z, solving the above determinant gives us the possible values of the refractive index n_w which are the eigenvalues. For each eigenvalue, using Eqs. (3.39)–(3.41), one can immediately determine the eigenvector $\mathbf{E}(E_x, E_y, E_z)$.

Opening the determinant in Eq. (3.43), appears to give us a cubic equation in n_w^2. In fact the coefficient of n_w^6 term vanishes identically leading to a quadratic equation in n_w^2 which can be shown to have two positive roots n_{w1} and n_{w2}. We first consider two specific examples.

3.5.1 Wave propagation along the z-direction
In such a case, we will have

$$\kappa_x = 0, \quad \kappa_y = 0, \quad \kappa_z = 1 \tag{3.44}$$

and Eq. (3.43) becomes

$$\begin{vmatrix} K_x/n_w^2 - 1 & 0 & 0 \\ 0 & K_y/n_w^2 - 1 & 0 \\ 0 & 0 & K_z/n_w^2 \end{vmatrix} = 0 \tag{3.45}$$

which gives us

$$\left(\frac{K_x}{n_w^2} - 1\right)\left(\frac{K_y}{n_w^2} - 1\right)\frac{K_z}{n_w^2} = 0$$

i.e.,

$$n_w = n_{w1} = K_x^{\frac{1}{2}} \tag{3.46}$$

$$n_w = n_{w2} = K_y^{\frac{1}{2}} \tag{3.47}$$

These are the two eigenvalues. The corresponding polarization directions can be determined by substituting $n_w = n_{w1}$ and $n_w = n_{w2}$ in Eqs. (3.39)–(3.4i) and using Eq. (3.44). Thus for a wave which has a refractive index n_{w1} Eq. (3.40) gives us

$$(K_y/K_x - 1)E_y = 0 \tag{3.48}$$

For $K_x \neq K_y$, we thus have $E_y = 0$. Similarly Eq. (3.41) gives us $E_z = 0$. Thus a wave with its **E** along x and propagating along z propagates with a phase velocity $c/K_x^{\frac{1}{2}}$. Similarly one can show that the other root $n_w = n_{w2} = K_y^{\frac{1}{2}}$ corresponds to a wave with its **E** lying along the y-direction.

Thus for waves propagating along the z-direction, the eigenwaves are (a) a wave with its **E** along x (and hence its **D** also along x since x, y, z is a principal axis system) which travels with a phase velocity $c/K_x^{\frac{1}{2}}$ and (b) a wave with its **E** along y (and hence **D** also along y) which travels with a phase velocity $c/K_y^{\frac{1}{2}}$. Any other polarization state propagating along the z-direction can be analysed by decomposing the wave into polarization components along the x and y-directions and then superposing them at any z-value with proper phases. Similarly one can show that a wave with its **E** (and **D**) along z and propagating along either the x or y directions travels with a phase velocity $c/K_z^{\frac{1}{2}}$.

3.5.2   *Wave propagation in the x–z plane*

We now consider a wave propagating in the x–z plane at an angle ψ with the z-direction (see Fig. 3.9). For such a case

$$\kappa_x = \sin \psi, \quad \kappa_y = 0, \quad \kappa_z = \cos \psi \tag{3.49}$$

Fig. 3.9 Here (a) and (b) represent the two eigen polarizations corresponding to waves propagating in the x–z plane at an angle ψ with the z-direction. Waves polarized with **D** along y travel with a velocity c/n_{w1} and waves with **D** perpendicular to the direction of propagation and lying in the x–z plane travel with a velocity c/n_{w2} where n_{w1} and n_{w2} are given by Eqs. (3.52) and (3.53).

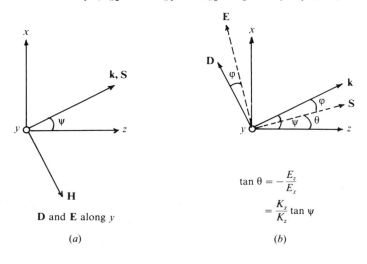

$$\tan \theta = -\frac{E_z}{E_x}$$

$$= \frac{K_z}{K_x} \tan \psi$$

D and **E** along y

(a) $\qquad\qquad\qquad\qquad$ (b)

and Eq. (3.43) becomes

$$
\begin{vmatrix}
K_x/n_w^2 - \cos^2\psi & 0 & \sin\psi\cos\psi \\
0 & K_y/n_w^2 - 1 & 0 \\
\sin\psi\cos\psi & 0 & K_z/n_w^2 - \sin^2\psi
\end{vmatrix} = 0
\tag{3.50}
$$

Expanding the determinant, we obtain

$$
\left(\frac{K_x}{n_w^2} - \cos^2\psi\right)\left(\frac{K_y}{n_w^2} - 1\right)\left(\frac{K_z}{n_w^2} - \sin^2\psi\right)
$$

$$
- \sin^2\psi\cos^2\psi\left(\frac{K_y}{n_w^2} - 1\right) = 0
$$

or

$$
\left(\frac{K_y}{n_w^2} - 1\right)\left[\frac{K_x K_z}{n_w^2} - (K_x\sin^2\psi + K_z\cos^2\psi)\right]\frac{1}{n_w^2} = 0
\tag{3.51}
$$

Thus the two eigenvalues are

$$
n_{w1} = K_y^{\frac{1}{2}}
\tag{3.52}
$$

and

$$
1/n_{w2}^2 = \cos^2\psi/K_x + \sin^2\psi/K_z
\tag{3.53}
$$

Corresponding to the solution n_{w1}, $\mathbf{D}$ will be along the y-direction (see Fig. 3.9(a)). For the solution $n_w = n_{w2}$, substituting $n_w = n_{w2}$ as given by Eq. (3.53) in Eq. (3.39), we obtain

$$
K_z E_z/K_x E_x = -\tan\psi
\tag{3.54}
$$

Since in the principal axis system the dielectric constant tensor is diagonal

$$
D_x = \epsilon_0 K_x E_x; \quad D_y = \epsilon_0 K_y E_y; \quad D_z = \epsilon_0 K_z E_z
\tag{3.55}
$$

Thus Eq. (3.54) implies

$$
D_z/D_x = -\tan\psi
\tag{3.56}
$$

i.e., the $\mathbf{D}$ of this wave is normal to y and to the direction of propagation (see Fig. 3.9(b)). Observe that $\mathbf{E}$ is *not* normal to the wave vector if $\psi \neq 0$ or $\pi/2$.

In this example, the refractive index seen by a wave polarized along the y-direction and propagating in the x–z plane at any arbitrary angle is $K_y^{\frac{1}{2}}$. On the other hand, for the other wave, the wave refractive index and the corresponding $\mathbf{D}$ change with the angle of propagation. The wave refractive index seen by a wave propagating along the z-axis ($\psi = 0$) is $K_x^{\frac{1}{2}}$ (with its $\mathbf{D}$ along x) and by a wave propagating along the x-axis ($\psi = \frac{1}{2}\pi$) it is $K_z^{\frac{1}{2}}$.

One can show, in general, that for a wave propagating along any direction, there are two possible refractive indices and hence two possible phase velocities corresponding to two definite linearly polarized waves. In Sec. 3.8 we will discuss a simple geometrical construction for determining the two possible polarization directions and their corresponding phase velocities.

As we have mentioned earlier, for biaxial crystals $K_x \neq K_y \neq K_z$ and the determination of the phase velocities and the corresponding polarization directions for a wave propagating along a general direction becomes complicated. For uniaxial crystals, two of the principal dielectric constants become equal and the analysis becomes simple. We will assume that the axes are oriented so that $K_x = K_y$ and the x and y-directions become arbitrary as long as they are perpendicular to z. For wave propagation along any direction, we choose the x-direction to lie in the plane containing the propagation direction and the z-axis. For such a case the two refractive indices are described by Eqs. (3.52) and (3.53). Thus these two equations describe completely wave propagation in uniaxial crystals, and ψ is the angle between the z-direction and the propagation direction. Also $n_w = n_{w1}$ corresponds to $\mathbf{D}$ lying perpendicular to $\mathbf{k}$ and to the z-direction and $n_w = n_{w2}$ corresponds to $\mathbf{D}$ lying perpendicular to $\mathbf{k}$ but in the plane containing $\mathbf{k}$ and the z-direction. Also along the z-direction, $\psi = 0$ and both polarization states travel with the same velocity $c/K_x^{\frac{1}{2}} = c/K_y^{\frac{1}{2}}$. Thus the z-direction corresponds to the optic axis of the crystal. As can be seen, one of the waves always has the same velocity $c/K_x^{\frac{1}{2}}$ and this wave is referred to as the o-wave. The o-wave or o-ray refractive index is simply $n_o = K_x^{\frac{1}{2}} = K_y^{\frac{1}{2}}$. The other wave has a velocity which depends on the direction of propagation. This wave is referred to as the e-wave. The velocities of the two waves differ most when travelling along $\psi = \frac{1}{2}\pi$ i.e., perpendicular to the optic axis. The corresponding phase or ray velocity is $c/K_z^{\frac{1}{2}}$ and the e-wave or e-ray refractive index is $n_e = K_z^{\frac{1}{2}}$.

3.6   Ray refractive index

In the above discussion, we considered the phase velocities of the waves and their polarization directions in an anisotropic medium. We will now show that there is an analogous analysis for ray velocity.

Since $\mathbf{D}$ and $\mathbf{E}$ are not along the same direction, the component of $\mathbf{D}$ along $\mathbf{E}$ is $(\mathbf{D} \cdot \mathbf{E})\mathbf{E}/E^2$. This can also be written as $\mathbf{D} - \hat{\mathbf{s}}(\mathbf{D} \cdot \hat{\mathbf{s}})$ where $\hat{\mathbf{s}}$ is the unit vector along the ray direction (i.e., along $\mathbf{S}$). Thus

$$\mathbf{D} - \hat{\mathbf{s}}(\mathbf{D} \cdot \hat{\mathbf{s}}) = (\mathbf{D} \cdot \mathbf{E}/E^2)\mathbf{E} \tag{3.57}$$

In order to evaluate the RHS of the above equation, we take a dot product of

Eq. (3.35) with $\mathbf{E}$ and observing that $\bar{\epsilon}\mathbf{E} = \mathbf{D}$, we obtain

$$(\mathbf{k}\cdot\mathbf{E})^2 - k^2 E^2 = -\omega^2\mu_0\mathbf{D}\cdot\mathbf{E}$$

or

$$\mathbf{D}\cdot\mathbf{E}/E^2 = k^2\cos^2\phi/\omega^2\mu_0 \tag{3.58}$$

where ϕ is the angle between $\mathbf{D}$ and $\mathbf{E}$ (which is the same as that between $\hat{\mathbf{s}}$ and $\hat{\boldsymbol{\kappa}}$) – see Fig. 3.8. Substituting from Eq. (3.58) in Eq. (3.57), we obtain

$$\mathbf{D} - \hat{\mathbf{s}}(\mathbf{D}\cdot\hat{\mathbf{s}}) = (k^2\cos^2\phi/\omega^2\mu_0)\mathbf{E} \tag{3.59}$$

Now, from Eq. (3.33), we get

$$k^2\cos^2\phi/\omega^2 = \cos^2\phi/v_w^2 = 1/v_r^2 \tag{3.60}$$

Thus

$$\mathbf{D} - \hat{\mathbf{s}}(\mathbf{D}\cdot\hat{\mathbf{s}}) = (1/v_r^2\mu_0)\mathbf{E}$$

or

$$(\hat{\mathbf{s}}\cdot\mathbf{D})\hat{\mathbf{s}} - \mathbf{D} = -(\epsilon_0 c^2/v_r^2)\mathbf{E} \tag{3.61}$$

We now define the ray refractive index n_r as the ratio of the ray velocity in free space to that in the anisotropic medium, i.e.,

$$n_r = c/v_r \tag{3.62}$$

Thus Eq. (3.61) becomes

$$(\hat{\mathbf{s}}\cdot\mathbf{D})\hat{\mathbf{s}} - \mathbf{D} = -n_r^2\epsilon_0\mathbf{E} \tag{3.63}$$

Comparing the above equation with Eq. (3.38), we see that Eq. (3.63) can be obtained from Eq. (3.38) by replacing $\hat{\boldsymbol{\kappa}}$ by $\hat{\mathbf{s}}$, $\mathbf{E}$ by $\mathbf{D}/\epsilon_0$, $\mathbf{D}$ by $\epsilon_0\mathbf{E}$ and $1/n_w$ by n_r. Referring again to the principal axis system the components of $\mathbf{D}$ and $\mathbf{E}$ are related through Eq. (3.55), the x, y and z-components of Eq. (3.63) become

$$D_x(n_r^2/K_x - s_y^2 - s_z^2) + s_x s_y D_y + s_x s_z D_z = 0 \tag{3.64}$$

$$s_x s_y D_x + D_y(n_r^2/K_y - s_x^2 - s_z^2) + s_z s_y D_z - 0 \tag{3.65}$$

$$s_x s_z D_x + s_y s_z D_y + D_z(n_r^2/K_z - s_x^2 - s_y^2) = 0 \tag{3.66}$$

(cf. Eqs. (3.39)–(3.41)). For nontrivial solutions the determinant of the coefficients of D_x, D_y and D_z must vanish. Following a procedure similar to the one used for obtaining the phase velocity and the corresponding refractive index n_w, the set of Eqs. (3.64)–(3.66) implies that to every ray direction there correspond two ray refractive indices and hence two ray velocities.

It easily follows with an analysis similar to the one used for the wave refractive index that for a ray propagating along the z-direction, there are two values of n_r: (*a*) for a ray with **E** along x, $n_r = K_x^{\frac{1}{2}}$ and the corresponding ray velocity is $c/K_x^{\frac{1}{2}}$; (*b*) for a ray with **E** along y, $n_r = K_y^{\frac{1}{2}}$ and the corresponding ray velocity is $c/K_y^{\frac{1}{2}}$. Similarly it can be shown that a ray with **E** along z travels with a velocity $c/K_z^{\frac{1}{2}}$. Observe that for propagation along any of the principal axes, **E** and **D** of the propagating wave and ray are parallel to each other and the ray and wave velocities are equal.

We now consider a ray propagating in the x–z plane at an angle θ with the z-axis (see Fig. 3.9). For such a case

$$s_x = \sin\theta, \quad s_y = 0, \quad s_z = \cos\theta \tag{3.67}$$

and the determinant from Eqs. (3.64)–(3.66) gives us two roots

$$n_{ro} = K_y^{\frac{1}{2}} \tag{3.68}$$

and

$$n_{re}^2 = K_x \cos^2\theta + K_z \sin^2\theta \tag{3.69}$$

(cf. Eqs. (3.52) and (3.53)). For the ray corresponding to $n_r = K_y^{\frac{1}{2}}$, the **D** and **E** are along the y-direction. For the ray having a refractive index given by Eq. (3.69), we obtain from Eqs. (3.64) and (3.69),

$$\frac{D_z/K_z}{D_x/K_x} = -\tan\theta$$

or

$$E_z/E_x = -\tan\theta \tag{3.70}$$

which implies that the **E** of the ray is normal to y and to the direction of propagation. Observe that the **D** for the ray is *not* parallel to **E** of the ray unless $\theta = 0$ or $\frac{1}{2}\pi$. Notice the difference in the equations describing ray propagation (Eq. (3.69)) and wave propagation (Eq. (3.53)).

As for wave propagation, Eqs. (3.68) and (3.69) completely describe ray propagation in uniaxial media with θ representing the angle between the ray propagation direction and the optic axis.

3.7   The ray velocity surface

Since $c/n_r = v_r$ (see Eq. (3.62)) and defining $v_r s_x = v_{rx}$, $v_r s_y = v_{ry}$, $v_r s_z = v_{rz}$, the three Cartesian components of the ray velocity, the determinant equation obtained from Eqs. (3.64)–(3.66) can be written as

$$\begin{vmatrix} c^2/K_x - v_{ry}^2 - v_{rz}^2 & v_{rx}v_{ry} & v_{rx}v_{rz} \\ v_{rx}v_{ry} & c^2/K_y - v_{rx}^2 - v_{rz}^2 & v_{rz}v_{ry} \\ v_{rx}v_{rz} & v_{ry}v_{rz} & c^2/K_z - v_{rx}^2 - v_{ry}^2 \end{vmatrix} = 0 \tag{3.71}$$

The above equation can be represented schematically as shown in Fig. 3.10. For a given direction of propagation (i.e., for given s_x, s_y, s_z) one obtains two ray velocities as solution to the determinant Eq. (3.71). These two solutions are plotted as vectors in the same direction $\hat{s}(s_x, s_y, s_z)$ and of lengths equal to the two ray velocities. For various directions, these end points lead to the surface shown in Fig. 3.10 which is called the ray velocity surface.

In order to understand how the surface is generated, we consider ray propagation in the x–z plane. We also assume (for specificity) that

$$K_x < K_y < K_z \tag{3.72}$$

We have seen earlier that in the x–z plane the two ray refractive indices are given by Eqs. (3.68) and (3.69). Thus in the x–z plane we would get two curves, one corresponding to

$$v_{rx}^2 + v_{rz}^2 = v_r^2 = c^2/n_r^2 = c^2/K_y \tag{3.73}$$

which is a circle of radius $c/K_y^{\frac{1}{2}}$. The other curve is obtained by replacing $\cos \theta = v_{rz}/v_r = v_{rz}n_r/c$ and $\sin \theta = v_{rx}/v_r = v_{rx}n_r/c$ in Eqs. (3.69) leading to

$$\frac{v_{rz}^2}{c^2/K_x} + \frac{v_{rx}^2}{c^2/K_z} = 1 \tag{3.74}$$

which represents an ellipse with axes $c/K_z^{\frac{1}{2}}$ and $c/K_x^{\frac{1}{2}}$ along x and z-directions respectively. Similarly one can obtain the curves in other planes. The intersections of the ray velocity surface and the three coordinate planes are shown in Fig. 3.11(a), (b) and (c). In every plane the intersection consists of a circle and an ellipse. Since $K_x < K_y < K_z$, the circle and ellipse intersect each

Fig. 3.10 The ray velocity surface in a biaxial medium. Along any direction two vectors are plotted whose lengths are the two ray velocities. The end points of these vectors form the ray velocity surface.

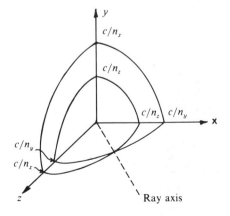

other at a pair of diametrically opposite points in the x–z plane. In the other planes, they are nonintersecting. As can be seen in the x–z plane, two special directions exist along which the two ray velocities are equal. These are called the ray axes which, in general, do not coincide with the optic axes. Such crystals are called biaxial crystals. In uniaxial crystals, two of the dielectric tensor components are equal and one has a single ray or optic axis.

We will show in Problem 3.1 that the angle between the ray axes and the z-direction is given by

$$\alpha = \tan^{-1}\left(\frac{K_y - K_x}{K_z - K_y}\right)^{\frac{1}{2}} \tag{3.75}$$

Fig. 3.11 The intersection curves of the ray velocity surface with the three coordinate planes. Since $K_x < K_y < K_z$ along two directions lying in the x–z plane, the two ray velocities are equal.

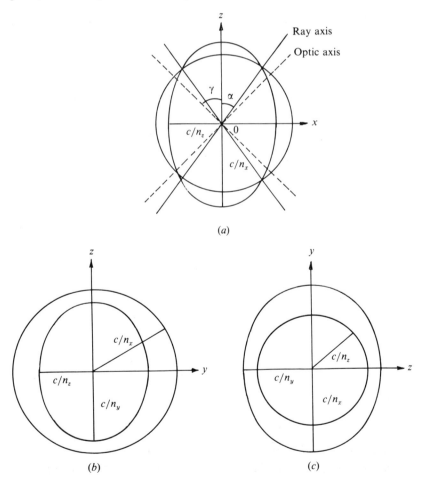

(a)

(b)   (c)

Thus as $K_x \rightarrow K_y$, $\alpha \rightarrow 0$ and there is only a single ray axis which is along the z-direction. Similarly in the limit $K_y \rightarrow K_z$, $\alpha \rightarrow \frac{1}{2}\pi$ and there is again a single ray axis now along x-direction. Thus $\alpha \rightarrow 0$ leads to a positive uniaxial crystal and $\alpha \rightarrow \frac{1}{2}\pi$ leads to a negative uniaxial crystal.

The ray velocity surface is, in general, a surface of fourth order but in the special case of uniaxial media, the ray surface becomes simple. For uniaxial media, we have $K_x = K_y$ assuming the optic axis to be along the z-direction, and thus the complete ray surface can be obtained by considering the intersection curve of the surface in, say, the x–z plane and then generating the surface by rotation about z-axis. We have already obtained the intersection curves in the x–z plane (see Eqs. (3.73) and (3.74)) as a circle and an ellipse. Thus the ray surface consists of two surfaces: (*a*) the o-ray surface which is a sphere described by

$$v_{rx}^2 + v_{ry}^2 + v_{rz}^2 = c^2/K_x \tag{3.76}$$

and (*b*) the e-ray surface which is an ellipsoid of revolution described by

$$\frac{v_{rx}^2 + v_{ry}^2}{c^2/K_z} + \frac{v_{rz}^2}{c^2/K_x} = 1 \tag{3.77}$$

These are the ray velocity surfaces which are used in discussing refraction at anisotropic interfaces using Huygens' construction (see Sec. 3.2). In fact, if we have a point source placed in an anisotropic medium, the wavefront emerging from the point source would correspond to these ray velocity surfaces.

3.8 The index ellipsoid

We saw in Sec. 3.5 that to every direction of propagation, there correspond two definite **D** directions which propagate with, in general, two different wave (phase) velocities. The phase velocities and the corresponding **D** directions are obtained by solving the determinant Eq. (3.43) and Eqs. (3.39)–(3.41) for a given $\hat{\boldsymbol{\kappa}}$.

We now introduce a geometrical construction which permits us to determine the two polarization directions and their velocities. We construct an ellipsoid with axes of half lengths $n_x (= K_x^{\frac{1}{2}})$, $n_y (= K_y^{\frac{1}{2}})$ and $n_z (= K_z^{\frac{1}{2}})$ along the x, y and z-directions respectively (see Fig. 3.12). The equation of such an ellipsoid will be

$$x^2/n_x^2 + y^2/n_y^2 + z^2/n_z^2 = 1 \tag{3.78}$$

The above ellipsoid is referred to as the index ellipsoid.

For a given direction of wave propagation i.e., for a given $\hat{\boldsymbol{\kappa}}$, one draws a plane perpendicular to $\hat{k}$ and passing through the centre of the ellipsoid. This plane intersects the ellipsoid in an ellipse; the directions of the major and minor axes correspond to the **D** directions of the two linearly polarized waves and their lengths give the respective wave refractive indices.

As an example let us consider a wave propagating with its **k** along the z-direction. To determine the phase velocities of the two waves we take a section of the ellipsoid perpendicular to **k** i.e., the x–y plane. The ellipse so obtained will have the equation

$$x^2/n_x^2 + y^2/n_y^2 = 1 \tag{3.79}$$

The principal axes of the ellipse are along x and y-directions and their lengths are n_x and n_y respectively. Thus along the z-direction the eigen waves are: (*a*) a wave polarized with its **D** along the x-direction and which propagates with a phase velocity c/n_x; and (*b*) a wave polarized with its **D** along y and propagating with a phase velocity c/n_y. The above result is the same as that obtained in Sec. 3.5.1.

One can similarly show that for a wave propagating with its **k** in the x–z plane making an angle ψ with the z-axis, the phase velocities of the two waves are: (*a*) c/n_y; and (*b*) c/n_{w2} with n_{w2} given by Eq. (3.53). Their corresponding

Fig. 3.12 The index ellipsoid described by Eq. (3.78). For any given direction of **k**, one finds the intersection curve of a plane normal to **k** and the index ellipsoid. This is, in general, an ellipse, the orientation of the axes of which gives the direction of the **D** corresponding to the two eigen polarizations and their lengths correspond to their respective refractive indices. For a uniaxial crystal, the index ellipsoid is an ellipsoid of revolution about the optic axis.

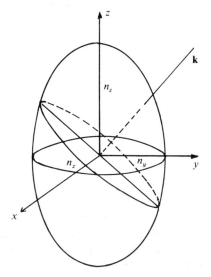

D vectors are along the y-direction and in the x–z plane perpendicular to **k** respectively. In Appendix B we show that, in general, one can use the index ellipsoid to find the phase velocities and the **D** corresponding to waves propagating along any general direction.

For an isotropic medium, $n_x = n_y = n_z$ and the index ellipsoid degenerates into a sphere. For a uniaxial medium with $n_x = n_y \neq n_z$, the index ellipsoid becomes an ellipsoid of revolution about z. In such a medium, it is obvious that if a wave propagates along the z-direction, then the section of the index ellipsoid perpendicular to z is a circle and thus there are no preferred polarization directions for waves travelling along z, and all polarization states travel along z with the same velocity. Thus the z-axis corresponds to the optic axis. Observe that there is only one section which is circular. For biaxial crystals $n_x \neq n_y \neq n_z$ and one can show that there are two circular sections of the ellipsoid. The directions perpendicular to these circular sections are the optic axes and along these directions the wave velocities are equal. The index ellipsoid will be used in Chapters 15 and 16 to study electrooptic effect and acoustooptic effect.

Problems

Problem 3.1: Show that the angle between the ray axes and the z-direction is

$$\alpha = \tan^{-1} \left(\frac{K_y - K_x}{K_z - K_y} \right)^{\frac{1}{2}}$$

Solution: Since the ray axes lie in the x–z plane, we have to find that direction for which v_{rx} and v_{rz} simultaneously satisfy Eqs. (3.73) and (3.74). From Eqs. (3.73) and (3.74) one easily obtains

$$\frac{v_{rx}}{c} = \left[\frac{K_y - K_x}{K_y(K_z - K_x)} \right]^{\frac{1}{2}} \tag{3.80}$$

$$\frac{v_{rz}}{c} = \left[\frac{K_z - K_y}{K_y(K_z - K_x)} \right]^{\frac{1}{2}} \tag{3.81}$$

and hence

$$\tan \alpha = \frac{v_{rx}}{v_{rz}} = \left(\frac{K_y - K_x}{K_z - K_y} \right)^{\frac{1}{2}} \tag{3.82}$$

Problem 3.2: The determinant equation, Eq. (3.43), can also be represented schematically in **k** space. Obtain the curves of intersection of this surface in the k_x–k_y, k_y–k_z and k_x–k_z planes. Show that for $K_x < K_y < K_z$ one has two directions lying in the k_x–k_z plane along which the phase velocities are equal. Show that the angle 2γ between these two directions (see Fig. 3.11(a)) is given by the following equation

$$\tan \gamma = \left[\frac{K_z (K_y - K_x)}{K_x (K_z - K_y)} \right]^{\frac{1}{2}} \tag{3.83}$$

(cf. Eq. (3.75)). These correspond to the optic axes of the crystal.

Problem 3.3: Fig. 3.13(a) shows a Wollaston prism consisting of two similar prisms of quartz, one having its optic axis perpendicular to the edge and the other parallel to the edge. Show that an unpolarized beam entering such a prism would split into two orthogonal linearly polarized waves. What is the difference between the prisms shown in Figs. 3.13(a) and (b); the latter one is referred to as a **Rochon prism**.

Problem 3.4: Fig. 3.14 shows a Babinet compensator consisting of two similar wedges of quartz, one having its optic axis parallel to the edge and the other perpendicular to the edge as shown. Show that if such a compensator is placed between two crossed Nicol prisms, with the pass axes of the Nicol prisms making an angle of 45° with the edge of the prism, one would obtain a system of equidistant dark and bright bands. The Babinet compensator is used for analysing polarized light.

Problem 3.5: Consider a uniaxial crystal with indices n_o and n_e cut so that the optic axis is normal to the surface. Show that for a light ray incident from air on the interface at an angle θ_i with the normal to the surface, the angle of refraction θ_e of the e-ray is given by

$$\tan \theta_e = \frac{n_o}{n_e} \frac{\sin \theta_i}{(n_e^2 - \sin^2 \theta_i)^{\frac{1}{2}}} \tag{3.84}$$

Solution: In Fig. 3.15(a) we have shown the incident wavefront AN corresponding to the rays incident at an angle θ_i. Let the distance $AB = d$. The time taken by the wave to travel from N to B is

$$\tau = NB/c = (d/c) \sin \theta_i \tag{3.85}$$

According to Huygens' construction, in order to determine the e-ray direction, we

Fig. 3.13 (*a*) The Wollaston prism, (*b*) the Rochon prism.

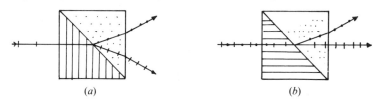

(*a*)   (*b*)

Fig. 3.14 The Babinet compensator.

draw an ellipse centred at A and with axes $(c/n_o)\tau$ and $(c/n_e)\tau$ along and normal to the optic axis which is chosen here as the z-direction. This would be the position of the secondary wavelet corresponding to the e-ray in the time the ray takes to travel from N to B. Now we draw a tangent from B to the elliptical wavefront which touches the ellipse at C. The line joining A to C will give us the e-ray direction.

In order to calculate θ_e, the angle of refraction, we find that point $C(x, z)$ on the ellipse where the tangent is parallel to the line joining the point C to B $(d, 0)$. The equation of the ellipse whose axes are along x and z with lengths $(c/n_e)\tau$ and $(c/n_o)\tau$ is

$$x^2 n_e^2 + z^2 n_o^2 = c^2 \tau^2 = d^2 \sin^2 \theta_i \qquad (3.86)$$

Fig. 3.15 (*a*) Refraction of a plane wave incident at an angle θ_i on an interface between air and a uniaxial medium whose optic axis is perpendicular to the interface. (*b*) The propagation directions corresponding to the refracted wave and ray in the uniaxial medium.

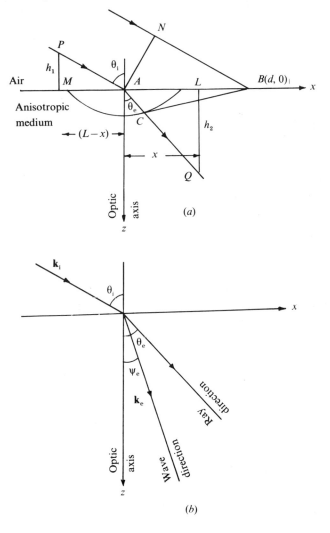

(*a*)

(*b*)

The slope of the tangent to the ellipse at any point (x, z) is

$$\frac{dz}{dx} = -\frac{x\,n_e^2}{z\,n_o^2} \tag{3.87}$$

This must be equal to the slope of BC which is $z/(x - d)$. Thus

$$\frac{x\,n_e^2}{z\,n_o^2} = \frac{z}{d - x} \tag{3.88}$$

or

$$x = d\sin^2\theta_i/n_e^2 \tag{3.89}$$

where we have used Eq. (3.86). Substituting this value of x in Eq. (3.86), we get

$$z = \frac{d}{n_o}\sin\theta_i\left(1 - \frac{\sin^2\theta_i}{n_e^2}\right)^{\frac{1}{2}} \tag{3.90}$$

Thus

$$\tan\theta_e = \frac{x}{z} = \frac{n_o}{n_e}\frac{\sin\theta_i}{(n_e^2 - \sin^2\theta_i)^{\frac{1}{2}}} \tag{3.91}$$

Problem 3.6: Derive Eq. (3.91) using Fermat's principle (see Ghatak (1977), Chapter 2).

Solution: The optical path length from P to Q is given by (see Fig. 3.15(a))

$$L_{OP} = PA + n_{re}(\theta_e)AQ$$
$$= [h_1^2 + (L - x)^2]^{\frac{1}{2}} + n_{re}(\theta_e)(h_2^2 + x^2)^{\frac{1}{2}} \tag{3.92}$$

where

$$n_{re}^2(\theta_e) = n_o^2\cos^2\theta_e + n_e^2\sin^2\theta_e \tag{3.93}$$

or

$$n_{re}^2(\theta_e) = n_o^2\frac{h_2^2}{h_2^2 + x^2} + n_e^2\frac{x^2}{h_2^2 + x^2} \tag{3.94}$$

and the distances h_1, h_2, L and x are shown in Fig. 3.15(a). According to Fermat's principle

$$dL_{OP}/dx = 0 \tag{3.95}$$

which readily gives us

$$\sin\theta_i = n_e^2\sin\theta_e/n_{re}(\theta_e) \tag{3.96}$$

If we now use Eq. (3.93), we get Eq. (3.91).

Problem 3.7: In the above problem if the optic axis is assumed to lie in the plane of incidence but making an angle Γ with the normal show that

$$\sin\theta_i = \frac{n_o^2\cos\theta_e\sin\Gamma + n_e^2\sin\theta_e\cos\Gamma}{n_{re}(\theta_e)} \tag{3.97}$$

where θ_e is now the angle that the refracted ray makes with the optic axis (see Ghatak (1977)).

Problem 3.8: In the above problem, assume normal incidence i.e., $\theta_i = 0$. If r represents the angle of refraction for the e-ray then show that for positive uniaxial crystals

$$\tan r = \tan(\theta_e + \Gamma)$$

$$= \frac{(n_e^2 - n_o^2)\cos\Gamma\sin\Gamma}{n_o^2 \sin^2\Gamma + n_e^2 \cos^2\Gamma} \tag{3.98}$$

Problem 3.9: Using the fact that the tangential component of the **k** of the incident and refracted waves must be continuous, derive Eq. (3.91).

Solution: If ψ_e is the angle made by the refracted wave vector $\mathbf{k}_e$ and the normal to the surface (see Fig. 3.15(b)) then

$$k_0 \sin\theta_i = k_0 n_{we}(\psi_e)\sin\psi_e \tag{3.99}$$

Now (see Eq. 3.53)

$$\frac{1}{n_{we}^2(\psi_e)} = \frac{\cos^2\psi_e}{n_o^2} + \frac{\sin^2\psi_e}{n_e^2} \tag{3.100}$$

Thus from the above two equations we obtain

$$\tan\psi_e = \frac{n_e}{n_o}\frac{\sin\theta_i}{(n_e^2 - \sin^2\theta_i)^{\frac{1}{2}}} \tag{3.101}$$

Using Eqs. (3.54) and (3.56), we note for the **E** of the refracted wave

$$\frac{E_z}{E_x} = -\frac{n_o^2}{n_e^2}\tan\psi_e$$

$$= -\frac{n_o}{n_e}\frac{\sin\theta_i}{(n_e^2 - \sin^2\theta_i)^{\frac{1}{2}}} \tag{3.102}$$

Since the refracted ray is perpendicular to **E**, Eq. (3.91) immediately follows.

Problem 3.10: In Fig. 3.3(a), the wave is assumed to be incident normally and the incident electric field is assumed to be described by the following equation:

$$E_y(x = 0) = \frac{1}{\sqrt{2}}E_0\cos\omega t$$

$$E_z(x = 0) = \frac{1}{\sqrt{2}}E_0\cos\omega t \tag{3.103}$$

(a) Show that E_y propagates as an o-wave and E_z as an e-wave.
(b) At $x = l$

$$E_y = \frac{1}{\sqrt{2}}E_0\cos(\omega t - k_0 n_0 l)$$

$$E_z = \frac{1}{\sqrt{2}}E_0\cos(\omega t - k_0 n_e l) \tag{3.104}$$

Show that for a quartz quarter wave plate (see Eq. 3.6)), the emergent light is right circularly polarized and for a calcite $\frac{1}{4}\lambda$ plate it is left circularly polarized.

Problem 3.11: A left circularly polarized beam at $\lambda_0 = 5893\,\text{Å}$ is incident normally on a calcite $\frac{1}{2}\lambda$ plate for which $n_o = 1.65836$, $n_e = 1.48641$. Calculate the thickness of the plate and show that the emergent beam is right circularly polarized.

Problem 3.12: Show that if a linearly polarized beam is incident on a half wave plate with the plane of polarization making an angle γ with the fast axis, then the output beam is linearly polarized with its plane of polarization making an angle $-\gamma$ with the fast axis.

Solution: Let us assume the half wave plate to be made of a positive uniaxial crystal i.e., $n_o < n_e$. Thus the direction perpendicular to the optic axis is the fast axis which we assume to be along x; thus the optic axis will be along z. The components along x and z of the incident linearly polarized waves will be

$$\left.\begin{aligned} E_x &= E_0 \cos\theta\, e^{i\omega t} \\ E_z &= E_0 \sin\theta\, e^{i\omega t} \end{aligned}\right\} \tag{3.105}$$

where θ is the angle made by the polarization direction with the x-axis. If the crystal thickness is l then at the output one has

$$\begin{aligned} E_x(l) &= E_0 \cos\theta\, e^{i(\omega t - k_0 n_o l)} \\ E_z(l) &= E_0 \sin\theta\, e^{i(\omega t - k_0 n_e l)} \end{aligned} \tag{3.106}$$

Thus

$$E_z(l)/E_x(l) = \tan\theta\, e^{i k_0 l (n_e - n_o)} \tag{3.107}$$

for a half wave plate $k_0(n_e - n_o)l = \pi$. Thus

$$E_z(l)/E_x(l) = -\tan\theta \tag{3.108}$$

i.e., the output is also linearly polarized with its **E** making an angle $-\theta$ with the x-direction.

Problem 3.13: A polarizer–quarter wave plate combination can be used as an isolator to prevent any normally reflected light returning along the incident direction. The construction consists of a polarizer followed by the $\frac{1}{4}\lambda$ plate with the polarizer pass axis making an angle of 45° with the axes of the $\frac{1}{4}\lambda$ plate. Show that light reflected normally from any subsequent surface after passing again through the $\frac{1}{4}\lambda$ plate becomes a linearly polarized wave, polarized at 90° to the pass axis of the polarizer and hence is cut off. Such isolators are used to prevent feedback from interferometers into lasers.

Problem 3.14: What is the difference in the output from a quarter wave plate when (a) a left circularly and (b) a right circularly polarized light is incident on it.

Problem 3.15: Using the fact that the ray propagation direction is perpendicular to

E and that the ray and wave velocities are related through Eq. (3.33), derive Eq. (3.69) from Eq. (3.53).

Solution: From Eq. (3.33) we may write

$$n_r^2 = n_w^2 \cos^2 \phi$$

$$= \frac{K_x K_z \cos^2 (\psi - \theta)}{(K_z \cos^2 \psi + K_x \sin^2 \psi)} \tag{3.109}$$

where we have used $\phi = \psi - \theta$ (see Fig. 3.9(*b*)) and Eq. (3.53). Now, since **E** is perpendicular to ray propagation direction, we may write

$$\tan \theta = -\frac{E_z}{E_x} = \frac{K_x}{K_z} \tan \psi \tag{3.110}$$

Expanding the cosine in Eq. (3.109) and using Eq. (3.110), we readily obtain Eq. (3.69).

Problem 3.16: Consider a uniaxial medium with ordinary and extraordinary refractive indices n_o and n_e respectively and consider the propagation of an e-wave at an angle ψ with the optic axis. Calculate the angle between the e-wave and the e-ray.

Solution: Since the angle between the e-wave and the e-ray is the same as that between **E** and **D**, if ϕ represents this angle, then we may write

$$\cos \phi = \frac{\mathbf{E \cdot D}}{|\mathbf{E}||\mathbf{D}|} = \frac{E_x D_x + E_z D_z}{(E_x^2 + E_z^2)^{\frac{1}{2}} (D_x^2 + D_z^2)^{\frac{1}{2}}} \tag{3.111}$$

where the propagation is chosen to be in the x–z plane. Since **D** is perpendicular to **k**, we have (see Eq. (3.56))

$$\frac{D_z}{D_x} = \frac{K_z E_z}{K_x E_x} = -\tan \psi \tag{3.112}$$

Substituting this in Eq. (3.111), we may simplify to obtain

$$\cos \phi = \frac{(n_e^2 + n_o^2 \tan^2 \psi)}{(n_e^4 + n_o^4 \tan^2 \psi)^{\frac{1}{2}}} \cos \psi \tag{3.113}$$

From above, we obtain after some algebra

$$\tan \phi = \frac{1}{2} \left(\frac{\cos^2 \psi}{n_o^2} + \frac{\sin^2 \psi}{n_e^2} \right)^{-1} \left(\frac{1}{n_e^2} - \frac{1}{n_o^2} \right) \sin 2\psi \tag{3.114}$$

As an example we consider KDP for which

$$n_o = 1.50737, \quad n_e = 1.46685$$

For propagation at an angle $\psi = \frac{1}{4}\pi$ from the optic axis we get

$$\tan \phi = 2.72 \times 10^{-2}$$

which corresponds to $\phi \simeq 1.56°$. For lithium niobate which is another important anisotropic medium,

$$n_o = 2.2967, \quad n_e = 2.2082$$

and for $\psi = \frac{1}{4}\pi$, we readily obtain $\phi \simeq 2.25°$. Thus for typical crystals, the angle between the ray and wave direction is about 1–3°. Obviously for $\psi = 0$ or $\frac{1}{2}\pi$, i.e., for propagation along or perpendicular to the optic axis, $\phi = 0$ implying that the e-wave and the e-ray propagate along the same direction.

4

Fraunhofer diffraction

4.1   Introduction

Consider a plane wave incident normally on a long narrow slit of width a as shown in Fig. 4.1. If we have a screen beyond the slit then according to geometrical optics, only the region LM will be illuminated and the remaining portion (which is known as the geometrical shadow) will be absolutely dark. However, experiments tell us that there is some light in the geometrical shadow; this spreading of light is due to the phenomenon of diffraction. In the experimental arrangement shown in Fig. 4.1, diffraction will occur along the x-direction and the amount of (diffraction) spreading increases with (*a*) increase in wavelength and (*b*) decrease in the width of the slit. Thus the smaller the slit width, the greater will be the spread due to diffraction.

The phenomenon of diffraction is usually divided into two categories:

Fig. 4.1 A plane wave is incident normally on a long narrow slit of width a. According to geometrical optics, only the region LM will be illuminated.

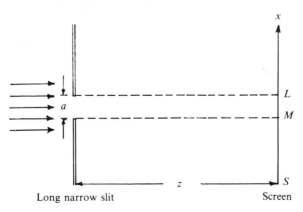

(i) Fresnel diffraction
(ii) Fraunhofer diffraction

In Fresnel diffraction, the source of light or the screen (or both) are at a finite distance from the diffracting aperture. For example, in the arrangement shown in Fig. 4.1, since the screen is at a finite distance from the diffracting aperture, it corresponds to Fresnel diffraction. In Fraunhofer diffraction, the source *and* the screen are at infinite distances from the aperture; this is easily achieved by placing the source on the focal plane of a convex lens and placing the screen on the focal plane of another convex lens (see Fig. 4.2). The two lenses effectively move the source and the screen to infinity because the first lens makes the light beam parallel and the second lens makes a point on the screen receive a parallel beam of light. Indeed in the arrangement shown in Fig. 4.1, if the distance z between the aperture and the screen satisfies

$$z \gg a^2/\lambda$$

then the diffraction pattern observed on the screen will be essentially of the Fraunhofer type.

It turns out that it is much easier to calculate the intensity distribution corresponding to Fraunhofer diffraction and fortuitously, it is the Fraunhofer diffraction pattern which is of greater importance in optics. Further, the Fraunhofer diffraction pattern is not difficult to observe; all that one needs is an ordinary laboratory spectrometer, the collimator gives out a parallel beam of light and the telescope receives a parallel beam of light on its focal plane. The diffracting aperture is placed on the prism table. We will discuss the Fraunhofer diffraction pattern in this chapter. The Fresnel diffraction pattern and its transition to Fraunhofer diffraction will be discussed in the next chapter.

Fig. 4.2 To observe the Fraunhofer diffraction pattern, the source and the screen are placed at the focal planes of two lenses. The beam diffracted along the direction θ is observed at the point P.

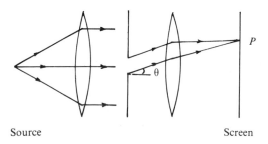

Source   Screen

4.2   The diffraction formula

We first consider a plane wave incident normally on a diffracting aperture $\mathscr{A}$ as shown in Fig. 4.3. The field at the point P can be calculated approximately by using the Huygens–Fresnel principle according to which each point on a wavefront is a source of secondary disturbance and the secondary wavelets emanating from different points mutually interfere. Now, the field at the point P due to spherical waves emanating from the area $d\xi\,d\eta$ around the point Q whose coordinates are $(\xi, \eta, 0)$ will be proportional to

$$A(e^{-ikr}/r)\,d\xi\,d\eta \qquad\qquad ? \tag{4.1}$$

where A represents the amplitude of the plane wave in the plane of the aperture and

$$r = [(x-\xi)^2 + (y-\eta)^2 + z^2]^{\frac{1}{2}} \tag{4.2}$$

represents the distance QP. In order to calculate the field at the point P due to the entire aperture we will have to sum over all the infinitesimal areas and if we replace the summation by an integral we will obtain

$$u(P) = C \iint_{\mathscr{A}} A(e^{-ikr}/r)\,d\xi\,d\eta \tag{4.3}$$

where C is the proportionality constant which can be determined by noting that in the absence of the aperture $u(P)$ must be equal to Ae^{-ikz}. Thus $\quad?$

$$C \iint_{-\infty}^{+\infty} A(e^{-ikr}/r)\,d\xi\,d\eta = Ae^{-ikz} \tag{4.4}$$

Fig. 4.3 A plane wave (propagating along the z-axis) is incident normally on a diffracting aperture $\mathscr{A}$.

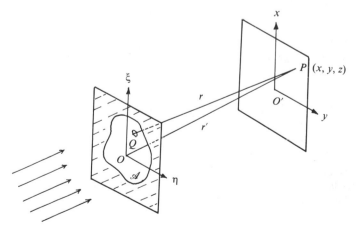

If we carry out the above integration under the approximation that most of the contribution to the integral comes from the domain $|x - \xi|$, $|y - \eta| \ll z$, we will obtain (see Problem 4.14):

$$C = i/\lambda \tag{4.5}$$

Thus

$$u(P) = \frac{iA}{\lambda} \int \int \frac{e^{-ikr}}{r} d\xi \, d\eta \tag{4.6}$$

It should be mentioned here that a more rigorous analysis based on the solution of the scalar wave equation gives the same result as Eq. (4.6) provided we neglect the obliquity factor (see, e.g., Ghatak and Thyagarajan (1978), Sec. 4.4).

In most problems of practical interest, the dimension of the aperture is small so that the quantity r does not vary appreciably over the area of the aperture and very little error will be involved if r in the denominator is replaced by r' and taken out of the integral,[†] i.e.,

$$u(P) \approx (iA/\lambda r') \int \int_{\mathscr{A}} e^{-ikr} d\xi \, d\eta \tag{4.7}$$

where r' represents the distance of the point P from a conveniently chosen origin in the aperture. Even when we have an infinitely extended aperture (such as a long narrow slit) we will see that the major contribution to the integral in Eq. (4.7) comes from a small domain of the aperture where r can be assumed to be a constant.

We choose a convenient point O (on the plane of the aperture) as the origin of the coordinate system. Let $(\xi, \eta, 0)$ and (x, y, z) represent the coordinates of arbitrary points Q and P on the planes of the aperture and of the screen respectively. Thus the distance QP is given by

$$r = QP = [(x - \xi)^2 + (y - \eta)^2 + z^2]^{\frac{1}{2}}$$

$$= r' \left[1 - \frac{2(x\xi + y\eta)}{r'^2} + \frac{\xi^2 + \eta^2}{r'^2} \right]^{\frac{1}{2}} \tag{4.8}$$

where $r' = (x^2 + y^2 + z^2)^{\frac{1}{2}}$ represents the distance of the point P from the origin O. Since the dimension of the aperture is, in general, very small

---

[†] It should be pointed out that although the quantity r does not vary appreciably over the aperture, the quantity e^{-ikr} could vary significantly because of the fact that k is very large. For example, for $\lambda = 6 \times 10^{-5}$ cm ($k \approx 10^5$ cm^{-1}), kr will change by $\frac{1}{2}\pi$ when r changes by $\frac{1}{4}\lambda = 1.5 \times 10^{-5}$ cm. Thus as r changes from 60 cm to 60.00015 cm, $\cos kr$ would change from 1.0 to 0.

compared to the distance r', we make a binomial expansion to obtain

$$r = r' - \frac{x\xi + y\eta}{r'} + \left[-\frac{(x\xi + y\eta)^2}{2r'^3} + \frac{\xi^2 + \eta^2}{2r'} \right] + \cdots \tag{4.9}$$

Now, if r' is so large that the quadratic (and higher order terms) in ξ and η can be neglected, then we have what is known as Fraunhofer diffraction. If it is necessary to retain terms that are quadratic in ξ and η then we have what is known as Fresnel diffraction (see Sec. 5.2). In Sec. 5.10.1, we will explicitly discuss the transition of the Fresnel diffraction pattern into the Fraunhofer pattern, and we will show that for a plane wave incident normally on a long narrow slit (whose width is a in the x-direction – see Fig. 4.1), the diffraction pattern along the x-direction will be Fraunhofer if

$$z \gg a^2/\lambda \tag{4.10}$$

In general, for an aperture of dimension a, the pattern will be Fraunhofer if

$$a^2/\lambda z \ll 1 \tag{4.11}$$

The quantity $a^2/\lambda z$ is known as the *Fresnel number* of the aperture (see Sec. 5.6).

Thus while discussing Fraunhofer diffraction we may write

$$r = r' - (l\xi + m\eta) \tag{4.12}$$

where

$$l = x/r' = \cos \alpha \quad \text{and} \quad m = y/r' = \cos \beta \tag{4.13}$$

represent the direction cosines of OP along the x and y-directions; α and β represent the angles that OP makes with the x and y-axes. Substituting in Eq. (4.7) we obtain

$$u(P) = C \iint_{\mathscr{A}} e^{ik(l\xi + m\eta)} \, d\xi \, d\eta \tag{4.14}$$

where

$$C = \frac{iA}{\lambda} \frac{e^{-ikr'}}{r'} \tag{4.15}$$

is a constant.

If we define an aperture function $P(\xi, \eta)$ by

$$\left. \begin{array}{ll} P(\xi, \eta) = 1 & \text{for } (\xi, \eta) \text{ inside the aperture} \\ = 0 & \text{otherwise} \end{array} \right\} \tag{4.16}$$

then Eq. (4.14) can be written as

$$u(P) = C \int\int_{-\infty}^{+\infty} P(\xi, \eta) e^{ik(l\xi + m\eta)} \, d\xi \, d\eta \tag{4.17}$$

Eq. (4.17) is nothing other than the Fourier transform of the aperture function $P(x, y)$. If instead of an aperture we have an amplitude distribution $A(x, y)$ across the wavefront, then the above equation would be modified to

$$u(P) = C \int\int_{-\infty}^{+\infty} A(\xi, \eta) e^{ik(l\xi + m\eta)} \, d\xi \, d\eta \tag{4.18}$$

Since the Fourier transform of a Gaussian function is another Gaussian function, if we consider a beam whose transverse intensity distribution is Gaussian, then the transverse intensity distribution associated with the Frauhnofer diffraction pattern will also be Gaussian. Indeed, even the Fresnel diffraction pattern is also Gaussian (see Sec. 5.4).

Now, in order to observe the Fraunhofer diffraction pattern, one can either place a screen at a distance $z \gg a^2/\lambda$ or one can put a corrected convex lens immediately after the aperture. Thus the diffracted wave corresponding to a particular set of direction cosines (l, m) will be focussed at a particular point P (see Fig. 4.2). Thus the intensity pattern on the focal plane of the lens will correspond to the Fraunhofer diffraction pattern of the aperture. We now consider some specific examples.

4.3   Rectangular aperture

We first consider a rectangular aperture of dimensions $a \times b$ as shown in Fig. 4.4(*a*). For such a case we have

$$u(P) = C \int_{-b/2}^{+b/2} d\eta \int_{-a/2}^{+a/2} d\xi \, e^{ik(l\xi + m\eta)} \tag{4.19}$$

where we have chosen the origin to be at the centre of the rectangular aperture. Now,

$$\int_{-a/2}^{+a/2} e^{ikl\xi} \, d\xi = \frac{2}{kl} \sin \frac{kal}{2} = a \frac{\sin \zeta}{\zeta}$$

where

$$\zeta = \tfrac{1}{2}kal = \pi al/\lambda \tag{4.20a}$$

Similarly

$$\int_{-b/2}^{+b/2} e^{ikm\eta} \, d\eta = b \sin \tau/\tau$$

where

$$\tau = \tfrac{1}{2}kbm = \pi bm/\lambda \tag{4.20b}$$

Thus

$$u(P) = u_0 \frac{\sin \zeta}{\zeta} \frac{\sin \tau}{\tau} \tag{4.21}$$

interference + diffraction

where

$$u_0 = Cab \tag{4.22}$$

represents the amplitude along $l = 0$, $m = 0$, i.e., at the axial point O'.

In order to relate l and m to the angles of diffraction along the x and

Fig. 4.4 (a) A plane wave propagating along the z-axis is incident normally on a rectangular aperture and the Fraunhofer diffraction pattern is observed on the focal plane of the lens L. (b) The direction cosines of the diffracted wave.

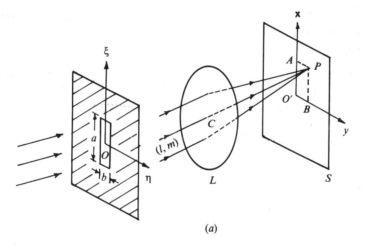

(a)

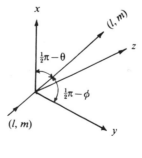

(b)

y-directions we refer to Fig. 4.4(a) and (b). As is obvious

$$l = \sin \theta = PB/CP \tag{4.23a}$$

$$m = \sin \phi = PA/CP \tag{4.23b}$$

Thus the intensity distribution in the Fraunhofer pattern will be given by

$$I = I_0 \left[\frac{\sin (\pi a \sin \theta/\lambda)}{\pi a \sin \theta/\lambda} \right]^2 \left[\frac{\sin (\pi b \sin \phi/\lambda)}{\pi b \sin \phi/\lambda} \right]^2 \tag{4.24}$$

A plot of

$$F(\tau) = \sin^2 \tau/\tau^2, \quad \tau = \pi b \sin \phi/\lambda \tag{4.25}$$

versus ϕ is shown in Fig. 4.5. As can be seen $F(\tau)$ has the maximum value

Fig. 4.5 Variation of $\sin^2 \tau/\tau^2$ with ϕ for (a) $b/\lambda = 139$ and (b) $b/\lambda = 278$. The intensity is zero when $\tau = \pi b \sin \phi/\lambda = m\pi$, $m = 1, 2, 3 \ldots$

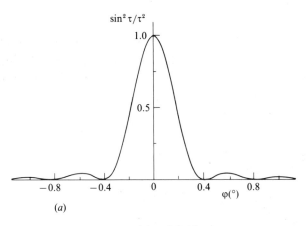

(a)

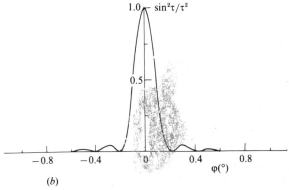

(b)

of unity for $\tau = 0$ (i.e., at $\phi = 0$) and is zero for

$$\left.\begin{array}{l} \tau = n\pi; \qquad n = 1, 2, 3, \ldots \\ \\ \text{i.e., when} \\ \\ \sin \phi = n\lambda/b; \quad n = 1, 2, 3 \ldots \end{array}\right\} \qquad (4.26)$$

Hence the intensity in the diffraction pattern given by Eq. (4.24) is zero when either

$$\left.\begin{array}{l} \zeta = \pi a \sin \theta/\lambda = m\pi \\ \\ \text{or} \\ \\ \tau = \pi b \sin \phi/\lambda = n\pi \end{array}\right\} n, m = 0, 1, 2, 3, \ldots \text{ except when } n = m = 0$$

$$(4.27)$$

i.e. when

$$\left.\begin{array}{l} \sin \theta = m\lambda/a \\ \\ \text{or} \\ \\ \sin \phi = n\lambda/b \end{array}\right\} n, m = 0, 1, 2, 3, \ldots \text{ except when } n = m = 0 \qquad (4.28)$$

The condition $n = m = 0$ i.e., $\theta = \phi = 0$ corresponds to the central maximum. For small values of θ and ϕ, minima will occur almost along straight lines parallel to the x and y-axes. A typical Fraunhofer diffraction pattern produced by a square aperture is shown in Fig. 4.6.

Fig. 4.6 The Fraunhofer diffraction pattern of a square aperture.

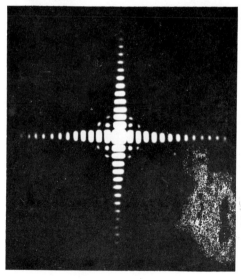

4.4   The single slit diffraction pattern

We next consider the diffraction by a long narrow slit in which $b \to \infty$. In Fig. 4.5(a) and (b) we have plotted

$$\left[\frac{\sin(\pi b \sin \phi / \lambda)}{(\pi b \sin \phi / \lambda)} \right]^2 \qquad (4.29)$$

as a function of ϕ for $b/\lambda = 139$ and 278. It is readily seen that as b/λ increases the intensity distribution becomes more sharply peaked around $\phi = 0$. Thus for $b \gg \lambda$, the intensity in the diffraction pattern will be appreciable only for small values of ϕ i.e., there will be very little diffraction along the y-direction. In this limit the intensity distribution is given by

$$\left. \begin{array}{ll} I = I_0 (\sin \zeta / \zeta)^2 & \text{for } \tau = 0 \\ = 0 & \text{for } \tau \neq 0 \end{array} \right\} \qquad (4.30)$$

which is shown schematically in Fig. 4.7. The intensity is zero when

$$\zeta = m\pi; \qquad m = \pm 1, \pm 2, \dots$$

i.e., $\qquad (4.31)$

$$a \sin \theta = m\lambda; \quad m = \pm 1, \pm 2, \dots$$

represents the conditions for minima. Thus for a plane wave incident normally on a long narrow slit, diffraction would occur only along the direction normal to the width of the slit as shown in Fig. 4.7. The intensity will be zero everywhere except on the x-axis. The intensity is zero when $\zeta = m\pi$ i.e.,

$$\sin \theta = m\lambda / a; \quad m = \pm 1, \pm 2, \dots$$

Fig. 4.7 For a plane wave incident normally on a long narrow slit, diffraction occurs only along the direction normal to the width of the slit.

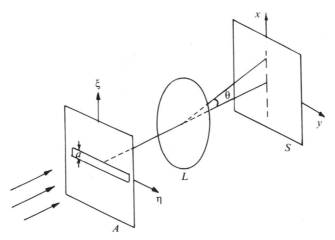

This represents the conditions for minima. Obviously the maximum value of m is the integer which is closest to and less than a/λ.

The positions of minima could have been obtained directly from very elementary considerations. For example, for $m = 1$, we divide the slit into two equal parts along the ξ-direction. For each point in the upper half there is a corresponding point in the lower half for which the path difference between the waves contributing to the diffracted wave along $\theta = \sin^{-1}(\lambda/a)$ is

$$\tfrac{1}{2}a\sin\theta = \tfrac{1}{2}\lambda \tag{4.32}$$

Thus the disturbance from each point in the upper half will be cancelled by a corresponding point in the lower half and the resultant diffracted wave amplitude will be zero. In a similar manner when $a\sin\theta = 2\lambda$, the slit has to be divided into four parts etc.

The positions of the maxima are readily obtained by setting

$$\mathrm{d}I/\mathrm{d}\zeta = 0$$

which gives us

$$\tan\zeta = \zeta \tag{4.33}$$

which represents the transcendental equation determining the positions of maxima. The roots of the above equation can readily be found by determining the points of intersections of the curve $y = \tan\zeta$ with the straight line $y = \zeta$. The roots are

$$\zeta = 0, 4.4934(\approx \tfrac{3}{2}\pi), 7.7253(\approx \tfrac{5}{2}\pi),\ldots \tag{4.34}$$

The corresponding intensities are given by

$$I/I_0 = 1, 0.0472, 0.0168, 0.0083 \tag{4.35}$$

Thus the intensity at the first and second maxima will be 4.96% and 1.68% of the central maximum which can also be seen from Fig. 4.5.

From the above discussion it is obvious that at large distances from the slit the beam diffracts with a divergence angle given by

$$\Delta\theta \sim \lambda/a \tag{4.36}$$

where we have assumed $\lambda/a \ll 1$ so that $\Delta\theta$ is small.

As can be seen, for a given wavelength the diffraction spreading becomes more predominant as a becomes smaller. Indeed one can write down an uncertainty relation

$$a\Delta\theta \sim \lambda \tag{4.37}$$

In the limit of $\lambda \to 0$, the diffraction effects will be completely absent and in this limit the geometrical optics is rigorously valid.

4.5   Circular aperture

We next consider the diffraction of a plane wave incident normally on a circular aperture of radius a (see Fig. 4.8). The field at the point P is again given by (see Eq. (4.14))

$$u(P) = C \iint_{\mathscr{A}} e^{ik(l\xi + m\eta)} \, d\xi \, d\eta \tag{4.38}$$

where the integration is now over the circular aperture. In order to carry out the integration we use the circular coordinates defined by the following equations

$$\xi = \sigma \cos \phi, \quad \eta = \sigma \sin \phi \tag{4.39}$$

Now because of the circular symmetry of the aperture, the diffraction pattern will consist of rings such that the intensity is the same on the circumference of a circle whose centre is at O'. Therefore, without any loss

Fig. 4.8 Fraunhofer diffraction of a plane wave incident normally on a circular aperture of radius a. On the screen S we will observe the Airy pattern.

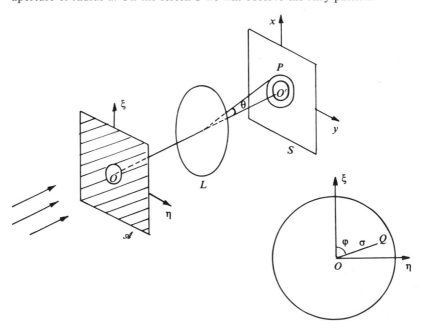

of generality we calculate the field on a point on the $O'-x$-axis for which

$$l = \sin \theta \quad \text{and} \quad m = 0$$

Thus

$$u(P) = C \int_0^a \sigma \, d\sigma \int_0^{2\pi} d\phi \, e^{ik\sigma \cos\phi \sin\theta} \tag{4.40}$$

where $\sigma \, d\sigma \, d\phi$ represents the elemental area on the area of the aperture. We introduce

$$\zeta = k\sigma \sin \theta \tag{4.41}$$

Thus

$$u(P) = \frac{C}{(k \sin \theta)^2} \int_0^{ka\sin\theta} \zeta \, d\zeta \left(\int_0^{2\pi} d\phi \, e^{i\zeta \cos\phi} \right)$$

Now, if J_n represents the Bessel function of the nth order then it is well known that (see e.g. Irving and Mullineux (1959))

$$J_0(\zeta) = (1/2\pi) \int_0^{2\pi} e^{i\zeta \cos\phi} \, d\phi \tag{4.42}$$

Thus

$$u(P) = \frac{2\pi C}{(k \sin \theta)^2} \int_0^{ka\sin\theta} \zeta J_0(\zeta) \, d\zeta \tag{4.43}$$

Using the identity

$$\frac{d}{d\zeta} [\zeta J_1(\zeta)] = \zeta J_0(\zeta) \tag{4.44}$$

we get

$$u(P) = \frac{2\pi C}{(k \sin \theta)^2} \zeta J_1(\zeta) \Big|_0^{ka\sin\theta}$$

$$= u_0 [2J_1(v)/v] \tag{4.45}$$

where

$$v = ka \sin \theta \tag{4.46}$$

and

$$u_0 = \pi C a^2$$

represents the amplitude at the central spot.[†] The corresponding intensity distribution is given by

$$I(\theta) = I_0 [2J_1(v)/v]^2 \tag{4.47}$$

which is known as the Airy pattern and is shown in Fig. 4.9. Now

$$J_1(v) = 0 \quad \text{when } v = 0, 3.8317, 7.0156, 10.1735, \ldots \tag{4.48}$$

Thus the dark rings in the diffraction pattern will occur when

$$v = ka \sin \theta = 3.8317, 7.0156, 10.1735, \ldots \tag{4.49}$$

(At $v = 0$, $2J_1(v)/v \to 1$ and the intensity has a maximum value.) The first dark ring appears when

$$\sin \theta = 3.832\lambda/2\pi a \approx 1.22\lambda/D \tag{4.50}$$

where $D = 2a$ represents the diameter of the circular aperture. In Sec. 4.5.1 we will show that about 84% of the energy is contained within the first dark ring; thus we may say that the angular spread of the beam is approximately given by

$$\Delta\theta \approx \lambda/D \tag{4.51}$$

From Eqs. (4.36) and (4.51) we may say that the angular divergence associated with the diffraction pattern can be written in the general form

$$\Delta\theta \sim \lambda/\text{linear dimension of the aperture} \tag{4.52}$$

Problem 4.1: Assume a plane wave ($\lambda = 6000$ Å) incident normally on a circular aperture of radius 0.1 mm. Immediately after the aperture we have a convex lens of focal length 20 cm. Calculate the radii of the first three dark rings observed on the focal plane of the lens.

Solution: The angles of diffraction for the first three dark rings would be given by

$$\sin \theta = \frac{3.832\lambda}{2\pi a}, \frac{7.016\lambda}{2\pi a} \quad \text{and} \quad \frac{10.174\lambda}{2\pi a}$$

$$= 3.66 \times 10^{-3}, 6.70 \times 10^{-3} \quad \text{and} \quad 9.72 \times 10^{-3}$$

[†] It should be noted that since

$$J_1(x) = \tfrac{1}{2}x - \tfrac{1}{2}(\tfrac{1}{2}x)^3 + \cdots$$

we have

$$\underset{x \to 0}{\text{Lt}} \; 2J_1(x)/x = 1$$

Fig. 4.9 (*a*) The intensity distribution in the Fraunhofer diffraction pattern produced by a circular aperture. (*b*) The observed Airy pattern.

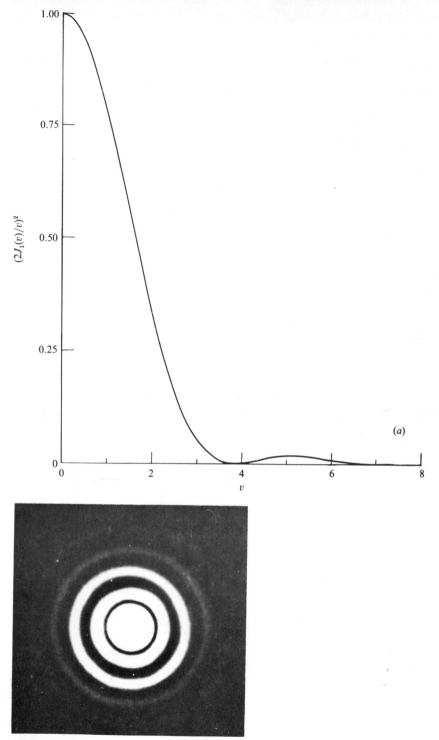

(*a*)

(*b*)

implying

$$\theta \approx 0.21°, 0.38° \quad \text{and} \quad 0.56°$$

Since the angles are small, the radii of the corresponding dark rings would be given by

$$r = f \tan \theta$$

$$\approx 0.732 \,\text{mm}, 1.34 \,\text{mm} \quad \text{and} \quad 1.94 \,\text{mm}$$

respectively.

4.5.1   Intensity distribution in the Airy pattern

We next calculate the energy contained inside the various rings of the Airy pattern. We will, for example, show that about 84% of the energy is contained within the circle bounded by the first dark ring and about 7% of energy is contained in the annular region bounded by the first two dark rings etc.

In the far field pattern the energy contained in the annular region between r and $r + dr$ is be proportional to

$$I(\theta) 2\pi \sin \theta \, d\theta$$

Thus if $f(\theta_0)$ represents the fraction of energy contained in the angular region 0 to θ_0 then

$$f(\theta_0) = \frac{\displaystyle\int_0^{\theta_0} I(\theta) 2\pi \sin \theta \, d\theta}{\displaystyle\int_0^{\pi/2} I(\theta) 2\pi \sin \theta \, d\theta} \tag{4.53}$$

Now, since

$$v = ka \sin \theta$$

we have

$$v \, dv = k^2 a^2 \sin \theta \cos \theta \, d\theta \approx k^2 a^2 \sin \theta \, d\theta \tag{4.54}$$

where we have assumed that most of the energy is contained around a small annular region where $\cos \theta \approx 1$; this is indeed true when $ka \gg 1$ which also represents the condition for the validity of the scalar wave theory.

If we use Eqs. (4.47) and (4.54) in Eq. (4.53) we obtain

$$f(\theta_0) = \frac{\displaystyle\int_0^{v_0} [2J_1(v)/v]^2 \, v \, dv}{\displaystyle\int_0^{ka} [2J_1(v)/v]^2 \, v \, dv} \tag{4.55}$$

where

$$v_0 = ka \sin \theta_0$$

Now

$$\left[\frac{J_1(v)}{v}\right]^2 v = J_1(v)\left[J_0(v) - \frac{dJ_1}{dv}\right]$$

$$= -\frac{dJ_0}{dv}J_0(v) - J_1(v)\frac{dJ_1(v)}{dv} \tag{4.56}$$

where we have used Eq. (4.44) and the relation[†]

$$dJ_0/dv = -J_1(v) \tag{4.57}$$

Thus

$$[J_1(v)/v]^2 v = -\tfrac{1}{2}(d/dv)[J_0^2(v) + J_1^2(v)] \tag{4.58}$$

The integrals in Eq. (4.55) can now readily be evaluated to give

$$f(v_0) = \frac{1 - J_0^2(v_0) - J_1^2(v_0)}{1 - J_0^2(ka) - J_1^2(ka)}$$

$$\approx 1 - J_0^2(v_0) - J_1^2(v_0) \tag{4.59}$$

where in the last step we have again assumed that $ka \gg 1$ which is indeed valid for most practical cases. In Fig. 4.10 we have plotted $f(v_0)$ as a function of v_0. Simple calculations show that

$$f(v_0) = 0.839, 0.910 \quad \text{and} \quad 0.938$$

for

$$v_0 = 3.832, 7.016 \quad \text{and} \quad 10.174$$

respectively. Thus about 84% of light energy is contained within the first dark ring and about 7% of energy is contained in the annular region between the first two dark rings etc.

[†] Eqs. (4.44) and (4.57) follow from the general relation

$$(d/dx)[x^{n+1}J_{n+1}(x)] = x^{n+1}J_n(x)$$

where we must remember that

$$J_{-n}(x) = (-1)^n J_n(x)$$

$(n = 0, \pm 1, \pm 2, \dots)$.

4.6   Directionality of laser beams

An ordinary source of light (like a sodium lamp) radiates in all directions. On the other hand, the output from a laser is usually such that its divergence is primarily due to diffraction effects. The output from a laser has usually a transverse intensity distribution. For example, a He–Ne laser oscillating in its fundamental mode has a Gaussian amplitude distribution:

$$u = u_0 e^{-r^2/w_0^2}, \quad r^2 = x^2 + y^2 \tag{4.60}$$

the propagation is along the z-axis and the quantity w_0 is usually referred to as the spot size of the beam. As we will show in the next chapter, when

Fig. 4.10 Variation of $f(v_0)$ with v_0.

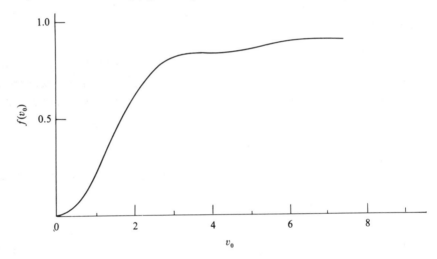

Fig. 4.11 Diffraction of a Gaussian beam.

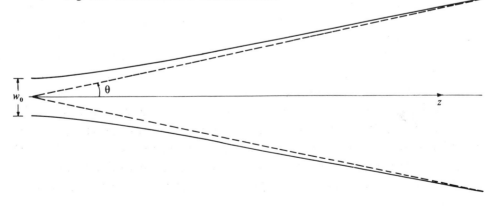

the beam propagates (along the z-direction) the spot size of the beam increases as (see Fig. 4.11)

$$w(z) = w_0(1 + \lambda^2 z^2/\pi^2 w_0^4)^{\frac{1}{2}} \qquad (4.61)$$

Thus for (cf. Eq. (4.10))

$$z \gg \pi w_0^2/\lambda \qquad (4.62)$$

the beam width increases as

$$w(z) \approx \lambda z/\pi w_0 \qquad (4.63)$$

Hence the diffraction divergence is given by

$$\Delta\theta \approx w(z)/z \approx \lambda/\pi w_0 \qquad (4.64)$$

As an example we consider a laser beam with $\lambda \approx 6 \times 10^{-5}$ cm and $w_0 \approx 1$ cm so that

$$\Delta\theta \approx 1.9 \times 10^{-5} \,\text{rad} \approx 3.9''$$

Due to the high directionality of the laser beam, it can be focussed to an extremely small spot (see Fig. 4.12(a)). If f is the focal length of the lens then the area of the focussed spot would be approximately

$$\pi(\lambda f/\pi w_0)^2 \sim 10^{-7} \,\text{cm}^2$$

for $f = 10$ cm. Thus a 1 kW laser beam will produce an intensity of about

$$I \sim \frac{P}{\pi(\lambda f/\pi w_0)^2} \approx 10^{10} \,\text{W/cm}^2 \qquad (4.65)$$

at the focus of the lens. Fig. 4.12(b) shows the spark created at the focus of a 3 MW peak power pulse from a ruby laser. The electric field strengths produced at the focus are of the order of 10^9 V/m. The large intensity is produced in an extremely small region whose radius is $\sim 2\,\mu$m. Such high intensities lead to numerous industrial applications of the laser such as welding, hole drilling, cutting materials, etc.; for further details, see, e.g., Thyagarajan and Ghatak (1981) Chapter 14 (and references therein).

From the above discussion it immediately follows that the greater the radius of the beam the smaller will be the size of the focussed spot and hence the greater will be the intensity at the focussed spot. Indeed one may use a beam expander (see Fig. 4.13) to produce a beam of greater size and hence a smaller focussed spot size. However, the beam would then have a greater divergence and therefore it would expand within a very short distance. One usually defines a *depth of focus* as the distance over which the intensity of the beam (on the axis) decreases by a certain factor of the value at the focal point. Thus a small focussed spot would lead to a small depth of focus. We will show in the next chapter that the depth of focus is given approximately by

$$\lambda(f/w_0)^2$$

Thus larger the value of w_0, the smaller will be the depth of focus. For

Fig. 4.12 (*a*) If a truncated plane wave of radius *a* is incident on an aberrationless lens of focal length f then the emerging spherical wave would be focussed to a spot whose radius is about $\lambda f/a$. (*b*) A spark created in air at the focus of a 3 MW peak power giant pulsed ruby laser. The electric field strengths produced at the focus are of the order of 10^9 V/m. (Photograph courtesy of Dr. R.W. Terhune.)

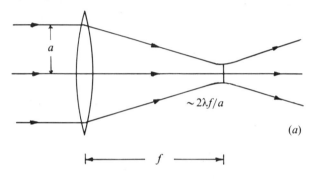

(*a*)

(*b*)

$\lambda = 6000\,\text{Å}$, $f = 10\,\text{cm}$ and $w_0 = 1\,\text{cm}$, the depth of focus would be about 0.006 cm.

4.7 Limit of resolution

Consider two point sources, such as stars (so that we can consider plane waves entering the aperture) being focussed by a telescope objective of diameter D (see Fig. 4.14). As discussed in the previous section, the system can be thought of as being equivalent to a circular aperture of diameter D, followed by a converging lens of focal length f, as shown in Fig. 4.8. As such, each point source will produce its Airy pattern as shown schematically in Fig. 4.14. The diameters of the Airy rings will be determined

Fig. 4.13 Two convex lenses separated by a distance equal to the sum of their focal lengths acts as a beam expander.

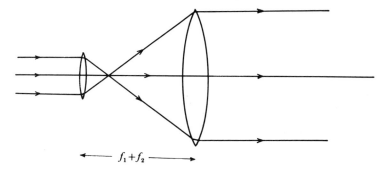

$$f_1 + f_2$$

Fig. 4.14 The image of two distant objects on the focal plane of a convex lens. If the diffraction patterns are well separated they are said to be resolved.

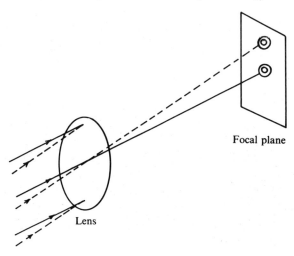

Focal plane

Lens

by the diameter of the objective, its focal length and the wavelength of light.

In Fig. 4.14, the Airy patterns are shown to be quite far away from each other and therefore, the two objects are said to be well resolved. This indeed happens when the angular separation between the two distant objects is much greater than the angular radius of the first dark ring, i.e.,

$$\Delta\theta \gg 1.22\lambda/D$$

According to the Rayleigh criterion, the two images are said to be just resolved when the first dark ring of one pattern falls on the central maximum

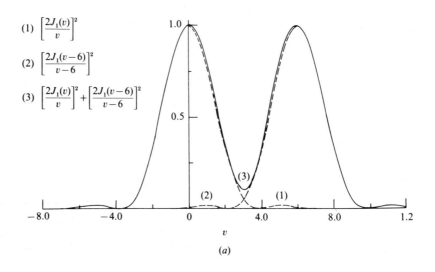

(1) $\left[\dfrac{2J_1(v)}{v}\right]^2$

(2) $\left[\dfrac{2J_1(v-6)}{v-6}\right]^2$

(3) $\left[\dfrac{2J_1(v)}{v}\right]^2 + \left[\dfrac{2J_1(v-6)}{v-6}\right]^2$

(a)

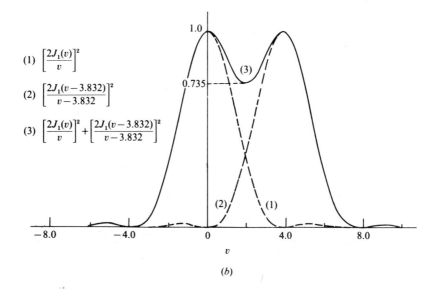

(1) $\left[\dfrac{2J_1(v)}{v}\right]^2$

(2) $\left[\dfrac{2J_1(v-3.832)}{v-3.832}\right]^2$

(3) $\left[\dfrac{2J_1(v)}{v}\right]^2 + \left[\dfrac{2J_1(v-3.832)}{v-3.832}\right]^2$

(b)

of the other pattern, i.e., when

$$\Delta\theta = 1.22\lambda/D \qquad (4.66)$$

The intensity distribution which results when $\Delta\theta$ is given by the above equation is plotted as the solid curve in Fig. 4.15(b) For such a case there is a drop in intensity of about 26% between the two maxima.

From the above it is obvious that for better resolution one requires a larger objective diameter. It is for this reason that a telescope is usually characterized by the diameter of the objective; for example, a 40 in telescope implies that the diameter of the objective is 40 in.

In Fig. 4.15 we have plotted the independent intensity distributions and their resultants produced by two distant point objects for various angular separations; in each case we have assumed that the two sources produce the same intensity at their respective central spots. Obviously, the resultant intensity distributions are quite complicated; what we plotted in Fig. 4.15 are the intensity distributions on the line joining the two centres of the Airy patterns. If we choose this line as x-axis then the parameter v in Figs. 4.15(a), (b) and (c) is given by

$$v = (2\pi a/\lambda f)x$$

Fig. 4.15 The dashed curves correspond to the individual intensity distributions produced by two distant point sources (producing the same intensity at the central spot); the solid curves represent the resultant; (a), (b) and (c) correspond to the cases when the angular separation between the two objects is $6\lambda/\pi D$ $1.22\lambda/D$ and $2\lambda/\pi D$ respectively.

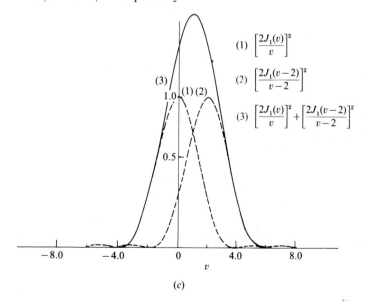

$$(1) \quad \left[\frac{2J_1(v)}{v}\right]^2$$

$$(2) \quad \left[\frac{2J_1(v-2)}{v-2}\right]^2$$

$$(3) \quad \left[\frac{2J_1(v)}{v}\right]^2 + \left[\frac{2J_1(v-2)}{v-2}\right]^2$$

(c)

Now, the intensity distributions given in Fig. 4.15(*a*) correspond to two distant point objects having an angular separation of $6\lambda/\pi D$ and as can be seen the two images are clearly resolved. Fig. 4.15(*c*) corresponds to

$$\Delta\theta \approx 2\lambda/\pi D$$

and as can be seen, the resultant intensity distribution has only one peak and therefore the two points cannot be resolved at all. Finally, if the angular separation of the two objects is $1.22\,\lambda/D$ then the central spot of the one pattern falls on the first minimum of the second and the objects are said to be just resolved.

In order to appreciate the above results numerically we consider a telescope objective whose diameter and focal length are 5 cm and 30 cm respectively. Assuming the light wavelength to be 6×10^{-5} cm, one finds that the minimum angular separation of two distant objects which can just be resolved will be

$$\frac{1.22\lambda}{D} = \frac{1.22 \times 6 \times 10^{-5}}{5} \approx 1.5 \times 10^{-5}\,\text{rad}$$

Further, the radius of the first dark ring of the Airy pattern will be

$$\frac{1.22\lambda}{D} \times \text{focal length} = \frac{1.22 \times 6 \times 10^{-5}}{5} \times 30$$

$$\approx 4.5 \times 10^{-4}\,\text{cm}$$

It is immediately obvious that the larger the diameter of the objective the better will be its resolving power. For example, the diameter of a large telescope objective is about 80 in and the corresponding angular separation of the objects that it can resolve is $\approx 0.07''$. This very low limit of resolution is never achieved in ground based telescopes due to the turbulence of the atmosphere. However, a larger aperture still provides a larger light gathering power and hence the ability to see deeper in space.

It is of interest to note that if we assume that the angular resolution of the human eye is primarily due to diffraction effects then it will be given by

$$\Delta\theta \sim \frac{\lambda}{D} \approx \frac{6 \times 10^{-5}}{2 \times 10^{-1}} = 3 \times 10^{-4}\,\text{rad}$$

where we have assumed the pupil diameter to be 2 mm. Thus, at a distance of 20 m, the eye should be able to resolve two points which are separated by a distance of

$$3 \times 10^{-4} \times 20 = 6 \times 10^{-3}\,\text{m} = 6\,\text{mm}$$

One can indeed verify that this result is qualitatively valid by finding the distance at which a millimetre scale becomes blurred.

In the above discussion we have assumed that the two object points produce identical (but displaced) Airy patterns. If that is not the case then the two central maxima will have different intensities; accordingly one has to set up a modified criterion for the limit of resolution such that the two maxima are well resolved. We have also assumed addition of intensities which is true for incoherent sources.

4.8   Resolving power of a microscope

We next consider the resolving power of a microscope objective of diameter D as shown in Fig. 4.16. Let P and Q represent two closely spaced self luminous point objects which are to be viewed through the microscope. Assuming the absence of any geometrical aberrations, rays emanating from the points P and Q will produce spherical wavefronts (after refraction through the lens) which will form Airy patterns around their paraxial image points P' and Q'. For the points P and Q to be just resolved, the point Q' should lie on the first dark ring surrounding the point P' and therefore we must have

$$\sin \theta' \approx 1.22\lambda/D = 1.22\lambda_0/n'D \tag{4.67}$$

where n and n' represent the refractive indices of the object and image spaces, and λ_0 and $\lambda(=\lambda_0/n')$ represent the wavelength of light in free space and in the medium of refractive index n' respectively. The angle θ' is defined in Fig. 4.16 and we have

$$\sin \theta' \approx \frac{y'}{OP'} = \frac{y' \tan i'}{D/2} \approx \frac{y' \sin i'}{D/2} \tag{4.68}$$

where we have assumed $\sin i' \approx \tan i'$, this is justified since the image distance

Fig. 4.16 Calculation of the resolving power of a microscope objective.

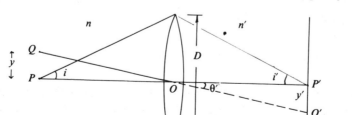

(OP') is large compared to D. Using Eqs. (4.67) and (4.68) we get

$$y' \approx 0.61 \lambda_0/n' \sin i' \tag{4.69}$$

If we now use the sine law $n'y' \sin i' = ny \sin i$ (see e.g., Ghatak and Thyagarajan (1978)) we get

$$y \approx 0.61 \lambda_0/n \sin i \tag{4.70}$$

which represents the smallest distance that the microscope can resolve. The quantity $n \sin i$ is the numerical aperture of the optical system and the resolving power increases with an increase in the numerical aperture. It is for this reason that in some microscopes the space between the object and the objective is filled with an oil – and they are referred to as 'oil immersion objectives'. Eq. (4.70) also tells us that the resolving power increases with a decrease in λ. As such, one often uses blue light (or even ultraviolet light) for the illumination of the object. It is of interest to mention that in an electron microscope the de- Broglie wavelength of electrons accelerated to $100 \, \text{keV}$ is about $0.03 \times 10^{-8} \, \text{cm}$ and therefore such a microscope has a very high resolving power.

In the above analysis, we have assumed that the two object points are self luminous so that the intensities can be added. However, in practice, the objects are illuminated by the same source and therefore, in general, there is some phase relationship between the waves emanating from the two object points; for such a case the intensities will not be strictly additive, nevertheless Eq. (4.70) will still give the correct order for the limit of resolution.

4.9   Annular aperture and apodization

We next consider Fraunhofer diffraction by an annular aperture bounded by two circles of radii a_1 and a_2 as shown in Fig. 4.17(a). Once again the intensity distribution on the focal plane of the lens will have circular symmetry and will consist of circular bright and dark rings. The field at the point P would obviously be given by (cf. Eq. (4.43))

$$u(P) = \frac{2\pi C}{(k \sin \theta)^2} \int_{ka_1 \sin \theta}^{ka_2 \sin \theta} \zeta J_0(\zeta) \, d\zeta \tag{4.71}$$

We can carry out the integration by using Eq. (4.44) and the resultant intensity distribution would be given by

$$I(P) = \frac{I_0}{(1 - \alpha^2)^2} \left[\frac{2J_1(ka_2 \sin \theta)}{ka_2 \sin \theta} - \alpha^2 \frac{2J_1(ka_1 \sin \theta)}{ka_1 \sin \theta} \right]^2 \tag{4.72}$$

Fig. 4.17 (a) An annular aperture bounded by circles of radii a_1 and a_2. (b) The corresponding Fraunhofer diffraction pattern for different values of α.

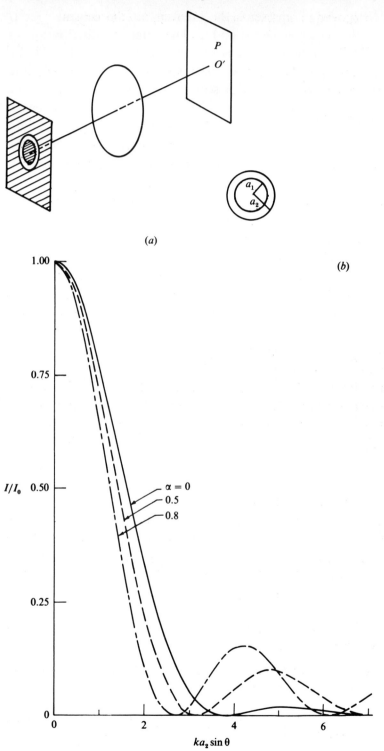

(a)

(b)

where $I_0 = \pi^2 C^2(a_2^2 - a_1^2)^2$ represents the intensity at the central spot O and $\alpha = a_1/a_2$ which lies between 0 and 1. The intensity distribution as given by Eq. (4.72) for different values of α is shown in Fig. 4.17(b), the curve $\alpha = 0$ corresponds to the circular aperture of radius a_2 and therefore it represents the Airy pattern. It may be seen that as $\alpha \to 1$, the central lobe of the diffraction pattern becomes sharper, therefore the resolution increases. This is known as *apodization*; obviously, as $\alpha \to 1$, the total amount of light reaching the plane P decreases and therefore the brightness of the image decreases. In addition, the intensity of the side lobes also increases.

Problem 4.2: In the limit $\alpha \to 1$, show that $I(P) = I(O)J_0^2(k_0 a_2 \sin \theta)$ and that the first dark ring occurs at $\theta \approx \sin^{-1}(2.405/ka_2)$.

4.10   Fraunhofer diffraction by a set of identical apertures

In this section we will consider the Fraunhofer diffraction of a plane wave incident normally on a set of identical apertures $S_1, S_2, \ldots$ as shown in Fig. 4.18. The analysis can readily be applied to a diffraction grating (see Sec. 4.10.2) which consists of a very large number of long parallel slits.

In the apertures shown in Fig. 4.18 we choose a convenient point inside one of the apertures as the origin. For example, for the rectangular apertures shown in Fig. 4.18 we choose the centre of the rectangular aperture S_1 as the origin O_1. Let $O_2, O_3, \ldots$ represent the corresponding points in apertures $S_2, S_3, \ldots$ and let the coordinates of $O_2, O_3, \ldots$ be (ξ_2, η_2), $(\xi_3, \eta_3), \ldots$ respectively. Now, the Fraunhofer diffraction pattern will be given by

$$u(P) = C \iint_{S_1 + S_2 + \ldots} e^{ik(l\xi + m\eta)} \, d\xi \, d\eta$$

$$= C(F_1 + F_2 + \cdots) \tag{4.73}$$

Fig. 4.18 A set of identical apertures.

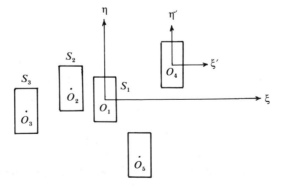

where

$$F_n = \iint_{S_n} e^{ik(l\xi + m\eta)} \, d\xi \, d\eta \tag{4.74}$$

In order to carry out the above integration we introduce the coordinates ξ', η':

$$\xi = \xi' + \xi_n, \quad \eta = \eta' + \eta_n$$

Obviously

$$\xi_1 = \eta_1 = 0$$

and

$$F_n = e^{ik(l\xi_n + m\eta_n)} \iint e^{ik(l\xi' + m\eta')} \, d\xi' \, d\eta' \tag{4.75}$$

The limits of ξ' and η' in the above integral are the same as the limits of ξ and η in the integral over S_1. This can be clearly understood by considering, for example the rectangular apertures shown in Fig. 4.18; thus

$$F_1 = \iint_{S_1} \cdots d\xi \, d\eta = \int_{-a/2}^{a/2} d\xi \int_{-b/2}^{b/2} d\eta \, e^{ik(l\xi + m\eta)}$$

and

$$F_n = \int_{\xi_n - a/2}^{\xi_n + a/2} d\xi \int_{\eta_n - b/2}^{\eta_n + b/2} d\eta \, e^{ik(l\xi + m\eta)}$$

$$= e^{ik(l\xi_n + m\eta_n)} \int_{-a/2}^{+a/2} d\xi' \int_{-b/2}^{+b/2} d\eta' \, e^{ik(l\xi' + m\eta')}$$

$$= e^{ik(l\xi_n + m\eta_n)} F_1$$

Returning to Eq. (4.73) we may therefore write it in the form

$$u(P) = \left[C \iint e^{ik(l\xi' + m\eta')} \, d\xi' \, d\eta' \right] \sum_{n=1}^{N} e^{ik(l\xi_n + m\eta_n)} \tag{4.76}$$

The term inside the square brackets represents the field produced by a single aperture and the summation represents the interference of N point sources placed at $(\xi_1, \eta_1), (\xi_2, \eta_2) \ldots$. The corresponding intensity distribution is given by

$$I = I_s P \tag{4.77}$$

where I_s represents the intensity distribution produced by a single aperture

and

$$P = \left| \sum_{n=1}^{N} e^{ik(l\xi_n + m\eta_n)} \right|^2 \tag{4.78}$$

represents the intensity distribution produced by N point sources. We will next consider the diffraction pattern produced by two slits and then the grating spectrum.

4.10.1   Two slit diffraction pattern

If we have only two apertues then

$$P = |1 + e^{ik(l\xi_2 + m\eta_2)}|^2 = |1 + e^{i\delta}|^2$$

where $\delta = k(l\xi_2 + m\eta_2)$. Simple calculations give

$$P = 4\cos^2(\delta/2) \tag{4.79}$$

which represents the interference pattern produced by two coherent sources (see, e.g., Ghatak (1977), Chapter 11). Thus for two long narrow slits (see Fig. 4.19)

$$I = (I_0 \sin^2 \zeta / \zeta^2) 4 \cos^2 (\delta/2) \tag{4.80}$$

where

$$\zeta = \pi a \sin \theta / \lambda \tag{4.81}$$

Fig. 4.19 Fraunhofer diffraction of a plane wave incident normally on a pair of long narrow slits each of width a.

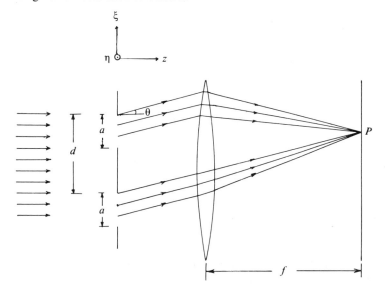

and[†]

$$\delta = 2\pi d \sin \theta / \lambda \tag{4.82}$$

a being the width of each slit and *d* being the distance between the two slits (see Fig. 4.19). Since the slits are infinitely long along the η-axis there will not be any diffraction in that direction and therefore, on the focal plane we will have an intensity distribution similar to that shown in Fig. 4.17. From Eq. (4.80) it is obvious that the intensity is zero when either

$$\zeta = \pi a \sin \theta / \lambda = \pi, 2\pi, 3\pi, \dots \tag{4.83}$$

or

$$\delta = 2\pi d \sin \theta / \lambda = \pi, 3\pi, 5\pi, \dots \tag{4.84}$$

which correspond to

$$a \sin \theta = m\lambda; \quad m = 1, 2, \dots \text{ diffraction minima} \tag{4.85}$$

and

$$d \sin \theta = (n + \tfrac{1}{2})\lambda; \quad n = 0, 1, 2, \dots \text{ interference minima} \tag{4.86}$$

In Fig. 4.20(*a*) we have plotted a typical double slit diffraction pattern for

$$a = 0.0070 \text{ cm}, \quad d = 0.0277 \text{ cm}$$

and

$$\lambda = 5000 \text{ Å} = 5 \times 10^{-5} \text{ cm}$$

Clearly, diffraction minima will occur when

$$\sin \theta = m\lambda / a = 7.1429 \times 10^{-3} m$$

or

$$\theta \approx 0.41°, 0.82°, 1.23°, 1.64°, \dots \tag{4.87}$$

The interference minima will occur when

$$\sin \theta = (n + \tfrac{1}{2})\lambda / d \approx 1.805 \times 10^{-3}(n + \tfrac{1}{2})$$

or

$$\theta \approx 0.052°, 0.155°, 0.259°, 0.362°, 0.465°, 0.569°, 0.672°, \dots$$

It should be noted that for

$$\theta = \sin^{-1}(4\lambda / d) \approx 0.414°$$

[†] From Fig. 4.19 it readily follows that $\xi_1 = 0$, $l = \sin \theta$, $\xi_2 - \xi_1 = \xi_2 = d$ and $\eta_1 = \eta_2 = 0$.

an interference maximum occurs; however around the same angle the first diffraction minimum also occurs (see Eq. 4.87) and the total intensity at this angle is extremely small (see Fig. 4.20(a)). Fig. 4.20(b) corresponds to a slit separation which is twice that for Fig. 4.20(a). Thus although the diffraction minima occur at the same angles there are now almost twice the number of interference minima between between two consecutive diffraction minima.

4.10.2　*The grating spectrum*

We next consider a diffraction grating which consists of N equidistant long parallel slits as shown in Fig. 4.21. Let the width of each slit be a and the distance between two consecutive slits be d (see Fig. 4.21). If we assume the centre of the first slit to be the origin O_1 then

$$\xi_1 = 0, \quad \xi_2 = d, \quad \xi_3 = 2d, \dots \quad \xi_N = (N-1)d$$

and

$$\eta_1 = \eta_2 = \dots = \eta_N = 0$$

Fig. 4.20 A typical intensity distribution in the Fraunhofer diffraction pattern corresponding to two slits with (a) $a = 0.0070$ cm, $d = 0.0277$ cm and (b) $a = 0.0070$ cm, $d = 0.0554$ cm.

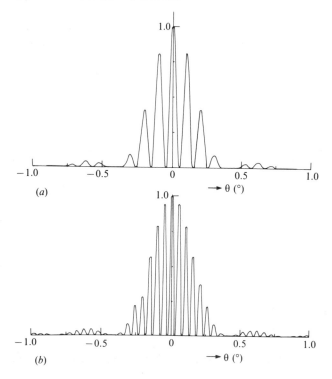

Thus

$$P = |1 + e^{2i\gamma} + e^{4i\gamma} + \cdots + e^{2i(N-1)\gamma}|^2$$

where

$$\gamma = \tfrac{1}{2}kdl = \pi d \sin \theta / \lambda \qquad (4.88)$$

The geometric series can easily be summed and one obtains

$$P = \left| \frac{1 - e^{2iN\gamma}}{1 - e^{2i\gamma}} \right|^2 = \frac{\sin^2 N\gamma}{\sin^2 \gamma} \qquad (4.89)$$

which represents the interference pattern produced by N equidistant point sources on a straight line. Thus the intensity distribution would be given by[†]

$$I = \left[I_0 \frac{\sin^2 \zeta}{\zeta^2} \right] \frac{\sin^2 N\gamma}{\sin^2 \gamma} \qquad (4.90)$$

where the term inside the square brackets represents the intensity distribution in the Fraunhofer pattern due to a single slit. It is easily seen that when

$$\gamma = n\pi; \quad n = 0, \pm 1, \pm 2, \ldots \qquad (4.91)$$

or, when

$$d \sin \theta = n\lambda \quad \text{grating equation} \qquad (4.92)$$

Fig. 4.21 Fraunhofer diffraction of a plane wave incident normally on a diffraction grating which consists of a large number of long narrow slits.

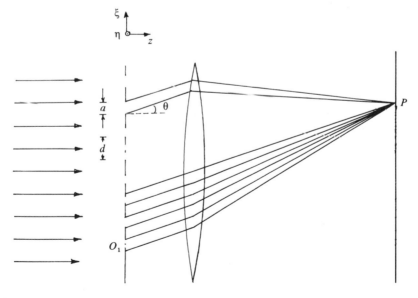

[†] It may be noted that for $N = 2$, since $\gamma = \delta/2$, Eq. (4.90) reduces to Eq. (4.80).

we get

$$I = N^2 \left[I_0 \frac{\sin^2 \zeta}{\zeta^2} \right] \tag{4.93}$$

Since N is usually a very large number, ($\approx 15\,000$) we get very intense maxima at the angles given by Eq. (4.92) unless the term inside the square brackets is <u>extremely small</u>; these intense maxima are referred to as principal maxima.

We next determine the positions of minima which occur when either

$$\zeta = \pi, 2\pi, 3\pi, \ldots \quad \text{diffraction minima} \tag{4.94}$$

or, when

$$\gamma = \frac{\pi}{N}, \frac{2\pi}{N}, \ldots \frac{(N-1)\pi}{N}, \frac{(N+1)\pi}{N}, \frac{(N+2)\pi}{N}, \ldots$$

secondary (or interference) minima (4.95)

i.e., when γ is an integral multiple of π/N but *not* an integral multiple of π. Indeed, when $\gamma = N\pi$ we obtain the principal maxima (see Eq. (4.91)). Physically, along these maxima, the fields produced by each of the slits are in phase and therefore the resultant field is N times the field produced by

Fig. 4.22 The intensity distribution corresponding to the Fraunhofer diffraction pattern consisting of four slits with $a = 0.003\,48$ cm, $d = 0.010\,44$ cm and $\lambda = 5000$ Å.

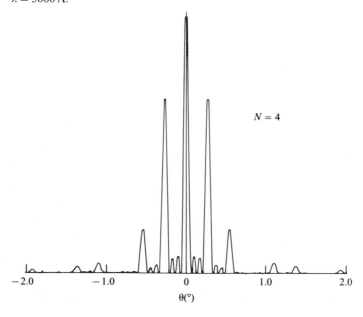

each of the slit. Eq. (4.92) is known as the grating equation and gives the angles at which principal maxima occur. Since N is usually very large ($\approx 15\,000$) the number of secondary minima ($= N - 1$) that occur between two principal maxima is extremely large and therefore the principal maxima are extremely sharp. A typical four slit diffraction pattern is shown in Fig. 4.22.

As mentioned earlier, an arrangement which essentially consists of a large number of equidistant slits is known as a diffraction grating. It is usually obtained by ruling grooves with a diamond point on an optically transparent sheet of material; the grooves act as opaque spaces. Since the distance between two consecutive rulings is extremely small and the rulings should be equally spaced, various sophisticated techniques exist to acheive such rulings. An ordinary laboratory grating may have about 15\,000 lines per inch whereas a good quality grating may have as many as 30\,000 lines per inch. As will be seen later, the greater the number of lines, the greater is the resolving power of the grating. We can observe the grating spectrum by using an ordinary laboratory spectrometer. The light beam from the collimator is allowed to fall normally on the diffraction grating and the diffraction pattern is observed on the cross wires of the rotating telescope. If we use a source like a mercury lamp, the zero order spectrum ($n = 0$) will have the same colour as the source because all wavelengths will overlap. This follows from the fact that for $n = 0$, $\theta = 0$ for all wavelengths (see Eq. (4.92)). The first and higher order spectra will show a number of coloured lines which are characteristic of the source used. If the value of d is known, then by measuring the angles of diffraction of various lines one can easily determine the various wavelengths.

4.10.3 *Dispersive power of a grating*

From the grating equation it readily follows that

$$\Delta\theta/\Delta\lambda = n/d\cos\theta \tag{4.96}$$

The quantity $\Delta\theta/\Delta\lambda$ is known as the dispersive power of the grating and only for small values of θ (i.e., when $\cos\theta \approx 1$), will $\Delta\theta$ be proportional to $\Delta\lambda$. When this happens we have what is known as a *normal spectrum*.

As an example, we calculate the angular separation of the D_1 and D_2 lines of sodium ($\lambda_1 \approx 5890\,\text{Å}$, $\lambda_2 \approx 5896\,\text{Å}$) in the first and second order spectrum of a grating with 15\,000 lines per inch. Now, for such a grating

$$d = 2.54/15\,000 \approx 1.69 \times 10^{-4}\,\text{cm}$$

and for the mean wavelength $\lambda = 5893\,\text{Å}$, the angle of diffraction θ is given by

$$\sin \theta = n\lambda/d$$

or

$$\theta = 20.4° \quad \text{and} \quad 44.2°$$

for the first and second order spectra respectively. Thus

$$\Delta\theta = (n/d \cos \theta)\Delta\lambda$$

$$\approx 0.0004 \, \text{rad} \, (\approx 1.30') \, \text{and} \, 0.0010 \, \text{rad} \, (\approx 3.40')$$

for the first and second order spectra. Thus if we are using a telescope of angular magnification 20, the two lines will appear to have an angular separation of $1°8'$ in the second order spectrum.

4.10.4 *Resolving power of a diffraction grating*
Fig. 4.23 shows the intensity distribution of the grating spectrum in the nth order for two nearby (equally intense) wavelengths λ and $\lambda + \Delta\lambda$. The nth order principal maximum corresponding to λ will occur at $\theta = \theta_n$ where

$$d \sin \theta_n = n\lambda \tag{4.97}$$

and the first minimum on either side of the principal maximum would occur at (see Eq. (4.95))

$$d \sin (\theta_n \pm \Delta\theta) = n\lambda \pm \lambda/N \tag{4.98}$$

Now, according to the Rayleigh criterion, the two wavelengths λ and $\lambda + \Delta\lambda$ are said to be just resolved if the principal maximum corresponding to the wavelength $\lambda + \Delta\lambda$ falls on the first minimum of λ (and conversely) as shown

Fig. 4.23 The intensity distribution around the nth order principal maximum for two nearby wavelengths λ and $\lambda + \Delta\lambda$ such that the principal maximum of one falls on the first minimum of the other.

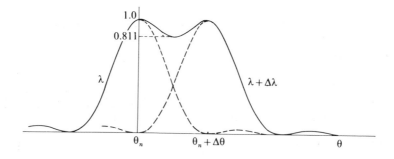

in Fig. 4.23. Thus for the two wavelengths to be just resolved we must have

$$d \sin (\theta_n + \Delta\theta) = n(\lambda + \Delta\lambda) \tag{4.99}$$

and

$$d \sin (\theta_n + \Delta\theta) = n\lambda + \lambda/N \tag{4.100}$$

From the above two equations it readily follows that if $\Delta\lambda$ is the separation of the two wavelengths which the grating can just resolve then

$$n\Delta\lambda = \lambda/N$$

or

$$R = \lambda/\Delta\lambda = nN \tag{4.101}$$

where R is known as the resolving power of the grating which is proportional to the order of the grating spectrum and the total number of (exposed) lines of the grating. Fig. 4.23 shows the exact variation of the total intensity for two nearby wavelengths of equal intensity. It can be seen that for the two wavelengths to be just resolved, the intensity dip between the two principal maxima would be about 19%.

In order to appreciate this numerically we once again consider D_1 and D_2 lines of sodium for which $\Delta\lambda \approx 6\,\text{Å}$. Thus

$$\frac{\lambda}{\Delta\lambda} = \frac{5893\,\text{Å}}{6\,\text{Å}} \approx 1000$$

and therefore a grating with 1000 lines should be just able to resolve the two lines of sodium in the first order. We should mention here that each wavelength has a spectral width of its own which may be due to the Doppler effect, collisions, etc. Even in the absence of these, the line has an intrinsic width, characteristic of the transition responsible for the radiation (see Sec. 8.8). Because of these line broadening mechanisms, the width of the principal maxima can be much greater than that shown in Fig. 4.23

Problem 4.3: Consider a plane wave incident obliquely (at an angle ϕ) on a diffraction grating as shown in Fig. 4.24. For such a case show that the field incident on the grating would be given by

$$A(\xi, \eta) = A_0 e^{-ik\xi\sin\phi} \tag{4.102}$$

Thus the entire analysis for normal incidence remains valid with l replaced by $(l - \sin\phi)$ and therefore the grating equation (Eq. 4.92) is modified to

$$d(\sin\theta - \sin\phi) = n\lambda \tag{4.103}$$

Problem 4.4: In continuation of the previous problem consider a grating with 15 000 lines per inch. Assuming $\lambda = 0.6\,\mu m$, show that for $\phi = 0°, 30°$ and $60°$, the highest order spectra that can be observed would be 2, 4 and 5 respectively. Thus by having oblique incidence one can have higher order spectra and consequently higher resolution.

It is of interest to mention that if we have a large number of identical apertures which are randomly distributed on the plane of the aperture, then the intensity distribution would still be given by Eq. (4.77) with

$$P = \left| \sum_{n=1}^{N} e^{i\phi_n} \right|^2 = \sum_{n=1}^{N} \sum_{m=1}^{N} e^{i\phi_n} e^{-i\phi_m}$$

$$= N + \sum_{n \neq m} \sum_{m} e^{i(\phi_n - \phi_m)} \tag{4.104}$$

where $\phi_n = k(l\xi_n + m\eta_n)$. Since the apertures are randomly distributed the second term (for a large value of N) would be extremely small. Thus the intensity distribution would be

$$I = NI_s \tag{4.105}$$

We may compare the above equation with Eq. (4.93) where we showed that the intensity (in certain directions) was proportional to N^2. Thus we may say that when a large number of identical apertures are randomly distributed, the intensities add.

Fig. 4.24 A plane wave incident obliquely at an angle ϕ on a diffraction grating.

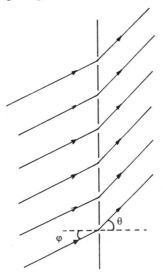

4.11 Resolving power of a prism

In this section, we shall calculate the resolving power of a prism. Fig. 4.25 gives a schematic description of the experimental arrangement for observing the prism spectrum which is determined through the following formula

$$n(\lambda) = \frac{\sin\{[A + \delta(\lambda)]/2\}}{\sin(A/2)} \qquad (4.106)$$

where A represents the angle of the prism and δ the angle of minimum deviation. We assume that the refractive index decreases with λ (which is usually the case) so that δ also decreases with λ. In Fig. 4.25 the points P_1 and P_2 represent the images corresponding to λ and $\lambda + \Delta\lambda$ respectively. We are assuming that $\Delta\lambda$ is small so that the same position of the prism corresponds to the minimum deviation position for both wavelengths. In an actual experiment one usually has a slit source (perpendicular to the plane of the paper) forming line images at P_1 and P_2. Since the faces of the prism are rectangular, the intensity distribution will be similar to that produced by a slit of width b (see Sec. 4.4). For the lines to be just resolved the first diffraction minimum of λ should fall at the central maximum of $\lambda + \Delta\lambda$, thus we must have

$$\Delta\delta \approx \lambda/b \qquad (4.107)$$

In order to express $\Delta\delta$ in terms of $\Delta\lambda$, we differentiate Eq. (4.106) to obtain

$$\Delta\delta = \frac{2\sin(A/2)}{\cos\{[A + \delta(\lambda)]/2\}}\frac{dn}{d\lambda}\Delta\lambda$$

Now from Fig. 4.25 we have

$$\theta = \tfrac{1}{2}[\pi - (A + \delta)]$$

Fig. 4.25 Schematic of the experimental arrangement to observe the prism spectrum. P_1 and P_2 represent the points around which the wavelengths λ and $\lambda + \Delta\lambda$ are focussed.

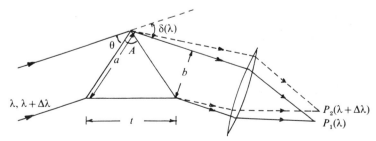

or

$$\sin \theta = b/a = \cos \left[(A + \delta)/2 \right]$$

where the length a is shown in the figure. Further

$$\sin \frac{A}{2} = \frac{t/2}{a}$$

where t is the length of the base of the prism. Thus

$$\Delta \delta \approx \frac{t}{b} \frac{dn}{d\lambda} \Delta \lambda \tag{4.108}$$

Substituting in Eq. (4.107) we get for the resolving power

$$R = \frac{\lambda}{\Delta \lambda} = t \frac{dn}{d\lambda} \tag{4.109}$$

Now, for most glasses, the wavelength dependence of the refractive index (in the visible region of the spectrum) can be accurately described by the Cauchy formula

$$n = A + B/\lambda^2 + C/\lambda^4 + \cdots \tag{4.110}$$

Thus

$$\frac{dn}{d\lambda} = - (2B/\lambda^3 + 4C/\lambda^5 + \cdots) \tag{4.111}$$

the negative sign implying that the refractive index n decreases with increase in wavelength. As an example we consider telescope crown glass for which

$$A = 1.51375, \quad B = 4.608 \times 10^{-11} \, \text{cm}^2, \quad C = 6.88 \times 10^{-22} \, \text{cm}^4$$

For $\lambda = 6 \times 10^{-5} \, \text{cm}$ we have

$$dn/d\lambda \approx - (4.27 \times 10^2 + 3.54) \, \text{cm}^{-1}$$

$$\approx - 4.30 \times 10^2 \, \text{cm}^{-1}$$

Thus for $t \approx 2.5 \, \text{cm}$ we have

$$R = \lambda/\Delta \lambda \approx 1000$$

which is an order of magnitude less than that for typical diffraction gratings with 15 000 lines. Although the resolving power of the prism is much lower than that of the grating, the intensities associated with the prism spectrum are much higher than the corresponding grating spectrum. This follows from the fact that in the prism spectrum, the entire incident energy is distributed in the spectrum. On the other hand, in the grating spectrum, a major portion of the energy appears in the zero order (where there is no dispersion) and the remaining energy is distributed in various orders. We should mention here that the resolving power of the Fabry–Perot interferometer (see Sec. 2.4) is much higher than that of a grating.

Additional problems

Problem 4.5: Consider a plane wave incident normally on a long narrow slit of width 0.02 cm. The Fraunhofer diffraction pattern is observed on the focal plane of a lens whose focal length is 20 cm. Assuming $\lambda = 6000$ Å determine the positions of the first and second minima. Also determine the positions and relative intensities of the first and second maxima.
(Answer: Minima: 0.06 cm and 0.12 cm from the central spot. Maxima: 0.086 cm and 0.148 cm from the central spot. Relative intensities: 4.96% and 1.68% of the central spot.)

Problem 4.6: In the above problem assume the length of the slit to be (a) 2 cm and (b) 0.02 cm. Obtain the positions of minima in the direction of the length of the slit and discuss qualitatively the diffraction pattern.
(Answer: (a) 6 μm and 12 μm (b) 600 μm and 1200 μm.)

Problem 4.7: The Fraunhofer diffraction pattern of an aperture is observed on the focal plane of a lens. The aperture is displaced laterally by distances d_1 and d_2 along the ξ and η directions (see Fig. 4.4). Show that there is no change in the diffraction pattern.

Problem 4.8: For a pinhole camera with 10 cm separation between the pinhole and the film plane, what pinhole diameter would you choose for optimum resolution for objects at infinity? Assume $\lambda = 5500$ Å.
(Answer: ~0.36 mm)
(Hint: For optimum resolution the diameter of the geometrical image must be roughly equal to the diameter of the first dark ring.)

Problem 4.9: A plane wave ($\lambda = 0.6$ μm) of 1 cm diameter is incident normally on a lens of focal length 15 cm. Calculate the radii of the first three dark rings on the focal plane of the lens.
(Answer: approximately 11 μm, 20 μm and 29 μm.)

Problem 4.10: Consider a double slit diffraction pattern with $a = 0.008$ cm, $b = 0.07$ cm and $\lambda = 6328$ Å (He–Ne laser wavelength). Show that there will be eighteen minima between the two first order diffraction minima on either side of the principal maximum. A photographic film is placed at a distance of 2 m. Assuming that at such a distance the diffraction is of the Fraunhofer type calculate the distance between two interference maxima.
(Answer: ≈ 0.18 cm)

Problem 4.11: The schematic of the diffraction pattern produced by an aperture consisting of N regularly spaced long narrow slits of widths a and periodicity b is shown in Fig. 4.26 ($\lambda = 5000$ Å)

(a) Calculate a and b corresponding to the grating
(b) Obtain the number of slits in the grating
(c) What will be the resolving power of the grating in the second order
(d) If the intensity of the central peak is I_0, what will be the intensity of the first order principal maximum.

(Answer: (a) $a = 0.002$ cm, $b = 0.005$ cm; (b) 5; (c) 10; (d) $0.573I_0$)

Problem 4.12: Consider a diffraction grating with 8000 lines per inch and assume that light of wavelength 5460 Å and 5460.072 Å illuminates the grating over a region of 2 in.

(a) Calculate the number of orders in the diffracted spectrum.
(b) Calculate the dispersion in the third order.
(c) In which diffraction orders will the two wavelength components be resolved?

(Answer: (a) 5, (b) $1.103\ \mu m^{-1}$, (c) Only in the fifth order.)

Problem 4.13: A diffraction grating with 15 000 lines per inch is illuminated by white light which may be assumed to range from 4000 Å (violet) to 7000 Å (red). Calculate the angular divergence of the second order spectrum and show that the third order spectrum overlaps with the second order.

(Answer: $\Delta\theta \approx 27.6°$)

Problem 4.14: Evaluate the integral in Eq. (4.4) under the approximation that most of the contribution to the integral comes from $|x - \xi|, |y - \eta| \ll z$ and show that

$$C = i/\lambda$$

Fig. 4.26 Schematic of the diffraction pattern produced by a grating.

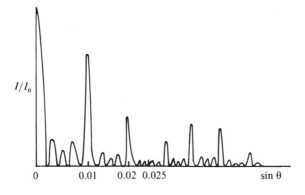

Hint:

$$\int\int \exp\frac{(-ikr)}{r}\,d\xi\,d\eta \approx \frac{1}{z}\exp(-ikz)$$

$$\times \int\int \exp\left\{-\frac{ik}{2z}[(x-\xi)^2+(y-\eta)^2]\right\}d\xi\,d\eta$$

which can readily be evaluated by using the standard integral

$$\int_{-\infty}^{+\infty} \exp(-\alpha x^2 + \beta x)\,dx = (\pi/\alpha)^{\frac{1}{2}}\exp(\beta^2/4\alpha)$$

5

Fresnel diffraction

5.1   Introduction

In the previous chapter we discussed in detail the Fraunhofer diffraction pattern produced by various apertures. As mentioned in Sec. 4.1, in Fraunhofer diffraction the source and the screen are at infinite distances from the aperture and in Fresnel diffraction, the source or the screen (or both) are at a finite distance from the diffracting aperture. Although the Fraunhofer diffraction pattern is of greater importance in optics, one must have a proper understanding of Fresnel diffraction and its transition to the Fraunhofer region. In this chapter, we will study the Fresnel diffraction pattern for some specific configurations and to make the analysis simple we will assume the source to be at infinity. We will also explicitly study the transition from the Fresnel to the Fraunhofer region of diffraction.

5.2   The diffraction integral

In the previous chapter we have shown that for a plane wave incident normally on an aperture, the diffraction pattern is given by (see Sec. 4.2)

$$u(P) = \frac{iA}{\lambda} \int\!\!\int_{\mathscr{A}} \frac{1}{r} e^{-ikr} \, d\xi \, d\eta \tag{5.1}$$

where the integration is over the area of the aperture (see Fig. 4.3) and

$$r = [(x - \xi)^2 + (y - \eta)^2 + z^2]^{\frac{1}{2}} \tag{5.2}$$

Now, if the amplitude and phase distribution on the plane $z = 0$ is given by $A(\xi, \eta)$ then the above integral is modified to

$$u(P) = \frac{i}{\lambda} \int\!\!\int A(\xi, \eta) \frac{e^{-ikr}}{r} \, d\xi \, d\eta \tag{5.3}$$

We may write

$$r = z[1 + (x - \xi)^2/z^2 + (y - \eta)^2/z^2]^{\frac{1}{2}}$$
$$\approx z + (x - \xi)^2/2z + (y - \eta)^2/2z \tag{5.4}$$

where we have assumed that most of the contribution to the integral comes from the domain

$$|x - \xi|, \quad |y - \eta| \ll z$$

so that terms higher than the quadratic term can be neglected (see also Problem 4.14). In the denominator we may safely replace r by z so that we may write

$$u(x, y, z)$$
$$\approx \frac{i}{\lambda z} \exp(-ikz) \int \int A(\xi, \eta) \exp \left\{ -\frac{ik}{2z} [(x - \xi)^2 + (y - \eta)^2] \right\} d\xi \, d\eta \tag{5.5}$$

We will apply the above formula to the calculation of Fresnel diffraction in various cases.

5.3 Uniform amplitude and phase distribution
For such a case, at $z = 0$

$$A(\xi, \eta) = A \quad \text{for all values of } \xi \text{ and } \eta \tag{5.6}$$

From Problem 4.14 it readily follows that

$$u(x, y, z) \approx A e^{-ikz} \tag{5.7}$$

as it indeed should for a uniform plane wave.

5.4 Diffraction of a Gaussian beam
In Chapter 9 we will show that when a laser oscillates in its fundamental transverse mode, the transverse amplitude distribution is Gaussian. Therefore the study of the diffraction of a Gaussian beam is of great importance. We consider a Gaussian beam propagating along the z-direction whose amplitude distribution on the plane $z = 0$ is given by

$$A(\xi, \eta) = a \exp[-(\xi^2 + \eta^2)/w_0^2] \tag{5.8}$$

implying that the phase front is plane at $z = 0$. From Eq. (5.8) it follows that at a distance w_0 from the z-axis, the amplitude falls by a factor $1/e$ (i.e., the intensity reduces by a factor $1/e^2$). This quantity w_0 is called the *spot size* of the beam. Substituting Eq. (5.8) in Eq. (5.5), we obtain

$$u(x, y, z) \approx \frac{ia}{\lambda z} \exp(-ikz) \int_{-\infty}^{\infty} \exp\left[-\frac{ik}{2z}(x - \xi)^2 - \frac{\xi^2}{w_0^2} \right] d\xi$$

$$\times \int_{-\infty}^{\infty} \exp\left[-\frac{ik}{2z}(y - \eta)^2 - \frac{\eta^2}{w_0^2} \right] d\eta$$

Using the integral

$$\int_{-\infty}^{\infty} \exp(-\alpha x^2 + \beta x) \, dx = (\pi/\alpha)^{\frac{1}{2}} \exp(\beta^2/4\alpha) \tag{5.9}$$

we get after simplification

$$u(x, y, z) = \frac{ia\pi}{\lambda} \frac{2w_0^2}{2z + ikw_0^2} \exp\left\{ -ik\left[z + \frac{x^2 + y^2}{2z(1 + \pi^2 w_0^4/\lambda^2 z^2)} \right] \right\}$$

$$\times \exp\left[-\frac{x^2 + y^2}{w^2(z)} \right] \tag{5.10}$$

where

$$w^2(z) = w_0^2 (1 + \lambda^2 z^2/\pi^2 w_0^4) \tag{5.11}$$

represents the width of the beam which increases with z (see Fig. 4.11). The intensity distribution will be proportional to

$$\exp\left[-2(x^2 + y^2)/w^2(z) \right] \tag{5.12}$$

which shows that the transverse intensity distribution remains Gaussian with the beam-width increasing with z which essentially implies diffraction

Fig. 5.1 Diffraction divergence of a Gaussian beam whose phase front is plane at $z = 0$. The dashed curves represent the phase fronts.

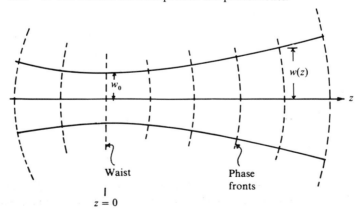

divergence. As can be seen from Eq. (5.11), for small values of z, the width increases quadratically with z but for large values of $z \gg w_0^2/\lambda$, we obtain

$$w(z) \approx w_0 \frac{\lambda z}{\pi w_0^2} = \frac{\lambda z}{\pi w_0} \tag{5.13}$$

which shows that the width increases linearly with z. This is schematically shown in Fig. 5.1. We should note from Eq. (5.13) that the rate of increase in the width is proportional to the wavelength and inversely proportional to the initial width of the beam (obviously!).

Now, for a spherical wave *diverging* from the origin, the field distribution is given by

$$u \sim e^{-ikr}/r \tag{5.14}$$

Thus on the plane $z = R$ (see Fig. 5.2), the phase distribution is given by

$$\exp\left[-ik(x^2 + y^2 + R^2)^{\frac{1}{2}}\right] \approx \exp(-ikR)\exp\left[-i\frac{k}{2R}(x^2 + y^2)\right] \tag{5.15}$$

where we have assumed $|x|, |y| \ll R$. On the plane $z = R$, the first factor is a constant and therefore from the above equation it follows that a phase variation of the type

$$\exp\left[-i\frac{k}{2R}(x^2 + y^2)\right] \tag{5.16}$$

(on the x–y plane) represents a *diverging* spherical wave of radius R. If we compare the above expression with the one inside the curly brackets in Eq. (5.10) we obtain the following approximate expression for the radius

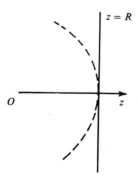

Fig. 5.2 A spherical wave diverging from the point O. The dashed curve represents a section of the spherical wavefront at a distance R from the source.

of curvature of the phase front

$$R(z) \approx z(1 + \pi^2 w_0^4 / \lambda^2 z^2) \tag{5.17}$$

which is shown in Fig. 5.1. Thus as the beam propagates, the phase front which was plane at $z = 0$ becomes curved. The plane $z = 0$, where the phase front is plane and the beam has the minimum spot size, is referred to as the waist of the Gaussian beam.

It should be mentioned that although in the derivation of Eq. (5.10) we have assumed z to be large (see Eq. (5.4)), Eq. (5.10) does give the correct field distribution even at $z = 0$.

5.5 Intensity distribution near the paraxial image point of a converging lens

We next consider a spherical wave emerging from a lens and converging to the paraxial image point I (see Fig. 5.3). For example, for a plane wave incident on the lens, the point I would represent the focal point. Since immediately after the lens we have a converging spherical wave the field distribution associated with this wave would be given by

$$u \sim e^{ikr} / r \tag{5.18}$$

where we have assumed the image point I to be the origin. With I as origin, on the plane P, $z = -s$ and

$$r = (x^2 + y^2 + s^2)^{\frac{1}{2}} \approx s + (x^2 + y^2)/2s$$

Fig. 5.3 A spherical wave emerging from a lens and converging to the paraxial image point I.

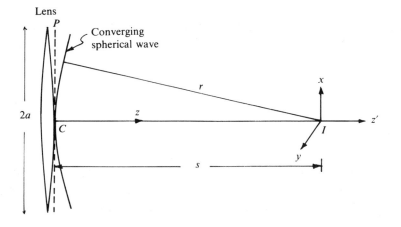

Thus the field distribution on the plane P would be given by

$$u \approx A/s \exp(iks) \exp\left[i\frac{k}{2s}(x^2 + y^2)\right] \tag{5.19}$$

Comparing with Eq. (5.16), we see that Eq. (5.19) does represent a converging spherical wave of radius s. Having obtained the field distribution on the plane P, we shift the origin to the point C and use Eq. (5.5) to determine the field at a distance z from the plane P:

$$u(x, y, z) \approx \frac{iA}{\lambda zs} \exp(iks) \exp(-ikz) \iint_{\xi^2 + \eta^2 \leqslant a^2} \exp\left[\frac{ik}{2s}(\xi^2 + \eta^2)\right]$$
$$\times \exp\left\{-\frac{ik}{2z}[(x - \xi)^2 + (y - \eta)^2]\right\} d\xi \, d\eta \tag{5.20}$$

where the integration has to be carried out over the area of the aperture which is defined by

$$\xi^2 + \eta^2 \leqslant a^2 \tag{5.21}$$

where a represents the radius of the lens (see Fig. 5.3). Straightforward simplifications give us

$$u(x, y, z)$$
$$\approx \frac{iA}{\lambda zs} \exp[ik(s - z)] \exp\left[-\frac{ik}{2z}(x^2 + y^2)\right]$$
$$\times \iint_{\xi^2 + \eta^2 \leqslant a^2} \exp\left[\frac{ik}{2}\left(\frac{1}{s} - \frac{1}{z}\right)(\xi^2 + \eta^2) + \frac{ik}{z}(x\xi + y\eta)\right] d\xi \, d\eta \tag{5.22}$$

We are interested in determining the intensity distribution around the image plane $z = s$. We first calculate the intensity distribution on the image plane $z = s$, so that Eq. (5.22) becomes

$$u(x, y, s) \approx \frac{iA}{\lambda s^2} \exp\left[-\frac{ik}{2s}(x^2 + y^2)\right] \iint_{\xi^2 + \eta^2 \leqslant a^2} \exp\left[\frac{ik}{s}(x\xi + y\eta)\right] d\xi \, d\eta \tag{5.23}$$

The above integral is of exactly the same form as the integral on the RHS of Eq. (4.38) which it indeed should be because on the image plane we would be observing the Fraunhofer diffraction pattern produced by a circular aperture. The final result would be[†]

$$u(x, y, z = s) \approx \frac{\pi i A a^2}{\lambda s^2} \left[\frac{2J_1(v)}{v} \right]$$
(5.24)

where

$$v = (ka/s)(x^2 + y^2)^{\frac{1}{2}}$$
(5.25)

where we have assumed that $(x^2 + y^2) \ll s/k$ so that the phase term can be neglected. Eq. (5.24) represents the Airy pattern which has been discussed in detail in Chapter 4.

We next calculate the intensity distribution along the z-axis ($x = y = 0$) near the image plane $z = s$. We introduce the variable

$$z' = z - s$$

to obtain

$$u(0, 0, z) \approx \frac{iA}{\lambda s^2} \exp(-ikz') \int\int_{\xi^2 + \eta^2 \leqslant a^2} \exp\left[\frac{iw}{2a^2} (\xi^2 + \eta^2) \right] d\xi \, d\eta$$
(5.26)

where we have assumed $z' \ll s$,

$$w = kz'a^2/s^2$$
(5.27)

We introduce the circular coordinates (ρ, ϕ)

$$\xi = \rho \cos \phi, \quad \eta = \rho \sin \phi$$

to obtain

$$u(0, 0, z) \approx \frac{iA}{\lambda s^2} e^{-ikz'} \int_0^{2\pi} d\phi \int_0^a \rho \, d\rho e^{iw\rho^2/2a^2}$$

and carrying out the elementary integration, we get

$$u(0, 0, z) \approx \frac{\pi i A a^2}{\lambda s^2} e^{-ikz'} e^{iw/4} \left[\frac{\sin(w/4)}{w/4} \right]$$

The corresponding intensity distribution will therefore be given by

$$I(0, 0, z) = I_0 \left[\frac{\sin(w/4)}{w/4} \right]^2$$
(5.28)

[†] Because of the circular symmetry of the problem, we may put $y = 0$ and calculate the intensity distribution along the x-axis. In the final result we should replace x by $(x^2 + y^2)^{\frac{1}{2}}$.

where I_0 represents the intensity at the point $I(z' = 0)$. Thus, on the axis, the intensity becomes zero when

$$w = \pm 4n\pi; \quad n = 1, 2, 3, \ldots \tag{5.29}$$

or when

$$z' = \pm \left(\frac{s}{a}\right)^2 2n\lambda; \quad n = 1, 2, 3, \ldots \tag{5.30}$$

The intensity drops by about 20% for $w \approx \pm 3.2$ or,

$$z' \approx \pm z_0 = \pm \frac{3.2}{k}\left(\frac{s}{a}\right)^2 \tag{5.31}$$

The quantity z_0 is usually referred to as the *focal tolerance*. For a plane wave incident on a lens of focal length $f = 25\,\mathrm{cm}$ and aperture radius $a = 2\,\mathrm{cm}$, the focal tolerance is

$$z_0 = \pm \frac{3.2 \times 5 \times 10^{-5}}{2\pi}\left(\frac{25}{2}\right)^2 \approx 0.004\,\mathrm{cm}$$

where we have assumed $\lambda = 5000\,\text{Å}$.

We should mention here that using the circular symmetry of the problem and straightforward algebra, Eq. (5.22) can be written in the following compact form (see, e.g., Ghatak and Thyagarajan (1978), Sec. 4.14)

$$u(x, y, z) = \frac{2\pi i A a^2}{\lambda s^2} \mathrm{e}^{-ikz'} \int_0^1 J_0(v\tau)\mathrm{e}^{\frac{1}{2}iw\tau^2}\tau \, \mathrm{d}\tau \tag{5.32}$$

where v is given by Eq. (5.25).

Fig. 5.4 *I* represents the image point of the point object *O*. If an aperture *A* is placed in front of the lens, then the intensity pattern observed on the image plane *QQ'* is the Fraunhofer diffraction pattern of the aperture.

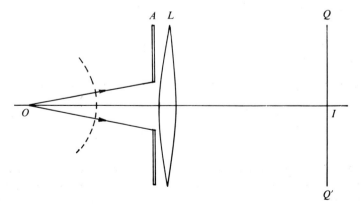

As a corollory of the above analysis, we consider the arrangement shown in Fig. 5.4 where we have spherical waves emanating from the axial point *O*. If the point *I* represents the image point corresponding to the aberrationless lens system *L*, then on the image plane *QQ'*, we will obtain the Fraunhofer diffraction pattern of the aperture *A* which is placed in front of the lens. Of course, we are assuming that the aperture size is much smaller than the size of the lens.

5.6   Fresnel diffraction by a circular aperture

We next consider the Fresnel diffraction of a plane wave incident normally on a circular aperture of radius *a* as shown in Fig. 5.5. The Fresnel diffraction pattern is observed on a screen *S* placed at a distance *z* from the aperture (see Fig. 5.5). Obviously, because of the circular symmetry of the problem the diffraction pattern will consist of concentric circles. The field pattern will be given by (see Eq. (5.5))

$$u(x, y, z) = \frac{iA}{\lambda z} \exp(-ikz)$$

Same A.

$$\times \int\int_{\xi^2 + \eta^2 \leqslant a^2} \exp\left\{ -\frac{ik}{2z} [(x - \xi)^2 + (y - \eta)^2] \right\} d\xi\, d\eta$$

(5.33)

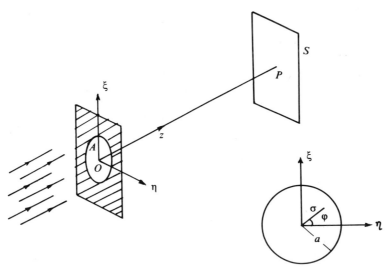

Fig. 5.5 A plane wave incident normally on a circular aperture of radius *a* and the Fresnel diffraction pattern is observed on the screen *S* which is at a finite distance *z* from the aperture.

where the integration is carried out over the area of the circular aperture. The integral is difficult to perform analytically; however, the intensity variation on the z-axis can easily be calculated. Now, on the z-axis, $x = y = 0$ so that

$$u = \frac{iA}{\lambda z} \exp(-ikz) \iint_{\xi^2 + \eta^2 \leqslant a^2} \exp\left[-\frac{ik}{2z}(\xi^2 + \eta^2) \right] d\xi\, d\eta \qquad (5.34)$$

We introduce the circular coordinates (σ, ϕ)

$$\xi = \sigma \cos \phi, \quad \eta = \sigma \sin \phi \qquad (5.35)$$

to obtain

$$u = \frac{iA}{\lambda z} \exp(-ikz) \int_0^{2\pi} d\phi \int_0^a \exp\left(-\frac{ik}{2z} \sigma^2 \right) \sigma\, d\sigma \qquad (5.36)$$

The integrations can easily be carried out and the final result is

$$u(0, 0, z) = A\,e^{-ikz}(1 - e^{-ip\pi}) \qquad (5.37)$$

or

$$u(0, 0, z) = 2iA\,e^{-ikz} e^{-ip\pi/2} \sin(p\pi/2) \qquad (5.38)$$

where

$$p = a^2/\lambda z \qquad (5.39)$$

is known as the Fresnel number of the aperture. The corresponding intensity variation (on the z-axis) is therefore given by

$$I = 4I_0 \sin^2(p\pi/2) \qquad (5.40)$$

which has been plotted in Fig. 5.6; the quantity I_0 represents the intensity in the absence of the aperture. In order to understand the intensity variation

Fig. 5.6 The variation of the intensity on the axis corresponding to the Fresnel diffraction produced by a circular aperture.

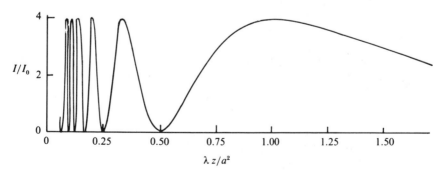

shown in Fig. 5.6 physically, we note that the quantity p represents the number of Fresnel half period zones that is contained in the circular aperture. This follows from the fact that if A represents an arbitrary point on the rim of the circular aperture then

$$(AP)^2 = a^2 + (OP)^2 = a^2 + z^2$$

(see Fig. 5.5), or

$$AP = z[1 + a^2/z^2]^{\frac{1}{2}} \approx z + a^2/2z$$

Thus

$$AP - OP \simeq a^2/2z \tag{5.41}$$

Now, if

$$AP - OP = n\lambda/2 \tag{5.42}$$

then the aperture is said to contain n half period zones. Obviously, the intensity (at the point P) is maximum or minimum depending whether the aperture contains an odd or even number of half period zones. This is consistent with Eq. (5.40). The intensity variation shown in Fig. 5.6 tells us that

$$I = 0 \quad \text{when } p = a^2/\lambda z = 2, 4, 6, \ldots \tag{5.43}$$

which respectively correspond to the aperture containing $2, 4, 6, \ldots$ half

Fig. 5.7 The zone plate in which alternate half period zones are blackened.

period zones. Similarly

$$I = 4I_0 \quad \text{when } p = a^2/\lambda z = 1, 3, 5, \ldots \tag{5.44}$$

which respectively correspond to the aperture containing $1, 3, 5, \ldots$ half period zones.

An interesting application of the above analysis is the zone plate in which alternate half period zones are blackened (see Fig. 5.7). Now, from Eqs. (5.41) and (5.42) it readily follows that the outer radius of the nth half period zone is given by

$$r_n^2/2z \approx n\lambda/2$$

or

$$r_n \approx n^{\frac{1}{2}}(\lambda z)^{\frac{1}{2}} \tag{5.45}$$

Thus the radii of the half period zones are proportional to the square root of natural numbers. If alternate zones are blackened, then the resultant field will be (cf. Eq. (5.36))

$$u = \frac{2\pi iA}{\lambda z} e^{-ikz} \left[\int_0^{r_1} + \int_{r_2}^{r_3} + \int_{r_4}^{r_5} + \cdots \right] e^{-(ik/2z)\sigma^2} \sigma \, d\sigma$$

where r_n is given by Eq. (5.45). Elementary calculations give us

$$\int_{r_n}^{r_{n+1}} e^{-(ik/2z)\sigma^2} \sigma \, d\sigma = 2z/ik \quad \text{for } n = 0, 2, 4, \ldots$$

Thus the field due to each of the odd zones is in phase so that

$$u = 2NA e^{-ikz} \tag{5.46}$$

where N represents the total number of odd half period zones exposed by the zone plate. The intensity at the point P is therefore very large and the zone plate acts as a converging lens.

It can easily be seen that if a plane wave is incident normally on a zone plate then it will have a large number of foci shown as $P_1, P_3, \ldots$ in Fig. 5.8.

Fig. 5.8 For a plane wave incident on a zone plate we will have a large number of foci $P_1, P_3, \ldots$. For the point P_n, the first exposed circle contains n half period zones.

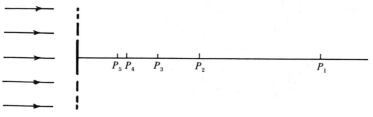

For the point P_1 the blackened rings correspond to the second, fourth, sixth half period zones. For the point P_2, the first exposed circle contains the first two half period zones etc, so that the intensity at P_2 is almost zero. For the point P_3, the first exposed circle contains the first three half period zones and the first blackened ring will contain the next three half period zones etc. Therefore we have alternate bright and dark points on the axis.

If the radii of the circles on the zone plate are written in the form

$$r_n = Kn^{\frac{1}{2}} \tag{5.47}$$

then the distance of the foci $P_1, P_3, \ldots$ from the zone plate would be

$$\frac{K^2}{\lambda}, \frac{K^2}{3\lambda}, \frac{K^2}{5\lambda}, \ldots \tag{5.48}$$

Thus for a zone plate whose radii are given by

$$r_n = 0.05\sqrt{n}\,\text{cm}$$

the focal lengths (for $\lambda = 5 \times 10^{-5}\,\text{cm}$) would be

$$50\,\text{cm}, 16.66\,\text{cm}, 10\,\text{cm}, \ldots$$

It may be noted that the zone plate has considerable chromatic aberration (see Problem 5.2).

5.7   Babinet's principle

We first define complementary apertures. Two apertures are said to be complementary if the open and opaque regions of one aperture correspond respectively to the opaque and open regions of the other aperture. For example, a circular aperture and a circular disc of the same radius are complementary apertures. Now if $u_1(P)$ and $u_2(P)$ represent the diffraction patterns due to two complementary apertures then

$$u_1(P) + u_2(P) = u(P) \tag{5.49}$$

where $u(P)$ represents the field at the point P in the absence of any aperture. The above equation represents Babinet's principle and follows immediately from the diffraction formula (see Eq. (5.1)). We will consider a simple application of Babinet's principle.

5.8   Fresnel diffraction due to a circular disc

We consider a plane wave incident normally on a circular disc of radius a (see Fig. 5.9). If $u_1(P)$ represents the corresponding diffraction pattern then according to Eq. (5.49)

$$u_1(P) = Ae^{-ikz} - u_2(P) \tag{5.50}$$

where $u_2(P)$ represents the diffraction pattern due to a circular hole of the same radius a and Ae^{-ikz} represents the field in the absence of any aperture. If we now use Eq. (5.37), the amplitude at the axial point P due to the circular disc will be given by

$$u_1(P) = Ae^{-ikz} - Ae^{-ikz}(1 - e^{-ip\pi})$$
$$= Ae^{-ikz}e^{-ip\pi} \tag{5.51}$$

Thus

$$|u_1(P)|^2 = |A|^2 \tag{5.52}$$

and therefore the intensity along the axis of the circular disc will be the same as in the absence of the disc! Thus we will always obtain a bright spot on the axis behind a circular disc[†]. This is called the Poisson spot. It may be of interest to mention that it was Poisson who first predicted from wave theory the existence of such a bright spot and used it as an argument against the validity of the wave theory. It was only later that Arago performed the experiment carefully and showed the bright spot which is now known as the Poisson spot.

Fig. 5.9 Fresnel diffraction of a normally incident plane wave by a circular disc of radius a. The central spot O is always bright; this spot is known as the Poisson spot.

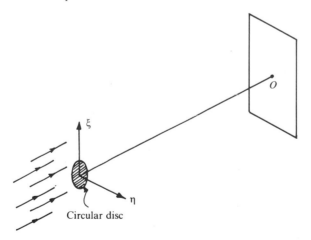

[†] It should be noted that Eq. (5.52) is not valid very close to the disc where the intensity becomes very small. This is because of the approximations made in calculating the circular aperture diffraction pattern.

5.9 Diffraction at a straight edge

We next consider a plane wave incident normally on a straight edge as shown in Fig. 5.10. We wish to find the intensity distribution on a screen S at a distance z from the straight edge. Obviously

$$A(\xi, \eta) = 0 \quad \xi < 0$$
$$= A \quad \xi > 0 \tag{5.53}$$

Thus the field distribution on the screen will be given by (see Eq. (5.5)):

$$u(x, y, z) = \frac{iA}{\lambda z} e^{-ikz} I_x I_y \tag{5.54}$$

where

$$I_x = \int_0^\infty \exp\left[-\frac{ik}{2z}(x - \xi)^2 \right] d\xi \tag{5.55}$$

$$I_y = \int_{-\infty}^\infty \exp\left[-\frac{ik}{2z}(y - \eta)^2 \right] d\eta \tag{5.56}$$

If we introduce the variable $\zeta = y - \eta$ then

$$I_y = \int_{-\infty}^\infty \exp\left(-\frac{ik}{2z}\zeta^2 \right) d\zeta = \left(\frac{\pi 2z}{ik} \right)^{\frac{1}{2}} = (\lambda z)^{\frac{1}{2}} \exp\left(-\frac{i\pi}{4} \right) \tag{5.57}$$

Fig. 5.10 A plane wave incident normally on a straight edge. $L'M'$ represents the edge of the geometrical shadow on the screen S.

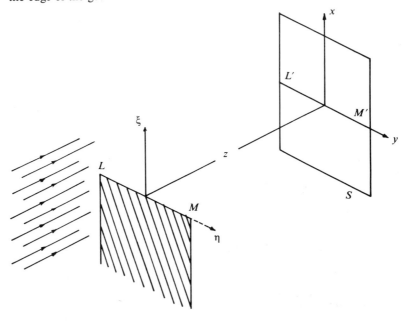

Table 5.1. *Fresnel integrals:* $C(\tau) = \int_0^\tau \cos \frac{1}{2}\pi v^2 \, dv; \; S(\tau) = \int_0^\tau \sin \frac{1}{2}\pi v^2 \, dv$

τ	$C(\tau)$	$S(\tau)$	τ	$C(\tau)$	$S(\tau)$
0.0	0.00000	0.00000	2.6	0.38894	0.54999
0.2	0.19992	0.00419	2.8	0.46749	0.39153
0.4	0.39748	0.03336	3.0	0.60572	0.49631
0.6	0.58110	0.11054	3.2	0.46632	0.59335
0.8	0.72284	0.24934	3.4	0.43849	0.42965
1.0	0.77989	0.42826	3.6	0.58795	0.49231
1.2	0.71544	0.62340	3.8	0.44809	0.56562
1.4	0.54310	0.71353	4.0	0.49843	0.42052
1.6	0.36546	0.63889	4.2	0.54172	0.56320
1.8	0.33363	0.45094	4.4	0.43833	0.46227
2.0	0.48825	0.34342	4.6	0.56724	0.51619
2.2	0.63629	0.45570	4.8	0.43380	0.49675
2.4	0.55496	0.61969	5.0	0.56363	0.49919

which is independent of y as indeed it should be because of the translational symmetry of the problem (along the y-direction). In order to evaluate I_x, we introduce the variable

$$\tau = (k/\pi z)^{\frac{1}{2}}(\xi - x) = (2/\lambda z)^{\frac{1}{2}}(\xi - x) \tag{5.58}$$

so that

$$I_x = (\lambda z/2)^{\frac{1}{2}} \int_{-\tau_0}^{\infty} e^{-i\pi\tau^2/2} \, d\tau \tag{5.59}$$

where

$$\tau_0 = (2/\lambda z)^{\frac{1}{2}} x \tag{5.60}$$

The above integral can be expressed in terms of the Fresnel integrals which are defined by the following equations

$$C(\tau) \equiv \int_0^\tau \cos(\pi x^2/2) \, dx \tag{5.61}$$

$$S(\tau) \equiv \int_0^\tau \sin(\pi x^2/2) \, dx \tag{5.62}$$

Both $C(\tau)$ and $S(\tau)$ are odd functions of τ, i.e.,

$$C(-\tau) = -C(\tau) \tag{5.63}$$

$$S(-\tau) = -S(\tau) \tag{5.64}$$

and have the following limiting values

$$C(0) = S(0) = 0 \tag{5.65}$$

$$C(\infty) = S(\infty) = \tfrac{1}{2} \tag{5.66}$$

and

$$C(-\infty) = S(-\infty) = -\tfrac{1}{2} \tag{5.67}$$

The Fresnel integrals are tabulated in Table 5.1 and in Fig. 5.11 we have plotted $C(\tau)$ and $S(\tau)$ with τ as a parameter. The resulting curve is known as Cornu's spiral and the values of τ are marked on the curve itself. We will demonstrate the use of Cornu's spiral later in this section. Now.

$$I_x = (\lambda z/2)^{\frac{1}{2}} \int_{-\tau_0}^{\infty} e^{-i\pi\tau^2/2} \, d\tau$$

$$= (\lambda z/2)^{\frac{1}{2}} \left(\int_0^{\infty} - \int_0^{-\tau_0} \right) e^{-i\pi\tau^2/2} \, d\tau$$

$$= (\lambda z/2)^{\frac{1}{2}} \left\{ [C(\infty) - C(-\tau_0)] - i[S(\infty) - S(-\tau_0)] \right\} \tag{5.68}$$

Substituting for I_y and I_x from Eqs. (5.57) and (5.68) in Eq. (5.54) we get

$$u(x, y, z) = \frac{iAe^{-ikz}}{\sqrt{2}} e^{-i\pi/4} \left\{ [\tfrac{1}{2} + C(\tau_0)] - i[\tfrac{1}{2} + S(\tau_0)] \right\} \tag{5.69}$$

Fig. 5.11 Cornu's spiral which represents a parametric plot of $C(\tau)$ and $S(\tau)$ with τ as a parameter.

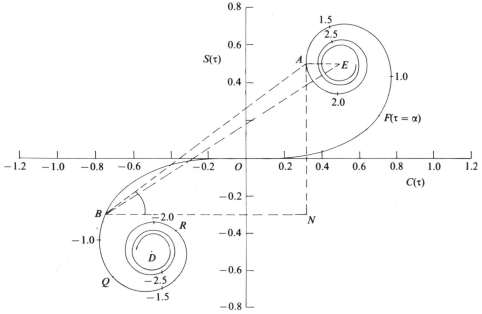

where we have used Eqs. (5.63)–(5.66). The corresponding intensity distribution would be given by

$$I(x, z) = \tfrac{1}{2} I_0 \{ [\tfrac{1}{2} + C(\tau_0)]^2 + [\tfrac{1}{2} + S(\tau_0)]^2 \} \tag{5.70}$$

where I_0 is the intensity in the absence of the straight edge. As mentioned earlier, the intensity distribution is independent of the y-coordinate. Furthermore, the dependence on x and z is through the parameter τ_0 (see Eq. (5.60)). Therefore the variation of I with τ_0 shown in Fig. 5.12 represents a universal curve. From the curve we may note the following:

(i) Deep inside the geometrical shadow, x (and therefore τ_0) tends to $-\infty$. Since $C(-\infty) = S(-\infty) = -\tfrac{1}{2}$, the intensity (deep inside the geometrical shadow) would tend to zero as indeed it should.

(ii) At the edge of the geometrical shadow (i.e., on the y-axis) $x = 0$ implying $\tau_0 = 0$. Therefore using Eq. (5.65) we get

$$I(x = 0, z) = \tfrac{1}{4} I_0 \tag{5.71}$$

Thus the intensity on the edge is quarter of the intensity that would have been in the absence of the edge.

Fig. 5.12 The intensity variation along the x-direction in the straight edge diffraction pattern.

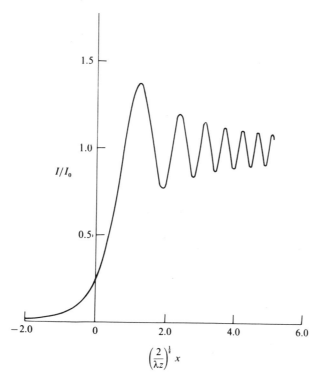

(iii) If we numerically evaluate the RHS of Eq. (5.70) we find that it attains maximum values for

$$\tau_0 \approx 1.22, 2.34, 3.08, \ldots \tag{5.72}$$

The corresponding intensities are given by

$$I \approx 1.37I_0, 1.20I_0, 1.15I_0, \ldots \tag{5.73}$$

The intensity minima occur at

$$\tau_0 \approx 1.87, 2.74, 3.39, \ldots \tag{5.74}$$

and the corresponding intensities are

$$I \approx 0.778I_0, 0.843I_0, 0.872I_0, \ldots \tag{5.75}$$

As an example, we assume

$$\lambda = 0.5 \, \mu\mathrm{m} \, (= 5000 \, \text{Å})$$
$$z = 100 \, \mathrm{cm}$$

so that Eq. (5.60) becomes

$$x = (\lambda z/2)^{\frac{1}{2}}\tau_0 = 0.05\tau_0$$

where x is now in centimetres. Thus the first three maxima occur at distances of

$$0.61 \, \mathrm{mm}, 1.17 \, \mathrm{mm} \quad \text{and} \quad 1.54 \, \mathrm{mm}$$

from the edge of the geometrical shadow. Similarly, the first three minima occur at distances of

$$0.935 \, \mathrm{mm}, 1.37 \, \mathrm{mm} \quad \text{and} \quad 1.695 \, \mathrm{mm}$$

from the edge of the geometrical shadow. One can physically interpret the appearance of successive maxima and minima with the help of Fresnel half period zones (see, e.g., Ghatak (1977) Sec. 16.4).

The intensity distribution on the screen can also be studied with the help of Cornu's spiral which is a geometrical representation of the Fresnel integrals $C(\tau)$ and $S(\tau)$ with τ as a parameter (see Fig. 5.11). The numbers written on the curve are the values of τ and as $\tau \to +\infty$ and $-\infty$, the curve spirals to the points $E(\frac{1}{2}, \frac{1}{2})$ and $D(-\frac{1}{2}, -\frac{1}{2})$ respectively. On Cornu's spiral, we consider two arbitrary points A and B corresponding to $\tau = \tau_1$, and $\tau = \tau_2$ respectively. Then

$$BN = C(\tau_1) - C(\tau_2) \tag{5.76}$$

and

$$AN = S(\tau_1) - S(\tau_2) \tag{5.77}$$

Thus

$$AB = \{[C(\tau_1) - C(\tau_2)]^2 + [S(\tau_1) - S(\tau_2)]^2\}^{\frac{1}{2}} \tag{5.78}$$

In order to illustrate the use of Cornu's spiral we consider a point inside the geometrical shadow for which x and therefore τ_0 is negative. Now using Eq. (5.68) we may write

$$I = \tfrac{1}{2}I_0\{[C(\infty) - C(\alpha)]^2 + [S(\infty) - S(\alpha)]^2\} \tag{5.79}$$

where

$$\alpha = -\tau_0 = -(2/\lambda z)^{\frac{1}{2}}x \tag{5.80}$$

is a positive quantity. Comparing with Eq. (5.78), we have $\tau_1 = \infty$ and $\tau_2 = \alpha$ and

$$I = \tfrac{1}{2}I_0(FE)^2$$

where the point F on the spiral is assumed to correspond to $\tau = \alpha$; since $\alpha > 0$, the point F lies in the upper portion of the spiral (see Fig. 5.11). As we go deep inside the geometrical shadow, $x \to -\infty$ and therefore $\alpha \to +\infty$ and the corresponding point F moves on the spiral towards E. Thus the length FE decreases uniformly and the intensity goes to zero.

At the edge of the geometrical shadow the point F coincides with the origin O and we get

$$I = \tfrac{1}{2}I_0(OE)^2$$
$$= \tfrac{1}{4}I_0$$

For a point on the illuminated region, $\tau_0 > 0$ and

$$I = \tfrac{1}{2}I_0\{[C(\infty) - C(-\tau_0)]^2 + [S(\infty) - S(-\tau_0)]^2\}$$
$$= \tfrac{1}{2}I_0(BE)^2 \tag{5.81}$$

where since $\tau_0 > 0$, the point B lies in the lower half of the spiral (see Fig. 5.11). As we go further away from the edge of the geometrical shadow, the value of $-\tau_0$ becomes more negative and the corresponding point moves on the spiral towards D. The length BE first increases till we reach the point $Q(\tau = -\tau_0 \approx -1.22)$ and then the length (and therefore the intensity) decreases till we reach the point $R(\tau = -\tau_0 \approx -1.87)$. Thus we will have maxima and minima in the intensity distribution consistent with

Fig. 5.12. As $\tau_0 \to \infty$, we reach the point D and

$$I = \tfrac{1}{2}I_0\{[C(\infty) - C(-\infty)]^2 + [S(\infty) - S(-\infty)]^2\}$$
$$= \tfrac{1}{2}I_0(DE)^2 = I_0 \tag{5.82}$$

Thus in the illuminated region as $x \to \infty$ the intensity tends to the value in the absence of the edge.

5.10   Fresnel diffraction by a long narrow slit

We next consider a plane wave incident normally on a long narrow slit. We wish to calculate the Fresnel diffraction pattern on a screen placed at a finite distance from the slit. We will also study the transition to the Fraunhofer region as the distance between the slit and the screen is made very large.

In Fig. 5.13, let LL' and MM' represent the edges of the geometrical shadow; we assume the slit width to be d and we choose the origin to be at the centre of the slit. Obviously,

$$A(\xi, \eta) = \begin{cases} A; & |\xi| < d/2 \\ 0; & |\xi| > d/2 \end{cases} \tag{5.83}$$

Thus the field distribution on the screen will be given by (cf. Eqs.

Fig. 5.13 A plane wave is incident normally on a long narrow slit of width d and the Fresnel diffraction pattern is observed on a screen at a finite distance z from the slit.

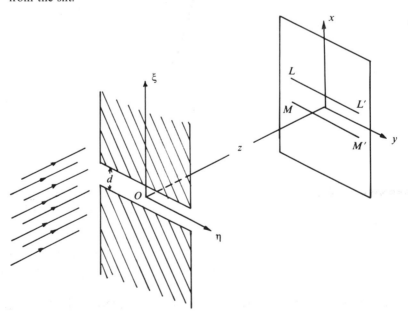

(5.54)–(5.57)]

$$u(x, y, z) = \frac{iA}{\lambda z} e^{-ikz} I_x I_y \tag{5.84}$$

where

$$I_x = \int_{-d/2}^{d/2} \exp\left[-\frac{ik}{2z}(x - \xi)^2 \right] d\xi \tag{5.85}$$

$$I_y = \int_{-\infty}^{\infty} \exp\left[-\frac{ik}{2z}(y - \eta)^2 \right] d\eta = (\lambda z)^{\frac{1}{2}} \exp(-i\pi/4) \tag{5.86}$$

Once again, I_y (and therefore the field distribution) is independent of y which is due to the translational symmetry of the problem. If we introduce the variable

$$\tau = (2/\lambda z)^{\frac{1}{2}}(\xi - x)$$

(see Eq. (5.58)) we would obtain

$$I_x = (\lambda z/2)^{\frac{1}{2}} \int_{-\tau_0 - \tau'}^{-\tau_0 + \tau'} e^{-i\pi\tau^2/2} d\tau \tag{5.87}$$

where

$$\tau_0 = (2/\lambda z)^{\frac{1}{2}} x \tag{5.88}$$

and

$$\tau' = \tfrac{1}{2}(2d^2/\lambda z)^{\frac{1}{2}} \tag{5.89}$$

Using a procedure similar to that in the previous section we obtain

$$
\begin{aligned}
I_x &= (\lambda z/2)^{\frac{1}{2}} \left(\int_0^{-\tau_0 + \tau'} e^{-i\pi\tau^2/2} d\tau - \int_0^{\tau_0 - \tau'} e^{-i\pi\tau^2/2} d\tau \right) \\
&= (\lambda z/2)^{\frac{1}{2}} \{ [C(-\tau_0 + \tau') - C(-\tau_0 - \tau')] \\
&\quad - i[S(-\tau_0 + \tau') - S(-\tau_0 - \tau')] \}
\end{aligned}
\tag{5.90}
$$

Obviously, the intensity distribution will be symmetric about the y-axis and therefore we may replace τ_0 by $-\tau_0$ to obtain the following expression for the intensity distribution

$$I(x, z) = \tfrac{1}{2} I_0 \{ [C(\tau_0 + \tau') - C(\tau_0 - \tau')]^2 + [S(\tau_0 + \tau') - S(\tau_0 - \tau')]^2 \} \tag{5.91}$$

For a given value of τ' a universal curve for the intensity distribution can

Fig. 5.14 The Fresnel diffraction pattern due to a single slit for $\tau' = (d^2/2\lambda z)^{\frac{1}{2}} = 5.0$. The vertical dashed lines represent the edges of the geometrical shadow.

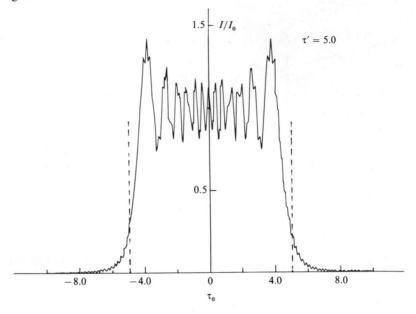

Fig. 5.15 The Fresnel diffraction pattern due to a single slit for $\tau' = (d^2/2\lambda z)^{\frac{1}{2}} = 1.0$. The vertical dashed lines correspond to the edge of the geometrical shadow and the dashed curve represents the corresponding Fraunhofer diffraction pattern (see Eq. (5.101)).

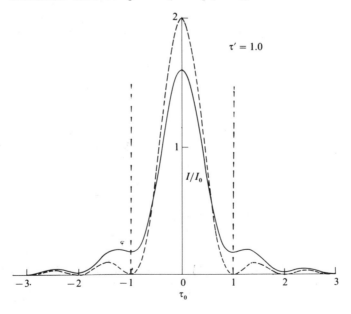

be plotted as a function of the parameter τ_0. Typical intensity distributions are plotted for

$$\tau' = 5.0, 1.0 \quad \text{and} \quad 0.7$$

in Figs. 5.14, 5.15 and 5.16 respectively. For $\tau' \gg 1$ i.e., for

$$d^2/\lambda z \gg 1 \tag{5.92}$$

the diffraction pattern resembles the pattern produced by two separate straight edges; indeed the intensity at the boundary of the geometrical shadow is very close to $I_0/4$ (see Fig. 5.14). As τ' becomes smaller the diffraction pattern starts to resemble the Fraunhofer diffraction pattern of the aperture. Indeed when τ' becomes small compared to unity, the diffraction pattern is almost of the Fraunhofer type as can be seen from Fig. 5.16. Thus we may say that when

$$\tau' = d^2/\lambda z \ll 1 \tag{5.93}$$

the diffraction is essentially of the Fraunhofer type as we will explicitly show in the following section.

5.10.1 Transition to the Fraunhofer region

In order to study the transition to the Fraunhofer region we first obtain the asymptotic forms of the Fresnel integrals. Now,

Fig. 5.16 The Fresnel diffraction pattern due to a single slit for $\tau' = (d^2/2\lambda z)^{\frac{1}{2}} = 0.7$. The vertical dashed lines correspond to the edge of the geometrical shadow and the dashed curve represents the corresponding Fraunhofer diffraction pattern (see Eq. (5.101)).

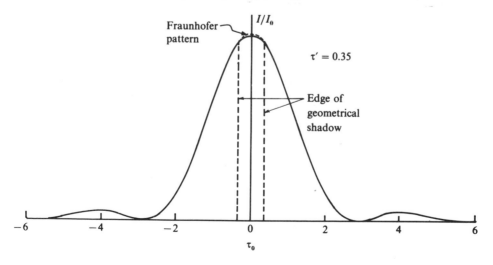

$$C(\tau) = \int_0^\tau \cos\frac{\pi x^2}{2}\,\mathrm{d}x$$

$$= \int_0^\infty \cos\frac{\pi x^2}{2}\,\mathrm{d}x - \int_\tau^\infty \cos\frac{\pi x^2}{2}\,\mathrm{d}x$$

$$= \tfrac{1}{2} - \int_\tau^\infty \frac{\mathrm{d}}{\mathrm{d}x}\left(\sin\frac{\pi x^2}{2}\right)\frac{1}{\pi x}\,\mathrm{d}x$$

$$= \tfrac{1}{2} + \frac{1}{\pi\tau}\sin\frac{\pi\tau^2}{2} + \int_\tau^\infty \frac{1}{\pi x^2}\sin\frac{\pi x^2}{2}\,\mathrm{d}x + \cdots$$

The integral can again be integrated by parts to obtain terms of order $1/\tau^2$ and $1/\tau^3$ etc. In the limit of $\tau \to \infty$ we may write

$$C(\tau) \approx \tfrac{1}{2} + \frac{1}{\pi\tau}\sin\frac{\pi\tau^2}{2} \tag{5.94}$$

and similarly

$$S(\tau) \approx \tfrac{1}{2} - \frac{1}{\pi\tau}\cos\frac{\pi\tau^2}{2} \tag{5.95}$$

We now consider the diffraction pattern for distances $z \gg d^2/\lambda$, i.e., for small values of τ'. Now, we can write

$$\tau_0 = (2/\lambda z)^{\frac{1}{2}}x \approx (2z/\lambda)^{\frac{1}{2}}\theta \tag{5.96}$$

where

$$\theta \approx x/z \tag{5.97}$$

represents the angle of diffraction (see Fig. 5.17). For a given angle of

Fig. 5.17 A screen placed at a distance $z \gg d^2/\lambda$ from the slit.

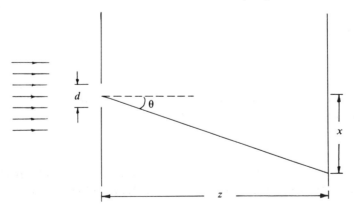

diffraction, as z increases τ_0 becomes very large and we may use the asymptotic expansions given by Eqs. (5.94) and (5.95). Thus

$$C(\tau_0 + \tau') - C(\tau_0 - \tau') \approx (2/\pi\tau_0) \cos\left[\tfrac{1}{2}\pi(\tau_0^2 + \tau'^2)\right] \sin \pi\tau_0\tau' \qquad (5.98)$$

Similarly

$$S(\tau_0 + \tau') - S(\tau_0 - \tau') \approx (2/\pi\tau_0) \sin\left[\tfrac{1}{2}\pi(\tau_0^2 + \tau'^2)\right] \sin \pi\tau_0\tau' \qquad (5.99)$$

If we square and add the above two equations and substitute in Eq. (5.91) we would obtain

$$I = \tfrac{1}{2}I_0(4/\pi^2\tau_0^2)\sin^2(\pi\tau_0\tau') \qquad (5.100)$$

If we now use Eq. (5.96), we get

$$I = I_{00}\sin^2\zeta/\zeta^2 \qquad (5.101)$$

where

$$I_{00} = 2I_0\tau'^2 = I_0 d^2/\lambda z \qquad (5.102)$$

and

$$\zeta = \pi\tau_0\tau' = (\pi/\lambda z)xd \approx (\pi d/\lambda)\theta \qquad (5.103)$$

where

$$\theta \approx x/z$$

represents the angle of diffraction. Eq. (5.101) represents the Fraunhofer diffraction pattern as is obvious from comparing it with Eq. (4.30).

Problems

Problem 5.1: Consider a circular aperture of radius 0.4 mm illuminated by a plane wave of wavelength 5000 Å. Calculate the positions of the brightest and darkest points on the axis.
(Answer: 32 cm, 16 cm)

Problem 5.2: Consider a zone plate whose first ring has a radius of 0.1 cm. Calculate the position of the principal focus for $\lambda = 5000$ Å and $\lambda = 6500$ Å and show that the chromatic aberration is large.
(Answer: 200 cm and 154 cm)

Problem 5.3: In Sec. 5.6 we had discussed the Fresnel diffraction of a plane wave incident normally on a circular aperture of radius a. If instead we have a point source on the axis, show that the intensity distribution is still given by Eq. (5.40) with

$$p \approx \frac{a^2}{\lambda}\left(\frac{1}{z_0} + \frac{1}{z}\right)$$

where z_0 and z represent the distances of the point object and the screen respectively from the the centre of the circular aperture. Hence physically interpret the intensity distribution in terms of the Fresnel half period zones (see Ghatak and Thyagarajan (1978)).

Problem 5.4: Consider a zone plate in which the the first, third, fifth,... zones are darkened. Discuss the resultant intensity distribution when a plane wave is incident normally and obtain the positions of the foci.

Problem 5.5: In the straight edge diffraction pattern, the distance between the first and second maxima is 1 mm. If the distance between the screen and the straight edge is 250 cm, calculate the wavelength of the lightwave.
(Answer: $\lambda \approx 6.38 \times 10^{-5}$ cm)

Problem 5.6: What is the primary focal length of a zone plate for $\lambda = 6500$ Å if the radius of the eighth ring is 4.5 mm. Compute the image distance for a point source at a distance of 7.788 m from the plate.
(Answer: ≈ 3.894 m; 7.788 m.)

Problem 5.7: An aperture containing an open portion between radii $\sqrt{2}$ mm and $\sqrt{3}$ mm is illuminated by a parallel beam of light of wavelength 0.6 μm. Calculate the distance of the brightest point on the axis of the aperture.
(Answer: ≈ 333 cm)

Problem 5.8: Consider a periodic object having a transmittance of the form

$$T(x) = A \cos(2\pi x/a)$$

Considering a normally incident plane wave, use Eq. (5.5) and show that the intensity distribution in the Fresnel diffraction pattern is given by

$$I(x, z) = A^2 \cos^2(2\pi x/a)$$

Observe that the intensity distribution in the transverse direction is independent of z.

Problem 5.9: In continuation with the last problem consider a general object with a periodicity x_0 being illuminated by normally incident plane parallel light. By making a Fourier series expansion show that for $z = 2mx_0^2/\lambda$, $m = 1, 2, 3, \ldots$, the original pattern is reproduced. Show also that for $z = (2m + 1)x_0^2/\lambda$, the original object distribution is reproduced but is displaced from the axis by $x_0/2$ (see Ghatak and Thyagarajan (1978)).

6

Spatial frequency filtering

6.1   Introduction

A very important application of lasers lies in the field of spatial frequency filtering which is the subject matter of this chapter. In Sec. 6.2 we will show that if $h(x, y)$ represents the field distribution on the front focal plane of a corrected lens (i.e., on the plane P_1 in Fig. 6.1) then the field distribution $g(x, y)$ on the back focal plane P_2 is the two-dimensional Fourier

Fig. 6.1 The plane P_2 is the Fourier transform plane where the spatial frequency components of the object (placed in the plane P_1) are displayed. The unwanted spatial frequency components of the object can be filtered out by putting suitable stops on plane P_2; in the absence of such a filter the original image (with an inversion) is displayed on the plane P_3.

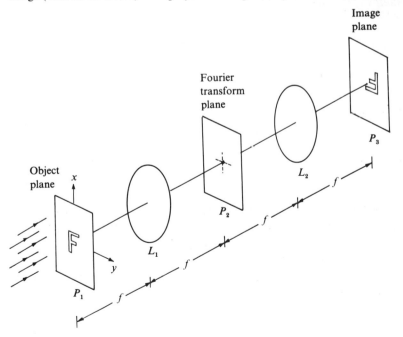

transform of $h(x, y)$:

$$g(x, y) = (1/\lambda f) \int \int h(x', y') e^{2\pi i(ux' + vy')} dx' dy' \tag{6.1}$$

where λ is the wavelength of light, f the focal length of the lens and

$$u \equiv x/\lambda f, \quad v \equiv y/\lambda f \tag{6.2}$$

are usually referred to as the spatial frequencies along the x and y-directions respectively. Thus on the back focal plane one obtains the spatial frequency spectrum of the field distribution on the front focal plane. We should mention here that just as the Fourier transform of a time varying signal gives its temporal frequency spectrum, similarly the spatial Fourier spectrum gives the spatial frequency spectrum of the field distribution.

Now, if we then put another corrected lens L_2, such that P_2 represents the front focal plane of L_2, on the back focal plane P_3 we will get the Fourier transform of the Fourier transform of the original field distribution $h(x, y)$ which will be the original field distribution (except for an inversion). Thus on the plane P_3 we will get a field distribution which will be similar [†] to the field distribution $h(x, y)$ on P_1. Thus by putting suitable stops and apertures on the plane P_2 we can filter out (or alter) certain spatial frequencies of the incident field distribution. This important result which allows us to filter out 'unwanted' spatial frequencies present in the object forms the basis of *spatial frequency filtering*.

6.2   The Fourier transform and some of its important properties

According to the Fourier integral theorem

$$h(x') = \int \int_{-\infty}^{\infty} h(x) e^{2\pi i u(x - x')} dx \, du \tag{6.3}$$

Thus if we write

$$H(u) = \int_{-\infty}^{+\infty} h(x) e^{2\pi i u x} dx = \mathscr{F}[h(x)] \tag{6.4}$$

then

$$h(x) = \int_{-\infty}^{+\infty} H(u) e^{-2\pi i u x} du \tag{6.5}$$

[†] If the focal lengths of the lenses L_1 and L_2 are f_1 and f_2 then the magnification of the final image will be $-f_2/f_1$.

The function $H(u)$ is known as the Fourier transform (often abbreviated as FT) of the function $h(x)$ and $h(x)$ is known as the inverse Fourier transform of $H(u)$. Elementary manipulations give us

$$\mathcal{F}\{\mathcal{F}[h(x)]\} = \int_{-\infty}^{+\infty} H(u)\,e^{2\pi iux}\,du = h(-x) \tag{6.6}$$

implying that the Fourier transform of the Fourier transform of a function is the original function itself except for an inversion. We also have the following important property

$$\mathcal{F}[h(x-a)] = \int_{-\infty}^{+\infty} h(x-a)\,e^{2\pi iux}\,dx = H(u)\,e^{2\pi iau} \tag{6.7}$$

As an example we consider the Fourier transform of the rectangle function

$$\operatorname{rect}(x/a) = \begin{cases} 1, & |x| < a/2 \\ 0, & |x| > a/2 \end{cases} \tag{6.8}$$

Thus

$$\mathcal{F}[\operatorname{rect}(x/a)] = \int_{-a/2}^{+a/2} e^{2\pi iux}\,dx = a(\sin \zeta/\zeta) \tag{6.9}$$

where $\zeta = \pi ua$.

In a similar manner we can define the two-dimensional Fourier transform:

$$H(u,v) = \int\!\!\int_{-\infty}^{+\infty} h(x,y)e^{2\pi i(ux+vy)}\,dx\,dy = \mathcal{F}[h(x,y)] \tag{6.10}$$

Properties similar to that given by Eqs. (6.6) and (6.7) readily follow:

$$\mathcal{F}\{\mathcal{F}[h(x,y)]\} = h(-x,-y) \tag{6.11}$$

$$\mathcal{F}\{h(x-a,y-b)\} = e^{2\pi i(ua+vb)}H(u,v) \tag{6.12}$$

6.3 The Fourier transforming property of a thin lens

Let $h(x,y)$ represent the field distribution on the plane P_1. We would first like to determine the field distribution on the plane P_4 i.e., at a distance f from the plane P_1 (see Fig. 6.2). Obviously the field will undergo Fresnel diffraction and on plane P_4 it will be given by (see Eq. (5.5))

$$u|_{P_4} = \frac{i}{\lambda f}\exp(-ikf)\int\!\!\int h(\xi,\eta)$$
$$\times \exp\left\{-\frac{ik}{2f}[(x-\xi)^2 + (y-\eta)^2]\right\}d\xi\,d\eta \tag{6.13}$$

Now the effect of a thin lens of focal length f is to multiply the incident field distribution by the factor p_L given by (see e.g., Ghatak and Thyagarajan (1978), Sec. 6.3)

$$p_L = \exp\left[\frac{ik}{2f}(x^2 + y^2)\right] \tag{6.14}$$

The above equation can be physically understood from the fact that for a normally incident plane wave the transmitted wave will be a converging spherical wave of radius of curvature f (cf. Eq. (5.19)). Thus on the plane P_5, the field distribution will be given by

$$u|_{P_5} = (i/\lambda f)\,e^{-ikf}\,e^{i\alpha(x^2+y^2)}\int\int h(\xi,\eta)\,e^{-i\alpha[(x-\xi)^2+(y-\eta)^2]}\,d\xi\,d\eta \tag{6.15}$$

where

$$\alpha = k/2f = \pi/\lambda f \tag{6.16}$$

From plane P_5 the field will again undergo Fresnel diffraction and therefore on plane P_2 it will be given by

$$u|_{P_2} = [(i/\lambda f)\,e^{-ikf}]^2 I \tag{6.17}$$

where

$$I = \int\int h(\xi,\eta)G(\xi,\eta)\,d\xi\,d\eta$$

and

$$G(\xi,\eta) = \int\int_{-\infty}^{+\infty} e^{i\alpha(\zeta^2+\tau^2)}e^{-i\alpha[(\zeta-\xi)^2+(\tau-\eta)^2]}$$
$$\times e^{-i\alpha[(x-\zeta)^2+(y-\tau)^2]}\,d\zeta\,d\tau$$

The integrals depending on ζ and τ separate out. Indeed we may write

$$G(\xi,\eta) = G_\xi G_\eta$$

Fig. 6.2 Demonstration of the Fourier transforming property of a thin lens.

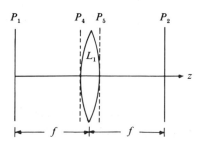

where

$$G_\xi = \int_{-\infty}^{\infty} e^{i\alpha[\zeta^2 - (\zeta - \xi)^2 - (x - \zeta)^2]} \, d\zeta$$

and a similar expression for G_η. Simple manipulations give us

$$G_\xi = e^{-i\alpha(x^2 + \xi^2)} e^{i\alpha(x + \xi)^2} \int_{-\infty}^{\infty} e^{-i\alpha\sigma^2} \, d\sigma$$

where $\sigma \equiv \zeta - (x + \xi)$. Thus

$$G_\xi = e^{2i\alpha x\xi} 2\left(\int_0^{\infty} \cos\alpha\sigma^2 \, d\sigma - i \int_0^{\infty} \sin\alpha\sigma^2 \, d\sigma \right)$$

$$= e^{2\pi i u\xi}(\lambda f/2)^{\frac{1}{2}}(1 - i) \tag{6.18}$$

where

$$u = (1/\pi)\alpha x = x/\lambda f \tag{6.19}$$

and use has been made of the properties of the Fresnel integrals $C(\infty) = S(\infty) = \frac{1}{2}$ (see Sec. 5.9). Similarly one can calculate G_η. We thus finally obtain

$$u|_{P_2} = \left(\frac{i}{\lambda f} e^{-ikf} \right)^2 \left[\left(\frac{\lambda f}{2} \right)^{\frac{1}{2}} (1 - i) \right]^2 \int\int e^{2\pi i(u\xi + v\eta)} h(\xi, \eta) \, d\xi \, d\eta \tag{6.20}$$

where u and v are the spatial frequencies given by Eq. (6.2) (see also Eq. (6.19)). Thus the field on the back focal plane P_2 is given by

$$g(x, y) = \frac{1}{\lambda f} \int\int_{-\infty}^{+\infty} h(\xi, \eta) e^{2\pi i(u\xi + v\eta)} \, d\xi \, d\eta \tag{6.21}$$

where we have neglected the unimportant constant phase factor $i e^{-2ikf}$. The above equation gives the important result that the field distribution on the back focal plane of a corrected lens is the Fourier transform of the field distribution on the front focal plane. We should mention here that in writing the limits in the integral from $-\infty$ to $+\infty$ we have assumed the lens to be of infinite extent; the error involved is usually very small because in almost all practical cases

$$a/\lambda \gg 1 \tag{6.22}$$

where a represents the aperture of the lens.

From Eq. (6.21) it follows that a point (x, y) on the back focal plane of the lens corresponds to spatial frequencies $u = x/\lambda f$ and $v = y/\lambda f$ of the object.

Therefore the amplitude distribution on the back focal plane actually gives the spatial frequency distribution of the object.

6.4 Some elementary examples of the Fourier transforming property of a lens

In order to have a physical understanding of the Fourier transforming property of a lens we will consider some simple examples:

Example: We first consider a uniform plane wave incident normally on the plane P_1 (see Fig. 6.2). Thus

$$h(x, y) = \text{constant} = A \, (\text{say}) \tag{6.23}$$

and

$$g(x, y) = (A/\lambda f) \int\int_{-\infty}^{+\infty} e^{2\pi i (u\xi + v\eta)} \, d\xi \, d\eta \tag{6.24}$$

Now an integral representation of the Dirac delta function is given by the following equation

$$\delta(x - a) = (1/2\pi) \int_{-\infty}^{+\infty} e^{\pm ik(x-a)} \, dk = \int_{-\infty}^{+\infty} e^{\pm 2\pi iu(x-a)} \, du \tag{6.25}$$

Thus Eq. (6.24) can be written as

$$g(x, y) = (A/\lambda f)\delta(u)\delta(v) \tag{6.26}$$

Since $\delta(u) = 0$ for $u \neq 0$, the above equation tells us that the field at the back focal plane is zero everywhere except at $u = 0$, $v = 0$ i.e., at $x = 0$, $y = 0$. This is indeed to be expected because a plane wave incident normally on a lens (of large aperture) is focussed almost to a point at the focal plane of a corrected lens. Eq. (6.26) also tells us that only zero frequencies are present in an infinitely extended plane wave (propagating along the z-direction) which is indeed true because there are no spatial variations in the field. We should point out here that the actual pattern is the Airy distribution (see Sec. 4.5) which approaches the delta function distribution (given by Eq. (6.26)) in the limit of the aperture radius a going to infinity.[†] Indeed the integral in Eq. (4.38) is very similar to the one in Eq. (6.24) except that the limits here are from $-\infty$ to $+\infty$.

Example: We next consider a long narrow slit of width a (along the x-axis) placed on the front focal plane. Thus

$$\begin{aligned} h(x, y) &= A & |x| < a/2 \\ &= 0 & |x| > a/2 \end{aligned} \tag{6.27}$$

for all values of y. (We have the same arrangement as shown in Fig. 4.7 with the planes A and S representing the front and back focal planes of the lens L.)

[†] When a becomes very large, the angles at which the dark rings appear become extremely small (see Eq. (4.50)). Thus the Airy pattern is concentrated very near to the axis.

Substituting Eq. (6.27) in Eq. (6.21) we obtain

$$g(x, y) = \frac{A}{\lambda f} \int_{-a/2}^{+a/2} e^{2\pi i u \xi} \, d\xi \int_{-\infty}^{+\infty} e^{2\pi i v \eta} \, d\eta$$

$$= \frac{Aa}{\lambda f} \left(\frac{\sin \zeta}{\zeta} \right) \delta(v) \tag{6.28}$$

where

$$\zeta = \pi u a = \pi a x / \lambda f \approx (\pi a / \lambda) \sin \theta$$

and θ is the angle of diffraction along the x-direction (see Sec. 4.4). We thus obtain the single slit diffraction pattern (cf. Eq. (4.30)) and the intensity is zero except on the x-axis.

Example: As a third example we consider a point source on the axis for which

$$h(x, y) = A\delta(x)\delta(y) \tag{6.29}$$

Substituting in Eq. (6.21), we obtain

$$g(x, y) = A/\lambda f \tag{6.30}$$

which simply implies that an impulse contains all frequencies with equal amplitudes.

Example: We next consider a field distribution on the plane P_1 of the form[†]

$$h(x, y) = A + B \cos 2\pi \alpha x \tag{6.31}$$

Thus

$$g(x, y) = (A/\lambda f)\delta(u)\delta(v) + p(x, y) \tag{6.32}$$

where

$$p(x, y) = (B/2\lambda f) \int\int_{-\infty}^{+\infty} (e^{2\pi i \alpha \xi} + e^{-2\pi i \alpha \xi}) e^{2\pi i (u\xi + v\eta)} \, d\xi \, d\eta$$

$$= (B/2\lambda f)[\delta(u + \alpha)\delta(v) + \delta(u - \alpha)\delta(v)] \tag{6.33}$$

Eqs. (6.32) and (6.33) imply that on the back focal plane we will have three spots corresponding to

$$u = 0, \quad \pm \alpha; \quad v = 0 \tag{6.34}$$

i.e., at

$$x = 0, \quad \pm \lambda f \alpha \tag{6.35}$$

on the x-axis. Thus the spatial frequency associated with the field distribution on P_1 is displayed on plane P_2.

The appearance of the spots can also be understood by noting that the field distribution on the plane P_1 would be proportional to

$$e^{i\omega t}(A + B \cos 2\pi \alpha x) = A e^{i\omega t} + \tfrac{1}{2} B(e^{i(\omega t + 2\pi \alpha x)} + e^{i(\omega t - 2\pi \alpha x)}) \tag{6.36}$$

[†] It is well known that one obtains a $\cos^2 \gamma x$ type of intensity distribution in the interference pattern produced in the two slit interference arrangement such as that using the Fresnel biprism (see, e.g., Ghatak (1977), Sec. 11.6). Since $\cos^2 \gamma x = \tfrac{1}{2}(1 + \cos 2\gamma x)$, the photographic plate recording the interference fringes will have a transmittance very similar to that given by Eq. (6.31).

Now a plane wave propagating along **k** (in vacuum) is described by (see Appendix A)

$$e^{i(\omega t - \mathbf{k \cdot r})} = e^{i(\omega t - k_x x - k_y y - k_z z)} \tag{6.37}$$

where

$$k = (k_x^2 + k_y^2 + k_z^2)^{\frac{1}{2}} = \omega/c \tag{6.38}$$

If ϕ_x, ϕ_y and ϕ_z are the angles that **k** makes with the x, y and z-axes then

$$\cos \phi_x = k_x/k, \quad \cos \phi_y = k_y/k, \quad \cos \phi_z = k_z/k \tag{6.39}$$

The phase variation on the plane $z = 0$ is given by

$$e^{i(\omega t - k_x x - k_y y)}$$

Fig. 6.3 If the amplitude distribution on the plane P_1 is $(A + B \cos 2\pi\alpha x)$, then on the back focal plane P_2 we will have three spots located at $(0,0)$, $(\lambda f \alpha, 0)$ and $(-\lambda f \alpha, 0)$.

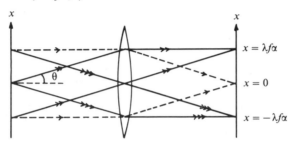

Fig. 6.4 The spatial frequency α present in the object can be filtered out by placing stops at $(\lambda f \alpha, 0)$ and $(-\lambda f \alpha, 0)$ on the back focal plane P_2.

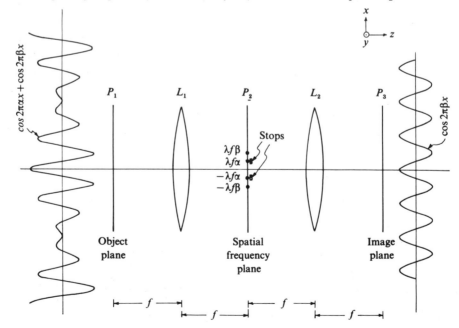

Comparing with Eq. (6.36) we have $k_y = 0$ and the three terms in Eq. (6.36) represent plane waves propagating along directions making angles 0, $-\theta$ and $+\theta$ with the z-axis (see Fig. 6.3) where

$$\sin\theta = \frac{k_x}{k} = \frac{2\pi\alpha}{2\pi/\lambda} = \alpha\lambda \qquad (6.40)$$

These three plane waves will produce spots at $x = 0$, $\pm\alpha\lambda f$ (on the x-axis); this is consistent with Eq. (6.35). The appearance of the spots is also consistent with the fact that the object described by Eq. (6.31) contains spatial frequencies $(0, 0)$ and $(\alpha, 0)$.

Example: We next consider the field variation on the front focal plane

$$h(x, y) = A\cos 2\pi\alpha x + B\cos 2\pi\beta x \qquad (6.41)$$

which implies the presence of two spatial frequencies α and β present in the incident field. Obviously, we will now have four spots on plane P_2 (along the x-axis) at

$$x = \pm\lambda f\alpha, \quad \pm\lambda f\beta; \quad y = 0 \qquad (6.42)$$

Now if we put stops at the points

$$x = \pm\lambda f\alpha, \quad y = 0$$

i.e., if we block the spatial frequency α then on plane P_3 we will only have the spatial frequency β (see Fig. 6.4). Thus the spatial frequency α has been filtered out. This is the basic principle of spatial frequency filtering.

6.5   Applications of spatial frequency filtering

As discussed in the previous section one can remove unwanted spatial frequencies by putting appropriate stops and apertures on the Fourier transform plane P_2. These stops and apertures are known as spatial frequency filters. Three simple filters are shown in Fig. 6.5. Fig. 6.5(a) shows a 'low pass filter' which allows only the low spatial frequency components to pass through. Fig. 6.5(b) is known as a 'high pass filter' because it blocks the low frequency components and allows only the high frequency components to pass through. Fig. 6.5(c) shows a 'band pass filter' as it blocks the low spatial frequencies as well as very high spatial frequencies.

Fig. 6.5 (a) Low pass filter, (b) high pass filter and (c) band pass filter.

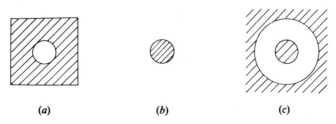

(a)   (b)   (c)

Fig. 6.6 Halftone dot structure present in (*a*) can be removed by using a low pass filter shown in Fig. 6.5(*a*) to form the filtered image shown in (*b*). In (*c*) we have shown the spatial frequency spectrum associated with (*a*). (Photograph courtesy of Dr. R.A. Philips.)

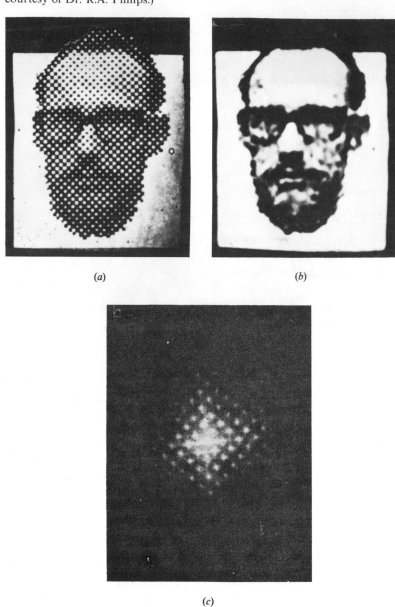

(*a*)   (*b*)

(*c*)

As a simple example we consider the removal of the dot structure present in a typical halftone photograph (see Fig. 6.6(*a*)); its corresponding spatial frequency spectrum (as observed on the Fourier transform plane) is shown in Fig. 6.6(*c*). Now the closely arranged dots represent a high spatial frequency which can be removed by putting the low pass filter (see Fig. 6.5(*a*)) on the back focal plane P_2. Thus if a small hole is placed on the plane P_2 then an image of the form shown in Fig. 6.6(*b*) is obtained. It may be seen that in Fig. 6.6(*b*) the missing parts of the eyeglasses, frame etc. appear.

We can similarly use spatial frequency filtering for contrast enhancement. For example, in the presence of a large amount of background light the contrast is poor. Since the background light corresponds to zero spatial frequency we can remove this by using a small opaque disc of the type shown in Fig. 6.5(*b*). Thus the stop will remove the low frequency components and the contrast will be enhanced.

6.6   Phase contrast microscope

Many objects in microscopy are phase objects which only change the phase of the incident wave without changing the amplitude. Thus if the thickness or the refractive index of an object varies across the transverse dimension then using an ordinary microscope it is difficult to observe such an object. Such objects can indeed be viewed through what is known as a phase contrast microscope first introduced by Zernike in 1942.

We consider a phase object whose transmittance is given by

$$h(x, y) = e^{-i\phi(x,y)} \tag{6.43}$$

Under normal observations, one observes the intensity of the transmitted wave which is proportional to $|h(x, y)|^2$, which for a phase object is a constant. If the phase variation $\phi(x, y)$ is small, then

$$h(x, y) \approx 1 - i\phi(x, y) \tag{6.44}$$

Observe that to first order in ϕ, the intensity variation which is proportional to $|h(x, y)|^2$ is still a constant. This is because of the additional factor of i (which corresponds to a phase difference of $\pi/2$) present in the $\phi(x, y)$ term.

In order to see how this phase factor can be nullified, we consider the object $h(x, y)$ to be placed in the front focal plane of a lens. The amplitude distribution in the back focal plane will be the Fourier transform of $h(x, y)$ (see Eq. (6.21)). Substituting from Eq. (6.44) we have

$$H(u, v) = \delta(u)\delta(v) - i\Phi(u, v) \tag{6.45}$$

where

$$\Phi(u, v) = \mathscr{F}[\phi(x, y)] \tag{6.46}$$

The first term on the RHS of Eq. (6.45) corresponds to zero spatial frequency and is centred on the axis of the back focal plane. The second term on the RHS of (6.45) would produce a distributed intensity pattern on the focal plane. We now place a phase filter on the axis so as to introduce a phase difference of $\frac{1}{2}\pi$ between the zero and other spatial frequency components. Such a filter could be a $\lambda_0/4(n-1)$ thick film of refractive index n placed axially on the back focal plane. The filtered spectrum would be

$$H_f(u, v) = -i\delta(u)\delta(v) - i\Phi(u, v) \tag{6.47}$$

If a second Fourier transform is performed by another lens then the intensity pattern on its back focal plane will be

$$|g(x, y)|^2 = |\mathscr{F}[H_f(u, v)]|^2$$
$$= 1 + 2\phi(x, y) \tag{6.48}$$

Thus the intensity distribution is linearly related to the phase variations in the object. The phase distribution is thus converted into intensity distribution in the image.

6.7   Image deblurring

We shall now discuss briefly another very interesting application of spatial frequency filtering namely image deblurring. Let $f(x, y)$ represent the image of an object. If during exposure of the film, the camera moves or is out of focus, then instead of the image $f(x, y)$ one would obtain a modified image $g(x, y)$ which is the blurred image corresponding to $f(x, y)$. We shall now show how the blurring in the image can be partially compensated i.e., the image can be deblurred from the blurred photograph.

If $h(x, y)$ represents the intensity distribution in the blurred image of a point object, then the intensity distribution of the blurred image can be written as

$$g(x, y) = \int\int f(x', y')h(x - x', y - y')\,dx'\,dy'$$
$$= f(x, y) * h(x, y) \tag{6.49}$$

where $*$ represents convolution and we have assumed imaging under incoherent light so that intensity addition takes place. If we assume the transmittance of the exposed film to be proportional to $g(x, y)$, then the amplitude transmittance of the recorded film would be proportional to $g(x, y)$.

If we place this film on the front focal plane of a lens and illuminate by a normally incident laser beam, then the amplitude distribution on the back focal plane would be the Fourier transform of $g(x, y)$. Using the fact that the Fourier transform of the convolution of two functions is the product of their Fourier transforms the amplitude distribution on the back focal plane would be

$$G(u, v) = F(u, v)H(u, v) \qquad (6.50)$$

where $u = x/\lambda f$, $v = y/\lambda f$ and G, F and H are the Fourier transforms of g, f and h respectively. If we place a filter whose transmittance is proportional to $1/H(u, v)$ on the back focal plane then the filtered spectrum would be

$$G(u, v)\frac{1}{H(u, v)} = F(u, v) \qquad (6.51)$$

Fig. 6.7 (*a*) Blurred photograph due to defocussing, (*b*) its spectrum (*c*) the filters used for deblurring and (*d*) the deblurred images. (After Stroke, Halioua and Srinivasan (1975).)

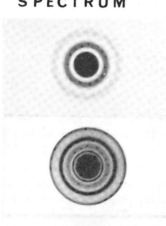
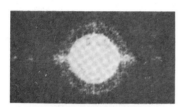

(*b*) **S P E C T R U M**

AMPLITUDE FILTERS

(*c*)

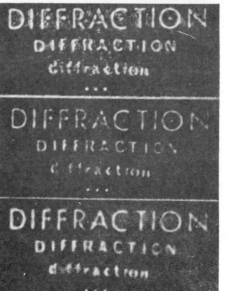

BLURRED PHOTOGRAPH　(*a*)

DEBLURRED IMAGES

(*d*)

A second lens can Fourier transform the filtered spectrum further to produce the deblurred image $f(x, y)$.

It must be noted that the filter $1/H(u, v)$ can only be obtained approximately since those spatial frequencies which have been degraded to zero in the original photograph (i.e., those spatial frequency values where $H(u, v) = 0$) cannot be restored. In addition, since $H(u, v)$ is, in general, a complex function, one may use holographic techniques to build the filter $1/H(u, v)$.

As an example let us consider the imaging of a point object by a camera and let us assume that the camera moves while imaging. The corresponding recorded image would be a line along the direction of motion (say along x) and let b represent the width of the line. The corresponding transmittance of the recorded pattern would be

$$g(x, y) = \begin{cases} A\delta(y); & |x| < b/2 \\ 0; & |x| > b/2 \end{cases} \tag{6.52}$$

The spectrum of $g(x, y)$ would be

$$G(u, v) = \int_{-\infty}^{\infty} dy \int_{-b/2}^{b/2} A\delta(y) e^{2\pi i(ux + vy)} dx$$

$$= A\frac{\sin \pi u b}{\pi u} \tag{6.53}$$

If the filter has a transmittance proportional to $T(u, v) = \pi u/\sin \pi u b$ then the filtered spectrum would be $G(u, v) T(u, v) = A$. A second lens can perform a second Fourier transform to give

$$f(x, y) = \mathscr{F}[A] = A\delta(x)\delta(y) \tag{6.54}$$

Notice that a filter with $T(u, v) = \pi u/\sin \pi u b$ cannot be realized exactly in practice since, for example, at $u = 1/b, 2/b$ etc. the filter transmittance has to be infinite.

Fig. 6.7 shows some results of image deblurring using spatial frequency techniques. The image blurring is due to defocussing and hence the corresponding filter should have a transmittance proportional to $W/J_1(W)$ where $W = ra/\lambda f$ where a corresponds to the radius of the defocussed image of a point. Although the deblurring is not perfect, it is significant. The deblurring operation described can be used for blurring caused by defocussing, aberrations of the optical system or even for imaging defects at microwave, ultrasonic or X-ray wavelengths.

7

Holography

7.1 Introduction

A photograph represents a two-dimensional recording of a three-dimensional scene. What is recorded is the intensity distribution that prevailed at the plane of the photograph when it was exposed. The light sensitive medium is sensitive only to the intensity variations and hence while recording a photograph, the phase distribution which prevailed at the plane of the photograph is lost. Since only the intensity pattern has been recorded, the three-dimensional character (e.g., parallax) of the object scene is lost. Thus one cannot change the perspective of the image in the photograph by viewing it from a different angle and one cannot refocus any unfocussed part of the image in the photograph. Holography is a method evolved by Gabor in 1948, in which one not only records the amplitude but also the phase of the light wave. Because of this the image produced by the technique of holography has a true three-dimensional form. Thus, as with the object, one can change one's position and view a different perspective of the image and one can focus at different distances. The capability to produce images as true as the object itself is what is responsible for the wide popularity gained by holography.

The basic technique in holography is the following: in the recording of the hologram, one superimposes on the object wave another wave called the reference wave (which is usually a plane wave) and the photographic plate is made to record the resulting interference pattern (see Fig. 7.1(a)). This recorded interference pattern forms the hologram and (as will be shown) contains information not only about the amplitude but also about the phase of the object wave. Unlike a photograph, the hologram bears little resemblance to the object; in fact information about the object is coded into the hologram. To view the image we illuminate the hologram with another wave, called the reconstruction wave (which in most cases is identical to the reference wave used during the formation of the hologram); this process is termed reconstruction (see Fig. 7.1(*b*) and (*c*)). The reconstruction process leads, in general, to a virtual and a real image of the object scene. The virtual image has all the characteristics of the object such as parallax etc. Thus one can move the position of the eye and look behind the object or one can focus at different distances. The real image can be photographed without the aid of lenses just by placing a light sensitive medium at the position where the real image is formed.

7.2   The basic principle

Fig. 7.1(*a*) shows a typical configuration used for recording a hologram. A portion of a coherent beam is allowed to be scattered by the object and the other portion is reflected by a mirror; the former corresponds to the object wave and the latter to the reference wave. The object wave and the reference wave are made to interfere and the resultant interference pattern is recorded on a photographic plate. Let

$$\psi_o(x, y) = A_o(x, y) e^{-i\phi_o(x,y)} \tag{7.1}$$

Fig. 7.1 (*a*) The recording of a hologram, (*b*) reconstruction with a wave identical to the reference wave and (*c*) reconstruction using a wave that is the conjugate of the reference wave.

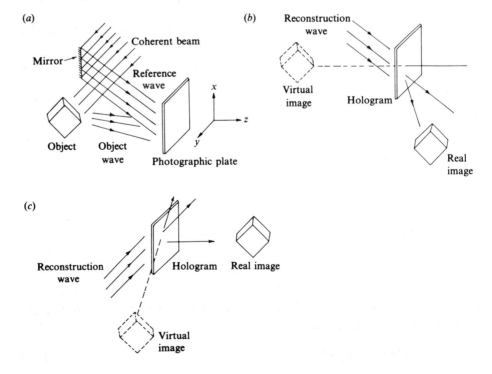

represent the field distribution of the object wave falling on the recording plate; here A_o and ϕ_o are the amplitude and phase distribution of the object wave. Similarly let

$$\psi_r(x, y) = A_r(x, y) e^{-i\phi_r(x,y)} \tag{7.2}$$

represent the field distribution of the reference wave on the plane of the recording medium; A_r and ϕ_r represent the corresponding amplitude and phase distribution. In Eqs. (7.1) and (7.2) we have assumed the plane of the recording medium to be the $z = 0$ plane. If the reference wave is a plane wave travelling in the x–z plane, then we have

$$\psi_r(x, y) = A_r e^{-ik_x x} \tag{7.3}$$

where k_x is the x-component of the propagation vector of the plane wave and A_r is a constant.

The object wave and the reference wave interfere on the recording medium which records the resultant intensity distribution given by

$$
\begin{aligned}
I(x, y) &= |\psi_o(x, y) + \psi_r(x, y)|^2 \\
&= A_o^2 + A_r^2 + A_r A_o e^{-i(\phi_o - \phi_r)} + A_r A_o e^{i(\phi_o - \phi_r)}
\end{aligned} \tag{7.4}
$$

or

$$I(x, y) = A_o^2 + A_r^2 + 2A_r A_o \cos(\phi_o - \phi_r) \tag{7.5}$$

As is obvious from Eq. (7.4) or (7.5) the recorded intensity pattern has also the phase of the object wave ϕ_o.

When the photographic plate which has recorded the intensity variation is developed one obtains a hologram of the object. The amplitude transmittance of the hologram i.e., the ratio of the transmitted amplitude to the incident field amplitude on the hologram is a function of $I(x, y)$. By a suitable developing process one can obtain a condition under which the amplitude transmittance is linearly related to $I(x, y)$. Thus we may write for the amplitude transmittance of the hologram

$$T(x, y) = I(x, y) \tag{7.6}$$

where we have omitted a constant of proportionality.

In order to reconstruct the object using the hologram, we illuminate it with a reconstruction wave as shown in Fig. 7.1(*b*). The reconstruction wave is in most cases identical to the reference wave and we assume the reconstruction wave to have a field distribution on the plane of the hologram (assumed to be $z = 0$) given by

$$\psi_c(x, y) = A_r(x, y) e^{-i\phi_r(x,y)} \tag{7.7}$$

Hence using Eqs. (7.4), (7.6) and (7.7), the field distribution of the transmitted wave on the plane of the hologram will be

$$\psi_t(x, y) = \psi_c(x, y) T(x, y)$$
$$= A_r(A_0^2 + A_r^2) e^{-i\phi_r(x,y)} + A_r^2 A_o e^{-i\phi_o} + A_r^2 A_o e^{i(\phi_o - 2\phi_r)} \quad (7.8)$$

We now assume the reference and reconstruction waves to be plane waves given by Eq. (7.3). If θ is the angle made by the waves with the z-axis then

$$k_x = k_o \sin \theta \quad (7.9)$$

Substituting from Eq. (7.3) in Eq. (7.8) we obtain

$$\psi_t(x, y) = A_r(A_0^2 + A_r^2) e^{-ik_x x} + A_r^2 A_o e^{-i\phi_o} + A_r^2 A_o e^{i\phi_o} e^{-2ik_x x} \quad (7.10)$$

We now consider the three terms on the RHS of Eq. (7.10). The first term represents the reconstruction wave itself but whose amplitude is modulated due to the term $A_o^2(x, y)$. The second term is identical within a constant multiplier of A_r^2 to the *object wave* which was present on the plane of the hologram when the hologram was recorded. If one views the object wave one would see a virtual image of the object exactly at the location where the object was placed during recording. In addition, since the original object wave itself has been reproduced, viewing the virtual image would have all the effects of depth and parallax. The last term is a conjugate of the object wave (due to $e^{i\phi_o}$ term) and is also phase modulated by a phase variation $e^{-2ik_x x}$. The conjugate wave would result in the reconstruction of a real image of the object since if the original wave is a diverging wave, the conjugate wave will be converging. The effect of the $e^{-2ik_x x}$ term is to rotate the direction of propagation of the conjugate wave (see Fig. 7.1(b)). It must be noted that in order to view the image of the object without overlap of the other waves, the three waves in Eq. (7.10) must be travelling along different directions. This was, in fact, the major problem in early holography due to the absence of sources with long coherence lengths and the arrival of the laser brought about a revival of holography in the early 1960s.

Example: As a specific example we consider the recording the reconstruction of a point object (see Fig. 7.2). We choose the plane of the recording medium to be the $z = 0$ plane. Let the point object be at a distance d from the recording medium and be lying on the z-axis. For such a case the object wave would be (cf. Eq. (5.14))

$$\psi_o(x, y) = \frac{A_o}{(x^2 + y^2 + d^2)^{\frac{1}{2}}} \exp\left[-ik_o(x^2 + y^2 + d^2)^{\frac{1}{2}}\right]$$
$$\approx \frac{A_o}{d} \exp(-ik_o d) \exp\left[-i\frac{k_o}{2d}(x^2 + y^2)\right] \quad (7.11)$$

where we have assumed $x, y \ll d$. We consider a reference wave of the form given by Eq. (7.3) with $k_x = k_0 \sin \theta$. Thus

$$\psi_r(x, y) = A_r e^{-ik_0 x \sin \theta} \qquad (7.12)$$

The transmittance of the recorded hologram will be

$$T(x, y) = |\psi_o(x, y) + \psi_r(x, y)|^2$$
$$= \frac{A_o^2}{d^2} + A_r^2 + \frac{2 A_r A_o}{d} \cos\left[k_0 d - k_0 x \sin \theta + \frac{k_0}{2d}(x^2 + y^2) \right] \qquad (7.13)$$

Fig. 7.2 (*a*) Recording of a hologram of a point object using an off-axis plane reference wave. (*b*) Reconstruction with a wave identical to the reference wave. The virtual image can be viewed without any interference with other waves.

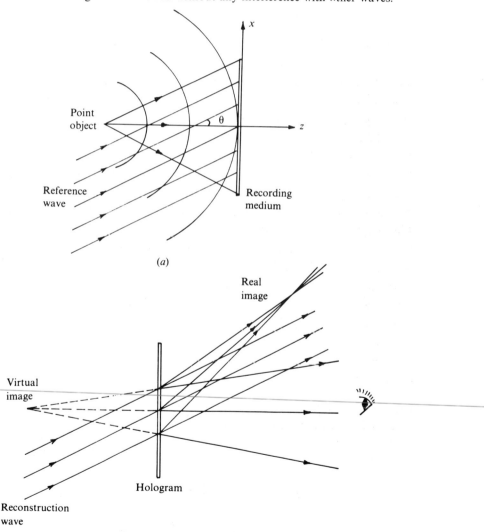

(*a*)

(*b*)

We first assume the reference wave to be normally incident on the recording medium so that $\theta = 0$ (see Fig. 7.3(a)). In such a case

$$T(x, y) = \left(\frac{A_o^2}{d^2} + A_r^2\right) + \frac{2A_rA_o}{d}\cos\left[k_od + \frac{k_0}{2d}(x^2 + y^2)\right] \tag{7.14}$$

As can be seen from Eq. (7.14), the recorded interference pattern is a set of circular fringes centred at $x = 0$, $y = 0$ and the radii of the bright fringes are given by

$$r_m = [(2d/k_0)2m\pi]^{\frac{1}{2}} = m^{\frac{1}{2}}(2\lambda d)^{\frac{1}{2}} \tag{7.15}$$

where we have omitted the constant phase factor k_0d. The radii of the fringes are proportional to the square roots of natural numbers and comparing with the zone plate discussed in Sec. 5.6 we see that the recorded pattern is nothing but a sinusoidal zone plate.

Fig. 7.3 (a) Recording of a hologram of a point object with an on-axis reference wave. (b) Reconstruction with a wave identical to the reference wave. The virtual image and the real image are formed on the z-axis and thus viewing any one of them one has the problem of a defocussed image of the other.

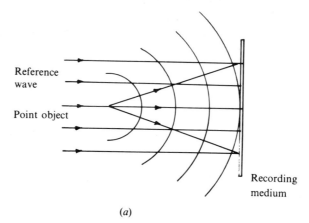

(a)

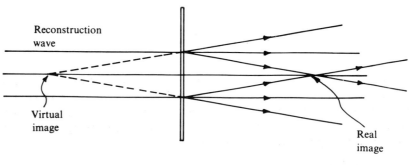

(b)

If the reconstruction is performed with the wave given by Eq. (7.12) with $\theta = 0$ (see Fig. 7.3(b)) we obtain for the transmitted wave on the hologram plane

$$\psi_t(x, y) = \psi_r(x, y) T(x, y)$$

$$= A_r\left(\frac{A_o^2}{d^2} + A_r^2\right) + \frac{A_r^2 A_o}{d}\exp\left(-ik_o d\right)\exp\left[-i\frac{k_o}{2d}(x^2 + y^2)\right]$$

$$+ \frac{A_r^2 A_o}{d}\exp\left(ik_o d\right)\exp\left[i\frac{k_o}{2d}(x^2 + y^2)\right] \tag{7.16}$$

The first term on the RHS is a plane wave propagating along the z-direction which is the reconstruction wave itself. The second term on the RHS is a diverging spherical wave of radius d and is the same as the object wave given by Eq. (7.11) (except for a constant multiplicative factor). The last term is a converging spherical wave of radius of curvature d and represents the conjugate wave. This wave would converge at a distance d from the hologram plane and form the real image (see Fig. 7.3(b)). Observe that in this case when the reference and reconstruction wave are travelling almost in the same direction as the object wave, the three reconstructed waves in Eq. (7.16) overlap and thus while viewing the virtual image one has the problem of a defocussed real image or vice versa (see Fig. 7.3(b)).

If $\theta \neq 0$, i.e., the reference wave falls obliquely on the recording medium, then Eq. (7.13) can be written as

$$T(x, y) = \frac{A_o^2}{d^2} + A_r^2 + \frac{2A_r A_o}{d}\cos\left\{k_o d - \frac{k_o d \sin^2\theta}{2} + \frac{k_o}{2d}[(x - d\sin\theta)^2 + y^2]\right\} \tag{7.17}$$

The resulting fringes are again circular but are now centred at

$$x = d\sin\theta, \quad y = 0 \tag{7.18}$$

If we now use a reconstruction wave given by Eq. (7.12), then the transmitted field at the plane of the hologram would be

$$\psi_t(x, y) = \left(\frac{A_o^2}{d^2} + A_r^2\right) A_r \exp\left(-ik_0 x \sin\theta\right)$$

$$+ \frac{A_r A_o}{d}\exp\left(-ik_o d\right)\exp\left[-i\frac{k_o}{2d}(x^2 + y^2)\right]$$

$$+ \frac{A_r A_o}{d}\exp\left(ik_o d\right)\exp\left[-2ik_0 x\sin\theta + \frac{ik_o}{2d}(x^2 + y^2)\right] \tag{7.19}$$

The first term on the RHS of the above equation propagates along the same direction as the reconstruction wave. The second term leads to a virtual image of the object point since it represents a diverging spherical wave and is the same as in Eq. (7.11). The last term can be shown to lead to a converging spherical wave converging to a point with coordinates

$$x = 2d\sin\theta, \quad y = 0, \quad z = d \tag{7.20}$$

Thus for sufficiently large θ, the three waves are propagating along different directions (see Fig. 7.2(*b*)) and thus the virtual image can be viewed without any interference. This leads to what is known as an off-axis hologram.

We should mention here that if we consider only a portion of the hologram discussed above, the reconstructed spherical wave will have a smaller transverse dimension. This results in a poorer resolution. Thus if a hologram is formed by a diffusing object, every portion of the hologram contains information about the complete object.

Problem 7.1 Consider the reconstruction by a wave which is the conjugate of the reference wave, i.e.,

$$\psi_c = \psi_r^* = A_r e^{ik_0 x \sin \theta} \tag{7.21}$$

Show that the real image is formed at $(0, 0, d)$ and the virtual image at $(2d \sin \theta, 0, d)$.

Problem 7.2 Consider a hologram formed by a set of inclined plane waves. Obtain the transmittance of the hologram. Discuss the reconstruction using another plane reconstruction wave. Does the hologram show any resemblance to the diffraction grating discussed in Sec. 4.10?

7.3   Coherence requirements

Since holography essentially involves an interference phenomenon, certain coherence requirements must be met. Thus the beam illuminating the object and from which a portion is taken as the reference beam (see Fig. 7.1) must be spatially coherent over the whole region. In addition, if stable interference fringes are to be formed (so that they are recordable) the maximum path difference between the object wave and the reference wave must be less than the coherence length.

During reconstruction if the original object wave is to be reconstructed with minimum aberration, then the reconstruction wave must be identical to the recording wave. For maximum resolution, the reconstructing light source must be as coherent as the recording source.

Another critical requirement in making holograms is the stability of the whole recording arrangement. Thus, the film, the object and any mirrors used in the recording process must be motionless with respect to one another during the exposure.

7.4   Resolution

The resolution of the reconstructed image is determined by the size of the hologram. The size of the hologram may be less than the actual size of the recording medium because of the resolution limit of the recording medium, i.e., the inability of the recording medium to record interference fringes with a spatial frequency greater than a certain maximum say l_m. If we recall Eq. (7.14) which describes the recording of the interference pattern of a point object, we see that the spacing between the interference fringes is a function of x (see also Eq. (7.15)). In order to define a local spatial frequency of the interference fringes we note that a variation of the form

$$f(x) = A \cos \alpha x \tag{7.22}$$

has a spatial frequency $\alpha/2\pi$ which can also be written as

$$\frac{1}{2\pi} \frac{d}{dx}(\alpha x)$$

Thus for a variation of the form

$$f(x) = A \cos \gamma(x) \tag{7.23}$$

we may define a local spatial frequency as

$$\beta = \frac{1}{2\pi} \left| \frac{d}{dx} \gamma(x) \right| \tag{7.24}$$

Hence for the interference pattern given by Eq. (7.14), we obtain for the local spatial frequency along x

$$\beta(x) = \frac{1}{2\pi} \left| \frac{d}{dx} \left(\frac{k_0}{2d} x^2 \right) \right|$$
$$= |x|/\lambda_0 d \tag{7.25}$$

Hence as x increases the spatial frequency of the fringe increases. If the maximum spatial frequency that the hologram can record is l_m, then the maximum radius of the recorded interference pattern would be

$$a \approx \lambda_0 d l_m \tag{7.26}$$

Thus during reconstruction one can imagine the hologram to be equivalent to a lens of diameter $2a$ forming an image at a distance d. Thus we may, in accordance with Rayleigh's resolution criterion, write *the limit of resolution* as

$$\Delta = \frac{1.22\lambda_0}{2a} d = \frac{0.61}{l_m} \tag{7.27}$$

Obviously if the size of the hologram is less than that given by Eq. (7.27) the resolution will be determined by the size of the hologram.

7.5   Fourier transform holograms

In the recording configuration discussed above the recording medium is assumed to be at a finite distance from the object and hence the hologram records the interference between the Fresnel diffracted object wave and the reference wave. We shall now discuss Fourier transform holograms in which one records the interference between the Fourier transform of the object and a reference wave. Such holograms find applications in spatial frequency filtering, character recognition (see Sec. 7.7.3) etc.

Fig. 7.4 shows an arrangement used for recording a Fourier transform hologram of an object transparency. The object $f(x, y)$ is placed in the front

Fig. 7.4 (*a*) Recording of a Fourier transform hologram of an object transparency. (*b*) Reconstruction.

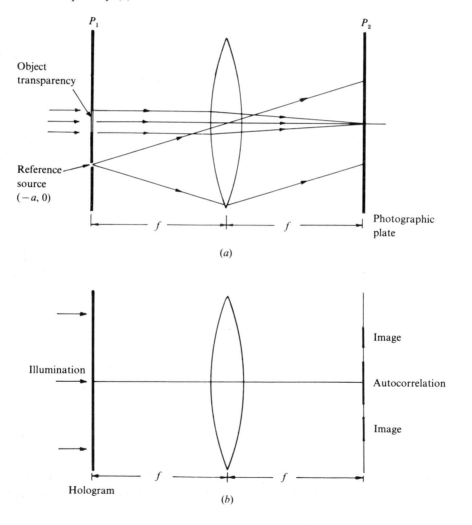

focal plane of a lens and the reference wave is provided by a point source also placed in the front focal plane. We have shown in Sec. 6.3 that on the back focal plane, the field distribution is the Fourier transform of the field distribution in the front focal plane. Using Eq. (6.1), we write the field produced on the plane P_2 due to $f(x, y)$ as

$$F(u, v) = (1/\lambda f) \int \int f(x', y') e^{2\pi i (ux' + vy')} \, dx' \, dy' \tag{7.28}$$

where $u = x/\lambda f$ and $v = y/\lambda f$ and x and y are measured in the plane P_2. The point reference source located at $x = -a$, $y = 0$ may be described by

$$g(x, y) = A\delta(x + a)\delta(y) \tag{7.29}$$

so that the reference wave on the back focal plane would be

$$G(u, v) = (A/\lambda f) \int \int \delta(x' + a)\delta(y') e^{2\pi i (ux' + vy')} \, dx' \, dy'$$

$$= (A/\lambda f) e^{-2\pi i ua} \tag{7.30}$$

which is nothing but an inclined plane wave as indeed it should be. Hence the recorded intensity pattern would be

$$I(x, y) = |F(u, v) + G(u, v)|^2$$

$$= \frac{A^2}{\lambda^2 f^2} + |F(u, v)|^2 + \frac{A}{\lambda f} F(u, v) e^{2\pi i ua} + \frac{A}{\lambda f} F^*(u, v) e^{-2\pi i ua} \tag{7.31}$$

Assuming the amplitude transmittance of the hologram to be proportional to $I(x, y)$, the transmittance of the hologram will again be given by Eq. (7.31). It may be noted that in the absence of a reference wave, one would have recorded $|F(u, v)|^2$ thus losing all information of phase. The presence of the reference wave helps in recording the complex Fourier transform $F(u, v)$ itself.

In the reconstruction, the hologram is placed on the front focal plane of a lens and illuminated by a parallel beam; the reconstructed image is viewed on the back focal plane of the lens (see Fig. 7.4(b)). The reconstruction lens takes a further Fourier transform of the function given in Eq. (7.31). Thus in the back focal plane we would find an amplitude distribution

$$h(x, y) = (1/\lambda f) \int \int I(x', y') e^{2\pi i (ux' + vy')} \, dx' \, dy' \tag{7.32}$$

where again $u = x/\lambda f$, $v = y/\lambda f$ and x and y are now measured in the back focal plane of the lens. Substituting from Eq. (7.31) and using the properties of Fourier transforms we obtain[†]

$$h(x, y) = \frac{A^2}{\lambda f} \delta(x)\delta(y) + \frac{1}{\lambda f} f(x, y) \circledast f(x, y)$$

$$+ \frac{A}{\lambda f} f(-x-a, -y) + \frac{A}{\lambda f} f^*(x-a, y) \qquad (7.33)$$

where $\circledast$ represents correlation defined by

$$f(x) \circledast g(x) = \int f(x')g^*(x+x')\,dx' \qquad (7.34)$$

The first term on the RHS of Eq. (7.33) produces a bright spot on the axis, the second term is the autocorrelation of $f(x, y)$ and is also centred on the axis. Third and last terms represent the reconstructed images and are centred at $x = -a$ and $x = a$ respectively (see Fig. 7.4(b)).

In order that the various reconstructed images be resolvable, we note that the autocorrelation of a function of width b has a dimension $2b$. For example, the autocorrelation of a rectangular function of width b is triangular with a base $2b$. Thus if b is the width of the object transparency then in order that the images be separate we must have

$$a \geqslant 3b/2 \qquad (7.35)$$

In order to understand the resolution in Fourier transform holograms we consider the object to be a point located in the front focal plane. The object wave for a point object will be a plane wave like the reference wave (Eq. (7.30)). Thus the hologram will be formed between a pair of inclined plane waves and the recorded hologram will consist of constant spatial frequency fringes. Unlike the Fresnel holograms discussed in Sec. 7.2, in the Fourier transform hologram every object point produces a constant spatial frequency fringe system and hence can be recorded over a large area. The finite cut-off frequency of the recording medium will only limit the field of view of the recorded object. For small objects, the Fourier transform hologram configuration is capable of producing a low frequency fringe system over a large aperture and thus images produced using this configuration have higher resolution.

Problem 7.3: Consider the hologram recording geometry shown in Fig. 7.5 (also referred to as a lensless Fourier transform hologram) in which a planar transparency $g(x, y)$ and a point reference source are placed at the same distance from the

---

[†] For details of the analysis see e.g., Ghatak and Thyagarajan (1978).

recording medium. Using Eq. (5.5) obtain the object and reference waves on the recording medium and calculate the transmittance of the hologram. Show that when such a hologram is reconstructed as a Fourier transform hologram, on the back focal plane we will obtain the reconstructed object (see, e.g., Ghatak and Thyagarajan (1978)).

Fig. 7.5 Recording of a lensless Fourier transform hologram.

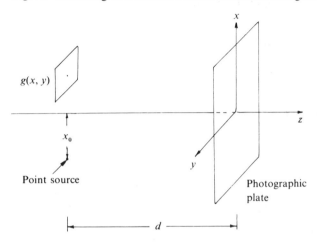

Fig. 7.6 A typical configuration for the recording of a volume reflection hologram.

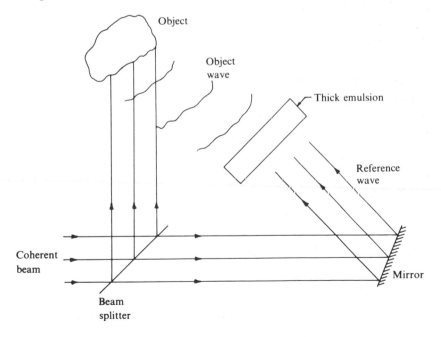

7.6   Volume holograms

In all the above discussions we have not considered the finite thickness of the recording medium. A hologram can be considered to be a plane hologram when the fringe spacing is very large compared to the thickness of the recording medium. If this is not the case then the whole volume of the recording medium takes part in the reconstruction process and finite thickness effects cannot be neglected. These are called volume holograms. The distinction between plane and volume holograms is very similar to that between Raman–Nath and Bragg acoustooptic diffractions (see Chapters 17 and 18); the former is essentially a thin phase grating and the latter a thick grating.

Fig. 7.6 shows a typical geometry used for recording one type of volume hologram, namely a volume reflection hologram. In this case the object wave scattered by the object and the reference wave are travelling in almost opposite directions through the recording medium. The resulting interference fringes are recorded in the depth of the recording medium and are almost parallel to the surface of the recording medium. The spacing between the fringes will be $\sim \lambda_0/2n_0$ where n_0 is the refractive index of the recording medium and λ_0 is the free space wavelength of the light wave used for recording. Such a recorded hologram becomes highly wavelength selective during reconstruction since constructive interference among the waves reflected by adjacent layers will occur only for an incident reconstruction wavelength λ_0. Hence such holograms can be reconstructed using white light and are also sometimes referred to as white light holograms. In Sec. 18.6.1 we have discussed the specific case of a reflection phase hologram and have shown the wavelength selectivity of such holograms.

7.7   Some applications

There are many diverse applications of holography. Here we shall discuss some of these applications.

7.7.1   *Microscopy*

The ability of holography to record information about depth finds application in studying transient microscopic events. Thus, if one has to study some transient phenomenon which occurs in a certain volume, then using ordinary microscopic techniques it becomes difficult first to locate the position and then make an observation. If a hologram of the scene is recorded, then the event is frozen into the hologram and hence one can focus through the depth of the reconstructed image and study the phenomenon at leisure. Holographic microscopy studies have been used in droplet and particle size analyses in cloud chambers, rocket engine exhaust, aerosols etc.

7.7.2 *Interferometry*

The most extensive use of holography appears to be in the field of nondestructive testing using interferometric methods. The ability of the holographic process to release the object wave when illuminated with a reconstruction wave allows us to have interference between different waves which exist at different times. Thus, in the technique called double exposure holographic interferometry, the photographic plate is first partially exposed to the object wave and the reference wave. Then, the object is stressed and the photographic plate is exposed again with the same reference wave. The photographic plate after development forms the hologram. When this hologram is illuminated with a reconstruction wave, then two object waves emerge from the hologram; one of them corresponds to the unstressed object and the other to the stressed object. Since the object waves themselves have been reconstructed, they interfere and produce interference fringes. These interference fringes are characteristic of the strain suffered by the body. A quantitative study of the fringe pattern produced in the body gives the distribution of strain in the object.

To understand the formation of the fringe pattern, we assume that the deformation of the object has been such as to alter only the phase distribution. Thus, let

$$\psi_o(x, y) = A_o(x, y)\,e^{-i\phi_o(x,y)} \tag{7.36}$$

and

$$\psi'_o(x, y) = A_o(x, y)\,e^{-i\phi'_o(x,y)} \tag{7.37}$$

represent the object waves before and after the object is stressed; here we are assuming that the only effect of the stress is to alter the phase distribution. On reconstruction the two object waves ψ_o and ψ'_o emerge together from the hologram and what one observes will be the intensity pattern due to interference between the two waves, which is given by

$$\begin{aligned}
I(x, y) &= |\psi_o(x, y) + \psi'_o(x, y)|^2 \\
&= 2A_o^2\{1 + \cos[\phi'_o(x, y) - \phi_o(x, y)]\}
\end{aligned} \tag{7.38}$$

Thus, whenever

$$\phi'_o - \phi_o = 2m\pi; \quad m = 0, 1, 2, \ldots \tag{7.39}$$

there will be constructive interference and whenever

$$\phi'_0 - \phi_0 = (2m + 1)\pi; \quad m = 0, 1, 2, \ldots \tag{7.40}$$

there will be destructive interference. Hence depending on $(\phi_o' - \phi_o)$ one will obtain on reconstruction, the object superimposed with bright and dark fringes. Fig. 7.7 shows a typical example of a double exposure hologram of a bullet in flight. The shock waves produced by the bullet in flight result in the interference fringes shown. Double exposure holography has been used in testing tyres, strain measurement, gas flow analysis etc.

In a variant of the above technique referred to as real time interferometry, instead of exposing the hologram twice, a hologram is first formed with the unstrained object. After development the hologram is replaced in exactly the same position. In such a case the virtual image produced by the hologram superimposes on the object and one receives the wave from the object as well as the reconstructed object wave. If the object now undergoes any strain, the interference between the object wave and the reconstructed wave from the hologram will produce fringes. Any real time motion of the object can also be viewed using such a technique.

Holographic methods also find application in nondestructive testing of vibrating objects. In this case the hologram of the object is recorded when the object undergoes vibration and completes many vibration periods during exposure. In such a case the reconstructed image is seen to contain fringes of equal amplitude (see Fig. 7.8). Such a method can be used in the study of turbine blades, musical instruments etc.

Fig. 7.7 Reconstruction from a double exposure hologram of a bullet in flight (after Weurker (1971)).

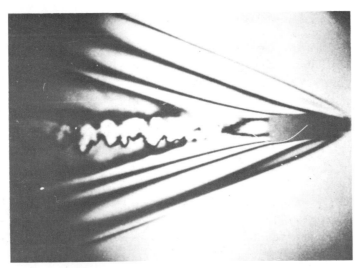

7.7.3   Character recognition

It is sometimes necessary to detect the presence of particular characters in an optical image. This is known as the character recognition problem. Such problems essentially involve cross correlation of the required character and the image. At the positions corresponding to the required character in the image one will obtain a strong correlation which will result in a bright spot of light.

Let $f(x, y)$ represent the function to be correlated with the character $h(x, y)$. Thus we are interested in performing the following operation

$$\phi = \int\int f(x', y')h^*(x + x', y + y')\,dx'\,dy'$$

$$= f(x, y) \circledast h(x, y) \tag{7.41}$$

where $\circledast$ stands for correlation. Consider the function

$$\Phi = \tilde{F}(u, v)\tilde{H}^*(u, v) \tag{7.42}$$

where $\tilde{F} = \mathscr{F}[f]$ and $\tilde{H} = \mathscr{F}[h]$. Now

$$\mathscr{F}[\Phi] = \int\int \tilde{F}(u, v)\tilde{H}^*(u, v)\,e^{2\pi i(ux + vy)}\,du\,dv$$

$$= \int\int \tilde{F}(u, v)\left(\int\int h^*(\xi, \eta)\,e^{-2\pi i(u\xi + v\eta)}\,d\xi\,d\eta\right)e^{2\pi i(ux + vy)}\,du\,dv$$

$$= \int\int h^*(\xi, \eta)\int\int \tilde{F}(u, v)e^{-2\pi i[u(\xi - x) + v(\eta - y)]}\,du\,dv\,d\xi\,d\eta$$

$$= \int\int h^*(\xi, \eta)f(\xi - x, \eta - y)\,d\xi\,d\eta$$

$$= \int\int f(x', y')h^*(x + x', y + y')\,dx'\,dy'$$

Thus the Fourier transform of $\tilde{F}(u, v)\tilde{H}^*(u, v)$ is the required cross correlation between $f(x, y)$ and $h(x, y)$. We have seen in Chapter 6 that a lens can perform a Fourier transformation. Thus if a field distribution of the form $\tilde{F}(u, v)\tilde{H}^*(u, v)$ can be produced on the front focal plane of a lens, then on its back focal plane one will obtain the cross correlation. This can be achieved as follows: first one fabricates a Fourier transform hologram of $h(x, y)$ using the procedure described in Sec. 7.5. Thus using a reference wave, as given by Eq. (7.30), the transmittance of the hologram will be given by (see Eq. 7.31)

$$T(x, y) = \frac{A^2}{\lambda^2 f^2} + |H(u, v)|^2 + \frac{A}{\lambda f} H(u, v) e^{2\pi i u a} + \frac{A}{\lambda f} H^*(u, v) e^{-2\pi i u a}$$

$$(7.43)$$

with

$$H(u, v) = (1/\lambda f) \int \int h(x, y) e^{2\pi i (ux + vy)} \, dx \, dy \qquad (7.44)$$

Eq. (7.43) contains a term proportional to the required term $H^*(u, v)$. In order to perform the cross correlation, we use a set up such as that shown in Fig. 6.1. A transparency with a transmittance proportional to $f(x, y)$ is kept on plane P_1 and the hologram whose transmittance is proportional to $T(x, y)$ given by Eq. (7.43) is placed on plane P_2. Lens L_1 takes a Fourier transform of $f(x, y)$ and since the transmittance of the filter placed on plane P_2 is $T(x, y)$, the effective field in the front focal plane of lens L_2 will be

$$g(x, y) = F(u, v) T(x, y)$$

$$= \frac{A^2}{\lambda^2 f^2} F(u, v) + F(u, v) |H(u, v)|^2 + \frac{A}{\lambda f} F(u, v) H(u, v) e^{2\pi i u a}$$

$$+ \frac{A}{\lambda f} F(u, v) H^*(u, v) e^{-2\pi i u a} \qquad (7.45)$$

Lens L_2 will take a Fourier transform of $g(x, y)$. The interesting terms are the third and the fourth terms on the RHS. In particular

Fig. 7.8 Reconstruction from a time averaged hologram of a can top vibrating in (a) the lowest resonance frequency and (b) the second resonance frequency. Fringes corresponds to loci of constant amplitude of vibration (after Powell and Stetson (1965)).

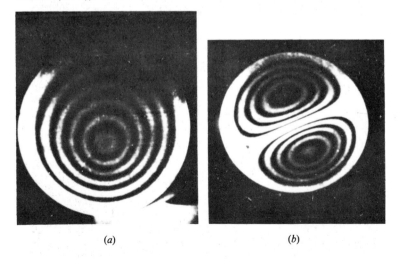

(a)   (b)

$$\mathscr{F}[F(u,v)H(u,v)\,\mathrm{e}^{2\pi i u a}]$$

$$= \int\int f(x',y')h(-x-a-x',-y-y')\,\mathrm{d}x'\,\mathrm{d}y' \tag{7.46}$$

which is the convolution between $f(x,y)$ and $h(x,y)$; the convolution is centred at $x = -a$, $y = 0$ and is also inverted. Similarly corresponding to the last term in Eq. (7.45) we would obtain

$$\mathscr{F}[F(u,v)H^*(u,v)\,\mathrm{e}^{-2\pi i u a}]$$

$$= \int\int f(x',y')h^*(x-a+x',y+y')\,\mathrm{d}x'\,\mathrm{d}y' \tag{7.47}$$

Fig. 7.9 Fingerprint identification using optical cross correlation techniques with a holographically produced spatial filter. A good correlation results in a bright concentrated spot of light (after Tsujiuchi *et al.* (1971)).

which is nothing but the cross correlation between $f(x, y)$ and $h(x, y)$; the cross correlation is centred at $x = a$, $y = 0$. It can also be shown that the first two terms on the RHS in Eq. (7.45) lead to distributions centred at $x = 0$, $y = 0$.

Thus on plane P_3 of Fig. 6.1 we will obtain a correlation of the object $f(x, y)$ with the character $h(x, y)$, and the correlation would be off-axis, centred at $x = a$, $y = 0$. In the region of correlation, wherever the character $h(x, y)$ is present in the original image one will obtain bright dots which is just a manifestation of high correlation. Fig. 7.9 shows the application of the technique to fingerprint identification; in the upper picture the suspect's fingerprint matches that at the scene of the crime, in the lower one it does not (Tsujiuchi, Matsuda and Takeya, 1971).

8

Lasers: I

8.1   Introduction

LASER is an acronym for Light Amplification by Stimulated Emission of Radiation. As the name implies, in a laser, the process of stimulated emission is used for amplifying light waves. It was as early as 1917 that Einstein first predicted the existence of two different kinds of processes by which an atom can emit radiation; these are called spontaneous and stimulated emissions. The fact that the stimulated emission process could be used in the construction of coherent optical sources was first put forward by Townes and Schawlow in the USA and Basov and Prochorov in the USSR. And finally in 1960 Maiman demonstrated the first laser. Since then the development of lasers has been extremely rapid and laser action has been demonstrated with gases, solids, liquids, free electrons, semiconductors etc.

The three main components of any laser are the amplifying medium, the pump and the optical resonator. The amplifying medium consists of a collection of atoms, molecules or ions which act as an amplifier for light waves. Under normal conditions, the number of atoms in the lower energy state is always larger than the number in the excited energy state; as such, a light wave passing through such a collection of atoms would cause more absorptions than emissions and therefore the wave will be attenuated. Thus in order to have amplification, it is necessary to have population inversion (between two atomic states) in which there is a large number of atoms in the higher energy state as compared to that in the lower energy state. When a wave passes through a collection of atoms which are in a state of population inversion, the wave will induce more emissions and will be amplified. The pump is the source of energy which maintains the medium in this population inverted state. The optical resonator which consists of a pair of mirrors facing each other provides optical feedback to the amplifier so that it can act as a source of radiation.

In this chapter we first obtain relationships between the absorption and emission processes and then show that in order to use a collection of atoms to amplify electromagnetic radiation, one must create a state of population inversion (see Sec. 8.3). In Sec. 8.5 we discuss three and four level laser systems. Finally in Sec. 8.8 we discuss the various phenomena which lead to broadening of spectral lines.

In the next chapter we shall discuss in detail, optical resonators and techniques for obtaining single transverse or single longitudinal mode oscillation and for obtaining short intense pulses of light using Q-switching and mode locking. In chapter 10 we shall briefly discuss some important laser systems.

8.2   The Einstein coefficients

We consider two levels of an atomic system as shown in Fig. 8.1 and let N_1 and N_2 be the number of atoms per unit volume present in the energy levels E_1 and E_2 respectively. If radiation at a frequency corresponding to the energy difference $(E_2 - E_1)$ falls on the atomic system, it can interact in three distinct ways:

(*a*) An atom in the lower energy level E_1 can absorb the incident radiation and be excited to E_2. This excitation process requires the presence of radiation. The rate at which absorption takes place from level 1 to level 2 will be proportional to the number of atoms present in the level E_1 and also to the energy density of the radiation at the frequency $\omega = (E_2 - E_1)/\hbar$. Thus if $u(\omega)\,d\omega$ represents the radiation energy per unit volume between ω and $\omega + d\omega$ then we may write the number of atoms undergoing absorptions per unit time per unit volume from level 1 to level 2 as

$$\Gamma_{12} = B_{12}u(\omega)N_1 \tag{8.1}$$

where B_{12} is a constant of proportionality and depends on the energy levels E_1 and E_2. Notice here that $u(\omega)$ has the units of energy density per frequency interval.

(*b*) For the reverse process, namely the deexcitation of the atom from E_2

Fig. 8.1 Two states of an atom with energies E_1 and E_2; their corresponding population densities are N_1 and N_2 respectively. At thermal equilibrium $N_2 < N_1$ and $N_2/N_1 = e^{-(E_2 - E_1)/k_B T}$.

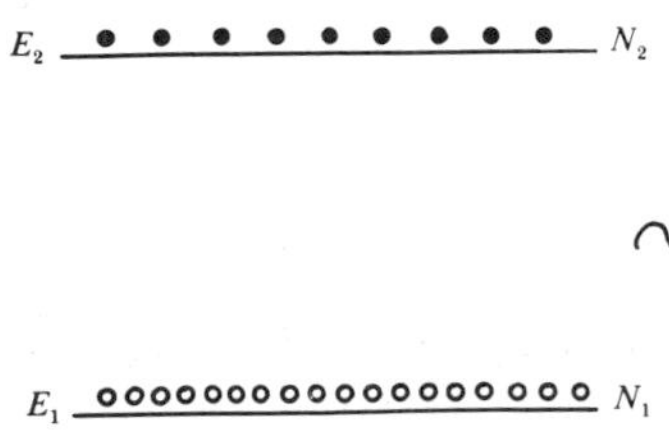

to E_1, Einstein postulated that an atom can make a transition from E_2 to E_1 through two distinct processes, namely stimulated emission and spontaneous emission. In the case of stimulated emission, the radiation which is incident on the atom stimulates it to emit radiation and the rate of transition to the lower energy level is proportional to the energy density of radiation at the frequency ω. Thus, the number of stimulated emissions per unit time per unit volume will be

$$\Gamma_{21} = B_{21}u(\omega)N_2 \tag{8.2}$$

where B_{21} is the coefficient of proportionality and depends on the energy levels.

(c) An atom which is in the upper energy level E_2 can also make a spontaneous emission; this rate will be proportional to N_2 only and thus we have for the number of atoms making spontaneous emissions per unit time per unit volume

$$U_{21} = A_{21}N_2 \tag{8.3}$$

At thermal equilibrium between the atomic system and the radiation field, the number of upward transitions must be equal to the number of downward transitions. Hence, at thermal equilibrium

$$N_1 B_{12}u(\omega) = N_2 A_{21} + N_2 B_{21}u(\omega)$$

or

$$u(\omega) = \frac{A_{21}}{(N_1/N_2)B_{12} - B_{21}} \tag{8.4}$$

Using Boltzmann's law, the ratio of the equilibrium populations of levels 1 and 2 at temperature T is

$$N_1/N_2 = e^{(E_2 - E_1)/k_B T} = e^{\hbar\omega/k_B T} \tag{8.5}$$

where k_B ($= 1.38 \times 10^{-23}\,\mathrm{J/K}$) is the Boltzmann's constant. Hence

$$u(\omega) = \frac{A_{21}}{B_{12}e^{\hbar\omega/k_B T} - B_{21}} \tag{8.6}$$

Now according to Planck's law, the radiation energy density per unit frequency interval is given by

$$u(\omega) = \frac{\hbar\omega^3 n_0^3}{\pi^2 c^3} \frac{1}{e^{\hbar\omega/k_B T} - 1} \tag{8.7}$$

where c is the velocity of light in free space and n_0 is the refractive index of the medium.

Comparing Eqs. (8.6) and (8.7), we obtain

$$B_{12} = B_{21} = B \tag{8.8}$$

and

$$A_{21}/B_{21} = \hbar\omega^3 n_0^3/\pi^2 c^3 \tag{8.9}$$

Thus the stimulated emission rate per atom is the same as the absorption rate per atom and the ratio of spontaneous to stimulated emission coefficients is given by Eq. (8.9). The coefficients A and B are referred to as the Einstein A and B coefficients.

At thermal equilibrium, the ratio of the number of spontaneous to stimulated emissions is given by

$$R = A_{21}N_2/B_{21}N_2 u(\omega) = e^{\hbar\omega/k_B T} - 1 \tag{8.10}$$

Thus at thermal equilibrium at a temperature T, for frequencies $\omega \gg k_B T/\hbar$, the number of spontaneous emissions far exceeds the number of stimulated emissions.

Example: Let us consider an optical source at $T = 1000$ K. At this temperature

$$\frac{k_B T}{\hbar} = \frac{1.38 \times 10^{-23}(\text{J/K}) \times 10^3(\text{K})}{1.054 \times 10^{-34}(\text{J s})} \approx 1.3 \times 10^{14}\,\text{Hz}$$

Thus for $\omega \gg 1.3 \times 10^{14}$ Hz, the radiation would be mostly due to spontaneous emissions. For $\lambda \simeq 5000$ Å, $\omega \approx 3.8 \times 10^{15}$ Hz and

$$R \approx e^{29.2} \approx 5.0 \times 10^{12}$$

Thus at optical frequencies the emission is predominantly due to spontaneous transitions and hence the light from usual light sources is incoherent.

We shall now obtain the relationship between the Einstein A coefficient and the spontaneous lifetime of level 2. Let us assume that an atom in level 2 can make a spontaneous transition only to level 1. Then since the number of atoms making spontaneous transitions per unit time per unit volume is $A_{21}N_2$, we may write the rate of change of population of level 2 with time due to spontaneous emission as

$$dN_2/dt = -A_{21}N_2 \tag{8.11}$$

the solution of which is

$$N_2(t) = N_2(0)e^{-A_{21}t} \tag{8.12}$$

Thus the population of level 2 reduces by $1/e$ in a time $t_{sp} = 1/A_{21}$ which is called the spontaneous lifetime associated with the transition $2 \to 1$.

Example: In the 2P $\to$ 1S transition in the hydrogen atom, the lifetime of the 2P state for spontaneous emission is given by (see, e.g., Thyagarajan and Ghatak (1981), Sec. 3.3):

$$t_{sp} = \frac{1}{A_{21}} \approx 1.6 \times 10^{-9}\,\text{s}$$

Thus

$$A_{21} \approx 6 \times 10^8\,\text{s}^{-1}$$

The frequency of the transition is given by

$$\omega \approx 1.55 \times 10^{16}\,\text{s}^{-1} \qquad (\hbar\omega \approx 10.2\,\text{eV})$$

Thus

$$B_{21} = \frac{\pi^2 c^3}{\hbar\omega^3 n_0^3} A_{21} \approx 4.1 \times 10^{20}\,\text{m}^3/\text{Js}^2$$

where we have assumed $n_0 \approx 1$. (Note the unit for B_{21}.)

Now, if one observes the spectrum of the radiation due to the spontaneous emission from a collection of atoms, one finds that the radiation is not strictly monochromatic but is spread over a certain frequency range. Similarly, if one measures the absorption by a collection of atoms as a function of frequency, one again finds that the atoms are capable of

Fig. 8.2 (*a*) Because of the finite lifetime of a state each state has a certain width $\Delta E (= \hbar\Delta\omega)$ so that the atom can absorb/emit radiation over a range of frequencies $\Delta\omega$ which is usually much less than ω. The numbers shown in the figure correspond to the 2P $\to$ 1S transition in the hydrogen atom. The corresponding lineshape function is shown in (*b*).

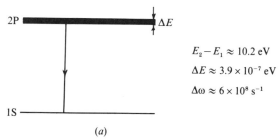

$E_2 - E_1 \approx 10.2\,\text{eV}$

$\Delta E \approx 3.9 \times 10^{-7}\,\text{eV}$

$\Delta\omega \approx 6 \times 10^8\,\text{s}^{-1}$

(*a*)

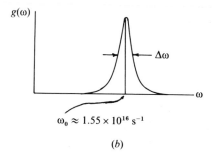

$\omega_0 \approx 1.55 \times 10^{16}\,\text{s}^{-1}$

(*b*)

absorbing not just a single frequency but radiation over a band of frequencies. This implies that energy levels have widths and the atoms can interact with radiation over a range of frequencies but the strength of interaction is a function of frequency (see Fig. 8.2). This function is called the lineshape function and is represented by $g(\omega)$. The function is usually normalized according to

$$\int g(\omega) \, d\omega = 1 \tag{8.13}$$

Explicit expressions for $g(\omega)$ will be obtained in Sec. 8.8.

From the above we may say that out of the total N_2 and N_1 atoms per unit volume, only $N_2 g(\omega) \, d\omega$ and $N_1 g(\omega) \, d\omega$ atoms per unit volume will be capable of interacting with radiation of frequency lying between ω and $\omega + d\omega$. Hence the number of stimulated emissions per unit time per unit

Fig. 8.3 (a) Atoms characterized by the line shape function $g(\omega)$ interacting with broadband radiation. (b) Atoms interacting with near-monochromatic radiation at $\omega = \omega'$.

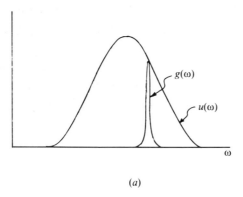

(a)

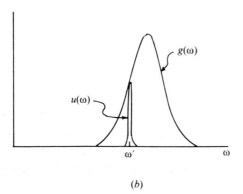

(b)

volume will now be given by

$$\Gamma_{21} = \int B_{21} u(\omega) N_2 g(\omega) \, d\omega$$

$$= N_2 \frac{\pi^2 c^3}{\hbar n_0^3 t_{sp}} \int \frac{u(\omega) g(\omega)}{\omega^3} \, d\omega \qquad (8.14)$$

where we have used Eq. (8.9) and $A_{21} = 1/t_{sp}$.

We now consider two specific cases.

(1) If the atoms are interacting with radiation whose spectrum is very broad compared to that of $g(\omega)$ (see Fig. 8.3(*a*)), then one may essentially assume that over the region of integration where $g(\omega)$ is appreciable $u(\omega)/\omega^3$ is essentially constant and thus may be taken out of the integral in Eq. (8.14). Using the normalization integral, Eq. (8.14) becomes

$$\Gamma_{21} = N_2 \frac{\pi^2 c^3}{\hbar \omega^3 n_0^3 t_{sp}} u(\omega) \qquad (8.15)$$

where ω now represents the transition frequency. Eq. (8.15) is consistent with Eq. (8.2) if we use Eq. (8.9) for B_{21}. Thus Eq. (8.15) represents the rate of stimulated emission per unit volume when the atom interacts with broadband radiation.

(2) We now consider the other extreme case in which the atom is interacting with near-monochromatic radiation. If the frequency of the incident radiation is ω', then the $u(\omega)$ curve will be extremely sharply peaked at $\omega = \omega'$ as compared to $g(\omega)$ (see Fig. 8.3(*b*)) and thus $g(\omega)/\omega^3$ can be taken out of the integral to obtain

$$\Gamma_{21} = N_2 \frac{\pi^2 c^3}{\hbar \omega'^3 n_0^3 t_{sp}} g(\omega') \int u(\omega) \, d\omega$$

$$= N_2 \frac{\pi^2 c^3}{\hbar \omega'^3 n_0^3 t_{sp}} g(\omega') u \qquad (8.16)$$

where

$$u = \int u(\omega) \, d\omega \qquad (8.17)$$

is the energy density of the incident near-monochromatic radiation. It may be noted that u has dimensions of energy per unit volume unlike $u(\omega)$ which has the dimensions of energy per unit volume per unit frequency interval. Thus when the atom described by a lineshape function $g(\omega)$ interacts with near-monochromatic radiation at frequency ω', the stimulated emission rate per unit volume is given by Eq. (8.16).

In a similar manner, the number of stimulated absorptions per unit time per unit volume will be

$$\Gamma_{12} = N_1 \frac{\pi^2 c^3}{\hbar \omega'^3 n_0^3 t_{sp}} g(\omega')u \tag{8.18}$$

We will use the above equations in later sections.

8.3   Light amplification

We next consider a collection of atoms and let a near-monochromatic radiation of energy density u at frequency ω' pass through it. We shall now obtain the rate of change of intensity of the radiation as it passes through the medium.

Let us consider two planes P_1 and P_2 of area S situated at z and $z + dz$; z being the direction of propagation of the radiation (see Fig. 8.4). If $I(z)$ and $I(z + dz)$ represent the intensity of the radiation at z and $z + dz$ respectively, then the net amount of energy entering the volume $S\,dz$ between P_1 and P_2 will be

$$[I(z) - I(z + dz)]S = [I(z) - I(z) - (dI/dz)\,dz]S$$
$$= -(dI/dz)S\,dz \tag{8.19}$$

This must be equal to the net energy absorbed by the atoms in the volume $S\,dz$. The energy absorbed by the atoms in going from level 1 to level 2 will be $\Gamma_{12} S\,dz\hbar\omega'$ where $\hbar\omega'$ is the energy absorbed when an atom goes from level 1 to level 2. Similarly the energy released through stimulated emissions from level 2 to level 1 will be $\Gamma_{21} S\,dz\hbar\omega'$. We shall neglect the energy arising from spontaneous emission since it appears over a broad frequency range and is also emitted in all directions. Thus the fraction of the spontaneous emission which would be at the radiation frequency ω' and which would be travelling along the z-direction will be very small (see also Sec. 8.5). Thus the net energy absorbed per unit time in the volume $S\,dz$ will be

Fig. 8.4 Propagation of radiation at frequency ω' through a medium; the intensities at z and $z + dz$ are $I(z)$ and $I(z + dz)$ respectively.

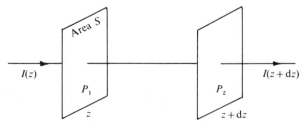

will be

$$(\Gamma_{12} - \Gamma_{21})\hbar\omega'S\,dz = \frac{\pi^2 c^3}{\hbar\omega'^3 n_0^3}\frac{1}{t_{sp}}ug(\omega')(N_1 - N_2)\hbar\omega'S\,dz$$

$$= \frac{\pi^2 c^3}{\omega'^2 n_0^3 t_{sp}}ug(\omega')(N_1 - N_2)S\,dz \qquad (8.20)$$

Now, the energy density u and the intensity of radiation I are related through the following equation[†]

$$I = vu = (c/n_0)u \qquad (8.21)$$

where $v(= c/n_0)$ is the velocity of the radiation in the medium of refractive index n_0.

Thus, using Eqs. (8.19)–(8.21) we obtain

$$dI/dz = -\alpha I \qquad (8.22)$$

where

$$\alpha = \frac{\pi^2 c^2}{\omega^2 n_0^2 t_{sp}}g(\omega)(N_1 - N_2) \qquad (8.23)$$

and we have removed the prime on ω with the understanding that ω represents the frequency of the incident radiation. Hence if $N_1 > N_2$, α is positive and the intensity decreases with z leading to an attenuation of the

Fig. 8.5 A typical variation of $\alpha(\omega)$ with ω for an amplifying medium corresponding to $N_2 > N_1$ (lower curve) and for an attenuating medium $N_1 > N_2$ (upper curve).

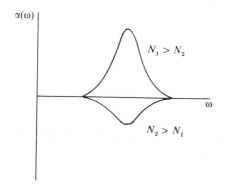

[†] This is analogous to the equation $\mathbf{J} = \rho\mathbf{v}$, where ρ represents the number of particles per unit volume propagating with velocity $\mathbf{v}$ and $\mathbf{J}$ represents the number of particles crossing a unit area normal to the direction of propagation per unit time. This can easily be seen from the fact that the number of particles crossing a unit area in a unit time will be the particles contained in a cylinder of length v units and having a unit area of cross section.

beam. On the other hand, if $N_2 > N_1$ then α is negative and the beam is amplified with z. Fig. 8.5 shows typical plots of α versus ω for $N_1 > N_2$ and $N_1 < N_2$. Obviously the frequency dependence of α will be almost the same as that of the lineshape function $g(\omega)$. The condition $N_2 > N_1$ is called population inversion and it is under this condition that one can obtain optical amplification.

In Eq. (8.23) if $(N_1 - N_2)$ is independent of I, then we have from Eq. (8.22)

$$I(z) = I(0)\,e^{-\alpha z} \tag{8.24}$$

i.e., an exponential attenuation when $N_1 > N_2$ and an exponential amplification when $N_1 < N_2$. We should mention that such an exponential decrease or increase of the intensity is obtained for low intensities; for large intensities saturation sets in and $(N_1 - N_2)$ is no longer independent of I (see Sec. 8.5).

Example: We consider a ruby laser (see Sec. 10.2) with the following characteristics:

$$n_0 = 1.76, \quad t_{sp} = 3 \times 10^{-3}\,\text{s}, \quad \lambda_0 = 6943\,\text{Å}$$
$$g(\omega_0) \approx 1/\Delta\omega \approx 1.1 \times 10^{-12}\,\text{s}$$

where we have assumed that for the normalized lineshape function[†]

$$g(\omega_0) \approx 1/\Delta\omega \tag{8.25}$$

where $\Delta\omega$ represents the full width at half maximum of the lineshape function and ω_0 represents the centre of the line. At thermal equilibrium at 300 K,

$$N_2/N_1 = e^{-h\nu/k_B T} \approx 10^{-30} \approx 0$$

A typical chromium ion density in a ruby laser is about $1.6 \times 10^{19}\,\text{cm}^{-3}$ and since at 300 K most atoms are in the ground level, the absorption coefficient at the centre of the line would be

$$\alpha = 1.4 \times 10^{-19}(N_1 - N_2) \approx 1.4 \times 10^{-19} \times 1.6 \times 10^{19}$$
$$\approx 2.2\,\text{cm}^{-1}$$

If a population inversion density of $5 \times 10^{16}\,\text{cm}^{-3}$ is generated (which represents a typical value) then the gain coefficient will be

$$-\alpha \approx 1.4 \times 10^{-19} \times 5 \times 10^{16}$$
$$\approx 7 \times 10^{-3}\,\text{cm}^{-1}$$

Example: As another example we consider the Nd:YAG laser (see Sec. 10.3.1) for which $n_0 = 1.82$, $t_{sp} = 0.23 \times 10^{-3}\,\text{s}$, $\lambda_0 = 1.06\,\mu\text{m}$

$$\Delta\nu = \frac{\Delta\omega}{2\pi} \approx \frac{1}{2\pi g(\omega_0)} \approx 1.95 \times 10^{11}\,\text{Hz}$$

[†] Indeed we will show in Sec. 8.8 that $g(\omega_0)\Delta\omega$ equals $(2/\pi)$ and $(4\ln 2/\pi)^{\frac{1}{2}}$ for Lorentzian and Gaussian lineshape functions respectively.

If we want a gain of $1\,\mathrm{m}^{-1}$, the inversion required can be calculated from Eq. (8.23) as

$$(N_2 - N_1) = 4v^2 n_0^2 t_{\mathrm{sp}} \alpha / c^2 g(\omega)$$
$$\approx 3.3 \times 10^{15}\,\mathrm{cm}^{-3}.$$

8.4 The threshold condition

In the last section we saw that in order that a medium should be capable of amplifying an incident radiation, one must create a state of population inversion in the medium. Such a medium will behave as an amplifier for those frequencies which fall within its linewidth. In order to generate radiation this amplifying medium is placed in an optical resonator which consists of a pair of mirrors facing each other much like in a Fabry–Perot etalon (see Fig. 8.6). Radiation which bounces back and forth between the mirrors is amplified by the amplifying medium and also suffers losses due to the finite reflectivity of the mirrors and other scattering and diffraction losses. If the oscillations have to be sustained in the cavity then the losses must be exactly compensated by the gain. Thus a minimum population inversion density is required to overcome the losses and this is called the threshold population inversion.

In order to obtain an expression for the threshold population inversion, let d represent the length of the resonator and let R_1 and R_2 represent the reflectivities of the mirrors (see Fig. 8.6). Let α_1 represent the average loss per unit length due to all loss mechanisms (other than the finite reflectivity) such as scattering loss, diffraction loss due to finite mirror sizes etc. Let us consider a radiation with an intensity I leaving mirror 1. As it propagates through the medium and reaches the second mirror, it is amplified by $e^{-\alpha d}$ and also suffers a loss of $e^{-\alpha_1 d}$; for an amplifying medium α is negative and $e^{-\alpha d} > 1$. The intensity of the reflected beam at the second mirror will be

Fig. 8.6 A typical optical resonator consisting of a pair of plane mirrors facing each other. The active medium is placed inside the cavity. One of the mirrors is made partially reflecting to couple out the laser beam.

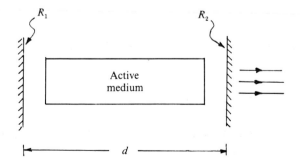

$IR_2 e^{-(\alpha_1 + \alpha)d}$. A second passage through the resonator and a reflection at the first mirror leads to an intensity for the radiation after one complete round trip of $IR_1 R_2 e^{-2(\alpha_1 + \alpha)d}$. Hence for laser oscillation to begin

$$R_1 R_2 e^{-2(\alpha_1 + \alpha)d} \geqslant 1 \tag{8.26}$$

the equality sign giving the threshold value for α (i.e., for population inversion). Indeed, when the laser is oscillating in a steady state with a continuous wave oscillation, then the equality sign in Eq. (8.26) must be satisfied. If the inversion is increased then the LHS becomes greater than unity; this implies that the round trip gain is greater than the round trip loss. This would result in an increasing intensity inside the laser till saturation effects take over, which would result in a decrease in the inversion (we shall explicitly show saturation effects in Sec. 8.5). Thus the gain is brought back to its value at threshold.

Eq. (8.26) can be written as

$$-\alpha \geqslant \alpha_1 - (1/2d) \ln R_1 R_2 \tag{8.27}$$

The RHS of Eq. (8.27) depends on the passive cavity parameters only. This can be related to the passive cavity lifetime t_c which is the time in which energy in the cavity reduces by a factor $1/e$. In the absence of the amplifying medium, the intensity at a point reduces by a factor $R_1 R_2 e^{-2\alpha_1 d} = e^{-(2\alpha_1 d - \ln R_1 R_2)}$ in a time corresponding to one round trip time. One round trip time corresponds to $t = 2d/(c/n_0) = 2dn_0/c$. Hence if the intensity reduces as e^{-t/t_c}, then in a time $t = 2dn_0/c$, the factor by which the intensity will be reduced is e^{-2dn_0/ct_c}. Thus

$$e^{-(2\alpha_1 d - \ln R_1 R_2)} = e^{-2dn_0/ct_c}$$

or

$$\frac{1}{t_c} = \frac{c}{2dn_0}(2\alpha_1 d - \ln R_1 R_2) \tag{8.28}$$

Using Eqs. (8.23) and (8.28), Eq. (8.27) becomes

$$(N_2 - N_1) \geqslant \frac{4v^2 n_0^3}{c^3} \frac{t_{sp}}{t_c} \frac{1}{g(\omega)} \tag{8.29}$$

Corresponding to the equality sign, we have the threshold population inversion density required for the oscillation of the laser.

According to Eq. (8.29), in order to have a low threshold value of population inversion, the following conditions must hold:

(a) The value of t_c should be large, i.e., the cavity losses must be small.

(b) Since $g(\omega)$ is normalized according to Eq. (8.13), the peak value of $g(\omega)$ will be inversely proportional to the width $\Delta\omega$ of the $g(\omega)$ function (see Eq. (8.25)). Thus smaller widths give larger values of $g(\omega)$ which implies smaller threshold values of $(N_2 - N_1)$. Also since the largest $g(\omega)$ appears at the line centre, the resonator mode which lies closest to the line centre will reach threshold first and begin to oscillate.

(c) Smaller values of t_{sp} (i.e., strongly allowed transitions) also lead to smaller values of threshold inversion. At the same time for smaller relaxation times (t_{sp}), larger pumping power will be required to maintain a given population inversion. In general, population inversion is more easily obtained on transitions which have longer relaxation times.

(d) The value of $g(\omega)$ at the centre of the line is inversely proportional to $\Delta\omega$ which, for example, in the case of Doppler broadening is proportional to ω (see Sec. 8.8.3). Thus the threshold population inversion increases approximately in proportion to ω^3. Thus it is much easier to obtain laser action at infrared wavelengths than in the ultraviolet region.

Example: We first consider a ruby laser[†] which has the following typical parameters:

$$\lambda_0 = 6943 \,\text{Å}, \quad t_{sp} \approx 3 \times 10^{-3}\,\text{s}, \quad n_0 = 1.76, \quad d = 5\,\text{cm},$$
$$R_1 = R_2 = 0.9, \quad \alpha_1 \approx 0$$
$$g(\omega_0) = \frac{1}{\Delta\omega} = \frac{1}{2\pi\Delta\nu} \approx 1.1 \times 10^{-12}\,\text{s}$$

Thus for the above values

$$t_c \approx 2.8 \times 10^{-9}\,\text{s}$$

and

$$(N_2 - N_1)_{th} \approx 1.5 \times 10^{17}\,\text{cm}^{-3}$$

Typical Cr^{3+} ion densities are about $1.6 \times 10^{19}\,\text{cm}^{-3}$. Thus the fractional excess population is very small. The above population inversion corresponds to a gain of $0.09\,\text{dB/cm}$.

Example: As another example, we consider a He–Ne laser with the following typical characteristics:

$$\lambda_0 = 6328 \,\text{Å}; \quad t_{sp} \approx 10^{-7}\,\text{s}, \quad n_0 \approx 1, \quad d = 20\,\text{cm}$$

[†] Ruby laser active medium consists of Cr^{3+} doped in Al_2O_3 and is an example of a three level laser. More details regarding the ruby laser are given in Sec. 10.2.

$$R_1 = R_2 = 0.98, \alpha_1 \simeq 0$$

$$\Delta v \approx 10^9 \, \text{Hz}$$

$$g(\omega_0) \approx \frac{1}{2\pi\Delta v} \approx 0.16 \times 10^{-9} \, \text{s}$$

For the above values

$$t_c \approx 3.3 \times 10^{-8} \, \text{s}$$

and

$$(N_2 - N_1)_{\text{th}} \approx 6.24 \times 10^8 \, \text{cm}^{-3}$$

8.5   Laser rate equations

In this section we shall develop laser rate equations which describe the rate at which the populations of various atomic levels change in the presence of external pumping and stimulated and spontaneous transitions. The rate equations provide a convenient approach for studying the variation of atomic populations, the steady state populations of various levels in the presence of radiation etc.

8.5.1   *The two level system*

We first consider a two level system consisting of energy levels E_1 and E_2 with N_1 and N_2 atoms per unit volume respectively. Let radiation at frequency ω with energy density u be incident on the system. The number of atoms per unit volume which absorb the radiation and are excited to the upper level will be (see Eq. 8.18)

$$\Gamma_{12} = \frac{\pi^2 c^3}{\hbar\omega^3 t_{\text{sp}} n_0^3} u g(\omega) N_1 = W_{12} N_1 \tag{8.30}$$

where

$$W_{12} = \frac{\pi^2 c^3}{\hbar\omega^3 t_{\text{sp}} n_0^3} u g(\omega) \tag{8.31}$$

The number of atoms undergoing stimulated emissions from E_2 to E_1 per unit volume per unit time will be (see Eqs. (8.16) and (8.18))

$$\Gamma_{21} = W_{21} N_2 = W_{12} N_2 \tag{8.32}$$

where we have used the fact that the absorption probability is the same as the stimulated emission probability. In addition to the above two transitions, atoms in the level E_2 would also undergo spontaneous transitions from E_2 to E_1. If A_{21} and S_{21} represent the radiative and nonradiative transition[†] rates

---

[†] In a nonradiative transition, when the atom deexcites, the energy is transferred to the translational, vibrational or rotational energies of the surrounding atoms or molecules.

from E_2 to E_1, then the number of atoms undergoing spontaneous transitions from E_2 to E_1 will be $T_{21}N_2$ where

$$T_{21} = A_{21} + S_{21} \tag{8.33}$$

Thus we may write the rate of change of population of energy levels E_2 and E_1 as

$$dN_2/dt = W_{12}(N_1 - N_2) - T_{21}N_2 \tag{8.34}$$

$$dN_1/dt = -W_{12}(N_1 - N_2) + T_{21}N_2 \tag{8.35}$$

As can be seen from Eqs. (8.34) and (8.35)

$$(d/dt)(N_1 + N_2) = 0$$

$$\Rightarrow N_1 + N_2 = \text{a constant} = N \quad \text{(say)} \tag{8.36}$$

which is nothing but the fact that the total number of atoms N per unit volume is constant. At steady state

$$dN_1/dt = 0 = dN_2/dt \tag{8.37}$$

which gives us

$$\frac{N_2}{N_1} = \frac{W_{12}}{W_{12} + T_{21}} \tag{8.38}$$

Since both W_{12} and T_{21} are positive quantities, Eq. (8.38) shows us that we can never obtain a steady state population inversion by optical pumping between just two levels.

Let us now have a look at the population difference between the two levels. From Eq. (8.38) we have

$$\frac{N_2 - N_1}{N_2 + N_1} = -\frac{T_{21}}{2W_{12} + T_{21}}$$

or if we write $\Delta N = N_2 - N_1$, we have

$$\frac{\Delta N}{N} = -\frac{1}{1 + 2W_{12}/T_{21}} \tag{8.39}$$

In order to put Eq. (8.39) in a slightly different form, we first assume that the transition from 2 to 1 is mostly radiative, i.e., $A_{21} \gg S_{21}$ and $T_{21} \approx A_{21}$. We also introduce a lineshape function $\tilde{g}(\omega)$ which is normalized to have unit value at $\omega = \omega_0$, the centre of the line, i.e.,

$$\tilde{g}(\omega) = g(\omega)/g(\omega_0) \tag{8.40}$$

Since $g(\omega) \leqslant g(\omega_0)$ for all ω, $0 < \tilde{g}(\omega) < 1$. Substituting the value of W_{12} in terms of u from Eq. (8.31) and observing that $u = n_0 I/c$ where I is the intensity of the incident radiation at ω, we have

$$\frac{W_{12}}{T_{21}} = \frac{\pi^2 c^3}{\hbar \omega^3 t_{\text{sp}} n_0^3} I \frac{n_0}{c} \tilde{g}(\omega) g(\omega_0) \frac{1}{A_{21}}$$

$$= \frac{\pi^2 c^2}{\hbar \omega^3 n_0^2} g(\omega_0) \tilde{g}(\omega) I \tag{8.41}$$

where we have used the fact that $A_{21} t_{\text{sp}} = 1$. Hence Eq. (8.39) becomes

$$\frac{\Delta N}{N} = -\frac{1}{1 + (I/I_s)\tilde{g}(\omega)} \tag{8.42}$$

where

$$I_s \equiv \hbar \omega^3 n_0^2 / 2\pi^2 c^2 g(\omega_0) \tag{8.43}$$

is called the saturation intensity. In order to see what I_s represents let us consider a monochromatic wave at frequency ω_0 interacting with a two level system. Since $\tilde{g}(\omega_0) = 1$, we see from Eq. (8.42) that for $I \ll I_s$, the density of population difference between the two levels ΔN is almost independent of the intensity of the incident radiation. On the other hand for I comparable to I_s, ΔN becomes a function of I and indeed for $I = I_s$, the value of ΔN is half the value at low incident intensities.

We showed in Sec. 8.3 that the loss/gain coefficient for a population difference $\Delta N = N_2 - N_1$ between two levels is given by [see Eq. (8.23)]

$$\alpha = -\frac{\pi^2 c^2}{\omega^2 t_{\text{sp}} n_0^2} g(\omega) \Delta N$$

$$= \frac{\alpha_0}{1 + (I/I_s)\tilde{g}(\omega)} \tag{8.44}$$

where

$$\alpha_0 = \frac{\pi^2 c^2}{\omega^2 t_{\text{sp}} n_0^2} g(\omega) N \tag{8.45}$$

corresponds to the small signal loss, i.e., the loss coefficient when $I \ll I_s$. We can see from Problem 8.1 that with α given by Eq. (8.44), the loss is exponential for $I \ll I_s$ while it becomes linear for $I \gg I_s$. Thus we see that the attenuation caused by a medium decreases as the incident intensity increases to values comparable to the saturation intensity. Organic dyes having reasonably low values of $I_s (\sim 5 \, \text{MW/cm}^2)$ are used as saturable absorbers in mode locking and Q-switching of lasers (see Sec. 9.7).

Problem 8.1: Using Eq. (8.44) in Eq. (8.22) obtain the variation of I with z. (Answer:

$$\ln\frac{I}{I_0} + \frac{\tilde{g}(\omega)}{I_s}(I - I_0) = -\alpha_0 z$$

where I_0 is the intensity at $z = 0$.)

8.5.2 *The three level laser system*

In the last section we saw that one cannot create a steady state population inversion between two levels just by using pumping between these levels. Thus in order to produce a steady state population inversion, one makes use of either a three level or a four level system. In this section we shall discuss a three level system.

We consider a three level system consisting of energy levels E_1, E_2 and E_3 all of which are assumed to be nondegenerate. Let N_1, N_2 and N_3 represent the population densities of the three levels (see Fig. 8.7). The pump is assumed to lift atoms from level 1 to level 3 from which they decay rapidly to level 2 through some nonradiative process. Thus the pump effectively transfers atoms from the ground level 1 to the excited level 2 which is now the upper laser level; the lower laser level being the ground state 1. If the relaxation from level 3 to level 2 is very fast, then the atoms will relax down to level 2 rather than to level 1. Since the upper level 3 is not a laser level, it can be a broad level (or a group of broad levels) so that a broadband light source may be efficiently used as a pump source (see e.g., the ruby laser discussed in Sec. 10.2).

If we assume that transitions take place only between these three levels then we may write

$$N = N_1 + N_2 + N_3 \tag{8.46}$$

where N represents the total number of atoms per unit volume.

Fig. 8.7 The three level laser system: the pump excites the atoms from level E_1 to E_3 and laser action takes place between levels E_2 and E_1.

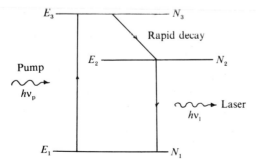

We may now write the rate equations describing the rate of change of N_1, N_2 and N_3. For example, the rate of change of N_3 may be written as

$$dN_3/dt = W_p(N_1 - N_3) - T_{32}N_3 \qquad (8.47)$$

where W_p is the rate of pumping per atom from level 1 to level 3 which depends on the pump intensity. The first term in Eq. (8.47) represents stimulated transitions between levels 1 and 3 and $T_{32}N_3$ represents the spontaneous transition from level 3 to level 2;

$$T_{32} = A_{32} + S_{32} \qquad (8.48)$$

A_{32} and S_{32} correspond respectively to the radiative and nonradiative transition rates between levels 3 and 2. In writing Eq. (8.47) we have neglected $T_{31}N_3$ which corresponds to spontaneous transitions between levels 3 and 1 since most atoms raised to level 3 make transitions to level 2 rather than to level 1.

In a similar manner, we may write

$$dN_2/dt = W_l(N_1 - N_2) + N_3T_{32} - N_2T_{21} \qquad (8.49)$$

and

$$dN_1/dt = W_p(N_3 - N_1) + W_l(N_2 - N_1) + N_2T_{21} \qquad (8.50)$$

where

$$W_l = \frac{\pi^2 c^2}{\hbar \omega^3 n_0^2} A_{21} g_l(\omega) I_1 \qquad (8.51)$$

represents the stimulated transition rate per atom between levels 1 and 2, I_1 is the intensity of the radiation in the $2 \to 1$ transition and $g_l(\omega)$ represents the lineshape function describing the transitions between levels 1 and 2. In writing Eq. (8.51) we have used Eqs. (8.16) and (8.21). Further,

$$T_{21} = A_{21} + S_{21} \qquad (8.52)$$

with A_{21} and S_{21} representing the radiative and nonradiative relaxation rates between levels 1 and 2. For good laser action since the transition must be mostly radiative, we shall assume $A_{21} \gg S_{21}$.

At steady state we must have

$$dN_1/dt = 0 = dN_2/dt = dN_3/dt \qquad (8.53)$$

From Eq. (8.47) we obtain

$$N_3 = \frac{W_p}{W_p + T_{32}} N_1 \qquad (8.54)$$

Using Eqs. (8.49), (8.50) and (8.54) we get

$$N_2 = \frac{W_1(T_{32} + W_p) + W_p T_{32}}{(W_p + T_{32})(W_1 + T_{21})} N_1 \tag{8.55}$$

Thus from Eqs. (8.46), (8.54) and (8.55), we get

$$\frac{N_2 - N_1}{N} = \frac{[W_p(T_{32} - T_{21}) - T_{32}T_{21}]}{[3W_p W_1 + 2W_p T_{21} + 2T_{32}W_1 + T_{32}W_p + T_{32}T_{21}]} \tag{8.56}$$

From the above equation, one may see that in order to obtain population inversion between levels 2 and 1, i.e., for $(N_2 - N_1)$ to be positive, a necessary (but not sufficient) condition is that

$$T_{32} > T_{21} \tag{8.57}$$

Since the lifetimes of levels 3 and 2 are inversely proportional to the relaxation rates, according to Eq. (8.57), the lifetime of level 3 must be smaller than that of level 2 for attainment of population inversion between levels 1 and 2. If this condition is satisfied then according to Eq. (8.56), there is a minimum pumping rate required to achieve population inversion which is given by

$$W_{pt} = \frac{T_{32}T_{21}}{T_{32} - T_{21}} \tag{8.58}$$

If $T_{32} \gg T_{21}$, then

$$W_{pt} \approx T_{21} \tag{8.58a}$$

and under the same approximation, Eq. (8.56) becomes

$$\frac{N_2 - N_1}{N} = \frac{(W_p - T_{21})/(W_p + T_{21})}{\left[1 + \dfrac{3W_p + 2T_{32}}{T_{32}(W_p + T_{21})} W_1\right]} \tag{8.59}$$

Below the threshold for laser oscillation, W_1 is very small and hence we may write

$$\frac{N_2 - N_1}{N} = \frac{W_p - T_{21}}{W_p + T_{21}} \tag{8.59a}$$

Thus when W_1 is small, i.e., when the intensity of the radiation corresponding to the laser transition is small (see Eq. (8.51)), then the population inversion is independent of I_1 and thus according to Eq. (8.22) there is an exponential amplification of the beam. As the laser starts oscillating, W_1 becomes large and from Eq. (8.59) we see that this reduces the inversion $N_2 - N_1$ which in turn reduces the amplification. When the laser oscillates under steady state conditions, the intensity of the radiation at the laser transition increases to such a value that the value of $N_2 - N_1$ is the same as the threshold value (see Eq. (8.29)).

Recalling Eq. (8.23), we see that for a population inversion $N_2 - N_1$, the gain coefficient of the laser medium is

$$\alpha = -\frac{\pi^2 c^2}{\omega^2 t_{sp} n_0^2} g(\omega)(N_2 - N_1)$$

$$= \frac{\alpha_0}{1 + \dfrac{3W_p + 2T_{32}}{T_{32}(W_p + T_{21})} W_1} \tag{8.60}$$

where

$$\alpha_0 = -\frac{\pi^2 c^2}{\omega^2 t_{sp} n_0^2} g(\omega) N \frac{W_p - T_{21}}{W_p + T_{21}} \tag{8.60a}$$

is the small signal gain coefficient. If we now carry out a similar analysis to that in Sec. 8.5.1, we may write

$$\alpha = \frac{\alpha_0}{1 + (I_1/I_s)\tilde{g}(\omega)} \tag{8.61}$$

where

$$\tilde{g}(\omega) = g(\omega)/g(\omega_0)$$

$$I_s = \frac{\hbar\omega^3 n_0^2}{\pi^2 c^2 A_{21} g(\omega_0)} \frac{T_{32}(W_p + T_{21})}{(3W_p + 2T_{32})} \tag{8.62}$$

I_s being the saturation intensity (see the discussion following Eq. (8.43)).

If T_{32} is very large then there will be very few atoms residing in level 3. Consequently, we may write

$$N = N_1 + N_2 + N_3 \approx N_1 + N_2 \tag{8.63}$$

Substituting in Eq. (8.59a), we get

$$\frac{N_2 - N_1}{N_2 + N_1} = \frac{W_p - T_{21}}{W_p + T_{21}}$$

or

$$W_p N_1 = T_{21} N_2 \tag{8.64}$$

The LHS of the above equation represents the number of atoms being lifted per unit volume per unit time from level 1 to level 2 via level 3 and the RHS corresponds to the spontaneous emission rate per unit volume from level 2 to level 1. These rates must be equal under steady state conditions for $W_1 \approx 0$, i.e. below threshold.

We shall now estimate the threshold pumping power required to start laser oscillation. In order to do this, we first observe that the threshold inversion required is usually very small compared to N (i.e., $N_2 - N_1 \ll N -$ see the example of the ruby laser discussed in Sec. 8.4). Thus from Eq. (8.64), we see that the threshold value of W_p required to start laser oscillation is also $\approx T_{21}$. Now the number of atoms being pumped per unit time per unit volume from level 1 to level 3 is $W_p N_1$. If ν_p represents the average pump frequency corresponding to excitation to E_3 from E_1, then the power required per unit volume will be

$$P = W_p N_1 h \nu_p \tag{8.65}$$

Thus the threshold pump power for laser oscillation is given by

$$P_t = T_{21} N_1 h \nu_p \tag{8.66}$$

Since $N_2 - N_1 \ll N$ and $N_3 \approx 0$, $N_1 \approx N_2 \approx N/2$. Also assuming the transition from level 2 to level 1 to be mainly radiative (i.e., $A_{21} \gg S_{21}$), we have

$$P_t \approx N h \nu_p / 2 t_{sp} \tag{8.67}$$

where we have used $A_{21} = 1/t_{sp}$.

As an example, we consider the ruby laser for which we have the following values of the various parameters:

$$N \approx 1.6 \times 10^{19} \, cm^{-3} \qquad t_{sp} \approx 3 \times 10^{-3} \, s \qquad \nu_p \approx 6.25 \times 10^{14} \, Hz \tag{8.68}$$

Substitution in Eq. (8.67) gives us

$$P_t \approx 1100 \, W/cm^3 \tag{8.69}$$

If we assume that the efficiency of the pumping source to be 25% and also that only 25% of the pump light is absorbed on passage through the ruby rod, then the electrical threshold power comes out to be about $18 \, kW/cm^3$ of the active medium. This is consistent with the threshold powers obtained experimentally.

8.5.3 The four level laser system

In the last section we found that since the lower laser level was the ground level, one has to lift more than 50% of the atoms in the ground level in order to obtain population inversion. This problem can be overcome by using another level of the atomic system and having the lower laser level also as an excited level. The four level laser system is shown in Fig. 8.8. Level 1 is the ground level and levels 2, 3 and 4 are excited levels of the system. Atoms from level 1 are pumped to level 4 from where they make a fast nonradiative relaxation to level 3. Level 3 which corresponds to the upper laser level is usually a metastable level having a long lifetime. The transition from level 3 to level 2 forms the laser transition. In order that atoms do not accumulate in level 2 and hence destroy the population inversion between levels 3 and 2, level 2 must have a very small lifetime so that atoms from level 2 are quickly removed to level 1 ready for pumping to level 4. If the relaxation rate of atoms from level 2 to level 1 is faster than the rate of arrival of atoms to level 2 then one can obtain population inversion between levels 3 and 2 even for very small pump powers. Level 4 can be a collection of a large number of levels or a broad level. In such a case an optical pump source emitting over a broad range of frequencies can be used to pump atoms from level 1 to level 4 effectively. In addition, level 2 is required to be sufficiently above the ground level so that, at ordinary temperatures, level 2 is almost unpopulated. The population of level 2 can also be reduced by lowering the temperature of the system.

We shall now write the rate equations corresponding to the populations of the four levels. Let N_1, N_2, N_3 and N_4 be the population densities of levels 1, 2, 3 and 4 respectively. The rate of change of N_4 can be written as

$$dN_4/dt = W_p(N_1 - N_4) - T_{43}N_4 \tag{8.70}$$

where, as before, $W_p N_1$ is the number of atoms being pumped per unit time

Fig. 8.8 The four level laser system: the pump lifts atoms from level E_1 to E_4 and laser action takes place between levels E_3 and E_2.

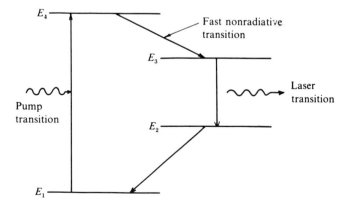

per unit volume, $W_p N_4$ is the stimulated emission rate per unit volume,

$$T_{43} = A_{43} + S_{43} \tag{8.71}$$

is the relaxation rate from level 4 to level 3 and is the sum of the radiative (A_{43}) and nonradiative (S_{43}) rates. In writing Eq. (8.70) we have neglected T_{42} and T_{41} in comparison to T_{43} i.e., we have assumed that the atoms in level 4 relax to level 3 rather than to levels 2 and 1.

Similarly, the rate equation for level 3 may be written as

$$dN_3/dt = W_l(N_2 - N_3) + T_{43}N_4 - T_{32}N_3 \tag{8.72}$$

where

$$W_l = \frac{\pi^2 c^2}{\hbar \omega^3 n_0^2} A_{32} g_l(\omega) I_1 \tag{8.73}$$

represents the stimulated transition rate per atom between levels 3 and 2 and the subscript 1 stands for laser transition; $g_l(\omega)$ is the lineshape function describing the $3 \leftrightarrow 2$ transition and I_1 is the intensity of the radiation at the frequency $\omega = (E_3 - E_2)/\hbar$. Also

$$T_{32} = A_{32} + S_{32} \tag{8.74}$$

is the net spontaneous relaxation rate from level 3 to level 2 and consists of the radiative (A_{32}) and nonradiative (S_{32}) contributions. Again we have neglected any spontaneous transition from level 3 to level 1. In a similar manner, we can write

$$dN_2/dt = - W_l(N_2 - N_3) + T_{32}N_3 - T_{21}N_2 \tag{8.75}$$

$$dN_1/dt = - W_p(N_1 - N_4) + T_{21}N_2 \tag{8.76}$$

where

$$T_{21} = A_{21} + S_{21} \tag{8.77}$$

is the spontaneous relaxation rate from $2 \rightarrow 1$.

Under steady state conditions

$$\frac{dN_1}{dt} = \frac{dN_2}{dt} = \frac{dN_3}{dt} = \frac{dN_4}{dt} = 0 \tag{8.78}$$

We will thus get four simultaneous equations in N_1, N_2, N_3 and N_4 and in addition we have,

$$N = N_1 + N_2 + N_3 + N_4 \tag{8.79}$$

for the total number of atoms per unit volume in the system.

From Eq. (8.70) we obtain, setting $dN_4/dt = 0$,

$$N_4/N_1 = W_p/(W_p + T_{43}) \tag{8.80}$$

If the relaxation from level 4 to level 3 is very rapid then $T_{43} \gg W_p$ and hence $N_4 \ll N_1$. Using this approximation in the remaining three equations we can obtain for the population difference,

$$\frac{N_3 - N_2}{N} \approx \frac{W_p(T_{21} - T_{32})}{W_p(T_{21} + T_{32}) + T_{32}T_{21} + W_1(2W_p + T_{21})} \tag{8.81}$$

Thus in order to be able to obtain population inversion between levels 3 and 2, we must have

$$T_{21} > T_{32} \tag{8.82}$$

i.e., the spontaneous rate of deexcitation of level 2 to level 1 must be larger than the spontaneous rate of deexcitation of level 3 to level 2.

If we now assume $T_{21} \gg T_{32}$, then from Eq. (8.81) we obtain

$$\frac{N_3 - N_2}{N} \approx \frac{W_p}{W_p + T_{32}} \frac{1}{1 + W_1(T_{21} + 2W_p)/T_{21}(W_p + T_{32})} \tag{8.83}$$

From the above equation we see that even for very small pump rates one can obtain population inversion between levels 3 and 2. This is contrary to what we found in a three level system, where there was a minimum pump rate, W_{pt}, required to achieve inversion. The first factor in Eq. (8.83) which is independent of W_1 (i.e., independent of the intensity of radiation corresponding to the laser transition – see Eq. (8.73)) gives the small signal gain coefficient whereas the second factor in Eq. (8.83) gives the saturation behaviour.

Just below threshold for laser oscillation, $W_1 \approx 0$, and hence from Eq. (8.83) we obtain

$$\Delta N/N \approx W_p/(W_p + T_{32}) \tag{8.84}$$

where $\Delta N = N_3 - N_2$ is the population inversion density. We shall now consider two examples of four level systems.

Example: The Nd:YAG laser corresponds to a four level laser system (see Sec. 10.3.1). For such a laser, typical values of various parameters are

$$\lambda_0 = 1.06\,\mu m(v = 2.83 \times 10^{14}\,Hz), \qquad \Delta v = 1.95 \times 10^{11}\,Hz,$$
$$t_{sp} = 2.3 \times 10^{-4}\,s, \qquad N = 6 \times 10^{19}\,cm^{-3}, \qquad n_0 = 1.82 \tag{8.85}$$

If we consider a resonator cavity of length 7 cm and $R_1 = 1.00$, $R_2 = 0.90$, neglecting other loss factors (i.e., $\alpha_l = 0$)

$$t_c = -\frac{2n_0 d}{c \ln R_1 R_2} \approx 8 \times 10^{-9}\,\text{s}$$

We now use Eq. (8.29) to estimate the population inversion density to start laser oscillation:

$$(\Delta N)_t \approx \frac{4v^2 n_0^3}{c^3} \frac{1}{g(\omega)} \frac{t_{sp}}{t_c}$$

$$= \frac{4v^2 n_0^3}{c^3} \pi^2 \Delta v \frac{t_{sp}}{t_c} \tag{8.86}$$

where for a homogeneous transition (see Sec. 8.8)

$$g(\omega_0) = 2/\pi\Delta\omega = 1/\pi^2\Delta v \tag{8.87}$$

Thus substituting various values, we obtain

$$(\Delta N)_t \approx 4 \times 10^{15}\,\text{cm}^{-3} \tag{8.88}$$

Since $(\Delta N)_t \ll N$, we may assume in Eq. (8.84) $T_{32} \gg W_p$ and hence we obtain for the threshold required to start laser oscillation

$$W_{pt} \approx \frac{(\Delta N)_t}{N} T_{32} \approx \frac{(\Delta N)_t}{N} \frac{1}{t_{sp}}$$

$$= \frac{4 \times 10^{15}}{6 \times 10^{19}} \times \frac{1}{2.3 \times 10^{-4}} \approx 0.3\,\text{s}^{-1}$$

At this pumping rate the number of atoms being pumped from level 1 to level 4 is $W_{pt}N_1$ and since N_2, N_3 and N_4 are all very small compared to N_1, we have $N_1 \approx N$. For every atom lifted from level 1 to level 4 an energy hv_p has to be given to the atom where v_p is the average pump frequency corresponding to the $1 \rightarrow 4$ transition. Assuming $v_p \approx 4 \times 10^{14}\,\text{Hz}$ we obtain for the threshold pump power required per unit volume of the laser medium

$$P_{th} = W_{pt}N_1 hv_p \approx W_{pt}Nhv_p$$

$$= 0.3 \times 6 \times 10^{19} \times 6.6 \times 10^{-34} \times 4 \times 10^{14}$$

$$\approx 4.8\,\text{W/cm}^3$$

which is about three orders of magnitude smaller than that obtained for ruby.

Example: As a second example of a four level laser system, we consider the He–Ne laser (see Sec. 10.4). We use the following data:

$$\lambda_0 = 0.6328 \times 10^{-4}\,\text{cm}\ (v = 4.74 \times 10^{14}\,\text{Hz}),$$

$$t_{sp} = 10^{-7}\,\text{s}, \qquad \Delta v \approx 10^9\,\text{Hz}, \qquad n_0 \approx 1 \tag{8.89}$$

If we consider the resonator to be of length 10 cm and having mirrors of reflectivities

$R_1 = R_2 = 0.98$, then assuming the absence of other loss mechanisms ($\alpha_l = 0$),

$$t_c = -2n_0 d/c \ln R_1 R_2$$
$$\approx 1.6 \times 10^{-8}\,\text{s} \tag{8.90}$$

For an inhomogeneously broadened transition (see Sec. 8.8)

$$g(\omega_0) = \frac{2}{\Delta\omega}\left(\frac{\ln 2}{\pi}\right)^{\frac{1}{2}}$$
$$\approx 1.5 \times 10^{-10}\,\text{s} \tag{8.91}$$

Thus the threshold population inversion required is

$$(\Delta N)_t \approx 1.4 \times 10^9\,\text{cm}^{-3} \tag{8.92}$$

Hence the threshold pump power required to start laser oscillation is

$$P_{th} = W_{pt} N_1 (E_4 - E_1)$$
$$\approx \frac{(\Delta N)_t}{t_{sp}} h\nu_p \tag{8.93}$$

where again we assume $(\Delta N)_t \ll N$ and $T_{32} \approx A_{32} = 1/t_{sp}$. Assuming $\nu_p \approx 5 \times 10^{15}$ Hz, we obtain

$$P_{th} = \frac{1.4 \times 10^9 \times 6.6 \times 10^{-34} \times 5 \times 10^{15}}{10^{-7}}$$
$$\approx 50\,\text{mW/cm}^3 \tag{8.94}$$

which again is very small compared to the threshold powers required for ruby laser.

8.6   Variation of laser power around threshold

In the earlier sections we considered the three level and four level laser systems and obtained conditions for the attainment of population inversion. In the present section we shall discuss the variation of the power in the laser transition as the pumping rate passes through threshold.

We consider the two levels involved in the laser transition in a four level

Fig. 8.9 The lower level of the laser transition is assumed to be unpopulated due to rapid relaxation to other lower levels. The upper level is being pumped at a rate R per unit volume and has a population density of N_2.

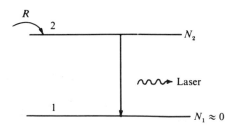

laser[†] and assume that the lower laser level has a very fast relaxation rate to lower levels so that it is essentially unpopulated. We will assume that only one mode has sufficient gain to oscillate and that the line is homogeneously broadened so that the same induced rate applies to all atoms (see Sec. 8.8). Let R represent the number of atoms that are being pumped into the upper level per unit time per unit volume (see Fig. 8.9). If the population density of the upper level is N_2, then the number of atoms undergoing stimulated emissions from level 2 to level 1 per unit time will be (see Eq. (8.16))

$$F_{21} = \Gamma_{21}V = \frac{\pi^2 c^3}{\hbar \omega^3 n_0^3} A_{21} u g(\omega) N_2 V \tag{8.95}$$

where u is the density of radiation at the oscillating mode frequency ω, V represents the volume of the active medium and n_0 is the refractive index of the medium.

Instead of working with the energy density u, we introduce the number of photons n in the oscillating mode. Since each photon carries an energy $\hbar \omega$, the number of photons n will be given by

$$n = uV/\hbar \omega \tag{8.96}$$

Thus

$$F_{21} = \frac{\pi^2 c^3}{\omega^2 n_0^3} A_{21} g(\omega) N_2 n = K n N_2 \tag{8.97}$$

where

$$K \equiv (\pi^2 c^3/\omega^2 n_0^3) A_{21} g(\omega) \tag{8.98}$$

The spontaneous relaxation rate from level 2 to level 1 in the whole volume will be $T_{21} N_2 V$ where

$$T_{21} = A_{21} + S_{21} \tag{8.99}$$

is the total relaxation rate consisting of the radiative (A_{21}) and nonradiative (S_{21}) components. Hence we have for the net rate of the change of population of level 2

$$\frac{d}{dt}(N_2 V) = -K n N_2 - T_{21} N_2 V + RV$$

or

$$\frac{dN_2}{dt} = -\frac{K n N_2}{V} - T_{21} N_2 + R \tag{8.100}$$

In order to write a rate equation describing the variation of photon

---

[†] A similar analysis can also be performed for a three level laser system but the general conclusions of this simple analysis remain valid.

number n in the oscillating mode in the cavity, we note that n change due to:

(a) All stimulated emissions caused by the n photons existing in the cavity mode result in a rate of increase of n of KnN_2 since every stimulated emission from level 2 to level 1 caused by radiation in that mode will result in the addition of a photon in that mode. There is no absorption since we have assumed the lower level to be unpopulated.

(b) In order to estimate the increase in the number of photons in the cavity mode due to spontaneous emission, we must note that not all spontaneous emissions occurring from the $2 \to 1$ transition will contribute to a photon in the oscillating mode. As we will show in Sec. 9.2 for an optical resonator which has dimensions which are large compared to the wavelength of light, there are an extremely large number of modes ($\sim 10^8$) that have their frequencies within the atomic linewidth. Thus when an atom deexcites from level 2 to level 1 by spontaneous emission it may appear in any one of these modes. Since we are only interested in the number of photons in the oscillating cavity mode, we must first obtain the rate of spontaneous emission into a mode of oscillation of the cavity. In order to obtain this we recall from Sec. 8.2 that the number of spontaneous emissions occurring between ω and $\omega + d\omega$ will be

$$G_{21}\, d\omega = A_{21} N_2 g(\omega)\, d\omega V \qquad (8.101)$$

We shall show in Appendix C that the number of oscillating modes lying in a frequency interval between ω and $\omega + d\omega$ is

$$N(\omega)\, d\omega = n_0^3 (\omega^2/\pi^2 c^3) V\, d\omega \qquad (8.102)$$

where n_0 is the refractive index of the medium. Thus the spontaneous emission rate per mode of oscillation at frequency ω is

$$S_{21} = \frac{G_{21}\, d\omega}{N(\omega)\, d\omega} = \frac{\pi^2 c^3}{n_0^3 \omega^2} g(\omega) A_{21} N_2$$

$$= KN_2 \qquad (8.103)$$

i.e., the rate of spontaneous emission into a particular cavity mode is the same as the rate of stimulated emission into the same mode when there is just one photon in that mode. This result can indeed be obtained by rigorous quantum mechanical derivation (see e.g., Thyagarajan and Ghatak (1981)).

(c) The photons in the cavity mode are also lost due to the finite cavity lifetime. Since the energy in the cavity reduces with time as e^{-t/t_c} (see Sec. 8.4), the rate of decrease of photon number in the cavity will also be n/t_c.

Thus we can write for the total rate of change of n

$$dn/dt = KnN_2 + KN_2 - n/t_c \tag{8.104}$$

Eqs. (8.100) and (8.104) represent the pair of coupled rate equations describing the variation of N_2 and n with time.

Under steady state conditions both time derivatives are zero. Thus we obtain from Eq. (8.104),

$$N_2 = \frac{n}{n+1} \frac{1}{Kt_c} \tag{8.105}$$

The above equation implies that under steady state conditions $N_2 \leqslant 1/Kt_c$. When the laser is oscillating under steady state conditions $n \gg 1$ and $N_2 \approx 1/Kt_c$. If we substitute the value of K from Eq. (8.98) we find that (for $n \gg 1$)

$$N_2 \approx \frac{\omega^2 n_0^3}{\pi^2 c^3} \frac{t_{sp}}{t_c} \frac{1}{g(\omega)} \tag{8.106}$$

which is nothing but the threshold population inversion density required for laser oscillation (cf. Eq. (8.29)). Thus Eq. (8.105) implies that when the laser oscillates under steady state conditions, the population inversion density is almost equal to and can never exceed the threshold value. This is also obvious since if the inversion density exceeds the threshold value, the gain in the cavity will exceed the loss and thus the laser power will start increasing. This increase will continue till saturation effects take over and reduce N_2 to the threshold value (see Eq. 8.106).

Substituting from Eq. (8.105) into Eq. (8.100) and putting $dN_2/dt = 0$, we get

$$\frac{K}{VT_{21}} n^2 + n\left(1 - \frac{R}{R_t}\right) - \frac{R}{R_t} = 0 \tag{8.107}$$

where

$$R_t = T_{21}/Kt_c \tag{8.108}$$

The solution of the above equation which gives a positive value of n is

$$n = \frac{VT_{21}}{2K} \left\{ \left(\frac{R}{R_t} - 1\right) + \left[\left(1 - \frac{R}{R_t}\right)^2 + \frac{4K}{VT_{21}} \frac{R}{R_t}\right]^{\frac{1}{2}} \right\} \tag{8.109}$$

The above equation gives the photon number in the cavity under steady state conditions for a pump rate R.

For a typical laser system, for example an Nd: glass laser (see Sec. 10.3.2),

$$V \approx 10 \, \text{cm}^3, \qquad n_0 \approx 1.5$$
$$\lambda \approx 1.06 \, \mu\text{m}, \qquad \Delta v \approx 3 \times 10^{12} \, \text{Hz}$$

so that

$$\frac{K}{V T_{21}} = \frac{c^3}{8v^2 n_0^3} \frac{1}{V \pi \Delta v} \approx 1.3 \times 10^{-13} \tag{8.110}$$

where we have used $T_{21} \approx A_{21}$ and Eq. (8.25). For such small values of $K/V T_{21}$, unless R/R_t is extremely close to unity, we can make a binomial expansion in Eq. (8.109) to get

$$n \approx \frac{R/R_t}{1 - R/R_t} \qquad \text{for} \qquad \frac{R}{R_t} < 1 - \Delta \tag{8.111}$$

$$n \approx \frac{V T_{21}}{K} \left(\frac{R}{R_t} - 1 \right) \qquad \text{for} \qquad \frac{R}{R_t} > 1 + \Delta \tag{8.112}$$

where $\Delta \gg (2K/V T_{21})^{\frac{1}{2}}$. Further

$$n = \left(\frac{V T_{21}}{K} \right)^{\frac{1}{2}} \qquad \text{for} \qquad \frac{R}{R_t} = 1 \tag{8.113}$$

Fig. 8.10 Variation of photon number n in the oscillating cavity mode as a function of pumping rate R; R_t is the threshold pump rate.

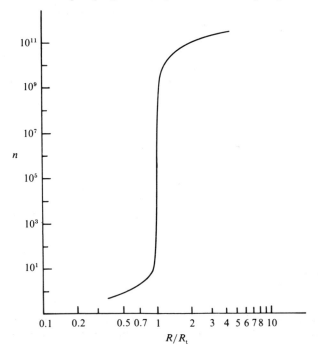

Fig. 8.10 shows a typical variation of n with R/R_t. As is evident $n \approx 1$ for $R < R_t$ and approaches 10^{12} for $R > R_t$. Thus R_t as given by Eq. (8.108) gives the threshold pump rate for laser oscillation.

Problem 8.2: Show that the threshold pump rate R_t given by Eq. (8.108) is consistent with that obtained in Sec. 8.4.

From the above analysis it follows that when the pumping rate is below threshold $(R < R_t)$ then the number of photons in the cavity mode is very small (~ 1). As one approaches the threshold, the number of photons in the preferred cavity mode (having higher gain and lower cavity losses) increases at a tremendous rate and as one passes the threshold, the number of photons in the oscillating cavity mode becomes extremely large. At the same time the number of photons in other cavity modes which are below threshold remains orders of magnitude smaller.

In addition to the sudden increase in the number of photons in the cavity mode and hence laser output power, the output also changes from an incoherent to a coherent emission. The output becomes an almost pure sinusoidal wave with a well defined wavefront, apart from small amplitude and phase fluctuations caused by the ever present spontaneous emission.[†] It is this spontaneous emission which determines the ultimate linewidth of the laser (see Sec. 9.4).

If the only loss mechanism in the cavity is that arising from output coupling due to the finite reflectivity of one of the mirrors, then the output laser power will be

$$P_{\text{out}} = (n/t_c)hv \tag{8.114}$$

where n/t_c is the number of photons escaping from the cavity per unit time and hv is the energy of each photon. Taking K/VT_{21} as given by Eq. (8.110) and $t_c \approx 10^{-8}$ s, for $R/R_t = 2$ we obtain

$$P_{\text{out}} = 144 \, \text{W}$$

Example: It is interesting to compare the number of photons per cavity mode in an oscillating laser and in a black body at a temperature T. The number of photons/mode in a black body is (see e.g., Thyagarajan and Ghatak (1981), Appendix C)

$$n = \frac{1}{e^{\hbar\omega/k_B T} - 1} \tag{8.115}$$

[†] In an actual laser system, the ultimate purity of the output beam is restricted due to mechanical vibrations of the laser, mirrors, temperature fluctuations etc.

Hence for $\lambda = 1.06\,\mu\mathrm{m}$, $T = 1000\,\mathrm{K}$, we obtain

$$n \approx \frac{1}{e^{13.5} - 1} \approx 1.4 \times 10^{-6}$$

which is orders of magnitude smaller than in an oscillating laser (see Fig. 8.10).

From Eq. (8.112) we may write for the change in number of photons dn for a change dR in the pump rate

$$\frac{\mathrm{d}n}{\mathrm{d}R} = \frac{VT_{21}}{K}\frac{1}{R_\mathrm{t}} = Vt_\mathrm{c}$$

or

$$V\,\mathrm{d}R = \mathrm{d}n/t_\mathrm{c} \tag{8.116}$$

where we have used Eq. (8.108). The LHS of Eq. (8.116) represents the additional number of atoms that are being pumped per unit time into the upper laser level and the RHS represents the additional number of photons that is being lost from the cavity. Thus above threshold all the increase in pump rate goes towards the increase in the laser power.

Example: Let us consider an Nd:glass laser (see Sec. 10.3.2) with the parameters given on page 230 and having

$$d = 10\,\mathrm{cm},$$
$$R_1 = 0.95, \qquad R_2 = 1.00$$

For these values of the parameters, using Eq. (8.28) we have

$$t_\mathrm{c} \approx -\frac{2n_0 d}{c \ln R_1 R_2} \approx 1.96 \times 10^{-8}\,\mathrm{s}$$

and

$$\frac{VT_{21}}{K} \approx \frac{V}{Kt_\mathrm{sp}} = \frac{4v^2 V n_0^3}{c^3 g(\omega)} \tag{8.117}$$

Thus for $R/R_\mathrm{t} = 2$, i.e., for a pumping rate twice the threshold value (see Eq. (8.112))

$$n = VT_{21}/K \approx 7.7 \times 10^{12}$$

Hence the energy inside the cavity is

$$E = nh\nu$$
$$\approx 1.4 \times 10^{-6}\,\mathrm{J} \tag{8.118}$$

If the only loss mechanism is the finite reflectivity of one of the mirrors, then the output power will be

$$P_\mathrm{out} = \frac{nh\nu}{t_\mathrm{c}} \approx 74\,\mathrm{W}$$

Problem 8.3: In the above example, if it is required that there be 1 W of power from the mirror at the left and 73 W of power from the right mirror, what should the reflectivities of the two mirrors be? Assume the absence of all other loss mechanisms in the cavity.

(Answer: $R_1 = 0.9993$, $R_2 = 0.9507$)

Example: In this example, we will obtain the relationship between the output power of the laser and the energy present inside the cavity by considering radiation to be making to and fro oscillations in the cavity. Fig. 8.11(*a*) shows the cavity of length l bounded by mirrors of reflectivities 1 and R and filled by a medium characterized by the gain coefficient α. Let us for simplicity assume absence of all other loss mechanisms. Fig. 8.11(*b*) shows schematically the variation of intensity along the length of the resonator when the laser oscillates under steady state conditions. For such a case, the intensity after one round trip I_4 must be equal to the intensity at the same point at the start of the round trip. Hence

$$R\,e^{2\alpha l} = 1 \tag{8.119}$$

Also, recalling the definition of cavity lifetime (see Eq. (8.28) with $\alpha_l = 0$), we have

$$t_c = -2l/c \ln R = 1/\alpha c \tag{8.120}$$

Now let us consider a plane P inside the resonator. Let the distance of the plane from mirror M_1 be x. Thus if I_1 is the intensity of the beam at mirror M_1, then assuming exponential amplification, the intensity of the beam going from left to right at P is

$$I_+ = I_1\,e^{\alpha x} \tag{8.121}$$

Fig. 8.11 (*a*) A resonator of length l bound by mirrors of reflectivities 1 and R and filled by a medium of gain coefficient α. (*b*) Curves 1 and 2 represent the qualitative variation of intensity associated with the waves propagating in the $+x$ and $-x$ directions respectively. The sudden drop in the intensity from I_2 to I_3 is due to the finite transmittivity of the mirror M_2.

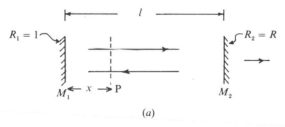

(*a*)

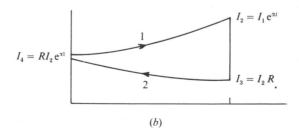

(*b*)

Similarly, the intensity of the beam going from right to left at P is

$$I_- = I_1 e^{\alpha l} R e^{\alpha(l-x)} = I_1 R e^{2\alpha l} e^{-\alpha x}$$
$$= I_1 e^{-\alpha x} \tag{8.122}$$

Hence the energy density at x is

$$u(x) = \frac{I_+ + I_-}{c} = \frac{I_1}{c}(e^{\alpha x} + e^{-\alpha x}) \tag{8.123}$$

If A is the area of cross section, then the total energy in the cavity is

$$W = \int \int u \, dA \, dx = A \int_0^l u \, dx$$
$$= \frac{AI_1}{\alpha c} e^{\alpha l}(1-R) \tag{8.124}$$
$$= AI_1 t_c e^{\alpha l}(1-R)$$

where we have used Eqs. (8.119) and (8.120) and have assumed, for the sake of simplicity, uniform intensity distribution in the transverse direction. Now the power emerging from mirror M_2 is

$$P_{\text{out}} = I_2 A(1-R)$$
$$= I_1 A e^{\alpha l}(1-R)$$
$$= W/t_c \tag{8.125}$$

which is consistent with Eq. (8.114).

8.7   Optimum output coupling

In the last section we obtained the steady state energy inside the resonator cavity as a function of the pump rate. In order to get an output laser beam, one of the mirrors is made partially transparent so that a part of the energy is coupled out. In this section we shall obtain the optimum reflectivity of the mirror so as to have a maximum output power.

The fact that an optimum output coupling exists can be understood as follows. If one has an almost zero output coupling (i.e., if both mirrors are almost 100% reflecting) then even though the laser may be oscillating, the output power will be almost zero. As one starts to increase the output coupling, the energy inside the cavity will start to decrease since the cavity loss is being increased but, since one is taking out a larger fraction of power, the output power starts increasing. The output power will start decreasing again if the reflectivity of the mirror is continuously reduced since if it is made too small, then for that pumping rate, the losses will exceed the gain and the laser will stop oscillating. Thus for a given pumping rate, there must be an optimum output coupling which gives the maximum output power.

In Sec. 8.4 we showed that the cavity lifetime of a passive resonator is

$$\frac{1}{t_c} = \frac{c}{2dn_0}(2\alpha_1 d - \ln R_1 R_2)$$

$$= \frac{1}{t_i} + \frac{1}{t_e} \tag{8.126}$$

where

$$\frac{1}{t_i} = \frac{c\alpha_1}{n_0}, \quad \frac{1}{t_e} = -\frac{c}{2dn_0}\ln R_1 R_2 \tag{8.127}$$

t_i accounts for all loss mechanisms except for the output coupling due to the finite mirror reflectivities and t_e for the loss due to output coupling only. Thus, the number of photons escaping the cavity due to finite mirror reflectivity will be n/t_e and hence the output power will be

$$P_{\text{out}} = \frac{nh\nu}{t_e}$$

$$= \frac{h\nu}{t_e}\frac{VT_{21}}{K}\left[\frac{RK}{T_{21}}\left(\frac{1}{t_i} + \frac{1}{t_e}\right)^{-1} - 1\right] \tag{8.128}$$

where we have used Eqs. (8.108), (8.112) and (8.126). The optimum output power will correspond to the value of t_e satisfying $\partial P_{\text{out}}/\partial t_e = 0$ which gives

$$\frac{1}{t_e} = \left(\frac{RK}{T_{21}t_i}\right)^{\frac{1}{2}} - \frac{1}{t_i} \tag{8.129}$$

Using Eqs. (8.126) and (8.108), the above equation can be simplified to

$$\frac{1}{t_e} = \frac{1}{t_i}\left(\frac{R}{R_t} - 1\right) \tag{8.130}$$

Substituting for t_e from Eq. (8.130) in Eq. (8.128) we obtain the maximum output power as

$$P_{\text{max}} = h\nu RV[1 - (T_{21}/KRt_i)^{\frac{1}{2}}]^2 \tag{8.131}$$

It is interesting to note that the optimum t_e and hence the optimum reflectivity is a function of the pump rate R.

Even though the output power passes through a maximum as the transmittivity $\mathcal{T}(= 1 - \mathcal{R})$ of the mirror is increased, the energy inside the cavity monotonically reduces from a maximum value as $\mathcal{T}$ is increased. This may be seen from the fact that the energy in the cavity is

$$E = nh\nu = \frac{VT_{21}}{K}\left(\frac{KRt_c}{T_{21}} - 1\right)h\nu \tag{8.132}$$

Thus as $\mathcal{T}$ is increased, t_c reduces and hence E reduces monotonically finally becoming zero when

$$t_c = T_{21}/KR \qquad (8.133)$$

beyond which the losses become more than the gain.

Problem 8.4: Using Eq. (8.130) calculate the optimum reflectivity of one of the mirrors of the resonator (assuming the other mirror to have 100% reflectivity) for $R = 2R_t$. Assume the length of the resonator to be 100 cm, $n_0 = 1$ and the intrinsic loss per unit length to be $3 \times 10^{-5} \, \mathrm{cm}^{-1}$.

8.8   Line broadening mechanisms

As we mentioned in Sec. 8.2 the radiation coming out of a collection of atoms making transitions between two energy levels is never perfectly monochromatic. This line broadening is described in terms of the lineshape function $g(\omega)$ that was introduced in Sec. 8.2. In this section, we shall discuss some important line broadening mechanisms and obtain the corresponding $g(\omega)$. A study of line broadening is extremely important since it determines the operation characteristics of the laser such as the threshold population inversion, the number of oscillating modes etc.

The various broadening mechanisms can be broadly classified as homogeneous or inhomogeneous broadening. In the case of homogeneous broadening (like natural or collision broadening) the mechanism acts to broaden the response of each atom in an identical fashion and for such a case the probability of absorption or emission of radiation of a certain frequency is the same for all atoms in the collection. Thus there is nothing which distinguishes one group of atoms from another in the collection. In the case of inhomogeneous broadening, different groups of atoms are distinguished by different frequency responses. Thus, for example, in Doppler broadening, group of atoms having different velocity components are distinguishable and they have different spectral responses. Similarly broadening caused by local inhomogeneities of a crystal lattice acts to shift the central frequency of the response of individual atoms by different amounts thereby leading to inhomogeneous broadening. In the following, we shall discuss natural, collision and Doppler broadening.

8.8.1   *Natural broadening*

We have seen earlier that an excited atom can emit its energy in the form of spontaneous emission. In order to investigate the spectral distribution of this spontaneous radiation, we recall that the rate of decrease of the number of atoms in level 2 due to transitions from level 2 to level 1 is (see Eq. (8.11))

$$dN_2/dt = -A_{21}N_2 \tag{8.134}$$

For every transition an energy $\hbar\omega_0 = E_2 - E_1$ is released. Thus the energy emitted per unit time per unit volume will be

$$W(t) = |dN_2/dt|\hbar\omega_0$$
$$= N_{20}A_{21}\hbar\omega_0 e^{-A_{21}t} \tag{8.135}$$

where we have used Eqs. (8.134) and (8.12). Since Eq. (8.135) describes the variation of the intensity of the spontaneously emitted radiation, we may write the electric field associated with the spontaneous radiation as

$$\mathscr{E}(t) = \mathscr{E}_0 e^{i\omega_0 t} e^{-t/2t_{sp}} \tag{8.136}$$

where $t_{sp} = 1/A_{21}$ and we have used the fact that intensity is proportional to the square of the electric field. Thus the electric field associated with spontaneous emission decreases exponentially.

In order to calculate the spectrum associated with the wave described by Eq. (8.136), we first take the Fourier transform:

$$\tilde{\mathscr{E}}(\omega) = \int_{-\infty}^{\infty} \mathscr{E}(t) e^{-i\omega t} dt$$

$$= \mathscr{E}_0 \int_0^{\infty} \exp\left[i(\omega_0 - \omega)t - t/2t_{sp}\right] dt$$

$$= \mathscr{E}_0 \frac{1}{1/2t_{sp} + i(\omega - \omega_0)} \tag{8.137}$$

where $t = 0$ is the time at which the atoms start emitting radiation. The power spectrum associated with the radiation will be proportional to $|\tilde{\mathscr{E}}(\omega)|^2$. Hence we may write the lineshape function associated with the spontaneously emitted radiation as

$$g(\omega) = K\frac{1}{(\omega - \omega_0)^2 + 1/4t_{sp}^2}$$

where K is a constant of proportionality which is determined such that $g(\omega)$ is normalized. Substituting for $g(\omega)$ in Eq. (8.13), and integrating, one can show that

$$K = 1/2\pi t_{sp} \tag{8.138}$$

Thus the normalized lineshape function is

$$g(\omega) = \frac{2t_{sp}}{\pi} \frac{1}{1 + 4(\omega - \omega_0)^2 t_{sp}^2} \qquad (8.139)$$

The above functional form is referred to as a Lorentzian and is plotted in Fig. 8.12. The full width at half maximum (FWHM) of the Lorentzian is

$$\Delta\omega_N = 1/t_{sp} \qquad (8.140)$$

Fig. 8.12 The Lorentzian and Gaussian lineshape functions having the same FWHM.

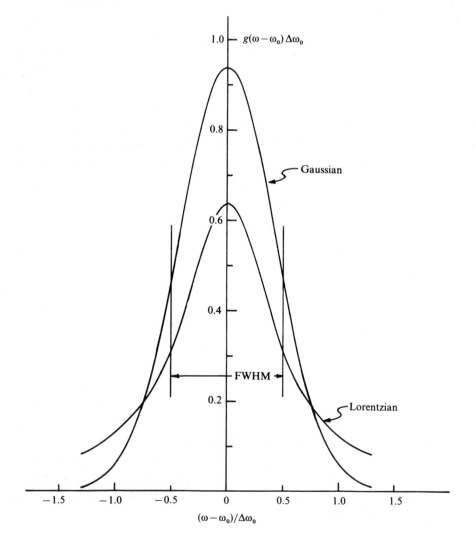

Thus, Eq. (8.139) can also be written as

$$g(\omega) = \frac{2}{\pi \Delta \omega_N} \frac{1}{1 + 4(\omega - \omega_0)^2/(\Delta \omega_N)^2} \tag{8.141}$$

A more precise derivation of Eq. (8.141) is given in Appendix F of Thyagarajan and Ghatak (1981).

Example: The spontaneous lifetime of the sodium level leading to a D_1 line ($\lambda = 5891 \text{ Å}$) is 16 ns. Thus the natural line width (FWHM) will be

$$\Delta \nu_N = 1/2\pi t_{sp} \approx 10 \text{ MHz} \tag{8.142}$$

which corresponds to $\Delta \lambda \approx 0.0001 \text{ Å}$.

8.8.2 *Collision broadening*

In a gas, random collisions occur between the atoms. In such a collision process, the energy levels of the atoms change when the atoms are very close due to their mutual interaction. Let us consider an atom which is emitting radiation and which collides with another atom. When the colliding atoms were far apart, their energy levels were unperturbed and the radiation was purely sinusoidal (if we neglect the decay in the amplitude due to spontaneous emission). As the atoms come close together their energy levels are perturbed and thus the frequency of emission changes during the collision time. After the collision the frequency returns to its original value.

If τ_c represents the time between collisions and $\Delta \tau_c$ the collision time then one can obtain order of magnitude expressions as follows:

$$\Delta \tau_c \approx \frac{\text{interatomic distance}}{\text{average thermal velocity}} \tag{8.143}$$

$$\approx \frac{1 \text{ Å}}{500 \text{ m/s}} \approx 2 \times 10^{-13} \text{ s}$$

$$\tau_c \approx \frac{\text{mean free path}}{\text{average thermal velocity}} \simeq \frac{5 \times 10^{-4} \text{ m}}{500 \text{ (m/s)}} \tag{8.144}$$

$$\approx 10^{-6} \text{ s}$$

Thus the collision time is very small compared to the time between collisions and thus the collisions may be taken to be almost instantaneous. Since the collision time $\Delta \tau_c$ is random, the phase of the wave after the collision is arbitrary with respect to the phase before the collision. Thus each collision may be assumed to lead to random phase changes as shown in Fig. 8.13. The wave shown in Fig. 8.13 is no longer monochromatic and this broadening is referred to as *collision broadening*.

In order to obtain the lineshape function for collision broadening, we note that the field associated with the wave shown in Fig. 8.13 can be represented by

$$\mathscr{E}(t) = \mathscr{E}_0\, e^{i(\omega_0 t + \phi)} \tag{8.145}$$

where the phase ϕ remains constant for $t_0 \lesssim t \lesssim t_0 + \tau_c$ and at each collision the phase ϕ changes randomly.

Since the wave is sinusoidal between two collisions, the spectrum of such a wave will be given by

$$\tilde{\mathscr{E}}(\omega) = \frac{1}{2\pi} \int_{t_0}^{t_0 + \tau_c} \mathscr{E}_0\, e^{i(\omega_0 t + \phi)} e^{-i\omega t}\, dt$$

$$= \frac{1}{2\pi}\mathscr{E}_0\, e^{i[(\omega_0 - \omega)t_0 + \phi]}\, \frac{e^{i(\omega_0 - \omega)\tau_c} - 1}{i(\omega_0 - \omega)} \tag{8.146}$$

The power spectrum of such a wave will be

$$I(\omega) \propto |\tilde{\mathscr{E}}(\omega)|^2 = \left(\frac{\mathscr{E}_0}{\pi}\right)^2 \frac{\sin^2\left[(\omega - \omega_0)\tau_c/2\right]}{(\omega - \omega_0)^2} \tag{8.147}$$

Now, at any instant, the radiation coming out of the atomic collection would be from atoms with different values of τ_c. In order to obtain the power spectrum we must multiply $I(\omega)$ by the probability $P(\tau_c)\, d\tau_c$ that the atom suffers a collision in the time interval between τ_c and $\tau_c + d\tau_c$ and integrate from 0 to ∞. It can be shown from kinetic theory that (see, e.g., Gopal (1974))

$$P(\tau_c)\, d\tau_c = \left(\frac{1}{\tau_0}\right) e^{-\tau_c/\tau_0}\, d\tau_c \tag{8.148}$$

Fig. 8.13 The wave coming out of an atom undergoing random collisions at which there are abrupt changes of phase.

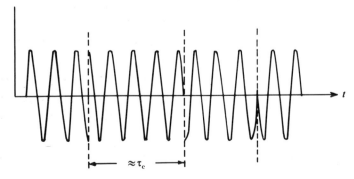

where τ_0 represents the mean time between two collisions. Notice that

$$\int_0^\infty P(\tau_c)\,\mathrm{d}\tau_c = 1, \qquad \int_0^\infty \tau_c P(\tau_c)\,\mathrm{d}\tau_c = \tau_0 \tag{8.149}$$

Hence the lineshape function for collision broadening will be

$$g(\omega) \propto \int_0^\infty I(\omega)P(\tau_c)\,\mathrm{d}\tau_c$$

$$= \left(\frac{\mathscr{E}_0}{\pi}\right)^2 \frac{1}{2} \frac{1}{(\omega - \omega_0)^2 + 1/\tau_0^2}$$

which is again a Lorentzian. The normalized lineshape function will thus be

$$g(\omega)\,\mathrm{d}\omega = \frac{\tau_0}{\pi} \frac{1}{1 + (\omega - \omega_0)^2\tau_0^2}\,\mathrm{d}\omega \tag{8.150}$$

and the FWHM will be

$$\Delta\omega_c = 2/\tau_0 \tag{8.151}$$

Thus a mean collision time of $\sim 10^{-6}$ s corresponds to a Δv of 0.3 MHz.

Problem 8.5: In the presence of both natural and collision broadening, in addition to the sudden phase changes at every collision, there will also be an exponential decay of the field as represented by Eq. (8.136). Show that in such a case, the FWHM is given by

$$\Delta\omega = 1/t_{sp} + 2/\tau_0 \tag{8.152}$$

8.8.3 Doppler broadening

In a gas, atoms move randomly and when a moving atom interacts with electromagnetic radiation, the apparent frequency of the wave is different from that seen from a stationary atom; this is called the Doppler effect and the broadening caused by this is termed Doppler broadening.

In order to obtain $g(\omega)$ for Doppler broadening, we consider radiation of frequency ω passing through a collection of atoms which have a resonant frequency ω_0 and which move randomly. (We neglect natural and collision broadening in this discussion.) In order that an atom may interact with the incident radiation, it is necessary that the apparent frequency seen by the atom in its frame of reference be ω_0. If the radiation is assumed to propagate along the z-direction, then the apparent frequency seen by the atom having a z-component of velocity v_z will be

$$\tilde{\omega} = \omega(1 - v_z/c) \tag{8.153}$$

Hence for a strong interaction, the frequency of the incident radiation must be such that $\tilde{\omega} = \omega_0$. Thus

$$\omega = \omega_0(1 - v_z/c)^{-1} \approx \omega_0(1 + v_z/c) \tag{8.154}$$

where we have assumed $v_z \ll c$. Thus the effect of the motion is to change the resonant frequency of the atom.

In order to obtain the $g(\omega)$ due to Doppler broadening, we note that the probability that an atom has a z-component of velocity lying between v_z and $v_z + dv_z$ is given by the Maxwell distribution

$$P(v_z)\,dv_z = \left(\frac{M}{2\pi k_B T}\right)^{\frac{1}{2}} \exp\left(-\frac{Mv_z^2}{2k_B T}\right) dv_z \tag{8.155}$$

where M is the mass of the atom and T the absolute temperature of the gas. Hence the probability $g(\omega)\,d\omega$ that the transition frequency lies between ω and $\omega + d\omega$ is equal to the probability that the z-component of the velocity of the atom lies between v_z and $v_z + dv_z$ where

$$v_z = \frac{(\omega - \omega_0)}{\omega_0} c$$

Thus

$$g(\omega)\,d\omega = \frac{c}{\omega_0}\left(\frac{M}{2\pi k_B T}\right)^{\frac{1}{2}} \exp\left[-\frac{Mc^2}{2k_B T}\frac{(\omega - \omega_0)^2}{\omega_0^2}\right] d\omega \tag{8.156}$$

which corresponds to a Gaussian distribution. The lineshape function is peaked at ω_0, and the FWHM is given by

$$\Delta\omega_D = 2\omega_0\left(\frac{2k_B T}{Mc^2}\ln 2\right)^{\frac{1}{2}} \tag{8.157}$$

In terms of $\Delta\omega_D$ Eq. (8.156) can be written as

$$g(\omega)\,d\omega = \frac{2}{\Delta\omega_D}\left(\frac{\ln 2}{\pi}\right)^{\frac{1}{2}} \exp\left[-4\ln 2\frac{(\omega - \omega_0)^2}{(\Delta\omega_D)^2}\right] d\omega \tag{8.158}$$

Fig. 8.12 shows a comparative plot of a Lorentzian and a Gaussian line having the same FWHM. It can be seen that the peak value of the Gaussian is more and that the Lorentzian has a much longer tail. As an example, for the D_1 line of sodium ($\lambda \simeq 5891$ Å) at $T = 500$ K, $\Delta v_D = 1.7 \times 10^9$ Hz which corresponds to $\Delta\lambda_D \approx 0.02$ Å. For neon atoms corresponding to $\lambda = 6328$ Å (the red line of the He–Ne laser) at 300 K, we have $\Delta v_D \approx 1600$ MHz where we have used $M_{Ne} \approx 20 \times 1.67 \times 10^{-27}$ kg. For the vibrational transition of the carbon dioxide molecule leading to the 10.6 μm radiation, at

$T = 300\,\text{K}$, we have

$$\Delta v_{\text{D}} \approx 5.6 \times 10^7 \,\text{Hz} \Rightarrow \Delta \lambda_{\text{D}} \approx 0.19\,\text{Å}$$

where we have used $M_{\text{CO}_2} \approx 44 \times 1.67 \times 10^{-27}\,\text{kg}$.

In all the above discussions we have considered a single broadening mechanism at a time. In general, all broadening mechanisms will be present simultaneously and the resultant lineshape function has to be evaluated by performing a convolution of the different lineshape functions.

Problem 8.6: Obtain the lineshape function in the presence of both natural and Doppler broadening.

Solution: From Maxwell's velocity distribution, the fraction of atoms with their centre frequency lying between ω' and $\omega' + d\omega'$ is given by

$$f(\omega')\,d\omega' = \left(\frac{M}{2\pi k_{\text{B}} T} \right)^{\frac{1}{2}} \frac{c}{\omega_0} \exp\left[-\frac{Mc^2}{2k_{\text{B}} T} \frac{(\omega' - \omega_0)^2}{\omega_0^2} \right] d\omega' \tag{8.159}$$

These atoms are characterized by a naturally broadened line-shape function described by

$$h(\omega - \omega') = \frac{2t_{\text{sp}}}{\pi} \frac{1}{1 + (\omega - \omega')^2 4 t_{\text{sp}}^2} \tag{8.160}$$

Thus the resultant lineshape function will be given by

$$g(\omega) = \int f(\omega') h(\omega - \omega')\,d\omega' \tag{8.161}$$

which is nothing but the convolution of $f(\omega)$ with $h(\omega)$.

Additional Problems

Problem 8.7: Consider the two level system shown in Fig. 8.1 with $E_1 = -13.6\,\text{eV}$ and $E_2 = -3.4\,\text{eV}$. Assume $A_{21} \approx 6 \times 10^8\,\text{s}^{-1}$. (a) What is the frequency of light emitted due to transitions from E_2 to E_1? Assuming the emission to have only natural broadening, what is the FWHM of the emission? What is the population ratio N_2/N_1 at $T = 300\,\text{K}$? (b) If the atomic system containing N_0 atoms/cm^3 of the above atoms is irradiated by a beam of intensity I_0 at the line centre frequency, calculate the value of I_0 required to produce $N_1 = 2N_2$.
(Answer: (a) $v \approx 2.5 \times 10^{15}\,\text{Hz}$, $\Delta v = A_{21}/2\pi \simeq 10^8\,\text{Hz}$, $N_2/N_1 \approx e^{-394}$. (b) $I_0 = \hbar\omega^3 n_0^2/\pi^2 c^2 g(\omega_0)$

Problem 8.8: Given that the gain coefficient in a Doppler broadened line is

$$\alpha(v) = \alpha(v_0) \exp\left[-4\ln 2(v - v_0)^2/(\Delta v_0)^2 \right]$$

where v_0 is the centre frequency and Δv_0 is the FWHM and that the gain coefficient

at the line centre is twice the loss averaged per unit length, calculate the bandwidth over which oscillation can take place.
(Answer: Δv_0)

Problem 8.9: Consider an atomic system as shown below:

$$
\begin{array}{ll}
3\text{————} & E_3 = 3\,\text{eV} \\
2\text{————} & E_2 = 1\,\text{eV} \\
1\text{————} & E_1 = 0\,\text{eV}
\end{array}
$$

The A coefficients of the various transitions are given by

$$A_{32} = 7 \times 10^7\,\text{s}^{-1}, \quad A_{31} = 10^7\,\text{s}^{-1}, \quad A_{21} = 10^8\,\text{s}^{-1}$$

(a) Show that this system cannot be used for continuous wave laser oscillation between levels 2 and 1.
(b) What is the spontaneous lifetime of level 3?
(c) Suppose at $t = 0$, N_0 atoms are lifted to level 3 by some external mechanism, describe the change of populations in levels 1, 2 and 3.
(d) If the steady state population of level 3 is 10^{15} atoms/cm^3, what is the power emitted spontaneously in the $3 \to 2$ transition?
(Answer: (b) $t_{\text{sp}} = 1.2 \times 10^{-8}\,\text{s}$ (d) $2.2 \times 10^{10}\,\text{W/m}^3$)

Problem 8.10: Given that the optimum external coupling occurs at

$$\frac{1}{t_e} = \frac{1}{t_i}\left(\frac{R}{R_t} - 1\right)$$

calculate the optimum reflectivity of one of the mirrors of the resonator (assuming the other mirror to be 100% reflecting) for $R = 2R_t$. Assume the length of the resonator to be 50 cm, $n_0 = 1$ and the intrinsic loss per unit length to be $3 \times 10^{-4}\,\text{m}^{-1}$. If the power output at the optimum coupling is 10 mW, what is the corresponding energy inside the cavity?
(Answer: $R \approx 0.9997$, Energy $\simeq 1.1 \times 10^{-7}\,\text{J}$)

9

Lasers: II

9.1 Introduction

In the last chapter we showed that for a medium to behave as an amplifier of light waves, one must create a state of population inversion between two energy levels of the system. In order to convert this amplifier to an oscillator i.e., a source of radiation, one must provide optical feedback to the amplifier. This is brought about by a pair of mirrors between which is enclosed the amplifying medium. This pair of mirrors forms what is known as the optical resonator. Fig. 9.1 shows a simple optical resonator formed by a pair of plane mirrors. Since the sides of the resonator cavity are open such resonators are also known as open resonators. The resonators can only support certain specific field configurations and certain specific oscillation frequencies. These are referred to as the modes of oscillation of the resonator. The need for an open resonator rather than a closed resonator (as in microwaves) arises because of the fact that the number of modes that a closed resonator can support within the frequency band in which the amplifying medium can provide gain is so large that the output from such a system would be far from monochromatic. By removing the side walls of a closed cavity and thus obtaining an open cavity, one increases the loss of most of the modes to such an extent that they cannot be sustained in the cavity. Thus one can achieve oscillation at a very few or even a

Fig. 9.1 A simple optical resonator formed by a pair of plane mirrors facing each other.

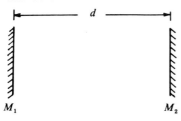

single frequency. In this chapter we first obtain the modes of a closed cavity and show that the number of modes within the oscillating linewidth is very large. In Sec. 9.3 we discuss the quality factor of a resonator and in Sec. 9.4 we obtain the ultimate linewidth of a laser. Sec. 9.5 discusses various mode selection techniques and Secs. 9.6 and 9.7 discusses Q-switching and mode locking in lasers. In Sec. 9.8 we obtain the modes in resonators formed by spherical mirrors.

9.2   Modes of a rectangular cavity and the open planar resonator

We consider a closed rectangular cavity of dimensions $2a \times 2b \times d$ with perfectly conducting walls (see Fig. 9.2). The field inside the cavity must satisfy the wave equation (see Eq. (1.16)) given by

$$\nabla^2 \mathscr{E} = \frac{1}{c^2} \frac{\partial^2 \mathscr{E}}{\partial t^2} \tag{9.1}$$

where c represents the velocity of light in the medium filling the cavity. Since the walls of the cavity are assumed to be perfectly conducting, the tangential component of the electric field must vanish at the walls. If $\hat{\mathbf{n}}$ represents the unit normal to the wall surface then we must have

$$\mathscr{E} \times \hat{\mathbf{n}} = 0 \tag{9.2}$$

In Appendix A we have obtained the solution of the wave equation by using the separation of variables technique. The time dependence is of the form $e^{i\omega t}$ where ω is the angular frequency of the wave. By considering each Cartesian component of Eq. (9.1), one can show that the x, y and z dependences are linear combinations of sine and cosine functions. Thus we

Fig. 9.2 A closed rectangular cavity bounded by perfectly conducting walls.

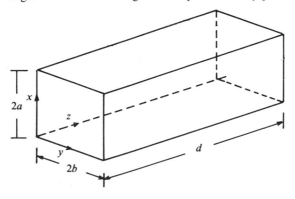

may write

$$\mathscr{E}_x = (A_1 \sin k_x x + B_1 \cos k_x x)(A_2 \sin k_y y + B_2 \cos k_y y)$$
$$\times (A_3 \sin k_z z + B_3 \cos k_z z)e^{i\omega t} \tag{9.3}$$

where k_x, k_y and k_z are the x, y and z-components of the propagation vector **k** and

$$k^2 = k_x^2 + k_y^2 + k_z^2 = \omega^2/c^2$$

Now, since $\mathscr{E}_x$ is a tangential component on the planes $y = 0$, $y = 2b$, $z = 0$ and $z = d$, it has to vanish on these planes. Thus we have from Eq. (9.3)

$$B_2 = 0, \quad B_3 = 0 \tag{9.4}$$

$$\sin (k_y 2b) = 0 \tag{9.5}$$

$$\sin (k_z d) = 0 \tag{9.6}$$

The last two equations imply

$$k_y = n\pi/2b, \quad k_z = q\pi/d; \quad n, q = 0, 1, 2, \ldots \tag{9.7}$$

where we have intentionally included the value 0, which, in this case, would lead to the trivial solution of $\mathscr{E}_x$ vanishing everywhere. In a similar manner we can show that the x and z dependences of $\mathscr{E}_y$ will be $\sin k_x x$ and $\sin k_z z$, respectively with

$$k_x = m\pi/2a; \quad m = 0, 1, 2, \ldots \tag{9.8}$$

and k_z again given by Eq. (9.7). Similarly the x and y dependences of $\mathscr{E}_z$ would be $\sin k_x x$ and $\sin k_y y$.

Now, due to the above forms of the x dependences of $\mathscr{E}_y$ and $\mathscr{E}_z$, $\partial \mathscr{E}_y/\partial y$ and $\partial \mathscr{E}_z/\partial z$ will vanish on the surface $x = 0$ and $x = 2a$. Hence on the planes $x = 0$ and $x = 2a$, the condition $\nabla \cdot \mathscr{E} = 0$ leads to $\partial \mathscr{E}_x/\partial x = 0$. Hence the x dependence of $\mathscr{E}_x$ must be of the form $\cos k_x x$ with k_x given by Eq. (9.8). Notice that the case $m = 0$ now corresponds to a nontrivial solution.

In a similar manner, one can obtain the solutions for $\mathscr{E}_y$ and $\mathscr{E}_z$. The complete solution inside the cavity can be written as

$$E_x = E_{0x} \cos k_x x \sin k_y y \sin k_z z \tag{9.9}$$

$$E_y = E_{0y} \sin k_x x \cos k_y y \sin k_z z \tag{9.10}$$

$$E_z = E_{0z} \sin k_x x \sin k_y y \cos k_z z \tag{9.11}$$

where E_{0x}, E_{0y} and E_{0z} are constants. Since k_x, k_y and k_z are given by

Eqs. (9.7) and (9.8), the allowed frequencies of oscillation in the cavity are

$$\omega = ck = c\pi \left(\frac{m^2}{4a^2} + \frac{n^2}{4b^2} + \frac{q^2}{d^2} \right)^{\frac{1}{2}} \tag{9.12}$$

The field configurations given by Eq. (9.9)–(9.11) are called the modes of the cavity and correspond to standing wave patterns in the cavity.

If we use Eqs. (9.9)–(9.11) along with the equation $\nabla \cdot \mathscr{E} = 0$, we will have

$$\mathbf{E}_0 \cdot \mathbf{k} = 0 \tag{9.13}$$

where $\mathbf{k} = k_x \hat{\mathbf{x}} + k_y \hat{\mathbf{y}} + k_z \hat{\mathbf{z}}$ and $\mathbf{E}_0 = E_{0x} \hat{\mathbf{x}} + E_{0y} \hat{\mathbf{y}} + E_{0z} \hat{\mathbf{z}}$. Since the coefficients E_{0x}, E_{0y} and E_{0z} have to satisfy Eq. (9.13), it follows that for a given mode, i.e., for given values of m, n and q, only two of the components of $\mathbf{E}_0$ can be chosen independently. Thus a given mode can have two independent states of polarization.

Example : As a specific example, we consider a mode with

$$m = 0, \quad n = 1, \quad q = 1$$

Thus $k_x = 0$, $k_y = \pi/2b$, $k_z = \pi/d$ and using Eqs. (9.9)–(9.11), we have

$$E_x = E_{0x} \sin k_y y \sin k_z z = E_{0x} \sin \left(\frac{\pi}{2b} y \right) \left(\sin \frac{\pi}{d} z \right)$$

$$E_y = 0 \tag{9.14}$$

$$E_z = 0$$

Using the time dependence of the form $e^{i\omega t}$ and expanding the sine functions into exponentials, we may write

$$\mathscr{E} = \frac{1}{(2i)^2} \hat{\mathbf{x}} E_{0x} (e^{i(\omega t - k_y y - k_z z)} + e^{i(\omega t - k_y y + k_z z)} + e^{i(\omega t + k_y y - k_z z)} + e^{i(\omega t + k_y y + k_z z)}) \tag{9.15}$$

Thus the total field inside the cavity has been broken up into four propagating

Fig. 9.3 A mode corresponding to $m = 0$, $n = 1$, $q = 1$ consists of four plane waves (whose directions of propagation are shown as 1–4) and whose mutual interference produces a standing wave pattern in the cavity.

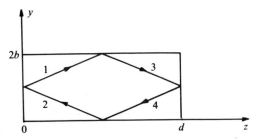

plane waves (see Fig. 9.3); in Eq. (9.15) the first term on the RHS represents a wave propagating along the $(+y, +z)$-direction, the second along the $(+y, -z)$ direction, the third along the $(-y, +z)$ direction and the fourth along the $(-y, -z)$ direction. These four plane waves interfere at every point inside the cavity to produce a standing wave pattern. However, since k_y and k_z take discrete values, the plane waves which constitute the mode make discrete angles with the axes.

As another example if we take $a = b = 1$ cm, $d = 20$ cm, and consider the mode with

$$m = 0, \quad n = 1, \quad q = 10^6$$

then

$$k_x = 0, \quad k_y = \pi/2 \, \text{cm}^{-1}, \quad k_z = 10^6 \pi/20 \, \text{cm}^{-1}$$

implying

$$k \approx 10^6 \pi/20 \, \text{cm}^{-1} \quad \text{and} \quad v = ck/2\pi = 7.5 \times 10^{14} \, \text{Hz}$$

which lies in the optical region. For such a case

$$\theta_y = \cos^{-1}(k_y/k) \approx 89.9994°$$
$$\theta_z = \cos^{-1}(k_z/k) = \cos^{-1}\{[1 + (k_y/k_z)^2]^{-\frac{1}{2}}\} \approx 0.0006°$$

and $\theta_x = 0$ because of which $\theta_y + \theta_z = 90°$. It may be noted that the component waves are propagating almost along the z-axis. In general

$$\cos^2 \theta_x + \cos^2 \theta_y + \cos^2 \theta_z = 1$$

Further, for $m \neq 0$, $n \neq 0$, $q \neq 0$, the cavity mode can be thought of as a standing wave pattern formed by eight plane waves with components of $\mathbf{k}$ given by $(\pm k_x, \pm k_y, \pm k_z)$.

Using the above formulation, we show in Appendix C that the number of modes per unit volume of the cavity in the frequency interval v to $v + dv$ is

$$p(v) \, dv = (8\pi v^2/c^3) \, dv \tag{9.16}$$

Thus, if we take for dv a typical value of the linewidth of an atomic system such as $dv = 3 \times 10^9$ Hz at $v = 3 \times 10^{14}$ Hz, then the number of modes per unit volume will be

$$p(v) \, dv = \frac{8 \times 3.14 \times (3 \times 10^{14})^2}{(3 \times 10^{10})^3} \times 3 \times 10^9 \approx 2 \times 10^8 \, \text{cm}^{-3} \tag{9.17}$$

Thus for cavities of practical dimensions, the number of oscillating modes which would lie within the atomic linewidth will be extremely large. Thus all these oscillating modes will draw energy from the atomic system and the resulting emission will be far from monochromatic. In order to have very few oscillating modes within the cavity, the dimensions of the cavity should be chosen to be of the order of the wavelength of the radiation. This is, of course, impractical at visible or infrared wavelengths.

The problem of the extremely large number of oscillating modes can be overcome by the use of open cavities which consist of two plane or curved mirrors facing each other. As we have seen earlier a mode can be considered to be a standing wave pattern formed between waves propagating within the cavity with **k** given by ($\pm k_x$, $\pm k_y$, $\pm k_z$); here k_x, k_y and k_z represent respectively the x, y and z-components of the propagation vectors of the component plane waves. Thus the angles made by the component plane waves with the x, y and z-directions will respectively be $\cos^{-1}(m\lambda/2a)$, $\cos^{-1}(n\lambda/2b)$ and $\cos^{-1}(q\lambda/d)$. Since in open resonators, the side walls of the cavity has been removed, those modes which are propagating almost along the z-direction (i.e., with a large value of q and small values of m and n) will have a loss which will be much less then the loss of the modes which make a large angle with the z-axis (i.e., large values of m and/or n). Thus on removing the side walls of the cavity, only modes having small values of m, n ($\sim 0, 1, 2, \ldots$) will have a small loss and thus as the amplifying medium placed inside the cavity is pumped, only these modes will be able to oscillate. Modes with large values of m and n will have a very large loss and thus will not be able to oscillate.

It should be noted here that since the resonator cavity is now open, all modes are lossy. Thus even the modes that have plane wave components travelling almost along the z-axis will suffer losses. Since m and n specify the field patterns along the transverse directions x and y, and q that along the longitudinal direction z, modes having different values of (m, n) are referred to as various transverse modes and modes differing in q values as various longitudinal modes.

The oscillation frequencies of the various modes of the closed cavity are given by Eq. (9.12). In order to obtain an approximate value for the oscillation frequencies of an open cavity, we may again use Eq. (9.12) with the condition $m, n \ll q$. Thus making a binomial expansion in Eq. (9.12), we obtain

$$v_{mnq} \approx \frac{c}{2}\left[\frac{q}{d} + \left(\frac{m^2}{a^2} + \frac{n^2}{b^2}\right)\frac{d}{8q}\right] \tag{9.18}$$

The difference in frequency between two adjacent modes having same values of m and n and differing in q by unity will be very nearly given by

$$\Delta v_q \approx c/2d \tag{9.19}$$

which corresponds to the longitudinal mode spacing. In addition if we completely neglect the term containing m and n, we will obtain

$$v_q \approx (c/2d)q \tag{9.20}$$

The above equation is similar to the frequencies of oscillation of a stretched string of length d.

Example: For a typical laser resonator $d \approx 100\,\text{cm}$ and assuming free space filling the cavity, the longitudinal mode spacing comes out to be $\sim 150\,\text{MHz}$.

Problem 9.1: Show that the separation between two adjacent transverse modes is much smaller than Δv_q.

Solution: The frequency separation between two modes differing in m values by unity will be

$$\Delta v_m \approx \frac{c}{2} \frac{d}{8a^2q} [m^2 - (m-1)^2]$$

$$\approx \Delta v_q \frac{\lambda d}{8a^2} (m - \tfrac{1}{2}) \tag{9.21}$$

where we have used $q \approx 2d/\lambda$ (see Eq. (9.20)). For typical values $\lambda = 6 \times 10^{-5}\,\text{cm}$, $d = 100\,\text{cm}$, $a = 1\,\text{cm}$; $\lambda d/8a^2 = 7.5 \times 10^{-3}$. Thus for $m \approx 1$, $\Delta v_m \ll \Delta v_q$.

It is of interest to mention that an open resonator consisting of two plane mirrors facing each other is, in principle, the same as a Fabry–Perot interferometer or etalon (see Sec. 2.4). The essential difference in respect of the geometrical dimensions is that in a Fabry–Perot interferometer the spacing between the mirrors is very small compared to the transverse dimension of the mirrors while in an optical resonator, the converse is true. In addition, in the former radiation is incident from outside while in the latter, the radiation is generated inside the cavity.

Earlier we showed that the modes in closed cavities are essentially superpositions of propagating plane waves. Because of diffraction effects, plane waves cannot represent modes in open cavities. Indeed, if we start with a plane wave travelling parallel to the axis from one of the mirrors, it will undergo diffraction as it reaches the second mirror and since the mirror is of finite transverse dimension, the energy in the diffracted wave which lies outside the mirror would be lost. The wave reflected from the mirror will again undergo losses when it is reflected from the first mirror. This loss is termed the diffraction loss. Fox and Li (1961) performed a numerical calculation of such a planar resonator. The analysis consisted of assuming a certain field distribution at one of the mirrors of the resonators and calculating the Fresnel diffracted field at the second mirror. The field reflected by the second mirror is used to calculate back the field distribution on the first mirror. It was shown that after many traversals the transverse field distribution settles down to a steady pattern i.e., it does not change between successive reflections but only the amplitude of the field decays exponentially in time due to diffraction losses. Such a field distribution represents a normal mode of the resonator and by changing the initial field distribution on the first mirror, other modes can also be obtained. For more details readers are referred to Fox and Li (1961); a brief analysis is also given in Thyagarajan and Ghatak (1981).

The use of curved mirrors instead of plane mirrors in the resonator has the advantage of having smaller diffraction losses. In fact, if the mirrors are sufficiently large in the transverse dimensions, the diffraction losses can almost be neglected. In Sec. 9.8 we will show that the modes of spherical mirror resonators are Hermite–Gauss functions.

9.3   The quality factor

Since an optical resonator is an open cavity, all modes suffer losses. These losses arise from the finite sizes of the mirrors (at the cavity ends) due to which there is diffraction spillover, the finite reflectivities of the mirrors and scattering and absorption by the medium filling the resonator cavity. In Sec. 8.4 we related the various losses to the cavity lifetime t_c. One can also describe the losses in the cavity by what is known as the quality factor. This is defined by

$$Q = \omega_0 \frac{\text{energy stored in the mode}}{\text{energy lost per unit time}} \qquad (9.22)$$

where ω_0 is the oscillation frequency of the mode. If $W(t)$ represents the energy in the mode at time t, then from Eq. (9.22), we obtain

$$Q = \omega_0 \frac{W(t)}{-dW/dt}$$

or

$$dW/dt = -(\omega_0/Q)W(t) \qquad (9.23)$$

whose solution is

$$W(t) = W(0)e^{-\omega_0 t/Q} \qquad (9.24)$$

Thus if t_c represents the cavity lifetime i.e., the time in which the energy in the mode decreases by a factor $1/e$, then

$$t_c = Q/\omega_0 = \frac{Q}{2\pi\nu_0} \qquad (9.25)$$

We can write for the field associated with the mode

$$E(t) = E_0 e^{i\omega_0 t} e^{-\omega_0 t/2Q} \tag{9.26}$$

The frequency spectrum of this wave train can be obtained in a manner similar to that used in Sec. 8.8 to obtain the spontaneous emission spectrum and it comes out to be

$$|\tilde{E}(v)|^2 = \frac{E_0^2}{4\pi^2} \frac{1}{(v - v_0)^2 + v_0^2/4Q^2} \tag{9.27}$$

which again represents a Lorentzian. The FWHM of the spectrum is

$$\Delta v_p = v_0/Q \tag{9.28}$$

Thus the linewidth of the passive mode depends inversely on the quality factor. The higher the quality factor (i.e., the longer the cavity lifetime, see Eq. (9.25)), the smaller will be the FWHM.

In Sec. 8.4 we derived an expression for t_c in terms of the length of the resonator and the reflectivity of the mirrors. Thus using Eqs. (8.28), (9.25) and (9.28), we have

$$Q = \frac{4\pi v_0 n_0 d}{c} \frac{1}{2\alpha_l d - \ln R_1 R_2} \tag{9.29}$$

and

$$\Delta v_p = \frac{c}{4\pi n_0 d}(2\alpha_l d - \ln R_1 R_2) \tag{9.30}$$

Example: Let us consider a typical cavity of a He–Ne laser with the following specifications:

$$d = 20\,\text{cm}, \quad n_0 \approx 1, \quad R_1 = 1, \quad R_2 = 0.98, \quad \alpha_l \approx 0 \tag{9.31}$$

For such cavity

$$\Delta v_p \approx 2.4\,\text{MHz} \tag{9.32}$$

For the same cavity, the frequency separation between adjacent longitudinal modes is

$$\delta v \approx c/2d = 750\,\text{MHz} \tag{9.33}$$

Thus the width of each mode is smaller than the separation between adjacent modes.

Example: As another example we consider a GaAs semiconductor laser (see Sec. 10.9) with the following values of the various parameters:

$$d = 500\,\mu\text{m}, \quad n_0 = 3.5, \quad R_1 = R_2 = 0.3, \quad \alpha_c \approx 0 \tag{9.34}$$

For such a cavity we obtain

$$\Delta v_p \approx 3.3 \times 10^{10}\,\text{Hz} \tag{9.35}$$

9.4   The ultimate linewidth of the laser

One of the most important properties of a laser is its ability to produce light of high spectral purity or high temporal coherence. The finite spectral width of a laser operating continuously in a single mode is caused by two mechanisms. One is the external factors which tend to perturb the cavity, for example, temperature fluctuations, vibrations, etc. randomly alter the oscillation frequency which results in a finite spectral width. The second more fundamental mechanism which determines the ultimate spectral linewidth of the laser is that due to the ever present random spontaneous emissions in the cavity. Since spontaneous emission is completely incoherent with respect to the existing energy in the cavity mode, it leads to a finite linewidth of the laser. In this section, we shall give a heuristic derivation for this ultimate linewidth of the laser (Gordon, Zeiger and Townes, 1955 and Maitland and Dunn, 1969).

In order to obtain a value for the ultimate laser linewidth, we assume that the radiation arising out of spontaneous emission represents a loss as far as the coherent output energy is concerned. This loss will then lead to a finite linewidth for the laser. We recall from Sec. 8.6 that the number of spontaneous emissions per unit time into a mode of the cavity is given by KN_2 (where K is defined by Eq. (8.98)) and N_2 represents the number of atoms/unit volume in the upper laser level. We are assuming that $N_1 \approx 0$. When the laser oscillates in a steady state then we know from Eq. (8.105) that $N_2 \approx 1/Kt_c$ where we are assuming $n \gg 1$ and t_c is the passive cavity lifetime. Thus above threshold, the number of spontaneous emissions per unit time will be $KN_2 = 1/t_c$. Thus the energy appearing per unit time in a mode due to spontaneous emission will be $h\nu_0/t_c$ where ν_0 is the oscillation frequency of the mode.

Now, the total energy contained in the mode is $nh\nu_0$ and since the output power P_{out} is given by $P_{out} = nh\nu_0/t_c$, the energy contained in the mode is $P_{out}t_c$.

We now use Eq. (9.28) and denote the linewidth of the oscillating laser caused by spontaneous emission by $\delta\nu_{sp}$ to obtain

$$Q = \frac{\nu_0}{\delta\nu_{sp}} = 2\pi\nu_0 \frac{P_{out}t_c}{h\nu_0/t_c} = \frac{2\pi P_{out}t_c^2}{h} \tag{9.36}$$

Now if $\Delta\nu_p$ is the passive cavity linewidth then $t_c = 1/2\pi\Delta\nu_p$ and thus from Eq. (9.36) we obtain

$$\delta\nu_{sp} = 2\pi(\Delta\nu_p)^2 h\nu_0/P_{out} \tag{9.37}$$

The above equation gives the ultimate linewidth of an oscillating laser and is similar to the one given by Schawlow and Townes (1958). It is interesting to note that δv_{sp} depends inversely on the output power P_{out}. This is physically due to the fact that for a given mirror reflectivity, an increase in P_{out} corresponds to an increase in the energy in the mode inside the cavity which in turn implies a greater dominance of stimulated emission over spontaneous emission. Fig. 9.4 shows a typical measured variation of the linewidth of a GaAlAs semiconductor laser which shows the linear increase of linewidth with inverse optical power.

The above derivation is rather heuristic; an analysis based on the random phase additions due to spontaneous emission is given by Jacobs (1979) which gives a result half that predicted by Eq. (9.37).

Example: Let us first consider the He–Ne laser given in Eq. (9.31) and we assume that it oscillates with an output power of 1 mW at $\lambda_0 = 6328$ Å. The spontaneous emission linewidth of the laser will be

$$\delta v_{sp} \approx 0.01 \text{ Hz} \tag{9.38}$$

which is extremely small. To emphasize how small these widths are, let us try to estimate the precision with which the length of the cavity has to be controlled in order that the oscillation frequency changes by 10^{-2} Hz. We know that the approximate oscillation frequency of a mode is $v = qc/2d$. Thus the change in frequency Δv caused by a change in length Δd is

$$\Delta v = (v/d)\,\Delta d \tag{9.39}$$

Using $d = 20$ cm, $\lambda_0 = 6328$ Å and $\delta v \approx 0.01$ Hz, we obtain

$$\Delta d \approx 4 \times 10^{-14} \text{ cm} \tag{9.40}$$

which corresponds to a stability of less than even nuclear dimensions!

Fig. 9.4 Experimentally obtained variations of laser linewidth in a single frequency GaAlAs semiconductor diode laser as a function of inverse output power. (After Mooradian (1985).)

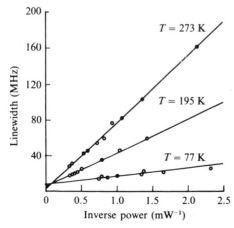

Example: As another example, we consider a GaAs semiconductor laser operating at $\lambda_0 = 0.85\,\mu$m with an output power of 1 mW with cavity dimensions as given by Eq. (9.34). For such a laser

$$\delta v_{sp} \approx 1.6\,\text{MHz} \tag{9.41}$$

which is much larger than that of a He–Ne laser. This is primarily due to the very large value of Δv_p in the case of semiconductor laser. Since Δv_p can be reduced by increasing the length of the cavity, one can use an external mirror for feedback and thus reduce the linewidth. For a more detailed discussion of semiconductor laser linewidth, readers are referred to Mooradian (1985).

9.5   Mode selection

Since optical resonators have dimensions which are large compared to the optical wavelength, there are, in general, a large number of modes which fall within the atomic linewidth and which can oscillate in the laser. Hence the output may consist of various transverse and longitudinal modes. In this section we shall describe some techniques which are used to select a single transverse and a single longitudinal mode to oscillate in the laser. For more details readers are referred to Smith (1972).

9.5.1   Transverse mode selection

We discussed in Sec. 9.2 that different transverse modes are characterized by different transverse amplitude distributions. In Secs 9.8 and 9.9 we shall show that for a stable spherical resonator, the transverse field distributions at the waist of the various modes are Hermite–Gauss functions given by

$$E_{mn}(x, y) = E_0 H_m(\sqrt{2}x/w_0) H_n(\sqrt{2}y/w_0) e^{-(x^2 + y^2)/w_0^2} \tag{9.42}$$

where m, n represent the transverse mode numbers, $H_m(\sqrt{2}x/w_0)$ and $H_n(\sqrt{2}y/w_0)$ represent Hermite polynomials (see Eq. (11.117)) and w_0 is the characteristic mode width which depends on the wavelength of operation and the resonator dimensions such as the radii of curvatures of the two mirrors and the distance between them. Fig. 9.5 shows a photograph of various transverse modes of a laser.

The fundamental transverse mode corresponds to $m = 0$ $n = 0$ for which we get

$$E_{00}(x, y) = E_0 e^{-(x^2 + y^2)/w_0^2} \tag{9.43}$$

Thus, the fundamental mode of spherical resonators is Gaussian with a 1/e amplitude radius w_0 at the waist. (In Sec. 5.4 we discussed in detail the propagation of a Gaussian beam through free space.) Most applications

Fig. 9.5 Photograph showing some of the lower order transverse modes of an optical resonator. TEM_{mn} corresponds to the (m, n)th transverse mode and TEM_{00} is the fundamental mode with Gaussian distribution (After Kogelnik and Li (1966). Photograph courtesy Dr. H. Kogelnik.)

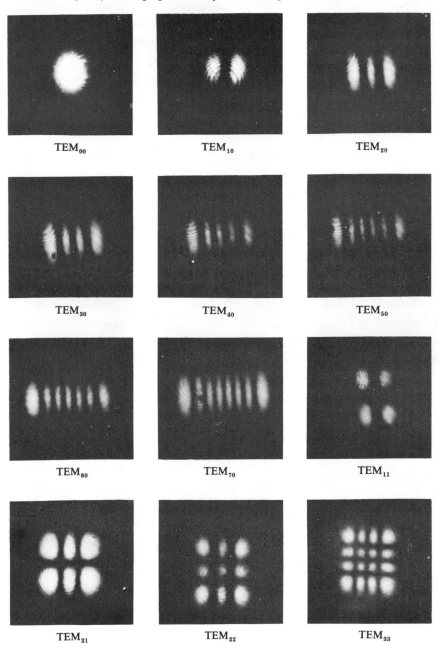

of lasers, such as holography, rangefinding, nonlinear optical experiments, etc., require the laser to oscillate in its fundamental transverse mode which is a Gaussian field distribution. This is because it is the fundamental mode which produces the minimum beam divergence, the highest power density and is also uniphase across its wavefront. In addition it is the fundamental transverse mode which produces the largest power per unit area in a focussed diffraction limited spot. In view of the above it is usually necessary to restrict oscillation of the laser only in its fundamental transverse mode.

Since the fundamental Gaussian mode is the mode with the narrowest transverse dimensions, an aperture placed inside the resonator will preferentially introduce higher losses for higher order modes. Thus if a circular aperture is introduced into the laser such that the loss suffered by all higher order modes is greater than the gain while the loss suffered by the fundamental mode is still lower than the gain, then one will have a single transverse mode oscillation.

Fig. 9.6 shows a typical configuration for producing single transverse mode oscillation (Davis, 1968) in a ruby laser. For the resonator shown in Fig. 9.6, the diffraction loss for a single pass for the fundamental mode is $\sim 20\%$ and for the first excited mode is about 50%. Thus the aperture introduces much higher losses for the higher order modes as compared to the fundamental mode.

It is interesting to note that specific higher order transverse modes can also be selected by choosing complex apertures which introduce high loss for all modes except the required mode or by profiling the reflectivity of one of the mirrors to suit the desired mode. Thus a thin wire placed normal to the axis passing through the resonator axis will select the TEM_{01} mode.

9.5.2 *Longitudinal mode selection*

We have seen in Sec. 9.2 that the various longitudinal modes corresponding to a transverse mode are approximately separated by $c/2d$,

Fig. 9.6 A typical configuration for producing only fundamental transverse mode operation of the laser.

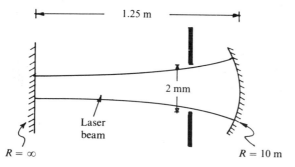

where d is the separation between the mirrors of the cavity. Hence if we consider a 50 cm long He–Ne laser operating at $\lambda = 6328$ Å and having an oscillating bandwidth of 1500 MHz (which is approximately the FWHM of the gain profile – see Sec. 8.8) then the longitudinal mode spacing will be

$$\Delta \nu = c/2d = 300 \text{ MHz} \tag{9.44}$$

Thus even if the laser is oscillating only in its fundamental transverse mode, there will still be five longitudinal modes which can oscillate in the laser (see Fig. 9.7). Thus the output will consist of five adjacent frequencies and will have only a short coherence length (see Problem 9.3). Thus in applications where a long coherence length is necessary (e.g., in holography, interferometry etc.) or where a well defined frequency is required (e.g., in spectroscopy) one will require single longitudinal mode operation of the laser in addition to its single transverse mode oscillation.

Fig. 9.7 (*a*) The longitudinal mode spacing of a resonator of length d is $c/2d$. For an oscillating bandwidth of 1500 MHz and an intermode spacing of 300 MHz, five different longitudinal modes can oscillate. (*b*) If $d < c/\delta \nu$ and there is a mode at the line centre then one can have single frequency oscillation of the laser.

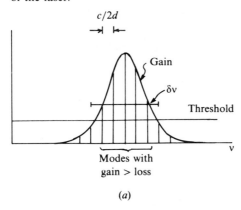

(*a*)

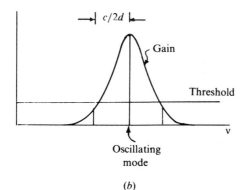

(*b*)

Referring to Fig. 9.7 we can have a simple method of obtaining single longitudinal mode oscillation by reducing the cavity length to a value such that the intermode spacing is larger than the spectral width over which oscillation can take place. Thus if the oscillating bandwidth is δv (see Fig. 9.7) then for single mode oscillation the cavity length must be such that

$$c/2d > \delta v \qquad (9.45)$$

For a He–Ne laser, $\delta v \approx 1500\,\text{MHz}$ and for single mode oscillation we must have

$$d < c/2\delta v \approx 10\,\text{cm}$$

We should note here that if one can ensure that a resonant mode exists at the centre of the gain profile then single mode oscillation can be obtained even with a cavity length of $\sim c/\delta v$ (see Fig. 9.7(*b*)).

One of the major drawbacks with the above method is that since the volume of the active medium gets very much reduced due to the restriction on the length of the cavity, the output power is small. In addition, in solid state lasers where the bandwidth is large, the above technique becomes impractical. Hence other techniques have been developed which can lead to single frequency oscillations without restricting the length of the cavity and hence are capable of high powers.

Oscillation of the laser in a given resonant mode can be achieved by introducing frequency selective elements such as Fabry–Perot etalons (see Sec. 2.4) into the laser cavity. The element should be so chosen that it introduces losses in all but the required mode so that their losses are larger than the gain. Fig. 9.8 shows a tilted Fabry–Perot etalon placed inside the resonator. The etalon consists of a pair of highly reflecting parallel surfaces which has a transmission versus frequency as shown in Fig. 9.9. As discussed in detail in Sec. 2.4 such an etalon has transmission peaks centred at frequencies given by

$$v_p = p\,\frac{c}{2nt\cos\theta}; \quad p = \text{any integer}$$

Fig. 9.8 A laser resonator with a Fabry–Perot etalon placed inside the cavity for the selection of single longitudinal mode oscillation.

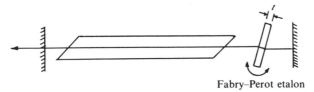

Fabry–Perot etalon

where t is the thickness of the etalon and n the refractive index of the medium between the reflecting surfaces. The width of each peak depends on the reflectivity of the surfaces; the higher the reflectivity, the sharper are the resonance peaks[†] (see Fig. 9.9). The frequency separation between two adjacent peak transmittances is

$$\Delta v = c/2nt \cos \theta \qquad (9.47)$$

which is also referred to as the free spectral range of the etalon.

If the etalon is so chosen that its free spectral range is greater than the spectral width of the gain profile then the Fabry–Perot etalon can be tilted inside the resonator so that one of the longitudinal modes of the resonator coincides with the peak transmittance of the etalon (see Fig. 9.10) and other modes are reflected away from the cavity. If in addition, the finesse of the etalon is so high as to introduce sufficiently high losses for the modes adjacent to the mode selected, then one may have a single mode oscillation (see Fig. 9.10).

Fig. 9.9 Transmittance versus frequency of a Fabry–Perot etalon for two different values of the reflectivity of the surfaces bounding the etalon. The peaks of transmission correspond to frequencies satisfying Eq. (9.46).

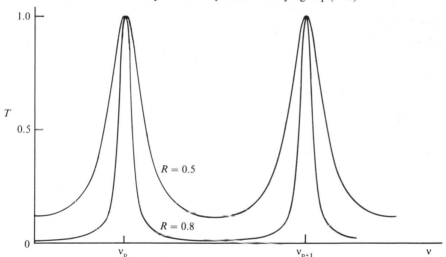

[†] The FWHM in transmittivity is approximately given by

$$\Delta v_{\frac{1}{2}} = c/\pi nt \cos \theta F^{\frac{1}{2}}$$

where F is the coefficient of finesse and is given by

$$F = 4R/(1 - R)^2$$

and R is the reflectivity of the two surfaces.

Example: Consider an argon ion laser for which the FWHM of the gain profile is about 8 GHz. Thus for near normal incidence ($\theta \approx 0$), the free spectral range of the etalon must be greater than about 10 GHz. Thus

$$c/2nt \gtrsim 10^{10}\,\text{Hz}$$

Taking fused quartz as the medium of the etalon, we have $n \approx 1.462$ (at $\lambda \sim 0.51\,\mu\text{m}$) and thus

$$t \lesssim 1\,\text{cm}$$

Another very important method used to obtain single frequency oscillation is to replace one of the mirrors of the resonator with a Fox–Smith

Fig. 9.10 The figure shows how by inserting a Fabry–Perot etalon inside the resonator, one can achieve single frequency oscillation. Without the etalon there are five modes above threshold. With the etalon only one of these modes is transmitted by the etalon and hence only that mode can oscillate. The other modes fall below threshold due to increased losses.

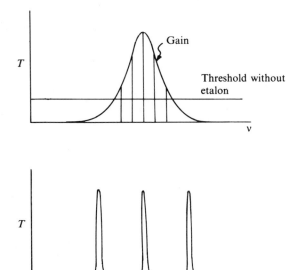

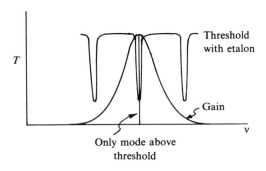

interferometer as shown in Fig. 9.11. Waves incident on the beam splitter BS from M_1 will suffer multiple reflections as follows:

$$\text{Reflection 1: } M_1 \to BS \to M_2 \to BS \to M_1$$
$$\text{Reflection 2: } M_1 \to BS \to M_2 \to BS \to M_3$$
$$\to BS \to M_2 \to BS \to M_1 \text{ etc.} \tag{9.48}$$

Thus the structure will behave much like a Fabry–Perot etalon if the beam splitter BS has a high reflectivity[†]. For constructive interference among waves reflected towards M_1 from the interferometer, the path difference between two consecutively reflected waves must be $m\lambda$, i.e.,

$$(2d_2 + 2d_3 + 2d_2 - 2d_2) = m\lambda$$

or

$$v = m\frac{c}{2(d_2 + d_3)}; \quad m = \text{any integer} \tag{9.49}$$

Thus frequencies separated by $\Delta v = c/2(d_2 + d_3)$ will have a low loss. Hence if Δv is greater than the bandwidth of oscillation of the laser, then one can achieve single mode oscillation. Since the frequencies of the resonator formed by mirrors M_1 and M_2 are

$$v = qc/2(d_1 + d_2); \quad q = \text{any integer} \tag{9.50}$$

for the oscillation of a mode one must have

$$m/(d_2 + d_3) = q/(d_1 + d_2) \tag{9.51}$$

which can be adjusted by varying d_3 by placing the mirror M_3 on a piezo electric movement.

Problem 9.2: If one wishes to choose a particular oscillating mode out of the possible resonator modes which are separated by 300 MHz, what is the approximate

Fig. 9.11 The Fox-Smith interferometer arrangement for selection of a single longitudinal mode.

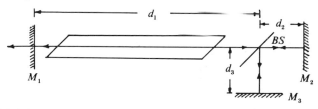

[†] It is interesting to note that if mirror M_3 is put above BS (in Fig. 9.11), then it would correspond to a Michelson interferometer arrangement and the transmittivity would not be sharply peaked.

change in d_3 required to change oscillation from one mode to another? Assume $\lambda_0 = 5000$ Å, $d_2 + d_3 = 5$ cm.

Solution: Differentiating Eq. (9.49), we have

$$\delta v = \frac{mc}{2(d_2 + d_3)^2}\,\delta d_3 = \frac{v}{(d_2 + d_3)}\,\delta d_3 \qquad (9.52)$$

which gives us

$$\delta d_3 = 0.025 \ \mu\text{m}.$$

Problem 9.3: Consider a laser which is oscillating simultaneously at two frequencies v_1 and v_2 ($v_2 - v_1 \approx c/2d$, where d is the resonator length). If this laser is used in an interference experiment, what is the minimum path difference between the interfering beams for which the interference pattern disappears?

Solution: The interference pattern disappears when the interference maxima produced by one wavelength fall on the interference minima produced by the other wavelength, i.e., when the path difference l is such that

$$l = mc/v_1 = (m + \tfrac{1}{2})c/v_2; \quad m = 0, 1, 2, \ldots \qquad (9.53)$$

we can write the above equations as

$$m = lv_1/c = lv_2/c - \tfrac{1}{2}$$

Thus the interference fringes disappear when

$$l = c/2(v_2 - v_1) = d \qquad (9.54)$$

i.e., the source can be considered coherent only for path differences $l < d$, the length of the laser. On the other hand if the laser was oscillating in just a single mode, the coherence length would have been much larger and would be determined by the frequency width of the oscillating mode only.

9.6   *Q*-switching

As discussed in Sec. 9.3, the quality factor (or Q factor) of a cavity is determined by the losses in the cavity; the smaller the losses, the larger is the Q-value. Consider a laser cavity in which a shutter is introduced in front of one of the mirrors as shown in Fig. 9.12. If the laser medium is

Fig. 9.12 A laser resonator with a shutter placed in front of one of the mirrors to achieve Q-switching.

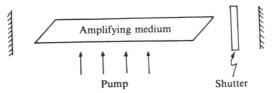

Amplifying medium

Pump

Shutter

continuously pumped, the population inversion in the cavity will go on increasing and will reach a very high value. This value could be much larger than the threshold inversion required for the same laser in the absence of the shutter. If the shutter is now suddenly opened, then the existing population inversion will correspond to a value much above the threshold value for oscillation. Thus the gain per round trip will be many times the loss per round trip and the radiation in the cavity mode will build up very rapidly. This rapid increase in the intensity will deplete the population inversion which will go below threshold. This results in the generation of an intense pulse of light from the cavity. Since the Q of the cavity is being switched from a small value to a large value, the above technique is referred to as Q-switching. Fig. 9.13 shows schematically the time variation of the

Fig. 9.13 Schematic representation of how the various quantities namely (*a*) loss, (*b*) Q, (*c*) population inversion ΔN and (*d*) laser output power vary with time when a laser is Q-switched.

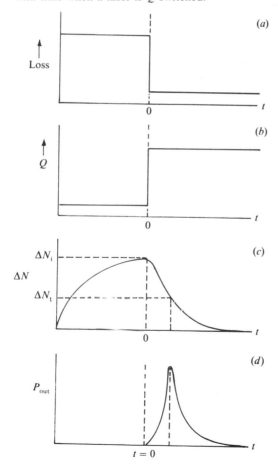

cavity loss, cavity Q, population inversion and the output power. As shown in the figure, an intense pulse is generated with the peak intensity appearing when the population inversion in the cavity is equal to the threshold value.

We will now write rate equations corresponding to Q-switching and obtain the most important parameters such as peak power, total energy, and duration of the pulse. We shall consider only one mode of the laser resonator and shall examine the specific case of a three level laser system such as that of ruby. In Sec. 8.6 while writing the rate equations for the population N_2 and the photon number n, we assumed the lower laser level to be essentially unpopulated. If this is not the case then instead of Eq. (8.100), we will have

$$d(N_2V)/dt = -KnN_2 + KnN_1 - T_{21}N_2V + RV \qquad (9.55)$$

where the second term on the RHS is the contribution due to absorption by N_1 atoms/unit volume in the lower level. Since the Q-switched pulse is of a very short duration, we will neglect the effect of the pump and spontaneous emission during the generation of the Q-switched pulse. It must, at the same time be noted that for the start of the laser oscillation, spontaneous emission is essential. Thus we get from Eq. (9.55)

$$dN_2'/dt = -(Kn/V)\Delta N' \qquad (9.56)$$

where

$$\Delta N' = (N_2 - N_1)V, \quad N_2' = N_2V \qquad (9.57)$$

and V is the volume of the amplifying medium. Similarly one can also obtain for the rate of change of population of the lower level

$$dN_1'/dt = (Kn/V)\Delta N' \qquad (9.58)$$

where $N_1' = N_1V$. Subtracting Eq. (9.58) from Eq. (9.57) we get

$$d(\Delta N')/dt = -(2Kn/V)\Delta N' \qquad (9.59)$$

We can also write the equation for the rate of change of the photon number n in the cavity mode in analogy to Eq. (8.104) as

$$dn/dt = Kn(N_2 - N_1) - n/t_c + KN_2$$
$$\approx (Kn/V)\Delta N' - n/t_c \qquad (9.60)$$

where we have again neglected the spontaneous emission term KN_2. From Eq. (9.60) we see that the threshold population inversion is

$$(\Delta N')_t = V/Kt_c \qquad (9.61)$$

when the gain represented by the first term on the RHS becomes equal to the loss represented by the second term (see Eq. (8.105)). Replacing V/K in Eqs. (9.59) and (9.60) by $(\Delta N')_t t_c$ and writing

$$\tau = t/t_c \tag{9.62}$$

we obtain

$$d(\Delta N')/d\tau = -2n\,\Delta N'/(\Delta N')_t \tag{9.63}$$

and

$$dn/d\tau = n[\Delta N'/(\Delta N')_t - 1] \tag{9.64}$$

Eqs. (9.63) and (9.64) give us the variation of the photon number n and the population inversion $\Delta N'$ in the cavity as a function of time. As can be seen the equations are nonlinear and solutions to the above set of equations can be obtained numerically by starting from an initial condition

$$\Delta N'(\tau = 0) = \Delta N_i' \quad \text{and} \quad n(\tau = 0) = n_i \tag{9.65}$$

where the subscript i stands for initial values. Here n_i represents the initial small number of photons excited in the cavity mode through spontaneous emission. This spontaneous emission is necessary to trigger laser oscillation.

From Eq. (9.64) we see that since the system is initially pumped to an inversion $\Delta N' > (\Delta N')_t$, $dn/d\tau$ is positive, thus the number of photons increases with time. The maximum number of photons in the cavity appear when $dn/d\tau = 0$ i.e., when $\Delta N' = (\Delta N')_t$. At such an instant n is very large and from Eq. (9.63) we see that $\Delta N'$ will further reduce below $(\Delta N')_t$ and thus will result in a decrease in n.

Although the time dependent solution of Eqs. (9.63) and (9.64) requires numerical computation, we can analytically obtain the variation of n with $\Delta N'$ and from this we can draw some general conclusions regarding the peak power, the total energy in the pulse and the approximate pulse duration. Indeed, dividing Eq. (9.64) by Eq. (9.63) we obtain

$$dn/d\Delta N' = \tfrac{1}{2}[(\Delta N')_t/\Delta N' - 1]$$

Integrating we get

$$n - n_i = \tfrac{1}{2}\{\Delta N_t' \ln [\Delta N'/(\Delta N')_i] + [(\Delta N')_i - \Delta N']\} \tag{9.67}$$

Peak power Assuming the only loss mechanism to be output coupling and recalling our discussion in Sec. 8.7, we have for the instantaneous power output

$$P_{\text{out}} = nhv/t_c \tag{9.68}$$

Thus the peak power output will correspond to maximum n which occurs when $\Delta N' = \Delta N'_t$. Thus

$$P_{\text{max}} = n_{\text{max}}hv/t_c$$

$$= \frac{hv}{2t_c}\left[(\Delta N')_t \ln\frac{(\Delta N')_t}{(\Delta N')_i} + ((\Delta N')_i - (\Delta N')_t) \right] \tag{9.69}$$

where we have neglected n_i (the small number of initial spontaneous photons in the cavity). This shows that the peak power is inversely proportional to cavity lifetime.

Total energy In order to calculate the total energy in the Q-switched pulse we come back to Eq. (9.64) and substitute for $\Delta N'/(\Delta N')_t$ from Eq. (9.63) to get

$$dn/d\tau = -\tfrac{1}{2}d(\Delta N')/d\tau - n \tag{9.70}$$

Integrating the above equation from $\tau = 0$ to ∞ we get

$$n_f - n_i = \tfrac{1}{2}[(\Delta N')_i - (\Delta N')_f] - \int_0^{\infty} n\,d\tau$$

or

$$\int_0^{\infty} n\,d\tau = \tfrac{1}{2}[(\Delta N')_i - (\Delta N')_f] - (n_f - n_i) \tag{9.71}$$

where the subscript f denotes final values. Since n_i and n_f are very small in comparison to the total integrated number of photons we may neglect them and obtain

$$\int_0^{\infty} n\,d\tau \approx \tfrac{1}{2}[(\Delta N')_i - (\Delta N')_f] \tag{9.72}$$

Thus the total energy of the Q-switched pulse is

$$E = \int_0^{\infty} P_{\text{out}}\,dt$$

$$= hv \int_0^{\infty} n\,d\tau$$

$$= \tfrac{1}{2}[(\Delta N')_i - (\Delta N')_f]hv \tag{9.73}$$

The above expression could also have been derived through physical arguments as follows: for every additional photon appearing in the cavity mode there is an atom making a transition from the upper to the lower energy level and for every atom making this transition the population inversion reduces by 2. Thus if the population inversion changes from $(\Delta N')_i$ to $(\Delta N')_f$, the number of photons emitted must be $\frac{1}{2}[(\Delta N')_i - (\Delta N)_f]$ and Eq. (9.73) follows immediately.

Pulse duration An approximate estimate for the duration of the Q-switched pulse can be obtained by dividing the total energy by the peak power. Thus

$$t_d = \frac{E}{P_{max}} = \frac{(\Delta N')_i - (\Delta N')_f}{\{(\Delta N')_t \ln[(\Delta N')_t/(\Delta N')_i] + (\Delta N')_i - (\Delta N')_t\}} t_c \qquad (9.74)$$

In the above formulas, we still have the unknown quantity $(\Delta N')_f$ the final inversion. In order to obtain this, we may use Eq. (9.67) for $t \to \infty$. Since the final number of photons in the cavity is small, we have

$$(\Delta N')_i - (\Delta N')_f = (\Delta N')_t \ln[(\Delta N')_i/(\Delta N')_f] \qquad (9.75)$$

from which we can obtain $(\Delta N')_f$ for a given set of $(\Delta N')_i$ and $(\Delta N')_t$.

As an example we consider the Q-switching of a ruby laser with the following characteristics:

$$\left.\begin{array}{l} \text{length of ruby rod} = 10\,\text{cm; area of cross section} = 1\,\text{cm}^2; \\ \text{resonator length} = 10\,\text{cm; mirror reflectivities} = 1 \\ \text{and } 0.7;\ Cr^{3+}\ \text{population density} = 1.58 \times 10^{19}\,\text{cm}^{-3}; \\ \lambda_0 = 6943\,\text{Å}; n_0 = 1.76, t_{sp} = 3 \times 10^{-3}\,\text{s}, g(\omega_0) = 1.1 \times 10^{-12}\,\text{s} \end{array}\right\} \quad (9.76)$$

The above parameters yield a cavity lifetime of 3.3×10^{-9} s and the required threshold population density of $1.25 \times 10^{17}\,\text{cm}^{-3}$. Thus

$$(\Delta N')_t = 1.25 \times 10^{18}$$

choosing

$$(\Delta N')_i = 4(\Delta N')_t = 5 \times 10^{18}$$

we get

$$P_{max} \approx 8.7 \times 10^7\,\text{W}$$

Solving Eq. (9.75) we obtain $(\Delta N')_f \approx 0.02\Delta N'_i$. Thus

$$E \approx 0.7\,\text{J}$$

and

$$t_d \approx 8 \, \text{ns}$$

9.6.1   *Techniques for Q-switching*

As discussed, for Q-switching the feedback to the amplifying medium must be initially inhibited and when the inversion is well past the threshold inversion, the optical feedback must be restored very rapidly. In order to perform this, various devices are available which include mechanical movements of the mirror or shutters which can be electronically controlled. The mechanical device may simply rotate one of the mirrors about an axis perpendicular to the laser axis which would restore the Q of the resonator once every rotation. Since the rotation speed cannot be made very large (typical rotation rates are $24\,000$ rpm), the switching of the Q from a low to a high values does not take place instantaneously and this leads to multiple pulsing.

In comparison to mechanical rotation, electronically controlled shutters employing the electrooptic or acoustooptic effect can be extremely rapid. A schematic arrangement of Q-switching using the electrooptic effect is shown in Fig. 9.14. As we shall discuss in Chapter 15, the electrooptic effect is the change in the birefringence of a material on application of an external electric field. Thus the electrooptic modulator (EOM) shown in Fig. 9.14

Fig. 9.14 A typical arrangement for achieving Q-switching using an electrooptic shutter. When a certain voltage is applied across the EOM it acts as a $\frac{1}{4}\lambda$ plate and thus the combination of polarizer–electrooptic modulator suppresses the reflection from M_2. When the voltage is removed, EOM does not change the state of polarization and thus Q becomes very high.

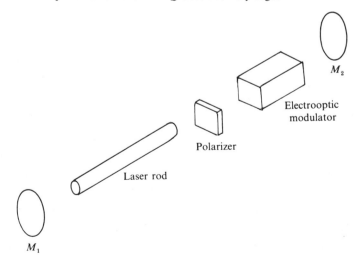

could be a crystal which is such that in the absence of any applied electric field the crystal does not introduce any phase difference between the two orthogonally polarized components travelling along the laser axis. On the other hand, if a voltage V_0 is applied, then the crystal introduces a phase difference of $\frac{1}{2}\pi$ between the orthogonal components, i.e., it behaves as a $\frac{1}{4}\lambda$ plate. If we now consider the polarizer–modulator–mirror system, when there is no applied voltage, the state of polarization (SOP) of the light incident on the polarizer after reflection by the mirror is along the pass axis of the polarizer and thus corresponds to a high Q state. When a voltage V_0 is applied, the linearly polarized light on passage through the EOM becomes circularly polarized (say right circularly). Reflection from the mirror converts this to left circularly polarized and passage through the EOM makes it linearly polarized but now polarized perpendicular to the pass axis of the polarizer. Thus there is essentially no feedback and this corresponds to the low Q state. Hence Q-switching can be accomplished by first applying a voltage across the crystal and removing it at the instant of highest inversion in the cavity. Some important electrooptic crystals used for Q-switching include potassium dihydrogen phosphate (KDP) and lithium niobate (LiNbO$_3$).

An acoustooptic Q-switch is based on the acoustooptic effect which is discussed in detail in Chapters 17 and 18. The acoustooptic effect is simply the diffraction of an incident light wave by an acoustic wave propagating in a medium. The acoustic wave generates a phase grating in the medium which is responsible for the diffraction. Thus if an acoustooptic cell is placed inside the resonator, it can be used to deflect the light beam out of the cavity thus leading to a low Q state. The Q can be switched to a high value by pulsing the acoustic wave. Details of electrooptic and acoustooptic Q-switches can be found in Koechner (1976).

Q-switching can also be obtained by using a saturable absorber inside the laser cavity. In a saturable absorber (which essentially consists of an organic dye dissolved in an appropriate solvent), the absorption coefficient of the material reduces with an increase in the incident intensity. This reduction in the absorption is caused by the saturation of a transition (see Sec. 8.5.1). In order to understand how a saturable absorber can be made to Q-switch, consider a laser resonator with the amplifying medium and the saturable absorber placed inside the cavity as shown in Fig. 9.15. As the amplifying medium is pumped, the intensity level inside the cavity is initially low since the saturable absorber does not allow any feedback from the mirror M_2. As the pumping increases, the intensity level inside the cavity increases which, in turn, starts to bleach the absorber. This leads to an increase in feedback which gives rise to an increased intensity and so on. Thus the energy stored inside the medium is released in the form of a giant pulse leading to Q-switching. If the relaxation time of the absorber is short compared to the cavity transit time then as we will discuss in Sec. 9.7, the saturable absorber would simultaneously mode lock and Q-switch the laser. For more details of Q-switching, readers may refer to Koechner (1976).

9.7   Mode locking in lasers

In this section we shall describe the technique of mode locking by which it is possible to produce ultrashort optical pulses (with pulse durations $\sim 10^{-12}$ s). In order to understand the concept of mode locking, we consider a laser formed by a pair of mirrors separated by a distance d. If the bandwidth over which gain exceeds losses in the cavity is Δv (see Fig. 9.16) then, since the intermodal spacing is $c/2d$, the laser will oscillate simultaneously in a large number of frequencies. The number of oscillating modes

Fig. 9.15 A laser resonator with a saturable absorber placed inside the cavity to achieve passive Q-switching.

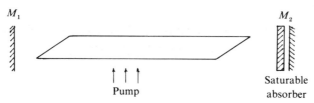

Fig. 9.16 The gain profile of an active medium centred at v_0.

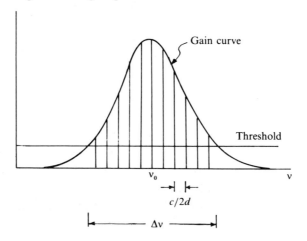

will be approximately (assuming that the laser is oscillating only in the fundamental transverse mode)

$$N + 1 \approx 1 + \text{integer closest to (but less than)} \frac{\Delta v}{(c/2d)} \qquad (9.77)$$

For example if we consider a ruby laser with an oscillating bandwidth of 6×10^{10} Hz with a cavity of length about 50 cm, the number of oscillating modes will be ~ 200. This large number of modes oscillates independently of each other and their relative phases are, in general, randomly distributed over the range $-\pi$ to $+\pi$. In order to obtain the output intensity of the laser when it oscillates in such a condition, we note that the total field E of the laser will be given by a superposition of the various modes of the laser. Thus

$$E(t) = \sum_{n=-N/2}^{N/2} A_n \exp(2\pi i v_n t + i\phi_n) \qquad (9.78)$$

where A_n and ϕ_n represent the amplitude and phase of the nth mode whose frequency is given by

$$v_n = v_0 + n\delta v; \quad n = -\tfrac{1}{2}N, \quad -\tfrac{1}{2}N + 1, \ldots, \tfrac{1}{2}N \qquad (9.79)$$

with $\delta v = c/2d$, the intermode spacing, and v_0 represents the frequency of the mode at the line centre.

In general the various modes represented by different values of n oscillate with different amplitudes A_n and also different phases ϕ_n. The intensity at the output of the laser will be given by

$$I = K|E(t)|^2 = K|\sum A_n \exp(2\pi i v_n t + i\phi_n)|^2$$
$$= K|\sum A_n \exp(2\pi i n\delta v t + i\phi_n)|^2 \qquad (9.80)$$

where K represents a constant of proportionality and we have used Eq. (9.79) for v_n. Eq. (9.80) can be rewritten as

$$I = K\sum_n |A_n|^2 + K\sum_n \sum_{m \neq n} A_n A_m^* \exp[2\pi i(n-m)\delta v t + i(\phi_n - \phi_m)] \qquad (9.81)$$

From the above equation the following observations can be made:

(a) Since the phases ϕ_n are randomly distributed in the range $-\pi$ to $+\pi$ for the various modes, if the number of modes is sufficiently large, the second term in Eq. (9.81) will have a very small value. Thus the intensity at the output would have an average value equal to the first term which is nothing but the sum of the intensities in various modes. (Compare with Eq. (4.105).)

(b) Although the output intensity has an average value of the sum of the individual mode intensities it is fluctuating with time (see Fig. 9.17) due to the second term in Eq. (9.81). It is obvious from Eq. (9.81) that if t is replaced by $t + q/\delta v$ where q is an integer, then the intensity value repeats itself. Thus the output intensity fluctuation repeats itself every $1/\delta v = 2d/c$ seconds which is nothing but the round trip transit time in the resonator.

(c) It also follows from Eq. (9.81) that, within this periodic repetition in intensity, the intensity fluctuates. The time interval of this intensity fluctuation (which is being caused by the beating between the two extreme modes) will be

$$t_f \simeq [(v_0 + \tfrac{1}{2}N\delta v) - (v_0 - \tfrac{1}{2}N\delta v)]^{-1} \approx 1/\Delta v \tag{9.82}$$

i.e., the inverse of the oscillation bandwidth of the laser medium.

When the laser is oscillating below threshold, the various modes are largely uncorrelated due to the absence of correlation among the various spontaneously emitting sources. The fluctuations become much less on passing above threshold but the different modes remain essentially un-correlated and the output intensity fluctuates with time.

Let us now consider the case in which the modes are locked in phase such that $\phi_n = \phi_0$, i.e., they are all in phase at some arbitrary instant of time $t = 0$. For such a case we have from Eq. (9.80),

$$I = K|\sum A_n e^{2\pi in\delta vt}|^2 \tag{9.83}$$

If we also assume that all modes have the same amplitude i.e., $A_n = A_0$, then we have

$$I = I_0|\sum e^{2\pi in\delta vt}|^2 \tag{9.84}$$

where $I_0 = KA_0^2$ is the intensity of each mode. The sum in Eq. (9.84) can

Fig. 9.17 Time variation of the output power from a laser oscillating in a number of longitudinal modes which have no phase relationship amongst themselves.

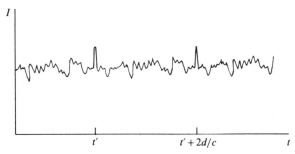

easily be evaluated and we obtain

$$I = I_0 \left\{ \frac{\sin\left[\pi(N+1)\delta vt\right]}{\sin(\pi\delta vt)} \right\}^2 \tag{9.85}$$

The variation of intensity with time as given by Eq. (9.85) is shown in Fig. 9.18. It is interesting to note that Eq. (9.85) represents a variation in time similar to that exhibited by a diffraction grating in terms of angle. Indeed the two situations are very similar; in the case of diffraction grating, the principal maxima are caused due to the constructive interference among waves diffracted by different slits and in the present case it is the interference in the temporal domain among modes of various frequencies.

From Eq. (9.85) we can conclude the following:

(1) The output of a mode locked laser will be in the form of a series of pulses and the pulses are separated by a duration

$$t_r = 1/\delta v = 2d/c \tag{9.86}$$

i.e., the cavity round trip time. Thus the mode locked condition can also be viewed as a condition in which a pulse of light is bouncing back and forth inside the cavity and every time it hits the mirror, a certain fraction is transmitted as the output pulse.

(2) From Eq. (9.85) it follows that the intensity falls off very rapidly around every peak (for large N) and the time interval between the zeroes of intensity on either side of the peak is

$$\Delta t \sim 2/(N+1)\delta v \approx 2/\Delta v \tag{9.87}$$

Fig. 9.18 The intensity variation as a function of time of a mode locked pulse train as described by Eq. (9.85). I_0 represents the intensity of each individual mode. The pulses are separated by a time interval of $2d/c$ which is just the cavity round trip time.

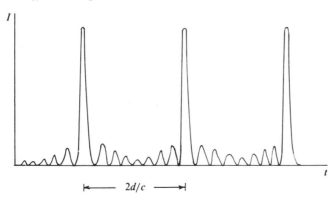

Table 9.1. *Some typical laser systems and their mode locked pulse widths*

Laser	$\lambda\,(\mu m)$	$\Delta v\,(\text{Hz})$	$t_D = (\Delta v)^{-1}\,(\text{s})$	Observed pulse width (s)
He–Ne	0.6328	$\sim 1.5 \times 10^9$	$\sim 6.7 \times 10^{-10}$	$\sim 6 \times 10^{-10}$
Argon ion	0.488	$\sim 7 \times 10^9$	$\sim 1.5 \times 10^{-10}$	$\sim 2.5 \times 10^{-10}$
Nd:YAG	1.060	$\sim 1.2 \times 10^{10}$	$\sim 8.3 \times 10^{-11}$	$\sim 7.6 \times 10^{-11}$
Ruby	0.6943	$\sim 6 \times 10^{10}$	$\sim 1.7 \times 10^{-11}$	$\sim 1.2 \times 10^{-11}$
Nd: glass	1.06	$\sim 3 \times 10^{12}$	$\sim 3.3 \times 10^{-13}$	$\sim 3 \times 10^{-13}$
Dye	0.6	$\sim 10^{13}$	$\sim 10^{-13}$	$\sim 10^{-13}$

Table adapted from Yariv (1985) and Demtroder (1981).

We may define the pulse duration as approximately (like the FWHM)

$$t_D \sim 1/\Delta v \tag{9.88}$$

i.e., the inverse of the oscillating bandwidth of the laser. Thus the larger the oscillating bandwidth the smaller will be the pulse width. Table 9.1 shows the bandwidth Δv, t_D and observed pulse width for some typical laser systems. As can be seen one can obtain pulses of duration 1 ps or shorter; such pulses, also referred to as ultrashort pulses, find wide applications in the study of ultrafast phenomena in physics, chemistry and biology (Shapiro, 1977).

(3) The peak intensity of each mode locked pulse is given by

$$I = (N + 1)^2 I_0 \tag{9.89}$$

which is $(N + 1)$ times the average intensity when the modes are not locked. Thus for typical solid state lasers which can oscillate simultaneously in 10^3–10^4 modes, the peak power enhancement due to mode locking can be very large.

Fig. 9.19(a) shows the output from a mode locked He–Ne laser operating at 6328 Å. Notice the regular train of pulses separated by the round trip time of the cavity. Fig. 9.19(b) shows an expanded view of one of the mode locked pulse; the pulse width is about 330 ps.

9.7.1   *Techniques for mode locking*

As we have seen above mode locking essentially requires that the various longitudinal modes be coupled to each other. In practice this can be achieved either by modulating the loss or optical path length of the cavity externally (active mode locking) or by placing saturable absorbers inside the laser cavity (passive mode locking).

In order to understand how a periodic loss modulation inside the resonator cavity can lead to mode locking, we consider a laser resonator having a loss modulator inside the cavity with the modulation frequency equal to the intermode frequency spacing δv. Consider one of the modes at a frequency v_q. Since the loss of the cavity is being modulated at a frequency δv, the amplitude of the mode will also be modulated at the same

Fig. 9.19 (*a*) Output pulse train of a mode locked He–Ne laser. (*b*) Expanded view of a single pulse from the pulse train. (After Fox, Schwarz and Smith (1968).)

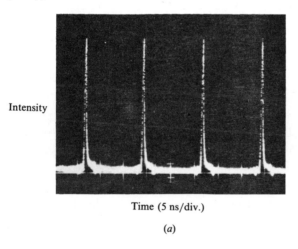

Intensity

Time (5 ns/div.)

(*a*)

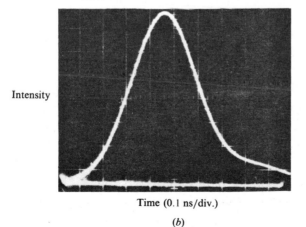

Intensity

Time (0.1 ns/div.)

(*b*)

frequency δv and thus the resultant field in the mode may be written as

$$(A + B\cos 2\pi\delta vt)\cos 2\pi v_q t = A\cos 2\pi v_q t + \tfrac{1}{2}B\cos[2\pi(v_q + \delta v)t]$$
$$+ \tfrac{1}{2}B\cos[2\pi(v_q - \delta v)t] \qquad (9.90)$$

Thus the amplitude modulated mode at a frequency v_q generates two waves at frequencies $v_q + \delta v$ and $v_q - \delta v$. Since δv is the intermode spacing, these new frequencies correspond to the two modes lying on either side of v_q. The oscillating field at the frequencies $v_q \pm \delta v = v_{q\pm1}$ forces the modes corresponding to these frequencies to oscillate such that a perfect phase relationship now exists between the three modes. Since the amplitudes of these new modes are also modulated at the frequency δv, they generate new side bands at $v_q + 2\delta v = v_{q+2}$ and $v_q - 2\delta v = v_{q-2}$. Thus all modes are forced to oscillate with a definite phase relationship and this leads to mode locking.

The above phenomenon of mode locking can also be understood in the time domain by noticing that the intermode frequency spacing $\delta v = c/2d$ corresponds to a time period of $2d/c$ which is exactly one round trip time inside the cavity. Hence considering the fluctuating intensity present inside the cavity (see Fig. 9.17), we observe that since the loss modulation has a period equal to a round trip time, the portion of the fluctuating intensity incident on the loss modulator at a given value of loss would after every round trip be incident at the same loss value. Thus the portion incident at the highest loss instant will suffer the highest loss at every round trip. Similarly, the portion incident at the instant of lowest loss will suffer the lowest loss at every round trip. This will result in the build up of narrow pulses of light which pass through the loss modulator at the instant of lowest loss. The pulse width must be approximately the inverse of the gain bandwidth since wider pulses would experience higher losses in the modulator and narrower pulses (which would have a spectrum broader than the gain bandwidth) would have lower gain. Thus the above process leads to mode locking.

The loss modulator inside the cavity could be an EOM or an acoustooptic modulator (AOM). The electrooptic and acoustooptic effects are discussed in detail in Chapters 15 and 19. Fig. 9.14 shows a typical arrangement for introducing loss modulation inside the cavity. The EOM changes the state of polarization (SOP) of the propagating light beam and the output SOP depends on the voltage applied across the modulator. Thus the SOP of the light passing through the modulator, reflected by the mirror and returning to the polarizer can be changed by the applied voltage and consequently the feedback provided by the polarizer–modulator–mirror system.

This leads to a loss modulation. Some of the electrooptic materials used include potassium dihydrogen phosphate (KDP) and lithium niobate (LiNbO$_3$).

Acoustooptic modulators can also be used for mode locking. Fig. 9.20 shows a typical arrangement in which an AOM is introduced into the cavity. Standing acoustic waves are generated inside the modulator which diffracts the light beam through Bragg diffraction effect. If the acoustic frequency is Ω, then the standing wave would oscillate at a frequency 2Ω and thus the loss would be modulated at a frequency 2Ω. When this frequency equals the intermode frequency spacing, one would obtain mode locking.

The above techniques in which an external signal is used to mode lock the laser are referred to as active mode locking. One can also obtain mode locking using a saturable absorber inside the laser cavity. This technique does not require an external signal to mode lock and is referred to as passive mode locking. As discussed in Sec. 9.6 in a saturable absorber, the absorption coefficient decreases with an increase in the incident light intensity (see Sec. 8.5.1). Thus, the material becomes more and more transparent as the intensity of the incident light increases. In order to understand how a saturable absorber can mode lock a laser, consider a laser cavity with a cell containing the saturable absorber placed in a cell adjacent to one of the resonator mirrors (see Fig. 9.15). Initially the saturable absorber does not transmit fully and the intensity inside the resonator has

Fig. 9.20 A typical configuration for mode locking of lasers using the acousto-optic effect. In the AOM standing acoustic waves are formed which through Bragg diffraction effects diffract the laser beam periodically out of the cavity leading to a mode locking.

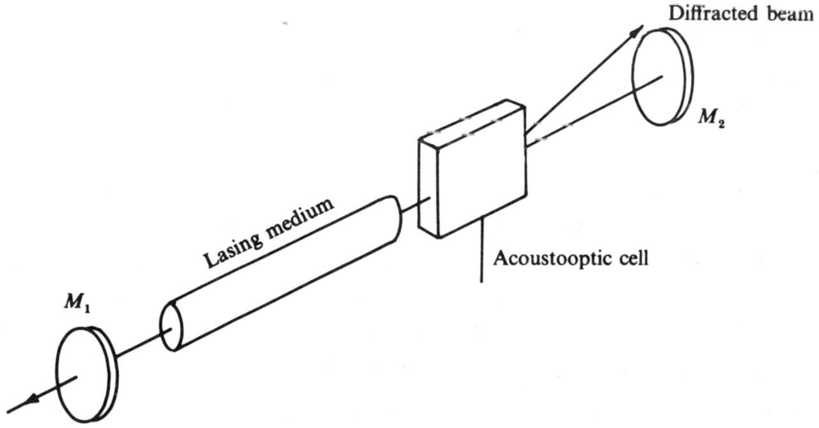

Table 9.2. *Some organic dyes used to mode lock laser systems*

	DDI[a]	Cryptocyanine	Eastman No. 9740	Eastman No. 9860
Laser	Ruby	Ruby	Nd	Nd
$I_s(\text{W/cm}^2)$	$\sim 2 \times 10^7$	$\sim 5 \times 10^6$	$\sim 4 \times 10^7$	$\sim 5.6 \times 10^7$
$\tau_D(\text{ps})$	~ 14	~ 22	~ 8.3	~ 9.3
Solvents	Methanol	Nitrobenzene		1, 2-Dichloroethane
	Ethanol	Acetone		Chlorobenzene
		Ethanol		
		Methanol		

[a]DDI stands for 1, 1'-diethyl-2, 2'-dicarbocyanine iodide. Table adapted from Koechner (1976).

a noiselike structure. The intensity peaks arising from this fluctuation bleach the saturable absorber more than the average intensity values. Thus the intensity peaks suffer less loss than the other intensity values and are amplified more rapidly as compared to the average intensity. If the saturable absorber has a rapid relaxation time (i.e., the excited atoms come back rapidly to the lower level) so that it can follow the fast oscillations in intensity in the cavity, one will obtain mode locking. Since the transit time of the pulse through the resonator is $2d/c$, the mode locked pulse train will appear at a frequency of $c/2d$. Table 9.2 gives some organic dyes used to mode lock ruby and Nd laser systems; I_s and τ_D represent the saturation intensity and relaxation time respectively.

9.8   Modes of a confocal resonator system

In Sec. 9.5 we mentioned that the amplitude distribution in the fundamental transverse mode is Gaussian. In this section we will first

Fig. 9.21 The symmetric confocal cavity consisting of two mirrors of radii of curvature R and separated by a distance R.

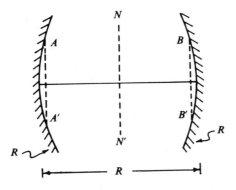

consider a confocal resonator cavity consisting of a pair of mirrors of equal radii of curvature R and separated by a distance R and obtain the modes of the cavity. Fig. 9.21 shows the symmetric confocal cavity. A mode of the cavity will be the field distribution which is such that after one complete round trip through the resonator, the resultant field distribution is the same as the distribution with which we started. Since the cavity shown in Fig. 9.21 is symmetric about the midplane NN', the modes can be simply obtained from the condition that the field distribution across the plane AA' must be the same as that across the plane BB' which is obtained after one half round trip. The allowed frequencies of oscillation will be those for which the phase change after one half round trip is an integral multiple of π.

In order to obtain the modes, we notice that a given field in the plane AA' propagating to the right first diffracts over a distance R and then suffers a reflection at the mirror M_2 to complete one half round trip. Hence in order to obtain the field across BB', we must know the effect of diffraction and also the effect of reflection by a mirror.

The effect of diffraction is given by Eq. (5.5) which we recall below:

$$g(x, y, z) = \frac{i}{\lambda z} \exp(-ikz) \int \int f(x', y', 0)$$

$$\times \exp\left\{-i\frac{k}{2z}[(x-x')^2 + (y-y')^2]\right\} dx' dy' \qquad (9.91)$$

where z is the distance between the observation plane and the initial plane where the field distribution is $f(z, y, 0)$.

In order to obtain the effect of the mirror on the incident field we use the same procedure as in Sec. 6.3 to obtain the effect of a lens. Fig. 9.22

Fig. 9.22 Diverging spherical waves from a point P converge after reflection from the mirror to the point Q.

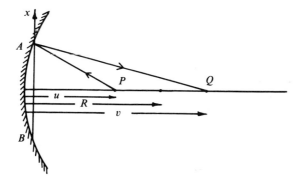

shows a point source P placed at a distance u in front of a concave mirror of radius of curvature R. Spherical waves emerging from P are focussed by the mirror to the image point Q at a distance from the mirror such that

$$\frac{1}{v} = \frac{2}{R} - \frac{1}{u} \tag{9.92}$$

Thus the mirror converts an incident diverging spherical wave of radius of curvature u to a converging spherical wave of radius of curvature v. Now as discussed in Sec. 5.4, the phase distribution in the transverse plane AB of the incident diverging spherical wave is

$$\exp\left[-i\frac{k}{2u}(x^2 + y^2) \right] \tag{9.93}$$

(apart from a constant phase term). Similarly the phase distribution of the reflected converging spherical wave on the same plane will be

$$\exp\left[i\frac{k}{2v}(x^2 + y^2) \right] \tag{9.94}$$

Hence if p_M represents the factor by which the incident phase distribution is to be multiplied to obtain the reflected phase distribution, then

$$p_M \exp\left[-i\frac{k}{2u}(x^2 + y^2) \right] = \exp\left[i\frac{k}{2v}(x^2 + y^2) \right]$$

or

$$p_M = \exp\left[i\frac{k}{R}(x^2 + y^2) \right] \tag{9.95}$$

The above equation may be compared with Eq. (6.14) for the effect of a lens.

Problem 9.4: Use Eq. (9.95) for an incident plane wave on (*a*) a concave mirror (represented by a positive R) and (*b*) a convex mirror (represented by a negative R) and discuss the result.

We will now use Eqs. (9.91) and (9.95) to obtain the modes of the resonator depicted in Fig. 9.21. Let $f(x, y)$ represent the field on the plane AA'. This field diffracts from mirror M_1 to M_2 to produce a field given by

$$g(x, y) = \frac{i}{\lambda R} \exp(-ikR) \iint_{\mathscr{A}} f(x', y')$$

$$\times \exp\left\{ -i\frac{k}{2R}[(x - x')^2 + (y - y')^2] \right\} dx' \, dy' \tag{9.96}$$

where we have used Eq. (9.91) and the fact that the distance between the mirrors is R; $\mathscr{A}$ represents the area of the mirror. Hence the field on the plane BB' after one half round trip will be

$$h(x, y) = g(x, y) \exp\left[i\frac{k}{R}(x^2 + y^2)\right] \tag{9.97}$$

If the field distribution $f(x, y)$ is to represent a mode then

$$h(x, y) = \sigma f(x, y) \tag{9.98}$$

where σ is a complex constant. The losses suffered by the field would be determined by the magnitude of σ and the phase shift in one half round trip by the phase of σ. We now substitute Eqs. (9.97) and (9.98) in Eq. (9.96) to obtain

$$\sigma f(x, y) = \frac{i}{\lambda R} \exp(-ikR) \int\int dx' \, dy' \, f(x', y')$$

$$\times \exp\left[-i\frac{k}{2R}(x'^2 + y'^2 - 2xx' - 2yy')\right]$$

$$\times \exp\left[i\frac{k}{2R}(x^2 + y^2)\right] \tag{9.99}$$

We now introduce a function $s(x, y)$:

$$s(x, y) = f(x, y) \exp\left[-i\frac{k}{2R}(x^2 + y^2)\right] \tag{9.100}$$

and also a set of dimensionless variables

$$\xi = (k/R)^{\frac{1}{2}}x, \quad \eta = (k/R)^{\frac{1}{2}}y \tag{9.101}$$

Using Eqs. (9.100) and (9.101), Eq. (9.99) becomes

$$\sigma s(\xi, \eta) = \frac{i}{2\pi} e^{-ikR} \int\int_{\mathscr{A}'} s(\xi', \eta') e^{i(\xi\xi' + \eta\eta')} \, d\xi' \, d\eta' \tag{9.102}$$

where $\mathscr{A}'$ represents the modified limits of integration.

In order to simplify the analysis we assume the mirrors to have a rectangular cross section with dimensions $2a \times 2b$. Thus Eq. (9.102) can be written as

$$\sigma s(\xi, \eta) = \frac{i}{2\pi} e^{-ikR} \int_{-\xi_0}^{\xi_0} \int_{-\eta_0}^{\eta_0} s(\xi', \eta') e^{i(\xi\xi' + \eta\eta')} \, d\xi' \, d\eta' \tag{9.103}$$

where

$$\xi_0 = (k/R)^{\frac{1}{2}}a, \quad \eta_0 = (k/R)^{\frac{1}{2}}b \tag{9.104}$$

We now try a separation of variables technique and write

$$s(\xi, \eta) = p(\xi)q(\eta) \tag{9.105}$$

$$\sigma = \kappa\tau$$

Substituting in Eq. (9.103) and separating the variables we get

$$\kappa p(\xi) = (i/2\pi)^{\frac{1}{2}}e^{-ikR/2} \int_{-\xi_0}^{\xi_0} p(\xi')e^{i\xi\xi'} \, d\xi' \tag{9.106}$$

$$\tau q(\eta) = (i/2\pi)^{\frac{1}{2}}e^{-ikR/2} \int_{-\eta_0}^{\eta_0} q(\eta')e^{i\eta\eta'} \, d\eta' \tag{9.107}$$

The integrals appearing in Eqs. (9.106) and (9.107) are similar to a Fourier transform and are referred to as finite Fourier transforms. They tend to a normal Fourier transform for $\xi_0 \to \infty$ and $\eta_0 \to \infty$. Slepian and Pollack (1961) have shown that the solutions of the above integral equations are prolate spheroidal functions. Here we will consider the case when

$$\xi_0 = \left(\frac{2\pi}{\lambda_0 R}\right)^{\frac{1}{2}} a \gg 1$$

$$\eta_0 = \left(\frac{2\pi}{\lambda_0 R}\right)^{\frac{1}{2}} b \gg 1 \tag{9.108}$$

We recall from Eq. (5.39) that $a^2/\lambda_0 R$ and $b^2/\lambda_0 R$ are nothing but the Fresnel numbers corresponding to the size of one mirror when viewed from the centre of the other mirror. Thus Eq. (9.108) essentially implies that the Fresnel numbers are large. We also assume that the functions $p(\xi)$ and $q(\eta)$ tend to zero as ξ and η tend to infinity; we shall later show the consistency of the assumption. Under the above assumptions we obtain

$$\kappa p(\xi) = (i/2\pi)^{\frac{1}{2}}e^{-ikR/2} \int_{-\infty}^{\infty} p(\xi')e^{i\xi\xi'} \, d\xi' \tag{9.109}$$

$$\tau q(\eta) = (i/2\pi)^{\frac{1}{2}}e^{-ikR/2} \int_{-\infty}^{\infty} q(\eta')e^{i\eta\eta'} \, d\eta' \tag{9.110}$$

Notice that Eq. (9.109) tells us that we are looking for functions $p(\xi)$ which are their own Fourier transforms. It is well known that Hermite–Gauss functions are a class of functions which are their own Fourier transforms. Eq. (9.109) can indeed be transformed to a differential equation for

Hermite–Gauss functions (see Thyagarajan and Ghatak (1981)). Here we assume the solution to be Gaussian and obtain the beam width which satisfies Eq. (9.109). Thus we let

$$p(\xi) = A e^{-\xi^2/\alpha^2} \tag{9.111}$$

where α is an unknown parameter to be determined such that $p(\xi)$ satisfies Eq. (9.109). Substituting in Eq. (9.109) and using Eq. (5.9) we obtain

$$\kappa e^{-\xi^2/\alpha^2} = \alpha \left(\frac{i}{2}\right)^{\frac{1}{2}} e^{-ikR/2} e^{-\alpha^2 \xi^2/4} \tag{9.112}$$

If the above equation is to be valid for all values of ξ, the exponents on the LHS and RHS must be equal i.e.,

$$\alpha = \sqrt{2} \tag{9.113}$$

Substituting this value of α in Eq. (9.112), we get

$$\kappa = e^{-i(kR/2 - \pi/4)} \tag{9.114}$$

where we have used $i = e^{i\pi/2}$. In an identical fashion, we can obtain

$$q(\eta) = B e^{-\eta^2/2} \tag{9.115}$$

with

$$\tau = e^{-i(kR/2 - \pi/4)} \tag{9.116}$$

Hence we obtain

$$f(x, y) = s(x, y) \exp\left[i\frac{k}{2R}(x^2 + y^2)\right]$$
$$= K \exp\left[-\frac{k}{2R}(x^2 + y^2)\right] \exp\left[i\frac{k}{2R}(x^2 + y^2)\right] \tag{9.117}$$

with $K = AB$ and

$$\sigma = \kappa\tau = e^{-i(kR - \pi/2)} \tag{9.118}$$

Eq. (9.117) gives the transverse distribution of the Gaussian mode of the confocal resonator at either mirror. We now note the following points:

(a) From Eq. (9.118) it follows that $|\sigma| = 1$, i.e., the amplitude of the field after one half round trip is the same as at the starting point, i.e., there is no loss. This is just due to the fact that we have assumed ξ_0 and $\eta_0 \to \infty$.

(b) From Eq. (9.118), we note that the phase shift after one half round trip is $kR - \frac{1}{2}\pi$. Hence only those frequencies are allowed for which this quantity is an integral multiple of π, i.e.,

$$kR - \tfrac{1}{2}\pi = q\pi; \quad q = 0, 1, 2, \ldots$$

or

$$v_q = \frac{c}{2R}(q + \tfrac{1}{2}) \tag{9.119}$$

which gives us the allowed frequencies of oscillation. Different value of q give us the different longitudinal modes of the cavity corresponding to a Gaussian transverse field distribution. The intermode frequency spacing is

$$\Delta v_q = c/2R \tag{9.120}$$

For a typical cavity with $R = 100\,\text{cm}$, $\Delta v_q = 150\,\text{MHz}$.

(c) Recalling the discussion in Sec. 5.4 we observe from Eq. (9.117) that the Gaussian mode has a curved phase front of radius of curvature R at the mirror. This is exactly equal to the radius of curvature of the mirror. Thus the Gaussian mode is incident normally on the mirror and will retrace its path after reflection. The spherical wavefront described by Eq. (9.117) is converging which is consistent with the fact that Eq. (9.117) represents the field after reflection.

(d) The $1/e$ width of the Gaussian field at the mirror can be obtained from Eq. (9.117) and is

$$w = (2R/k)^{\frac{1}{2}} \tag{9.121}$$

(e) We have shown in Sec. 5.4 that a Gaussian beam remains always Gaussian when it diffracts in space. Since the mode field inside the resonator cavity is formed by the diffraction of the field given by Eq. (9.117) into the cavity, it follows that the mode field is Gaussian everywhere both inside and outside the cavity. Since a Gaussian beam diffracts symmetrically on either direction of the waist, it follows from the symmetry of the resonator shown in Fig. 9.21 that the waist must lie at the centre of the resonator. Recalling the equation representing the change of the beam width of a Gaussian beam as it propagates (see Eq. (5.11))

$$w^2(z) = w_0^2(1 + \lambda^2 z^2/\pi^2 w_0^4) \tag{9.122}$$

we can easily obtain the waist size w_0 of the Gaussian mode by using the fact that at $z = R/2$, $w(z) = (2R/k)^{\frac{1}{2}}$ – see Eq. (9.121). Thus we get

$$w_0 = (R/k)^{\frac{1}{2}} \tag{9.123}$$

which is independent of the transverse dimension of the mirror and depends only on R and λ. Fig. 9.23 shows the variation of $w(z)$ and the phase front of a Gaussian mode in a confocal resonator.

The $1/e$ beam radius of the Gaussian mode at the mirror can be obtained from Eq. (9.117) as $(2R/k)^{\frac{1}{2}}$. Obviously if the transverse dimension of the mirrors is very large compared to $(2R/k)^{\frac{1}{2}}$, then the mirrors will behave practically as infinite size mirrors. Thus if the transverse dimension of the mirror is $2a \times 2b$, then for the approximation of extending the limits of integration in Eqs. (9.106) and (9.107) to be valid, we must have

$$a \gg (\lambda_0 R/\pi)^{\frac{1}{2}}; \quad b \gg (\lambda_0 R/\pi)^{\frac{1}{2}} \tag{9.124}$$

which is almost the same as Eq. (9.108).

As an example we consider a symmetric confocal resonator $R = 1\,\mathrm{m}$ operating at $\lambda_0 = 1\,\mu\mathrm{m}$. For such a case we have

$$w_0 = \left(\frac{10^{-3} \times 10^3}{2\pi}\right)^{\frac{1}{2}} \approx 0.4\,\mathrm{mm}$$

and the beam width at the mirrors is $\sqrt{2}w_0 \approx 0.57\,\mathrm{mm}$.

Problem 9.5: Obtain the angle of divergence of the Gaussian mode discussed above.

9.9 General spherical resonator

In Sec. 9.8 we have shown that the fundamental mode of a symmetric confocal resonator has a Gaussian field distribution. As we discussed in Sec. 9.2, in general, one can form a resonator with a pair of spherical mirrors of radii of curvature R_1 and R_2 and separated by a distance d (see Fig. 9.24). If we can find a Gaussian beam which has a phase front with radius of curvature R_1 on mirror M_1 and a radius of curvature R_2 on mirror M_2,

Fig. 9.23 Variation of $w(z)$ and the phase front of a Gaussian mode in a confocal resonator.

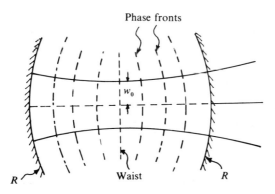

and if the beam widths at both the mirrors are small compared to the mirror sizes, then such a Gaussian beam can resonate in the resonator and will correspond to the fundamental mode of the resonator (see Fig. 9.24).

Before we consider a general spherical resonator, we first consider a resonator consisting of a plane mirror and a concave mirror of radius of curvature R (see Fig. 9.25). In order to find the Gaussian mode of the resonator, we use the fact that the Gaussian mode must have radii of curvature equal to those of the mirrors at the position of the mirrors. Hence it must have an infinite radius of curvature at the plane mirror and a radius of curvature R on the concave mirror. Since the radius of curvature of a Gaussian beam at its waist is infinite, the Gaussian mode must have its

Fig. 9.24 A Gaussian beam resonating between two spherical mirrors of radii of curvature R_1 and R_2 and separated by a distance d; $z = 0$ is the position of the waist of the beam where the phase front is plane.

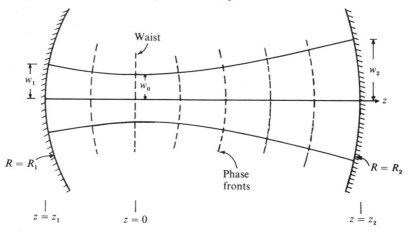

Fig. 9.25 A resonator consisting of a plane mirror and a concave mirror of radius of curvature R.

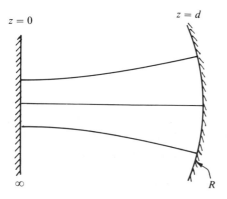

waist at the plane mirror. It should also have a radius of curvature R at a distance d from the plane mirror. Hence using Eq. (5.17) we may write

$$R(z = d) = R = d(1 + \pi^2 w_0^4/\lambda^2 d^2)$$

which can be solved to give

$$w_0 = \left(\frac{\lambda d}{\pi}\right)^{\frac{1}{2}} \left(\frac{R}{d} - 1\right)^{\frac{1}{4}} \tag{9.125}$$

This gives us the Gaussian mode of the resonator shown in Fig. 9.25.

Problem 9.6: If $d = R/2$ in Fig. 9.25 then it will correspond to a symmetric confocal resonator. Show from Eq. (9.125) that w_0 is consistent with Eq. (9.123).

From Eq. (9.125) we see that if $d > R$, then w_0 becomes complex. Such a resonator cannot support a fundamental Gaussian mode and is unstable. It also follows from Eq. (9.125) that w_0 can be increased by increasing R while keeping d fixed. Thus if we consider $d = 50\,\text{cm}$, $\lambda = 1\,\mu\text{m}$ then $w_0 \approx 0.4\,\text{mm}$ for $R = 1\,\text{m}$, $w_0 \approx 1.5\,\text{mm}$ for $R = 100\,\text{m}$. The effect of increasing w_0 is to increase the volume of the region occupied by the mode in the cavity. This volume which is also referred to as the cavity mode volume essentially determines the volume over which interaction between the beam and the amplifying medium is taking place. Thus for higher laser powers, one must have larger cavity mode volumes which can be obtained by using larger R values. Eq. (9.125) also gives the waist size for symmetric spherical resonators but d then represents half the distance between the mirrors.

We now consider a general spherical resonator as shown in Fig. 9.24. We have to find a Gaussian beam having a phase front of radius of curvature R_1 at mirror M_1 and R_2 at mirror M_2. If we let the waist lie at some plane $z = 0$ as shown in Fig. 9.24 and let the coordinates of the poles of the mirrors M_1 and M_2 be z_1 and z_2, then we have

$$R(z_1) = z_1 + \kappa^2/z_1, \quad R(z_2) = z_2 + \kappa^2/z_2 \tag{9.126}$$

where

$$\kappa^2 = \pi^2 w_0^4/\lambda^2 \tag{9.127}$$

We choose a sign convention such that the radius of curvature is positive if the mirror is concave towards the resonator. Then for the structure shown in Fig. 9.24, we have $R(z_2) = R_2$ and $R(z_1) = -R_1$. Thus

$$-R_1 = z_1 + \kappa^2/z_1, \quad R_2 = z_2 + \kappa^2/z_2 \tag{9.128}$$

Since the distance between the mirrors is d, we also have

$$z_2 - z_1 = d \tag{9.129}$$

We can solve for κ^2 from Eqs. (9.128) and (9.129) and using Eq. (9.127) obtain

$$w_0^4 = \frac{\lambda^2 d^2}{\pi^2} \frac{(1 - g_1 g_2) g_1 g_2}{(g_1 + g_2 - 2 g_1 g_2)^2} \tag{9.130}$$

where

$$g_1 = 1 - d/R_1, \quad g_2 = 1 - d/R_2 \tag{9.131}$$

Knowing w_0 and z_1 and z_2 from Eq. (9.128), we obtain

$$w^2(z_1) = \frac{\lambda d}{\pi} \left[\frac{g_2}{g_1(1 - g_1 g_2)} \right]^{\frac{1}{2}} \tag{9.132}$$

$$w^2(z_2) = \frac{\lambda d}{\pi} \left[\frac{g_1}{g_2(1 - g_1 g_2)} \right]^{\frac{1}{2}} \tag{9.133}$$

where $w(z_1)$ and $w(z_2)$ give the beam radii at the two mirrors. For the above analysis to be valid, the transverse dimensions of the mirrors must be large compared to $w(z_1)$ and $w(z_2)$. From Eqs. (9.132) and (9.133) it follows that

Fig. 9.26 The stability diagram for optical resonators. The shaded regions correspond to stable resonator configurations.

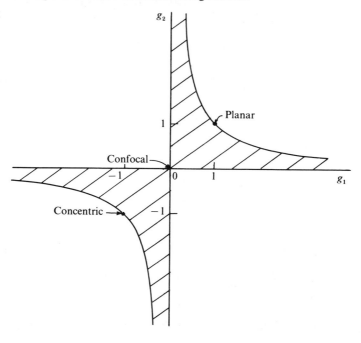

for the existence of a stable Gaussian mode we must have

$$0 \leqslant g_1 g_2 = (1 - d/R_1)(1 - d/R_2) \leqslant 1 \qquad (9.134)$$

The above condition is referred to as the stability condition for an optical resonator. Resonators satisfying $0 \leqslant g_1 g_2 \leqslant 1$ are termed stable; outside this range, the resonators become unstable. Fig. 9.26 depicts the stability diagram which is a plot of g_1 versus g_2. A point in the diagram represents a resonator. Resonators falling in the shaded region satisfy the condition (9.134) and hence are stable. Resonators in the unshaded region correspond to unstable resonators.

9.10 Higher order modes

In Sec. 9.8 we mentioned that the mode of a resonator is such that after every complete round trip in the resonator it repeats itself. Applying this condition we have shown that the modal field distribution must satisfy Eqs. (9.109) and (9.110) under the approximation that the mirrors are of large transverse extent. We have also shown that a Gaussian field distribution satisfies Eqs. (9.109) and (9.110) and thus they represent a mode of the resonator. It can be shown that Eq. (9.109) is actually satisfied by an infinite set of functions known as Hermite–Gauss functions represented by $H_n(\xi)e^{-\xi^2/2}$; here $H_n(\xi)$ represents the Hermite polynomial of order n. A few lower order Hermite polynomials are given below:

$$H_0(\xi) = 1, \quad H_1(\xi) = 2\xi, \quad H_2(\xi) = 4\xi^2 - 2 \qquad (9.135)$$

The Hermite–Gauss functions satisfy the following equation

$$i^m H_m(\xi)e^{-\xi^2/2} = \frac{1}{(2\pi)^{\frac{1}{2}}} \int_{-\infty}^{\infty} H_m(\xi')e^{-\xi'^2/2}e^{i\xi\xi'} \, d\xi' \qquad (9.136)$$

Comparing with Eq. (9.109) we can write

$$p(\xi) = H_m(\xi)e^{-\xi^2/2} \qquad (9.137)$$

$$\kappa = \exp\left\{ -i\left[\frac{kR}{2} - (m + \tfrac{1}{2})\frac{\pi}{2} \right] \right\} \qquad (9.138)$$

Similarly we can write

$$q(\eta) = H_n(\eta)e^{-\eta^2/2} \qquad (9.139)$$

$$\tau = \exp\left\{ -i\left[\frac{kR}{2} - (n + \tfrac{1}{2})\frac{\pi}{2} \right] \right\} \qquad (9.140)$$

Substituting for $p(\xi)$ and $q(\eta)$ from Eqs. (9.137) and (9.139) in Eq. (9.105) and using Eq. (9.100) we obtain the field distribution corresponding to the (m, n)th mode on the mirror as

$$f(x, y) = CH_m\left(\frac{x}{w_0}\right)H_n\left(\frac{y}{w_0}\right)\exp\left[-\frac{(x^2 + y^2)}{2w_0^2}\right]\exp\left[i\frac{k}{2R}(x^2 + y^2)\right]$$

(9.141)

where C is a constant. The above field distribution corresponds to what is known as the TEM_{mn} mode; TEM stands for Transverse Electric and Magnetic. As discussed earlier, the last factor represents the curvature of the wavefront at the mirrors. (For a more detailed analysis see, e.g., Ghatak and Thyagarajan (1978), Sec. 4.15.)

The fundamental mode corresponds to $m = 0, n = 0$ and using Eqs. (9.135) in Eq. (9.141), we find that it is indeed Gaussian. The mode TEM_{10} will correspond to a field pattern given by

$$f(x, y) = 2C\frac{x}{w_0}\exp\left[-\frac{(x^2 + y^2)}{2w_0^2}\right]\exp\left[-i\frac{k}{2R}(x^2 + y^2)\right]$$

(9.142)

The corresponding intensity pattern will be given by

$$I(x, y) = I_0 x^2 \exp\left[-(x^2 + y^2)/w_0^2\right]$$

(9.143)

where I_0 is a constant. The intensity distribution along the y-direction is Gaussian and along the x-direction is of the form $x^2 e^{-x^2/w_0^2}$. The intensity peaks at $x = \pm w_0$ and is zero at $x = 0$. Thus if a laser is oscillating in the TEM_{10} mode and we let it fall on a screen we will observe two distinct blobs of light as shown in Fig. 9.5. Fig. 9.5 also shows the observed pattern corresponding to other higher order modes. It is evident from Fig. 9.5 that m corresponds to the number of zeroes in intensity along the x-direction and n to the number of zeroes in intensity along the y-direction.

In Sec. 9.9 we have shown that the fundamental mode of a stable general spherical resonator is Gaussian. The higher order modes corresponding to the general resonator are also Hermite–Gaussian.

Substituting for κ and τ in Eq. (9.105) we get

$$\sigma = \exp\{-i[kR - (m + n + 1)\pi/2]\}$$

(9.144)

Thus the phase shift in one half round trip is $kR - (m + n + 1)\pi/2$ which for a standing wave pattern must be an integral multiple of π. Thus we get

$$kR - (m + n + 1)\pi/2 = q\pi$$

or

(9.145)

$$v_{mnq} = (2q + m + n + 1)c/4R$$

The frequency separation between two adjacent longitudinal modes differing in q by unity will be

$$\Delta v_q = c/2R \qquad (9.146)$$

Eq. (9.145) gives the frequencies of oscillation of a mode characterized by (m, n, q) in a confocal resonator.

10

Some laser systems

10.1   Introduction

In this chapter we shall discuss some specific laser systems and their important operating characteristics. The systems that we shall consider are some of the more important lasers that are in widespread use today for different applications. The lasers considered are:

    (*a*) solid state lasers: ruby, Nd:YAG, Nd:glass;
    (*b*) gas lasers: He–Ne, argon ion and CO_2;
    (*c*) liquid lasers: dyes;
    (*d*) excimer lasers;
    (*e*) semiconductor lasers.

10.2   Ruby lasers

The first laser to be operated successfully was the ruby laser which was fabricated by Maiman in 1960. Ruby, which is the lasing medium, consists of a matrix of aluminium oxide in which some of the aluminium ions are replaced by chromium ions. It is the energy levels of the chromium ions which take part in the lasing action. Typical concentrations of chromium ions are $\sim 0.05\%$ by weight. The energy level diagram of the chromium ion is shown in Fig. 10.1. As is evident from the figure this is a three level laser.[†] The pumping of the chromium ions is performed with the help of flash lamp (e.g., a xenon or krypton flashlamp) and the chromium ions in the ground state absorb radiation around wavelengths of 5500 Å and 4000 Å and are excited to the levels marked E_1 and E_2. The chromium ions excited to these levels relax rapidly through a nonradiative transition (in a time $\sim 10^{-8}$–10^{-9} s) to the level marked M which is the upper laser level. The

---

[†] The level M actually consists of a pair of levels corresponding to wavelengths of 6943 Å and 6929 Å. However laser action takes place only on the 6943 Å line because of higher inversion.

level M is a metastable level with a lifetime of ~ 3 ms. Laser emission occurs between level M and the ground state G at an output wavelength of $\lambda_0 = 6943$ Å.

The flashlamp operation of the laser leads to a pulsed output of the laser. As soon as the flashlamp stops operating the population of the upper level is depleted very rapidly and lasing action stops till the arrival of the next flash. Even during the short period of a few tens of microseconds in which the laser is oscillating, the output is a highly irregular function of time with the intensity having random amplitude fluctuations of varying duration as shown in Fig. 10.2. This is called laser spiking, the formation of which can be understood as follows: when the pump is turned on, the intensity of light at the laser transition is small and hence the pump builds up the inversion rapidly. Although under steady state conditions the inversion cannot exceed the threshold inversion, on a transient basis it can go beyond the threshold value due to the absence of sufficient laser radiation in the cavity which causes stimulated emission. Thus the inversion goes beyond threshold when the radiation density in the cavity builds up rapidly. Since the inversion is greater than threshold, the radiation density goes beyond the steady state value which in turn depletes the upper level population and reduces the

Fig. 10.1 The energy levels of the chromium ions in the ruby laser.

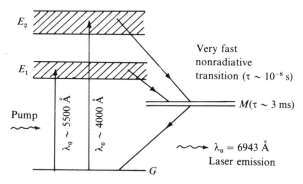

Fig. 10.2 Temporal output power variations of a ruby laser beam leading to what is referred to as laser spiking.

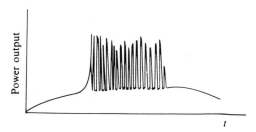

inversion below threshold. This leads to an interruption of laser oscillation till the pump can again create an inversion beyond threshold. This cycle repeats itself to produce the characteristic spiking in lasers.

Fig. 10.3 shows a typical set up of a flashlamp pumped pulsed ruby laser. The helical flashlamp is surrounded by a cylindrical reflector to direct the pump light onto the ruby rod efficiently. The ruby rod length is typically 2–20 cm with diameters of 0.1–2 cm. As we have seen in Sec, 8.5.2, typical input electrical energies are in the range of 10–20 kJ. In addition to the helical flashlamp pumping scheme shown in Fig. 10.3, one may use other pumping schemes such as that shown in Fig. 10.4 in which the pump lamp and the laser rod are placed along the foci of an elliptical cylindrical reflector. It is well known that the elliptical reflector focusses the light emerging from one focus into the other focus of the ellipse thus leading to an efficient focussing of pump light on the laser rod.

Fig. 10.3 A typical set up of a flashlamp pumped pulsed ruby laser. The flashlamp is covered by a cylindrical reflector for efficient coupling of the pump light to the ruby rod.

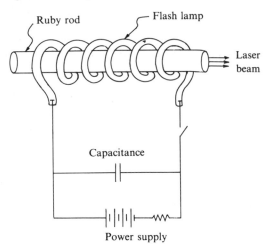

Fig. 10.4 Elliptical pump cavity in which the lamp and the ruby rod are placed along the foci of the elliptical cylindrical reflector.

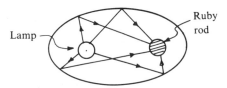

In spite of the fact that the ruby laser is a three level laser, it still is one of the important practical lasers. The absorption bands of ruby are very well matched with the emission spectra of practically available flashlamps so that an efficient use of the pump can be made. It also has a favourable combination of a long lifetime and a narrow linewidth. The ruby laser is also attractive from an application point of view since its output lies in the visible region where photographic emulsions and photodetectors are much more sensitive than they are in the infrared region. Ruby lasers find applications in pulsed holography, in laser ranging etc.

10.3 Neodymium based lasers

The Nd:YAG laser (YAG stands for yttrium aluminium garnet which is $Y_3Al_5O_{12}$) and the Nd:glass laser are two very important solid state laser systems in which the energy levels of the neodymium ion take part in laser emission. They both correspond to a four level laser. Using neodymium ions in a YAG or glass host has specific advantages and applications.

(a) Since glass has an amorphous structure the flourescent linewidth of emission is very large leading to a high value of the laser threshold. On the other hand YAG is a crystalline material and the corresponding linewidth is much smaller which implies much lower thresholds for laser oscillation.

(b) The fact that the linewidth in the case of the glass host is much larger than in the case of the YAG host can be made use of in the production of ultrashort pulses using mode locking since as discussed in Sec. 9.7, the pulsewidth obtainable by mode locking is the inverse of the oscillating linewidth.

(c) The larger linewidth in glass leads to a smaller amplification coefficient and thus the capability of storing a larger amount of energy before the occurence of saturation. This is especially important in obtaining very high energy pulses using Q-switching.

(d) Other advantages of the glass host are the excellent optical quality and excellent uniformity of doping that can be obtained and also the range of glasses with different properties that can be used for solving specific design problems.

(e) As compared to YAG, glass has a much lower thermal conductivity which may lead to induced birefringence and optical distortion.

From the above discussion we can see that for continuous or very high pulse repetition rate operation the Nd:YAG laser will be preferred over

Nd:glass. On the other hand for high energy pulsed operation, Nd:glass lasers may be preferred. In the following we discuss some specific characteristics of Nd:YAG and Nd:glass laser systems.

10.3.1 *Nd:YAG laser*

The Nd:YAG laser is a four level laser and the energy level diagram of the neodymium ion is shown in Fig. 10.5. The laser emission occurs at $\lambda_0 \approx$ 1.06 μm. Since the energy difference between the lower laser level and the ground level is ~ 0.26 eV, the ratio of its population to that of the ground state at room temperature $(T = 300 \, \text{K})$ is $e^{-\Delta E/k_B T} \approx e^{-9} \ll 1$. Thus the lower laser level is almost unpopulated and hence inversion is easy to achieve. The main pump bands for excitation of the neodymium ions are in the 0.81 μm and 0.75 μm wavelength regions and pumping is done using arc lamps (e.g., the Krypton arc lamp). Typical neodymium ion concentrations used are $\sim 1.38 \times 10^{20}$ cm^{-3}. The spontaneous lifetime corresponding to the laser transition is 550 μs and the emission line corresponds to homogeneous broadening and has a width $\Delta v \sim 1.2 \times 10^{11}$ Hz which corresponds to $\Delta \lambda \sim 4.5$ Å. We have shown in Sec. 8.5.3 that the Nd:YAG laser has a much lower threshold of oscillation than a ruby laser.

Nd:YAG lasers find many applications in range finders, illuminators with Q-switched operation giving about 10–50 pulses per second with output energies in the range of 100 mJ per pulse and pulse width ~ 10 ns. They also find applications in resistor trimming, scribing, micromachining operations as well as welding, hole drilling etc.

Fig. 10.5 The energy levels of neodymium ion in the Nd:YAG laser.

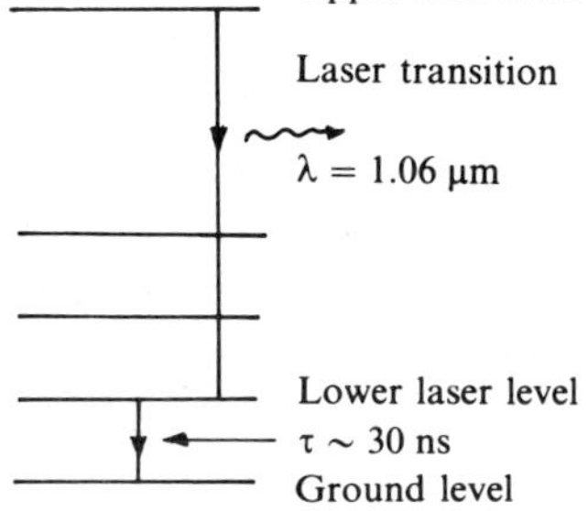

Table 10.1 *Comparison of ruby, Nd:YAG and Nd:glass laser systems.*

Laser	Ruby	Nd:YAG	Nd:glass
Wavelength (Å)	6943	10,641	10,623
Spontaneous lifetime (μs)	3000	240	300
Active ion concentration (cm^{-3})	1.58×10^{19}	1.38×10^{20}	2.83×10^{20}
Linewidth (GHz)	330	120	7500
(Å)	5.5	4.0	260
Population inversion density for 1% gain/cm (cm^{-3})	4×10^{7} $+ 7.6 \times 10^{18}$	1.1×10^{16}	3.3×10^{17}
Index of refraction (n) (at laser λ)	$n_o = 1.763$ $n_e = 1.755$	1.82	1.55
Major pump bands (Å)	4040 5540	5800 7500 8100	5800 7500 8100

Table adapted from Koechner (1976).

10.3.2 Nd:glass lasers

The Nd:glass laser is again a four level laser system with a laser emission around 1.06 μm. Typical neodymium ion concentrations are $\sim 2.8 \times 10^{20}$ cm^{-3} and various silicate and phosphate glasses are used as the host material. Since glass has an amorphous structure different neodymium ions situated at different sites have slightly different surroundings. This leads to an inhomogeneous broadening and the resultant linewidth is $\Delta v \sim 7.5 \times 10^{12}$ Hz which corresponds to $\Delta \lambda \sim 260$ Å. This width is much larger than in Nd:YAG lasers and consequently the threshold pump powers are also much higher. The spontaneous lifetime of the laser transition is $\sim 300 \, \mu$s.

Nd:glass lasers are more suitable for high energy pulsed operation such as in laser fusion where the requirement is of subnanosecond pulses with an energy content of several kilojoules (i.e., peak powers of several tens of terawatts – see e.g., Thyagarajan and Ghatak (1981)). Other applications are in welding or drilling operations requiring high pulse energies.

Table 10.1 gives a comparison of some important characteristics of ruby, Nd:YAG and Nd:glass laser systems.

10.4   The He–Ne laser

The first gas laser to be operated successfully was the He–Ne laser. As we discussed earlier in solid state lasers, the pumping is usually done using a flashlamp or a continuous high power lamp. Such a technique is efficient if the lasing system has broad absorption bands. In gas lasers since the atoms are characterized by sharp energy levels as compared to those in solids, one generally uses an electrical discharge to pump the atoms.

The He–Ne laser consists of a long and narrow discharge tube (diameter ~ 2–8 mm and length 10–100 cm) which is filled with helium and neon with typical pressures of 1 torr[†] and 0.1 torr. The actual lasing atoms are the neon atoms and as we shall discuss helium is used for a selective pumping of the upper laser level of neon. The laser resonator may consist of either internal or

Fig. 10.6 A typical He–Ne laser with external mirrors. The ends of the discharge tube are fitted with Brewster windows.

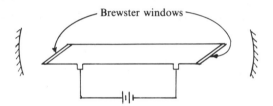

Fig. 10.7 Low lying energy levels of helium and neon taking part in the He–Ne laser.

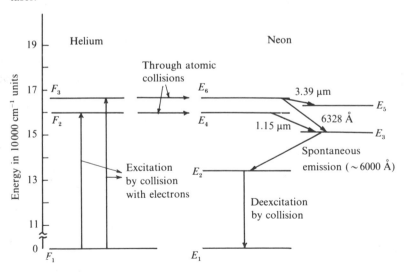

[†] Torr is a unit of pressure and 1 torr = 1 mm Hg.

external mirrors (see Fig. 10.6). Fig. 10.7 shows the energy levels of helium and neon. When an electrical discharge is passed through the gas, the electrons which are accelerated down the tube collide with helium and neon atoms and excite them to higher energy levels. The helium atoms tend to accumulate at levels F_2 and F_3 due to their long lifetimes of $\sim 10^{-4}$ and 5×10^{-6} s respectively. Since the levels E_4 and E_6 of neon atoms have almost the same energy as F_2 and F_3, excited helium atoms colliding with neon atoms in the ground state can excite the neon atoms to E_4 and E_6. Since the pressure of helium is ten times that of neon, the levels E_4 and E_6 of neon are selectively populated as compared to other levels of neon.

Transition between E_6 and E_3 produces the very popular 6328 Å line of the He–Ne laser. Neon atoms deexcite through spontaneous emission from E_3 to E_2 (lifetime $\sim 10^{-8}$ s). Since this time is shorter than the lifetime of level E_6 (which is $\sim 10^{-7}$ s) one can achieve steady state population inversion between E_6 and E_3. Level E_2 is metastable and thus tends to collect atoms. The atoms from this level relax back to the ground level mainly through collisions with the walls of the tube. Since E_2 is metastable it is possible for the atoms in this level to absorb the spontaneously emitted radiation in the $E_3 \rightarrow E_2$ transition to be reexcited to E_3. This will have the effect of reducing the inversion. It is for this reason that the gain in this laser transition is found to increase with decreasing tube diameter.

The other two important wavelengths from the He–Ne laser are 1.15 μm and 3.39 μm, which correspond to the $E_4 \rightarrow E_3$ and $E_6 \rightarrow E_5$ transitions. It is interesting to observe that both 3.39 μm and 6328 Å transitions share the same upper laser level. Now since the 3.39 μm transition corresponds to a much lower frequency than the 6328 Å line, the Doppler broadening is much smaller at 3.39 μm and also since gain depends inversely on v^2 (see Eq. (8.23)), the gain at 3.39 μm is much higher than at 6328 Å. Thus due to the very large gain, oscillations will normally tend to occur at 3.39 μm rather than at 6328 Å. Once the laser starts to oscillate at 3.39 μm, further build up of population in E_6 is not possible. The laser can be made to oscillate at 6328 Å by either using optical elements in the path which strongly absorb the 3.39 μm wavelength or increasing the linewidth through the Zeeman effect by applying an inhomogeneous magnetic field across the tube.

If the resonator mirrors are placed outside the discharge tube then reflections from the ends of the discharge tube can be avoided by placing the windows at the Brewster angle (see Fig. 10.6). In such a case the beam polarized in the plane of incidence suffers no reflection at the windows while the perpendicular polarization suffers reflection losses. This leads to a polarized output of the laser.

10.5   The argon ion laser

In an argon ion laser, one uses the energy levels of the ionized argon atom and the laser emits various discrete lines in the 3500–5200 Å wavelength region. Fig. 10.8 shows some of the energy levels taking part in the laser transition. The argon atoms have to be first ionized and then excited to the higher energy levels of the ion. Because of the large energies involved in this, the argon ion laser discharge is very intense; typical values being 40 A at 165 V. A particular wavelength out of the many possible lines is chosen by placing a dispersive prism inside the cavity close to one of the mirrors. Rotation of the prism–mirror system provides feedback only at the wavelength which is incident normally on the mirror. Typical output power in a continuous wave argon ion laser is 3–5 W. Some of the important emission wavelengths include 5145 Å, 4965 Å, 4880 Å, 4765 Å and 4579 Å.

10.6   The CO_2 laser

The lasers discussed above use transitions among the various excited electronic states of an atom or an ion. In a CO_2 laser one uses the transitions occuring between different vibrational states of the carbon dioxide molecule.

Fig. 10.9 shows the carbon dioxide molecule consisting of a central carbon atom with two oxygen atoms attached one on either side. Such a molecule

Fig. 10.8 Some of the levels taking part in the laser transition corresponding to the argon ion laser.

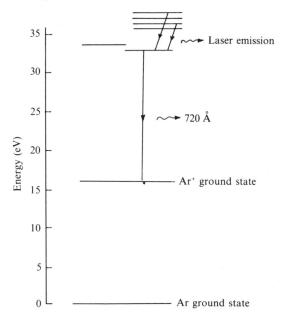

can vibrate in the three independent modes of vibration shown in Fig. 10.9. These correspond to the symmetric stretch, the bending and the asymmetric stretch modes. Each of these modes is characterized by a definite frequency of vibration. According to basic quantum mechanics these vibrational degrees of freedom are quantized, i.e., when a molecule vibrates in any of the modes it can have only a descrete set of energies. Thus if we call v_1 the frequency corresponding to the symmetric stretch mode then the molecule can have energies of only

$$E_1 = (m + \tfrac{1}{2})hv_1; \qquad m = 0, 1, 2, \ldots \tag{10.1}$$

when it vibrates in the symmetric stretch mode. Thus the degree of excitation is characterized by the integer m when the carbon dioxide molecule vibrates in the symmetric stretch mode. In general, since the carbon dioxide molecule can vibrate in a combination of the three modes the state of vibration can be described by three integers (mnq); the three integers correspond respectively to the degree of excitation in the symmetric stretch, bending and asymmetric stretch modes respectively. Fig. 10.10 shows the various vibrational energy levels taking part in the laser transition.

The laser transition at 10.6 μm occurs between the (001) and (100) levels of carbon dioxide. The excitation of the carbon dioxide molecules to the long lived level (001) occurs both through collisional transfer from nearly resonant excited nitrogen molecules and also from the cascading down of carbon dioxide molecules from higher energy levels.

Fig. 10.9 The three independent modes of vibration of the carbon dioxide molecule.

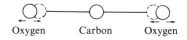

Oxygen   Carbon   Oxygen

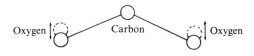

Oxygen   Carbon   Oxygen

Oxygen   Carbon   Oxygen

The CO_2 laser possesses an extremely high efficiency of $\sim 30\%$. This is because of efficient pumping to the (001) level and also because all the energy levels involved are close to the ground level. Thus the atomic quantum efficiency which is the ratio of the energy difference corresponding to the laser transition to the energy difference of the pump transition i.e.,

$$\eta = \frac{E_5 - E_4}{E_5 - E_1}$$

is quite high ($\sim 45\%$). Thus a large portion of the input power can be converted into useful laser power.

Output powers of several watts to several kilowatts can be obtained from CO_2 lasers. High power CO_2 lasers find applications in materials processing, welding, hole drilling, cutting, etc., because of their very high output power. In addition, the atmospheric attenuation is low at $10.6\,\mu m$ which leads to some applications of CO_2 lasers in open air communications.

10.7   Dye lasers

One of the most widely used tunable lasers in the visible region is the organic dye laser. The dyes used in the lasers are organic substances which are dissolved in solvents such as water, ethyl alcohol, methanol,

Fig. 10.10 The low lying vibrational levels of nitrogen and carbon dioxide molecules. Energy transfer from excited nitrogen molecules to carbon dioxide molecules results in the excitation of carbon dioxide molecules. Important lasing transitions occur at $9.6\,\mu m$ and $10.6\,\mu m$.

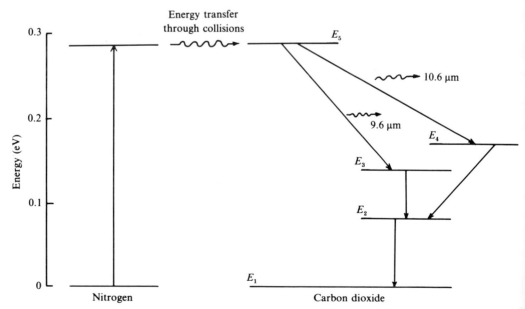

ethylene glycol etc. These dyes exhibit strong and broad absorption and flourescent spectra and because of this they can be made tunable. By choosing different dyes one can obtain tunability from 3000 Å to 1.2 μm.

The levels taking part in the absorption and lasing correspond to the various vibrational sublevels of different electronic states of the dye molecule. Fig. 10.11 shows a typical energy level diagram of a dye in which S_0 is the ground state, S_1 is the first excited singlet state, and T_1, T_2 are the excited triplet states of the dye molecule. Each state consists of a large number of closely spaced vibrational and rotational sublevels. Because of strong interaction with the solvent, the closely spaced sublevels are collision broadened to such an extent that they almost form a continuum.

When dye molecules in the solvent are irradiated by visible or ultraviolet radiation then the molecules are excited to the various sublevels of the state S_1. Due to collisions with the solvent molecules, the molecules excited to higher vibrational and rotational states of S_1 relax very quickly (in times $\sim 10^{-11}$–10^{-12} s) to the lowest level V_2 of the state S_1. Molecules from this level emit spontaneously and deexcite to the different sublevels of S_0. Thus the fluorescent spectrum is found to be red shifted against the absorption spectrum. Fig. 10.12 shows a typical absorption and flourescence spectrum of one of the most commonly used dye Rhodamine 6G. Observe the extremely broad fluorescence spectrum.

Fig. 10.11 Typical energy level diagram of a dye molecule.

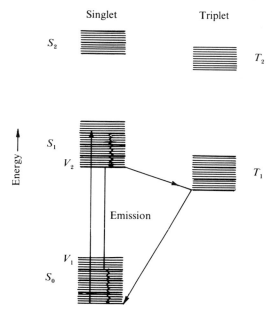

If the pump intensity is very high, then it is possible to generate population inversion between V_2 and some higher sublevel V_1 of S_0. As soon as the gain at this transition exceeds the loss, laser oscillation can begin. The lower laser level V_1 is depleted rapidly by collisions with the solvent molecules. Thus the dye laser corresponds to a four level laser system.

Some important points to be noted are: molecules from S_1 can also make a nonradiative relaxation to level T_1; this is called intersystem crossing. This is deterimental to laser action for two reasons. Firstly it results in a decrease in the population of V_1 which is the upper laser level. Secondly the absorption spectrum corresponding to $T_1 \to T_2$ usually overlaps with the emission spectrum of $S_1 \to S_0$. This would lead to an absorption of the laser radiation and if T_1 contains sufficient population this loss can inhibit laser oscillation completely. Thus care must be taken to remove these molecules from level T_1 which can be achieved either by adding what are called triplet quenching additives or by mechanically transporting the dye solution rapidly (in about 10^{-6} s) through the active zone in the cavity by using free flowing jets.

Experimental dye lasers use flashlamps, pulsed lasers or continuous wave lasers as pumping sources. Pump lasers include nitrogen lasers, argon lasers, krypton lasers, frequency doubled YAG lasers etc.

The broad flouresence spectrum as shown in Fig. 10.12 suggests a broad tunability of these lasers. Fig. 10.13 shows a typical arrangement used in a continuous wave dye laser pumped by a continuous wave argon laser. The pump beam is focussed on a flowing stream of dye through a cell (of typically 1 mm thickness) which may be placed at the Brewster angle to eliminate any reflection losses in the cavity. The dye laser cavity is formed around the dye

Fig. 10.12 Typical absorption and fluorescence spectrum of Rhodamine 6G; the quantities $E(\lambda)$ and $\epsilon_{ss}(\lambda)$ represent the spontaneous fluorescence lineshape function and the molecular extinction function respectively (Snavely, 1969).

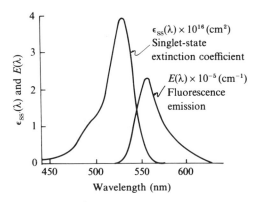

Table 10.2 *Typical characteristics of dye lasers with different pump sources.*

Pump	Tuning range (nm)	Average output power (W)	Peak output power (W)	Pulse duration (ns)
Nitrogen laser	350–1000	0.1–1	10^4–10^5	1–10
Flashlamp	400–960	0.1–100	10^5–10^6	10^2–10^5
Continuous wave argon laser	400–800	0.1–10	Max. reported 40 W	CW
Frequency doubled YAG laser	400–800	0.1–1	10^4–10^6	5–30

Adapted from Demtroder (1981).

cell and at one end one may have a prism–mirror combination for wavelength tuning. As can be seen from Fig. 10.13 only that wavelength which falls normally on the mirror will have feedback and the laser will oscillate at that wavelength. Wavelength selection can also be achieved by replacing the prism–mirror combination by a reflection grating. The threshold pump power depends on the resonator losses and the spot size at the focus of the pump but is typically between 1 mW and 1 W. Output powers of 1–4 W can be obtained from these lasers. Table 10.2 shows typical characteristics of dye lasers with different pump sources.

Fig. 10.13 A typical arrangement used in a tunable continuous wave dye laser pumped by an argon laser.

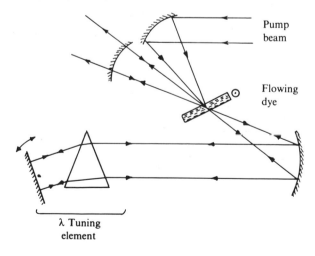

Pump beam

Flowing dye

λ Tuning element

10.8   Excimer lasers

Excimers are molecules which are bound in their excited electronic states but are unstable in their ground states. Examples of excimers are argon fluoride, krypton chloride, krypton fluoride, xenon bromide, etc. Since excimers are unstable in their ground state they are ideal for forming active media for lasers since the lower level can be the ground state in which the molecule is unstable (dissociation time $\sim 10^{-12}$–10^{-13} s) and hence population inversion can be automatically maintained. Pumping is usually accomplished either by fast transverse discharges or through high voltage, high current electron sources. Excimer lasers are tunable and emit mostly in the ultraviolet region (120–350 nm).

10.9   Semiconductor lasers

The semiconductor laser (which is also sometimes referred to as the junction laser or the diode laser) is today one of the most important types of lasers with its very important application in fibre optic communication (see Chapter 13). These lasers use semiconductors as the lasing medium and are characterized by specific advantages such as the capability of direct modulation into the gigahertz region, small size and low cost, the capability of monolithic integration with electronic circuitry, direct pumping with conventional electronic circuitry and compatibility with optical fibres.

The basic mechanism responsible for light emission from a semiconductor is the recombination of electrons and holes at a p–n junction when a current is passed through the diode. Just like in other laser systems, there can be three interaction processes: (a) an electron in the valence band can absorb the incident radiation and be excited to the conduction band leading to the generation of electron–hole pair; (b) an electron can make a spontaneous transition in which it combines with a hole, i.e., it makes a transition from the conduction to the valence band and in the process it emits radiation; (c) a stimulated emission may occur in which the incident radiation stimulates an electron in the conduction band to make a transition to the valence band and in the process emit radiation. If now by some mechanism a large density of electrons is created in the bottom of the conduction band and simultaneously in the same region of space a large density of holes is created at the top of the valence band (see Fig. 10.14) then an optical beam with a frequency slightly greater than $E_g/\hbar$, where E_g is the bandgap energy, will cause a larger number of stimulated emissions as compared to absorptions and thus can be amplified. In order to convert the amplifying medium into a laser, one must provide optical feedback which is usually done by cleaving or polishing the ends of the p–n junction diode at right angles to the junction.

Thus, when a current is passed through a *p–n* junction under forward bias, the injected electrons and holes will increase the density of electrons in the conduction band and holes in the valence band and at some value of current, the stimulated emission rate will exceed the absorption rate and amplification will begin. As the current is further increased, at some threshold value of current, the amplification will overcome the losses in the cavity and the laser will begin to emit coherent radiation.

The early semiconductor lasers were based on *p–n* junctions formed on the same material by proper doping and these are referred to as homojunction lasers. Due to the absence of any potential barriers for the confinement of carriers or abrupt refractive index discontinuities for the confinement of optical radiation these laser structures required large threshold current densities ($\sim 50\,000\,\mathrm{A/cm^2}$). The absence of carrier confinement resulted in a diffusion of the carriers near the *p–n* junction plane due to which a significant optical gain was available only over a very small region around the junction. The absence of any strong optical confinement resulted in the optical energy penetrating beyond the gain region where it was absorbed. Thus larger current densities were required for laser operation.

A significant reduction in threshold current densities was achieved by forming what are referred to as heterojunctions. A heterojunction is a junction formed between two dissimilar semiconductors. The present day lasers are based on the double heterojunction in which a thin active layer of a semiconductor with a narrow bandgap is sandwiched between two larger bandgap semiconductors as shown in Fig. 10.15. As seen from Fig. 10.15 in this configuration the regions in which the electrons and holes recombine is bound on either side by potential barriers and thus they are confined to the thin active region. Fortuitously the refractive index of the semiconductor decreases with an increase in the bandgap. Thus the refractive index of the central active layer is higher than the two surrounding regions (typically

Fig. 10.14 If a large density of electrons is created at the bottom of the conduction band and simultaneously in the same region a large density of holes is produced at the top of the valence band, then such a semiconductor can amplify optical radiation at a frequency slightly greater than $E_g/\hbar$.

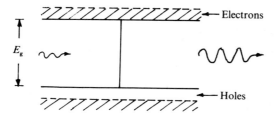

~ 5–10% higher). Such a refractive index profile can confine the emitted optical radiation to the active region by the mechanism of waveguidance which occurs due to total internal reflections taking place at the boundaries. In addition, since the layers surrounding the active central region have larger bandgaps, the optical field which penetrates into the surrounding region is also not absorbed. Use of such double heterojunctions results in a reduction of the threshold current densities to ~ 2000–$4000 \, \text{A/cm}^2$. Taking a typical cavity length of $300 \, \mu\text{m}$ and a width of $100 \, \mu\text{m}$ (see Fig. 10.16), the required threshold current for a threshold current density of $4000 \, \text{A/cm}^2$ will be
$$I_{\text{th}} \approx 300 \times 100 \times 10^{-8} \times 4000 = 1.2 \, \text{A}.$$

In the laser structure discussed above the carriers and the optical waves are confined only along one direction. One can provide in addition a

Fig. 10.15 Energy diagram of a double heterostructure laser.

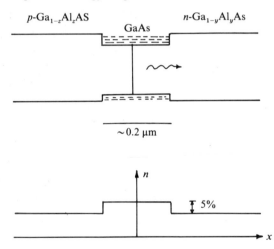

Fig. 10.16 A typical broad area p–n junction laser showing the dimensions of the laser.

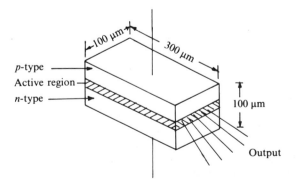

confinement in the lateral direction also. In such a laser, the active region will have an approximately rectangular cross section and will be surrounded by higher bandgap materials on all sides. Such a laser, referred to as a buried heterostructure laser (see Fig. 10.17) can operate with threshold currents which are much smaller. Thus if the current is restricted to flow across a lateral dimention of $\sim 5\,\mu$m, then the threshold current for a threshold current density of 4000 A/cm^2 will be

$$I_{\text{th}} \approx 5 \times 300 \times 10^{-8} \times 4000 = 60\,\text{mA}$$

which is a significant reduction in the threshold current.

Most semiconductor lasers operate either in the 0.8–0.9 μm or in the 1–1.7 μm spectral region. Since the wavelength of emission is determined by the bandgap, different semiconductor materials are used for the two different spectral regions. Lasers operating in the 0.8–0.9 μm spectral region are based on gallium arsenide. By replacing a fraction of gallium atoms by aluminium, the bandgap can be increased. Thus one can form heterojunctions by proper combinations of GaAlAs and GaAs which can provide both carrier confinement and optical waveguidance. For example, the bandagap of GaAs is 1.424 eV and that of $Ga_{0.7}Al_{0.3}As$ is ≈ 1.798 eV; the corresponding refractive index difference is $\Delta n \approx 0.19$. Thus surrounding the GaAs layer on either side with $Ga_{0.7}Al_{0.3}As$, one can achieve confinement of both carriers and light waves. For lasers operating in the 1.0–1.7 μm band the semiconductor material is InP with gallium and arsenic used to replace fractions of indium and phosphorous respectively. The above wavelength region is extremely important in connection with fibre optic communication since silica based optical fibres exhibit both low loss and very high bandwidth around 1.55 μm (see Chapter 13).

Typical output characteristics Fig. 10.18 shows a typical light output versus current characteristic of a GaAs semiconductor laser. As can be seen the

Fig. 10.17 Cross section of a typical buried heterostructure laser.

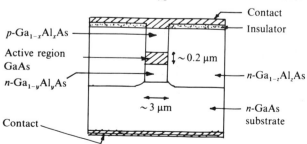

output optical power starts to increase very rapidly around a threshold current which essentially represents the beginning of laser oscillation. For digital modulation of the laser diodes, they are biased at slightly above the threshold and on this bias is superposed current pulses corresponding to the digital data. Thus the electrical signal can be directly encoded into an optical signal. For analog modulation the laser is usually biased above threshold and the analog signal is fed in the form of current variations. Lasers with modulation bandwidths greater than 6 GHz are available commerically.

An important characteristic of the laser is its mode of oscillation and as seen in Chapter 9, this can be divided into the transverse modal field pattern and the longitudinal mode. The mode field pattern consists of the transverse mode distribution (perpendicular to the junction plane, and the lateral mode distribution (parallel to the junction plane). The fundamental mode distribution of the laser exhibits no zeroes in the near field pattern and is characterized by a centrally peaked far field pattern. Due to the very small thickness of the active region ($\sim 0.2\,\mu$m) the laser usually oscillates in the fundamental transverse mode. The fundamental lateral mode oscillation can also be obtained by restricting the width of the active region in the junction plane to small values (~ 1–$2\,\mu$m) in the case of buried heterostructure lasers. For typical lasers the full angular width at half power in the far field is ≈ 35–$50°$ perpendicular to the junction plane and ≈ 6–$10°$ parallel to the junction plane.

Another important characteristic of a laser diode is the spectral width of emission. The spontaneous emission spectrum is usually very broad with a

Fig. 10.18 A typical light output vs. current characteristic of a semiconductor laser. (Adapted from the Ortel Corporation Report on High Speed Laser Diodes.)

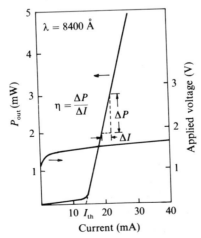

typical width of ~ 200–$300\,\text{Å}$. As discussed in Sec. 9.2, the frequencies of oscillation are approximately given by

$$v \approx (c/2nd)q; \qquad q = \text{integer} \tag{10.2}$$

where n is the refractive index of the laser medium and d is the resonator length. In semiconductor lasers, the variation of n with v is very significant and one can show that the intermode spacing is (see Problem 10.1)

$$\Delta v \approx \frac{c}{2nd}\left(1 + \frac{v}{n}\frac{dn}{dv}\right)^{-1} \tag{10.3}$$

Typically $n \approx 3.6$, $d \approx 250\,\mu\text{m}$, $(v/n)(dn/dv) \approx 0.38$ (Kressel and Bulter, 1977) and we have

$$\Delta v \approx 125\,\text{GHz} \tag{10.4}$$

For $\lambda_0 \approx 0.9\,\mu\text{m}$, the corresponding mode separation is $\approx 3.4\,\text{Å}$. Fig. 10.19 shows a typical output spectrum of a multilongitudinal mode laser with a FWHM of about 3 nm. In some applications (for example, in fibre optic communications) one would like to have single longitudinal mode oscill-ation of the laser so that its spectral width is $\ll 1\,\text{Å}$. One can achieve this either by using a cleaved coupled cavity configuration or by using distributed feedback. In the former case the laser device essentially consists of two independent cavities which are optically coupled. The mode which

Fig. 10.19 A typical output spectrum of: (*a*) a multilongitudinal mode laser; (*b*) a single longitudinal mode laser. (Adapted from Ortel Corporation Report on High Speed Laser Diodes).

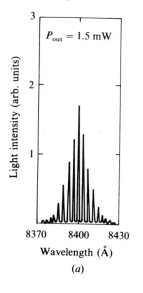
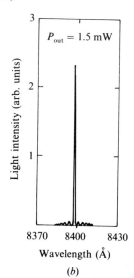

(*a*)　　　　　　　(*b*)

can oscillate is the one which is a mode of either of the cavities and also has the lowest loss. The second concept uses a periodic variation of the thickness of the layer surrounding the active region of the laser. This essentially acts as a Bragg reflector which is highly wavelength selective (see Chapter 18). Thus the feedback which, in this case is distributed throughout the length of the cavity, is wavelength selective and this leads to a single longitudinal mode oscillation of the laser (see e.g., Agarwal and Dutta (1986), Suematsu, Kishino, Arai and Koyama, (1985)).

Problems

Problem 10.1: In Chapter 11 we will show that for a symmetric planar waveguide to be single moded, (i.e., support only the fundamental transverse mode) one must have

$$V = (2\pi/\lambda_0)d(2n\Delta n)^{\frac{1}{2}} < \pi \tag{10.5}$$

where d is the width of the waveguide, n is the refractive index and Δn is the index step. Estimate the maximum d for a single mode oscillation in a GaAlAs–GaAs laser for $\Delta n \sim 0.2$.

Solution: For GaAs,

$$n \approx 3.5, \quad \lambda_0 \approx 0.85\,\mu m$$

and Eq. (10.5) gives us

$$d \lesssim 0.36\,\mu m$$

which gives a typical order of the spacing of the heterojunctions in a GaAs laser.

Problem 10.2: Starting from Eq. (10.2) obtain Eq. (10.3).

Solution: We write

$$v = \frac{c}{2n(v)d}q$$

$$v + \Delta v = \frac{c}{2n(v + \Delta v)d}(q + 1)$$

If we write

$$n(v + \Delta v) = n(v) + (dn/dv)\Delta v$$

we have

$$(v + \Delta v)[n + (dn/dv)\Delta v] = (c/2d)(q + 1)$$

or

$$\Delta v = \frac{c}{2nd}\left(1 + \frac{v}{n}\frac{dn}{dv}\right)^{-1}$$

11

Electromagnetic analysis of the simplest optical waveguide

11.1 Introduction

An optical waveguide is a structure which confines and guides the light beam by the process of total internal reflection. The most extensively used optical waveguide is the step index optical fibre which consists of a cylindrical central core, clad by a material of slightly lower refractive index (see Fig. 11.1). If the refractive indices of the core and cladding are n_1 and n_2

Fig. 11.1 (a) A typical optical fibre waveguide consists of a thin cylindrical glass rod of radius a and refractive index n_1 clad by glass of slightly lower refractive index n_2. Light guidance takes place through the phenomenon of total internal reflection at the core–cladding interfaces. (b) The corresponding refractive index variation; for a typical (multimode) optical fibre $n_1 \approx 1.50$, $n_2 \approx 1.49$, $a \approx 25\,\mu m$.

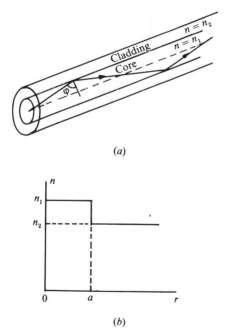

(a)

(b)

respectively, then for a ray entering the fibre, if the angle of incidence (at the core–cladding interface) ϕ is greater than the critical angle

$$\phi_c = \sin^{-1}(n_2/n_1), \tag{11.1}$$

then the ray will undergo total internal reflection at that interface. Further, because of the cylindrical symmetry in the fibre structure, this ray will suffer total internal reflection at the lower interface also and will therefore be guided through the core by repeated total internal reflections. This is the basic principle of light guidance through the optical fibre. We will present a detailed electromagnetic analysis of the waveguiding action in optical fibres in Chapter 13.

The simplest optical waveguide to analyse is probably the planar waveguide which consists of a thin dielectric film (of refractive index n_1) sandwiched between materials of slightly lower refractive indices. Such planar waveguides are important components in integrated optics which will be discussed in Chapter 14.

In this chapter we will present a detailed electromagnetic analysis of the symmetric planar waveguide for which the refractive indices of the materials on the top and bottom of the film are assumed to be the same. Although

Fig. 11.2 (*a*) The simplest planar optical waveguide consists of a planar film (of refractive index n_1) sandwiched between two materials of lower refractive indices. Light guidance takes place by the phenomenon of total internal reflection. (*b*) The refractive index distribution for a symmetric planar waveguide.

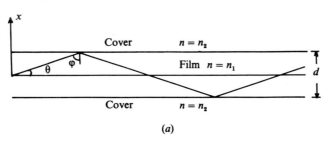

(*a*)

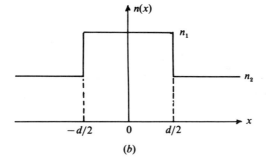

(*b*)

almost all waveguides used in integrated optics are asymmetric in nature, we felt that the electromagnetic analysis of a symmetric waveguide is much easier to understand and at the same time it brings out almost all the salient points associated with the modes of a waveguide, therefore making it easier to understand the physical principles of more complicated guiding structures.

In Fig. 11.2 we show the light guidance through a symmetric planar waveguide. The film is assumed to extend to infinity in the y-direction. The propagation is assumed to be in the z-direction. We consider a bundle of rays launched in the film of the waveguide. If ϕ represents the angle that a ray makes with the x-axis then for total internal reflection to occur at the interface of the film and the cover we must have

$$n_2/n_1 < \sin \phi < 1. \tag{11.2}$$

Thus if θ is the angle that the ray makes with the z-axis (see Fig. 11.2), then for total internal reflection to occur (or for light guidance to take place) we must have

$$n_2/n_1 < \cos \theta < 1 \tag{11.3}$$

On the other hand, when

$$\cos \theta < n_2/n_1 \tag{11.4}$$

the angle of incidence at the film–cover interface will be less than the critical angle and the beam will be partially reflected and partially transmitted. After undergoing several such partial reflections, the beam will 'leak away' from the waveguide.

The above considerations are valid in the geometric optics approximation. In the following sections we will give rigorous solutions of Maxwell's equations for the refractive index profile shown in Fig. 11.2 and discuss the concept of guided modes of the waveguide and their relationship to rays. In particular we will show that for the refractive index distribution depending only on the x-coordinate, i.e., for

$$n^2 = n^2(x) \tag{11.5}$$

Maxwell's equations reduce to two independent set of equations: the first set corresponding to what are known as TE (transverse electric) modes where the electric field does not have a longitudinal component and the second set corresponding to what are known as TM (transverse magnetic) modes where the magnetic field does not have a longitudinal component. In particular, for a symmetric waveguide (like the one shown in Fig. 11.2) for which

$$n^2(-x) = n^2(x) \tag{11.6}$$

both TE and TM modes can always be classified under two categories: one symmetric in x and the other antisymmetric in x. We will derive equations from which the propagation characteristics are determined, and will have a detailed discussion on the qualitative characteristics of modes which will be valid for all waveguiding structures.

In summary, this chapter is completely devoted to the quantitative understanding of a simple optical waveguide – but concepts derived from this will be building blocks for more complicated structures.

11.2   Classification of modes for a planar waveguide

In this section we will discuss the broad classification of modes for a planar waveguide – this will be used in the next two sections where we will have a detailed modal analysis for a specific profile.

In Sec. 11.9 we will show that when the refractive index depends only on the x-coordinate, the electric and magnetic fields associated with a propagating electromagnetic wave can be written in the form

$$\mathcal{E}_j = E_j(x)\,e^{i(\omega t - \beta z)}; \qquad j = x, y, z \tag{11.7}$$

$$\mathcal{H}_j = H_j(x)\,e^{i(\omega t - \beta z)}; \qquad j = x, y, z \tag{11.8}$$

where ω represents the angular frequency of the wave and β is known as the propagation constant. Corresponding to a specific value of β, there is a specific field distribution described by $\mathbf{E}(x)$ and $\mathbf{H}(x)$ and for these specific distributions, the nature of the distribution remains unchanged with propagation along the guide; such distributions are referred to as *modes* of the waveguide. A study of the propagation characteristics and the corresponding field distributions of these modes is of primary importance in the design of efficient integrated optic devices.

If we substitute the above forms of the electric and magnetic fields in Maxwell's equations (Eqs. (11.128) and (11.129)), it can easily be shown that the different components of the electric and magnetic fields are related through the following equations (see Sec. 11.9):

$$H_x = -\frac{\beta}{\omega\mu_0}E_y \tag{11.9}$$

$$H_z = \frac{i}{\omega\mu_0}\frac{\partial E_y}{\partial x} \qquad\left.\right\} \quad \text{TE modes} \tag{11.10}$$

$$-i\beta H_x - \frac{\partial H_z}{\partial x} = i\omega\epsilon_0 K(x)E_y \tag{11.11}$$

$$E_x = \frac{\beta}{\omega\epsilon_0 K(x)} H_y \qquad (11.12)$$

$$E_z = \frac{1}{i\omega\epsilon_0 K(x)} \frac{\partial H_y}{\partial x} \quad \left.\right\} \quad \text{TM modes} \qquad (11.13)$$

$$i\beta E_x + \frac{\partial E_z}{\partial x} = i\omega\mu_0 H_y \qquad (11.14)$$

where $K(x) = n^2(x)$, ϵ_0 and μ_0 are the dielectric permittivity and magnetic permeability of free space. As can be seen, the first three equations involve only E_y, H_x and H_z and the last three equations involve only E_x, E_z and H_y. Thus for such a waveguide configuration, Maxwell's equations reduce to two independent sets of equations. The first set corresponds to nonvanishing values of E_y, H_x and H_z and with E_x, E_z and H_y vanishing, giving rise to what are known as transverse electric (TE) modes because the electric field has only a transverse component. The second set corresponds to the nonvanishing values of E_x, E_z and H_y with E_y, H_x and H_z vanishing, giving rise to what are known as transverse magnetic (TM) modes because the magnetic field now has only a transverse component. The propagation of waves in such planar waveguides may thus be described in terms of TE and TM modes. In the next two sections we will discuss the TE and TM modes of a symmetric step index planar waveguide.

11.3 TE modes of a symmetric step index planar waveguide

In this and the following section we will carry out a detailed modal analysis of a symmetric step index planar waveguide. We first consider TE modes: we substitute H_x and H_z from Eqs. (11.9) and (11.10) in Eq. (11.11) to obtain

$$d^2 E_y/dx^2 + [k_0^2 n^2(x) - \beta^2]E_y = 0 \qquad (11.15)$$

where

$$k_0 = \omega(\epsilon_0\mu_0)^{\frac{1}{2}} = \omega/c \qquad (11.16)$$

is the free space wave number and $c \,(= 1/(\epsilon_0\mu_0)^{\frac{1}{2}})$ is the speed of light in free space.

Until now our analysis has been valid for an arbitrary x dependent profile. We now assume a specific profile given by (see Fig. 11.2)

$$n(x) = \begin{cases} n_1; & |x| < d/2 \\ n_2; & |x| > d/2 \end{cases} \qquad (11.17)$$

Using the above equations we will solve Eq. (11.15) subject to the appropriate boundary conditions at the discontinuities. Since E_y and H_z represent tangential components on the planes $x = \pm d/2$, they must be continuous at $x = \pm d/2$ and since H_z is proportional to dE_y/dx (see Eq. (11.10)) we must have

$$E_y \text{ and } dE_y/dx \text{ continuous at } x = \pm d/2 \tag{11.18}$$

The above condition represents the boundary conditions that have to be satisfied.[†] Substituting for $n(x)$ in Eq. (11.15) we obtain

$$d^2 E_y/dx^2 + (k_0^2 n_1^2 - \beta^2)E_y = 0; \quad |x| < d/2 \text{ film} \tag{11.19}$$

$$d^2 E_y/dx^2 + (k_0^2 n_2^2 - \beta^2)E_y = 0; \quad |x| > d/2 \text{ cover} \tag{11.20}$$

For guided modes we require that the field should decay in the cover (i.e., in the region $|x| > d/2$) so that most of the energy associated with the mode lies inside the film. Thus we must have[‡]

$$\beta^2 > k_0^2 n_2^2 \tag{11.21}$$

Furthermore, we must also have $\beta^2 < k_0^2 n_1^2$, otherwise the boundary conditions cannot be satisfied[§] at $x = \pm d/2$. Thus for guided modes we must have

$$k_0^2 n_2^2 < \beta^2 < k_0^2 n_1^2 \tag{11.22}$$

We therefore write Eqs. (11.19) and (11.20) in the form

$$d^2 E_y/dx^2 + \kappa^2 E_y = 0; \quad |x| < d/2 \text{ film} \tag{11.23}$$

$$d^2 E_y/dx^2 - \gamma^2 E_y = 0; \quad |x| > d/2 \text{ cover} \tag{11.24}$$

where

$$\kappa^2 = k_0^2 n_1^2 - \beta^2 \tag{11.25}$$

[†] The very fact that E_y satisfies Eq. (11.15) also implies that E_y and dE_y/dx are continuous unless $n^2(x)$ has an infinite discontinuity. This follows from the fact that if E_y' is discontinuous then E_y'' will be a delta function and Eq. (11.15) will lead to an inconsistent equation. Thus the continuity conditions are imbedded in Maxwell's equations.

[‡] When $\beta^2 < k_0^2 n_2^2$, the solutions are oscillatory in the region $|x| > d/2$ and they correspond to what are known as radiation modes of the waveguide. These modes correspond to rays which undergo refraction (rather than total internal reflection) at the film–cover interface and when these are excited, they quickly leak away from the core of the waveguide. Some aspects of radiation modes will be discussed in Chapter 12.

[§] It is left as an exercise for the reader to show that if we assume $\beta^2 > k_0^2 n_1^2$ and also assume decaying fields in the region $|x| > d/2$ then the boundary conditions at $x = +d/2$ *and* at $x = -d/2$ can *never* be satisfied (see also Problem 11.8).

and

$$\gamma^2 = \beta^2 - k_0^2 n_2^2 \tag{11.26}$$

The solution of Eq. (11.23) can be written in the form

$$E_y(x) = A\cos\kappa x + B\sin\kappa x; \qquad |x| < d/2 \tag{11.27}$$

where A and B are constants. In the region $x > d/2$ and $x < -d/2$ the solutions are $e^{\pm\gamma x}$ and if we neglect the exponentially amplifying one, we will obtain

$$E_y(x) = \begin{cases} Ce^{\gamma x}; & x < -d/2 \\ De^{-\gamma x}; & x > d/2 \end{cases} \tag{11.28}$$

If we now apply the boundary conditions (*viz.*, continuity of E_y and dE_y/dx at $x = \pm d/2$) we will get four equations from which we can get the transcendental equation, which will determine the allowed values of β. This is indeed the general procedure for determining the propagation constants (see e.g., Sec. 14.2); however, when the refractive index distribution is symmetric about $x = 0$, i.e., when

$$n^2(-x) = n^2(x) \tag{11.29}$$

the solutions are either symmetric or antisymmetric functions of x; thus we must have

$$E_y(-x) = E_y(x) \text{ symmetric modes} \tag{11.30}$$

$$E_y(-x) = -E_y(x) \text{ antisymmetric modes} \tag{11.31}$$

(The proof of this theorem is discussed in Problem 11.10.) For the symmetric mode, we must have

$$E_y(x) = \begin{cases} A\cos\kappa x; & |x| < d/2 \\ Ce^{-\gamma|x|}; & |x| > d/2 \end{cases} \tag{11.32}\tag{11.33}$$

Continuity of $E_y(x)$ and dE_y/dx at $x = \pm d/2$ gives us

$$A\cos(\kappa d/2) = Ce^{-\gamma d/2} \tag{11.34}$$

and

$$-\kappa A\sin(\kappa d/2) = -\gamma Ce^{-\gamma d/2} \tag{11.35}$$

respectively. Dividing Eq. (11.35) by Eq. (11.34) we get

$$(\kappa d/2)\tan(\kappa d/2) = (\gamma d/2) \tag{11.36}$$

Since

$$\gamma^2 = \beta^2 - k_0^2 n_2^2 = k_0^2(n_1^2 - n_2^2) - \kappa^2 \tag{11.37}$$

we may write

$$\gamma d/2 = (V^2/4 - \xi^2)^{\frac{1}{2}} \tag{11.38}$$

where

$$\xi = \kappa d/2 = (k_0^2 n_1^2 - \beta^2)^{\frac{1}{2}} d/2 \tag{11.39}$$

and

$$V = k_0 d(n_1^2 - n_2^2)^{\frac{1}{2}} \tag{11.40}$$

is known as the dimensionless waveguide parameter. Thus Eq. (11.36) can be put in the following form:

$$\xi \tan \xi = (V^2/4 - \xi^2)^{\frac{1}{2}} \tag{11.41}$$

Similarly, for the antisymmetric mode we will have

$$E_y(x) = \begin{cases} B \sin \kappa x; & |x| < d/2 \tag{11.42} \\ \dfrac{x}{|x|} D e^{-\gamma|x|}; & |x| > d/2 \tag{11.43} \end{cases}$$

Following an exactly similar procedure we get

$$-\xi \cot \xi = (V^2/4 - \xi^2)^{\frac{1}{2}} \tag{11.44}$$

Fig. 11.3 The variation of $\xi \tan \xi$ (solid curves) and $-\xi \cot \xi$ (dotted curves) as a function of ξ. The points of intersection of the solid and dotted curves with the quadrant of a circle of radius $V_0 (= V/2)$ determine the propagation constants of the optical waveguide corresponding to symmetric and antisymmetric modes respectively.

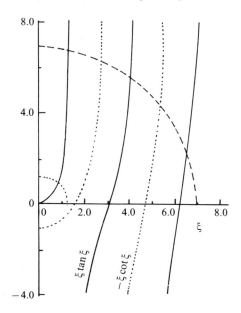

Thus, we have

$$\xi \tan \xi = [V_0^2 - \xi^2]^{\frac{1}{2}} \text{ symmetric case} \tag{11.45}$$

$$-\xi \cot \xi = [V_0^2 - \xi^2]^{\frac{1}{2}} \text{ antisymmetric case} \tag{11.46}$$

where

$$V_0 = V/2 = k_0 d(n_1^2 - n_2^2)^{\frac{1}{2}}/2 \tag{11.47}$$

Since the equation

$$\eta = (V_0^2 - \xi^2)^{\frac{1}{2}} \tag{11.48}$$

represents a portion of a circle (of radius V_0) in the (ξ, η) plane, the numerical evaluation of the allowed values of ξ (and hence of the propagation constants) is quite simple. In Fig. 11.3 we have plotted the functions $\xi \tan \xi$ (solid curve) and $-\xi \cot \xi$ (dotted curve) as a function of ξ. Their points of intersection with the quadrant of the circle determine the allowed values of ξ and if we use Eq. (11.39) we can determine the corresponding values of β.

11.3.1 *Some general comments about the modes*

From Fig. 11.3 we can derive the following conclusions:

(*a*) If $0 < V_0 < \pi/2$, i.e., when

$$0 < V < \pi \tag{11.49}$$

we have only one discrete (TE) mode of the waveguide and this mode is symmetric in x. When this happens, we refer to the waveguide as a 'single moded waveguide'. For example, if $\lambda_0 \approx 1.5 \, \mu\text{m}$, $n_1 = 1.50$, $n_2 = 1.48$ then for single mode operation we must have

$$\frac{2\pi}{1.5} d[(1.50)^2 - (1.48)^2]^{\frac{1}{2}} < \pi$$

where d is measured in microns. Solving we get

$$d < 3.07 \, \mu\text{m}$$

We will show below that if the operating wavelength is made smaller, the same waveguide will be able to support more than one mode.

(*b*) From Fig. 11.3 it is easy to see that if $\pi/2 < V_0 < \pi$ (or, $\pi < V < 2\pi$) we will have one symmetric and one antisymmetric mode. In general, if

$$2m\pi < V < (2m + 1)\pi \tag{11.50}$$

we will have $(m + 1)$ symmetric modes, and m antisymmetric modes and if

$$(2m + 1)\pi < V < (2m + 2)\pi \tag{11.51}$$

we will have $(m + 1)$ symmetric modes, and $(m + 1)$ antisymmetric modes where $m = 0, 1, 2, \ldots$. Thus the total number of modes will be the integer closest to (and greater than) V/π. Thus for the waveguide considered above, if the operating wavelength is made $0.6 \, \mu m$ then $V = 2.5\pi$ and therefore we will have three modes (two symmetric and one antisymmetric).

(c) When the waveguide supports many modes (i.e., when $V \gg 1$) the points of intersection (in Fig. 11.3) will be very close to $\xi = \pi/2, \pi, 3\pi/2$, etc; thus the propagation constants corresponding to the first few modes will be approximately given by the following equation

$$\xi = \xi_m = (k_0^2 n_1^2 - \beta_m^2)^{\frac{1}{2}} d/2 \approx (m + 1)\pi/2; \qquad V \gg 1 \tag{11.52}$$

where

$m = 0, 2, 4, \ldots$ correspond to symmetric modes

and

$m = 1, 3, 5, \ldots$ correspond to antisymmetric modes

(d) It is obvious from Fig. 11.3 that for the fundamental mode (which we will refer to as the zero order mode), $\xi \, (= \kappa d/2)$ will *always* lie between 0 and

Fig. 11.4 The solid curves represent the modal fields for the symmetric step index planar waveguide with $V = 4.4 \pi$; even and odd values of m correspond to symmetric and antisymmetric modes respectively. The dashed curve represents the fundamental mode for $V = 0.8 \pi$. All modes have been normalized to carry the same power.

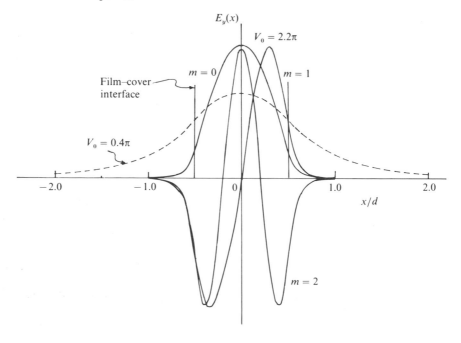

$\pi/2$ and the corresponding field variation $E_y(x)$ will have no zeroes. For the next mode (which will be antisymmetric in x) ξ ($= \kappa d/2$) will always lie between $\pi/2$ and π and therefore the corresponding $E_y(x)$ will have only one zero (at $x = 0$). It is easy to extend the analysis and prove that

$$\text{the } m^{\text{th}} \text{ mode will have } m \text{ zeroes} \qquad (11.53)$$

The above statement is valid for an arbitrary waveguiding structure. The actual plot of the modal pattern for the first few modes is shown in Fig. 11.4. It may be noted that the field spreads out more as the wavelength increases or V number decreases.

(*e*) We define a dimensionless parameter

$$b \equiv \frac{\beta^2/k_0^2 - n_2^2}{n_1^2 - n_2^2} = 1 - \frac{\xi^2}{V^2/4} \qquad (11.54)$$

(The quantity b is usually referred to as the normalized propagation constant). The allowed values of ξ (and hence of b) are calculated using Eqs. (11.45) and (11.46) for different values of V. The corresponding variations of b with V are plotted in Fig. 11.5 for the first few modes. The curves are universal, i.e., for a given waveguide and a given operating wavelength we have first to determine V and then to 'read off' from the curves the exact values of b from which we can determine the values of β by using the

Fig. 11.5 The variation of the normalized propagation constant b as a function of V for a symmetric step index planar waveguide. The solid and the dashed curves correspond to TE and TM modes respectively. The value of V at which $b = 0$ corresponds to what is known as the cutoff frequency.

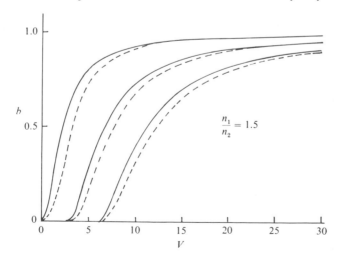

following equation:

$$\beta^2 = k_0^2[n_2^2 + b(n_1^2 - n_2^2)] \tag{11.55}$$

It may be noted that since for guided modes

$$n_2^2 < \beta^2/k_0^2 < n_1^2 \tag{11.56}$$

We must have

$$0 < b < 1 \tag{11.57}$$

Since for a guided mode β cannot be less than $n_2 k_0$, when (for a particular mode) β reaches the value equal to $n_2 k_0$ (i.e., when b becomes equal to zero), the mode is said to have reached 'cutoff'. Thus at cutoff

$$\beta = n_2 k_0, \qquad \gamma = 0 \qquad \text{and} \qquad b = 0$$

For the symmetric waveguide that we have been discussing, at cutoff $\xi = V/2 = V_0$ and hence the cutoff of TE modes is determined by

$$\frac{V}{2} \tan\left(\frac{V}{2}\right) = 0 \qquad \text{symmetric modes}$$

$$\frac{V}{2} \cot\left(\frac{V}{2}\right) = 0 \qquad \text{antisymmetric modes}$$

The above equation implies that the cutoff V values for various modes are given by

$$V_c = m\pi; \qquad m = 0, 1, 2, 3, \ldots \tag{11.58}$$

where even and odd values of m correspond to symmetric and antisymmetric modes respectively. Notice that the fundamental mode has no cutoff and therefore there will always be at least one guided mode. The physical understanding of cutoff will be discussed at the end of next section.

11.3.2   *Physical understanding of the modes*

In order to have a physical understanding of the modes we consider the electric field pattern inside the film $(-d/2 < x < d/2)$. For example, for a symmetric TE mode, this is given by (see Eq. (11.32)):

$$E_y(x) = A \cos \kappa x$$

Thus the complete field inside the film is given by

$$\mathscr{E}_y = A \cos \kappa x \, e^{i(\omega t - \beta z)}$$
$$= \tfrac{1}{2} A \, e^{i(\omega t - \beta z - \kappa x)} + \tfrac{1}{2} A \, e^{i(\omega t - \beta z + \kappa x)} \tag{11.59}$$

Now

$$e^{i(\omega t - k_x x - k_y y - k_z z)} \tag{11.60}$$

represents a wave propagating along the direction of **k** whose x, y and z-components are k_x, k_y and k_z respectively. Thus for the two terms on the RHS of Eq. (11.59) we have

$$k_x = \pm \kappa, k_y = 0 \qquad \text{and} \qquad k_z = \beta \tag{11.61}$$

which represent plane waves with propagation vectors parallel to the x–z plane making angles $\pm \theta$ with the z-axis where

$$\tan \theta = k_x/k_z = \kappa/\beta \tag{11.62}$$

or

$$\cos \theta = \beta/(\beta^2 + \kappa^2)^{\frac{1}{2}} = \beta/k_0 n_1 \tag{11.63}$$

Thus a guided mode can be considered to be a superposition of a pair of plane waves which are propagating at angles $\pm \theta \, (= \pm \cos^{-1}(\beta/k_0 n_1))$ with the z-axis (see Fig. 11.6). Since only discrete values of β are allowed (which we designate as β_m), only discrete angles of propagation of waves (or of the rays) are allowed. Each mode is characterised by a *discrete* angle of propagation θ_m. We will use this concept to derive the eigenvalue equation in Sec. 14.3.

The concept of the cutoff of a mode can also be easily understood from the above discussion. Since the guided waves correspond to

$$n_2 < \beta/k_0 < n_1 \tag{11.64}$$

we have

$$n_2/n_1 < \cos \theta < 1 \tag{11.65}$$

The condition that β cannot be less than $n_2 k_0$ implies that $\cos \theta$ should be greater than n_2/n_1 which is nothing but the condition for total internal reflection at the core–cladding interface (see Eq. (11.3)). Thus beyond cutoff, i.e., for $V < V_c$, the component waves no longer undergo total internal reflections at the boundaries.

Fig. 11.6 A guided mode in a step index waveguide corresponds to the superposition of two plane waves (inside the film) propagating at particular angles $\pm \theta$ with the z-axis. Different modes will correspond to different (discrete) values of θ.

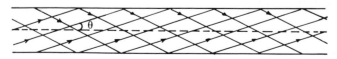

11.4   TM modes of a symmetric step index planar waveguide

In the above discussion we have considered the TE modes of the waveguide. An exactly similar analysis can also be performed for the TM modes which are characterized by field components E_x, E_z and H_y (see Eqs. (11.12)–(11.14)). If we substitute for E_x and E_z from Eqs. (11.12) and (11.13) in Eq. (11.14) we will get

$$n^2(x)\frac{d}{dx}\left[\frac{1}{n^2(x)}\frac{dH_y}{dx}\right]+(k_0^2 n^2-\beta^2)H_y=0 \tag{11.66}$$

which can be rewritten as

$$\frac{d^2 H_y}{dx^2}-\left[\frac{1}{n^2(x)}\frac{dn^2}{dx}\right]\frac{dH_y}{dx}+[k_0^2 n^2(x)-\beta^2]H_y(x)=0 \tag{11.67}$$

The above equation is of a form which is somewhat different from the equation satisfied by E_y for TE modes (see Eq. (11.15)); however, for the step index waveguide shown in Fig. 11.2, the refractive index is constant in each region and therefore we have

$$d^2 H_y/dx^2+(k_0^2 n_1^2-\beta^2)H_y(x)=0;\quad |x|<d/2 \tag{11.68}$$

and

$$d^2 H_y/dx^2-(\beta^2-k_0^2 n_2^2)H_y(x)=0;\quad |x|>d/2 \tag{11.69}$$

We must be careful about the boundary conditions. Since H_y and E_z are tangential components on the planes $x=\pm d/2$, we must have (see Eq. (11.13))

$$H_y \text{ and } \frac{1}{n^2}\frac{dH_y}{dx} \text{ continuous at } x=\pm\frac{d}{2} \tag{11.70}$$

This is also obvious from Eq. (11.66)[†]. The solutions of Eqs. (11.68) and (11.69) can be written immediately. Considering first the symmetric modes, we have

$$H_y(x)=\begin{cases} A\cos \kappa x; & |x|<d/2 \\ Be^{-\gamma|x|}; & |x|>d/2 \end{cases} \qquad\begin{matrix}(11.71)\\(11.72)\end{matrix}$$

where the symbols κ and γ are the same as given by Eqs. (11.25) and (11.26).

[†] Once again the condition that H_y and $(1/n^2)dH_y/dx$ should be continuous at $x=\pm d/2$ follows from Eq. (11.66) because if $(1/n^2)dH_y/dx$ was discontinuous $d/dx[(1/n^2)H_y']$ would be a delta function and Eq. (11.66) would lead to an inconsistent equation. Thus the continuity of H_y and $(1/n^2)H_y'$ are contained in Eq. (11.66).

The boundary conditions given by Eq. (11.70) give us

$$A \cos(\kappa d/2) = B \mathrm{e}^{-\gamma d/2} \tag{11.73}$$

$$\frac{1}{n_1^2}\left(-A\kappa \sin\frac{\kappa d}{2}\right) = \frac{1}{n_2^2}(-B\gamma \mathrm{e}^{-\gamma d/2}) \tag{11.74}$$

Dividing we get

$$\kappa \tan(\kappa d/2) = (n_1^2/n_2^2)\gamma \tag{11.75}$$

which can be rewritten in the form

$$\xi \tan \xi = (n_1^2/n_2^2)(V_0^2 - \xi^2)^{\frac{1}{2}} \qquad \text{symmetric TM modes} \tag{11.76}$$

A similar derivation gives us

$$-\xi \cot \xi = (n_1^2/n_2^2)(V_0^2 - \xi^2)^{\frac{1}{2}} \qquad \text{antisymmetric TM modes} \tag{11.77}$$

where, as before,

$$\xi = \kappa d/2 = (k_0^2 n_1^2 - \beta^2)^{\frac{1}{2}} d/2$$

and

$$V_0 = V/2 = k_0 d(n_1^2 - n_2^2)^{\frac{1}{2}}/2$$

The numerical solutions of Eqs. (11.76) and (11.77) can be discussed in a manner exactly similar to the TE case with the difference that the RHS of Eqs. (11.76) and (11.77) now represents an ellipse whose semimajor axis (along the η-direction) is of magnitude $(n_1/n_2)V_0$ and whose semi-minor axis (along the ξ-direction) is of magnitude V_0. All qualitative conclusions discussed in Sec. 11.3 for TE modes (*viz.*, the cutoff frequencies and number of zeroes of various modes, the physical interpretation of modal fields etc.) will remain valid. We should also mention the following three points:

(a) Although a waveguide for which $0 < V < \pi$ is referred to as a single moded waveguide, we actually have two modes (one TE and one TM) characterized by slightly different propagation constants. However, the incident field is usually linearly polarized and if **E** is along the y-axis, the TE mode is excited and if **E** is along the x-axis, the TM mode is excited. *This result is quite general and is valid for all planar waveguides.* On the other hand, if the incident field has a polarization which makes an angle with the x-axis (or, if the field is elliptically polarized) then both TE and TM modes will be excited and because they have slightly different propagation constants, they will superpose with different phases at different values of z changing the state of the resultant polarization. As an example, we consider the incidence of a linearly polarized wave with the electric vector making an angle of 45° with the x and y-axes. Thus at $z = 0$ we have

$$\left.\begin{array}{l} \mathscr{E}_x = \mathscr{E}_0 \cos 45 \cos \omega t \\ \mathscr{E}_y = \mathscr{E}_0 \sin 45 \cos \omega t \end{array}\right\} \begin{array}{l} \text{Field distribution} \\ \text{at } z = 0 \end{array} \qquad (11.78)$$

where $\mathscr{E}_0$ represents the transverse variation of the modal field which we have assumed to be the same for the TE and TM modes (see Eqs. (11.32) and (11.71) – the values of β are assumed to be nearly equal). If the propagation constants for the TE and TM modes are denoted by β_0 and $(\beta_0 - \Delta\beta_0)$ respectively, then the field distributions for $z > 0$ will be

$$\left.\begin{array}{ll} \text{TE:} & \mathscr{E}_y = \dfrac{\mathscr{E}_0}{\sqrt{2}} \cos [\omega t - \beta_0 z] \\[4mm] \text{TM:} & \mathscr{E}_x = \dfrac{\mathscr{E}_0}{\sqrt{2}} \cos [\omega t - \beta_0 z + \Delta\beta_0 z] \end{array}\right\} \begin{array}{l} \text{Field distribution} \\ \text{at } z > 0 \end{array} \qquad (11.79)$$

It can be readily seen that at $z = \pi/(2\Delta\beta_0)$ the beam will be circularly polarized and at $z = \pi/(\Delta\beta_0)$ the beam will be linearly polarized (with the electric vector now at right angles to the original direction) – for intermediate values of z, the beam will be elliptically polarized. For a distance $z = L_b = 2\pi/\Delta\beta_0$, the original polarization state is restored and this characteristic length is referred to as the beat length.

(b) Similarly, for $\pi < V < 2\pi$, although the waveguide is referred to as a 'two-moded waveguide' there are actually four modes (two TE and two TM) etc.

(c) For most practical waveguides, $n_1 \approx n_2$ and the propagation constants (and the field patterns) for the TE and TM modes are very nearly equal.

Problem 11.1: Consider a planar symmetric waveguide with $n_1 = 1.5$, $n_2 = 1.0$ and $V = 3.0$. Assuming $\lambda_0 = 1.3\,\mu\text{m}$ calculate $\Delta\beta_0$ and obtain the corresponding beat length.
(Answer: $\beta_{TE} = 6.4574\,\mu\text{m}^{-1}$, $\beta_{TM} = 6.0393\,\mu\text{m}^{-1}$, $L_b \approx 15\,\mu\text{m}$.)

11.5   The relative magnitude of the longitudinal components of the E and H fields

We first consider the TE modes. Using Eqs. (11.9) and (11.10) we get

$$\left| \frac{H_z}{H_x} \right| = \frac{1}{\beta} \left| \frac{\partial E_y/\partial x}{E_y} \right| \qquad (11.80)$$

Now from Eq. (11.32) we have (inside the film)

$$\left| \frac{E_y}{\partial E_y/\partial x} \right| \sim \frac{1}{\kappa} \qquad (11.81)$$

which represents the characteristic distance for the spatial variation in the *x*-direction. Thus

$$|H_z/H_x| \sim \kappa/\beta \tag{11.82}$$

Since

$$\kappa^2 = k_0^2 n_1^2 - \beta^2$$

we readily have

$$\kappa/\beta = (k_0^2 n_1^2/\beta^2 - 1)^{\frac{1}{2}} \tag{11.83}$$

Now, the guided modes correspond to

$$n_2^2 < \beta^2/k_0^2 < n_1^2$$

therefore

$$0 < \frac{\kappa}{\beta} < \left(\frac{n_1^2 - n_2^2}{n_2^2}\right)^{\frac{1}{2}} \tag{11.84}$$

Thus

$$\left|\frac{H_z}{H_x}\right| \lesssim \left(\frac{n_1^2 - n_2^2}{n_2^2}\right)^{\frac{1}{2}} \tag{11.85}$$

For $n_1 \approx 1.50$ and $n_2 \approx 1.49$, the RHS of the above equation is about 0.1 which shows that the longitudinal component is very weak in comparison to the transverse component. Thus, as long as $n_1 \approx n_2$, the mode can be approximately assumed to be a 'transverse electromagnetic mode'. The same is also true for the TM modes.

The above discussion is valid, in general, as long as $n_1 \approx n_2$. In Sec. 14.2 we will discuss asymmetric waveguides where the refractive index of the material on top of the film (n_c) is different from that of the material below the film (n_s). However, as long as n_c and n_s have values which are close to the refractive index of the film, the mode will be approximately transverse in both **E** and **H** fields. In Sec. 13.5 we will discuss round optical waveguides with cylindrical symmetry – even there, as long as the core and cladding refractive indices are nearly equal (which is true for *all* practical fibres) the longitudinal components of the electric and magnetic fields are usually negligible in comparison to the corresponding transverse components. When this is the case, the waveguide is referred to as '*weakly guiding*'.

11.6 Power associated with a mode

In this section we will calculate the power associated with the TE mode. The power flow is given by (see Eq. (1.92))

$$\langle \mathbf{S} \rangle = \tfrac{1}{2} \operatorname{Re}(\mathscr{E} \times \mathscr{H}^*) \tag{11.86}$$

Now for the TE mode we have (see Eqs. (11.7)–(11.11))

$$\mathscr{E}_y = E_y(x)\, e^{i(\omega t - \beta z)} \tag{11.87}$$

$$\mathscr{H}_x = -\frac{\beta}{\omega\mu_0}\mathscr{E}_y = -\frac{\beta}{\omega\mu_0}E_y(x)\, e^{i(\omega t - \beta z)} \tag{11.88}$$

and

$$\mathscr{H}_z = \frac{i}{\omega\mu_0}\frac{\partial \mathscr{E}_y}{\partial x} = \frac{i}{\omega\mu_0}\frac{dE_y}{dx}\, e^{i(\omega t - \beta z)} \tag{11.89}$$

Using the above equations we readily get $\langle S_y \rangle = 0$ and

$$\langle S_x \rangle = \tfrac{1}{2}\,\mathrm{Re}\,(\mathscr{E}_y \mathscr{H}_z^*) = 0 \tag{11.90}$$

Since E_y is real. Further,

$$\langle S_z \rangle = -\tfrac{1}{2}\,\mathrm{Re}\langle \mathscr{E}_y \mathscr{H}_x^* \rangle \tag{11.91}$$

or

$$\langle S_z \rangle = \frac{\beta}{2\omega\mu_0}|E_y|^2 \tag{11.92}$$

Although the above expression is rigorously valid only for the TE mode in a slab waveguide, it is approximately valid for all waveguides in the weakly guiding approximation.

The power associated with the mode (per unit length in the y-direction) is given by

$$P = \frac{1}{2}\frac{\beta}{\omega\mu_0}\int_{-\infty}^{+\infty}|E_y|^2\, dx \tag{11.93}$$

We consider the symmetric mode (see Eqs. (11.32) and (11.33)) for which

$$P = \frac{1}{2}\frac{\beta}{\omega\mu_0}2\Bigl(A^2\int_0^{d/2}\cos^2\kappa x\, dx + C^2\int_{d/2}^{\infty}e^{-2\gamma x}\, dx\Bigr) \tag{11.94}$$

or

$$P = \frac{\beta}{2\omega\mu_0}A^2\Bigl(\frac{d}{2} + \frac{1}{2\kappa}\sin\kappa d + \frac{C^2}{A^2}\frac{1}{\gamma}e^{-\gamma d}\Bigr)$$

If we now use Eq. (11.34) for C/A we get

$$P = \frac{\beta A^2}{4\omega\mu_0}\Bigl\{d + \frac{2\sin(\kappa d/2)\cos(\kappa d/2)}{\kappa} + \frac{2}{\gamma}[1 - \sin^2(\kappa d/2)]\Bigr\}$$

$$= \frac{\beta A^2}{4\omega\mu_0}\Bigl\{d + \frac{2}{\gamma} + \frac{2\sin(\kappa d/2)\cos(\kappa d/2)}{\gamma\kappa}[\gamma - \kappa\tan(\kappa d/2)]\Bigr\}$$

$$P = \frac{\beta A^2}{4\omega\mu_0}\Bigl(d + \frac{2}{\gamma}\Bigr) \tag{11.95}$$

where we have used Eq. (11.36).

Problem 11.2: Show that even for the antisymmetric TE mode, the power associated (per unit length in the y-direction) is given by Eq. (11.95).

Problem 11.3: Carry out the analysis of power flow of TM modes and show that

$$P = \frac{A^2 \beta}{2\omega\epsilon_0 n_1^2}\left[\frac{d}{2} + \frac{(n_1 n_2)^2}{\gamma} \frac{k_0^2(n_1^2 - n_2^2)}{(n_2^4 \kappa^2 + n_1^4 \gamma^2)}\right] \tag{11.96}$$

for both symmetric as well as antisymmetric modes.

Problem 11.4: Consider a symmetric planar waveguide with the following parameters:

$$n_1 = 1.50, \quad n_2 = 1.48, \quad d = 3.912\,\mu\text{m}$$

At $\lambda_0 = 1\,\mu$m, (a) show that there will be only two TE modes, the corresponding propagation constants being $\beta_0 = 9.4058\,\mu\text{m}^{-1}$ and $\beta_1 = 9.3525\,\mu\text{m}^{-1}$. (b) At $z = 0$ assume that the field in the core is given by

$$E_y(x) = 1.375 \times 10^4 \cos\kappa_0 x e^{i\omega t} + 1.309 \times 10^4 \sin\kappa_1 x e^{i\omega t}\,V/\text{m}$$

Show that equal power of 1 W is carried by the two modes. Calculate the transverse intensity distribution at

$$z = 0, \quad \pi/\Delta\beta, \quad 2\pi/\Delta\beta$$

where $\Delta\beta = \beta_0 - \beta_1$. Interpret the results physically.

11.7 Radiation modes

Till now we have considered the guided modes of the waveguide for which

$$n_2^2 < \beta^2/k_0^2 < n_1^2.$$

There exists another class of modes for which[†]

$$\beta^2/k_0^2 < n_2^2 \tag{11.97}$$

These are referred to as the *radiation modes* of the waveguide. It can be immediately seen that for $(\beta^2/k_0^2) < n_2^2$ the wave equation (say for the TE modes) in the region $|x| > d/2$ takes the form

$$d^2 E_y/dx^2 + \delta^2 E_y = 0 \tag{11.98}$$

where

$$\delta^2 = k_0^2 n_2^2 - \beta^2 \tag{11.99}$$

which is now a positive quantity, thus the solutions in the region $|x| > d/2$ will be wavelike of the form

$$e^{\pm i\delta x} \tag{11.100}$$

[†] It is impossible to have $\beta/k_0 > n_1$ (see Problem 11.8).

We may recall that in the region $|x| > d/2$, the field associated with guided modes decayed exponentially in the x-direction (see Eq. (11.33)). On the other hand, Eq. (11.100) tells us that the radiation modes correspond to oscillatory solutions in the cover and will be briefly discussed in the next chapter.

11.8 Excitation of guided modes

We start with the wave equation satisfied by TE modes (see Eq. (11.15))

$$d^2 u_m/dx^2 + [k_0^2 n^2(x) - \beta_m^2] u_m(x) = 0; \quad m = 0, 1, 2, \ldots \qquad (11.101)$$

where $u_m(x)$ represents the field pattern corresponding to the propagation constant β_m; we have used the symbol $u_m(x)$ instead of the more complicated symbol $E_y^{(m)}(x)$. Using the condition that for guided modes $u_m(x)$ will go to zero as $x \to \pm \infty$, we can readily show that (see Problem 11.7)

$$\int_{-\infty}^{+\infty} u_m^*(x) u_k(x)\, dx = 0 \qquad \text{for} \qquad m \neq k \qquad (11.102)$$

which is known as the *orthogonality condition*. Since Eq. (11.101) is a linear equation, a constant multiple of $u_m(x)$ is also a solution and we can always choose the constant such that

$$\int_{-\infty}^{+\infty} |u_m(x)|^2\, dx = 1 \qquad (11.103)$$

which is known as the *normalization condition*. Combining Eqs. (11.102) and (11.103) we get the *orthonormality condition*

$$\int_{-\infty}^{+\infty} u_m^*(x) u_k(x)\, dx = \delta_{mk} \qquad (11.104)$$

where δ_{mk} is the Kronecker delta function defined by the following equation

$$\delta_{mk} = \begin{cases} 0 & \text{for} \quad m \neq k \\ 1 & \text{for} \quad m = k \end{cases} \qquad (11.105)$$

Eq. (11.104) represents the orthonormality condition satisfied by the discrete (guided) modes. The radiation modes (which form a continuum) also form an orthogonal set in the sense that

$$\int_{-\infty}^{+\infty} u_{\beta'}^*(x) u_\beta(x)\, dx = 0, \qquad \text{for} \qquad \beta \neq \beta' \qquad (11.106)$$

However, the integral for $\beta = \beta'$ is not defined and the orthonormality condition is in terms of the Dirac delta function; we will discuss this in the next chapter.

The important point is that the finite number of guided modes along with the continuum of radiation modes form a complete set of functions in the sense that any 'well behaved' function of x can be expanded in terms of these functions, i.e.,

$$\phi(x) = \sum_m a_m u_m(x) + \int a(\beta) u_\beta(x) \, d\beta \tag{11.107}$$

where the first term on the RHS represents a sum over discrete (guided) modes and the second term represents an integral over the continuum (radiation) modes. If we multiply Eq. (11.107) by $u_k^*(x)$ and integrate we will readily obtain

$$a_k = \int_{-\infty}^{+\infty} u_k^*(x) \phi(x) \, dx \tag{11.108}$$

where we have used Eq. (11.104). Now, let $E_y(x, z = 0)$ represent the actual incident field (polarized in the y-direction) at the entrance aperture of the waveguide $(z = 0)$. The power launched in the mth mode will be (see Eq. (11.93))

$$P_m = (1/2\omega\mu_0)\beta_m |a_m|^2 \int_{-\infty}^{+\infty} |u_m(x)|^2 \, dx$$

$$= (1/2\omega\mu_0)\beta_m \left| \int u_m^*(x) E_y(x, z = 0) \, dx \right|^2 \tag{11.109}$$

As the beam propagates through the waveguide, the field in the region $(z > 0)$ will be given by

$$E_y(x, z) = \sum_m a_m u_m(x) e^{-i\beta_m z} + \int a(\beta) u_\beta(x) e^{-i\beta z} \, d\beta \tag{11.110}$$

One can easily see that at an arbitrary value of z, the power in the mth mode will be proportional to

$$|a_m e^{-i\beta_m z}|^2 = |a_m|^2 \tag{11.111}$$

which remains constant with z. Different modes superpose with different phases at different values of z as a consequence of which – considering guided modes only – the transverse intensity distribution will vary with z (see Problem 11.4).

As a very beautiful (and practically important[†]) example we consider the infinitely extended parabolic profile for which the refractive index distribution can be written in the form[‡]

$$n^2 = n_1^2 [1 - (2\Delta/a^2)x^2] \tag{11.112}$$

where Δ and a are constants. For such a profile, we have *only* discrete modes. The normalized modal patterns (see Eq. (11.103) and propagation constants are given by (see Appendix D)

$$u_m = N_m H_m(\xi) e^{-\xi^2/2} \qquad m = 0, 1, 2, \dots \tag{11.113}$$

$$\beta_m = n_1 k_0 \left[1 - (2m + 1) \frac{(2\Delta)^{\frac{1}{2}}}{ak_0 n_1} \right]^{\frac{1}{2}} \tag{11.114}$$

$$\xi = \frac{x\sqrt{2}}{w_0}, \qquad w_0 = \left[\frac{2a}{(2\Delta)^{\frac{1}{2}} n_1 k_0} \right]^{\frac{1}{2}} \tag{11.115}$$

and

$$N_m = [2^m m! \pi^{\frac{1}{2}} (w_0/\sqrt{2})]^{-\frac{1}{2}} \tag{11.116}$$

represents the normalization constant[§]. The functions $H_m(\xi)$ are the well known Hermite polynomials given by

$$\left. \begin{aligned} &H_0(\xi) = 1, \qquad H_1(\xi) = 2\xi \\ &H_2(\xi) = 4\xi^2 - 2, \qquad H_3(\xi) = 8\xi^3 - 12\xi, \dots \\ \text{and} \\ &H_{m+1}(\xi) = 2\xi H_m(\xi) - 2m H_{m-1}(\xi) \end{aligned} \right\} \tag{11.117}$$

A few interesting points may be noted:

(a) Eqs. (11.113) and (11.114) represent *exact* solutions of Eq. (11.101) for the profile given by Eq. (11.112) which can be verified by direct substitution.

[†] Fibres with parabolic refractive index variation are of great practical importance in optical communication systems and therefore a lot of work has been carried out on the analysis of parabolic index waveguides (see, e.g., Sodha and Ghatak (1977), Ghatak and Thyagarajan (1981)).

[‡] We may mention here that Eq. (11.112) is rather hypothetical in the sense that for large values of $|x|$, Eq. (11.112) predicts negative values of n^2 – a realistic profile is truncated, i.e., Eq. (11.112) is valid for $|x| < a$ beyond which the refractive index is constant. However, for a highly multimoded waveguide, when a large number of low order modes are excited, the analysis of the infinitely extended profile gives results which are close to the results obtained by using the truncated profile.

[§] The normalization constants are such that $\int_{-\infty}^{+\infty} |u_m(x)|^2 \, dx = 1$ and *not* $\int_{-\infty}^{\infty} |u_m(\xi)|^2 \, d\xi = 1$.

(b) Since the profile is symmetric in x (i.e., $n^2(-x) = n^2(x)$) the modal fields are either symmetric in x ($m = 0, 2, 4, \ldots$) or antisymmetric in x ($m = 1, 3, 5, \ldots$).

(c) The function $H_0(\xi)$ (and hence $u_0(\xi)$) has no zeroes; the function $H_1(\xi)$ (and hence $u_1(\xi)$) has only one zero (at $\xi = 0$); the function $H_2(\xi)$ (and hence $u_2(\xi)$) has two zeroes (at $\xi = \pm\sqrt{\frac{1}{2}}$), the function $H_3(\xi)$ (and hence $u_3(\xi)$) has three zeroes (at $\xi = 0, \pm\sqrt{\frac{2}{3}}$) etc. In general, $H_m(\xi)$ (and hence $u_m(\xi)$) will have m zeroes – consistent with the remarks made in Sec. 11.3.

(d) Eq. (11.114) tells us that for large values of m, β_m becomes imaginary. Thus if we write $\beta_m = -i\gamma_m$ then the z-dependence will be $e^{-\gamma_m z}$ which represents the attenuating modes.

Now, for an incident Gaussian field we may assume

$$E(x, z = 0) = E_0 e^{-x^2/w_i^2} \tag{11.118}$$

where w_i represents the spot size of the beam. Expanding it in terms of the modes we get

$$E_0 e^{-x^2/w_i^2} = \sum_m a_m u_m(x) \tag{11.119}$$

with

$$a_m = E_0 \int_{-\infty}^{+\infty} e^{-x^2/w_i^2} u_m(x)\, dx$$

$$= E_0 N_m \frac{w_0}{\sqrt{2}} \int_{-\infty}^{+\infty} e^{-w_0^2 \xi^2/2w_i^2} H_m(\xi) e^{-\frac{1}{2}\xi^2}\, d\xi$$

If we use the following relation (Gradshtein and Ryzhik, 1965)

$$\int_{-\infty}^{\infty} e^{-\xi^2/a^2} H_{2m}(\xi)\, d\xi = \pi^{\frac{1}{2}} a(2m!/m!)(a^2 - 1)^m$$

we will obtain

$$a_m = \begin{cases} E_0 N_m \dfrac{w_0}{\sqrt{2}} (2\pi)^{\frac{1}{2}} \dfrac{m!}{(m/2)!} \left(\dfrac{w_0^2}{w_i^2} + 1\right)^{-\frac{1}{2}} \left(\dfrac{1 - w_0^2/w_i^2}{1 + w_0^2/w_i^2}\right)^{m/2}; & m = 0, 2, 4 \ldots \\ 0 & m = 1, 3, 5, \ldots \end{cases} \tag{11.120}$$

The quantity $|a_m|^2$ is proportional to the power excited in the mth mode. Notice that if $w_i = w_0$ then $a_m = 0$ for all m except $m = 0$ implying that only the fundamental mode is excited which should indeed be the case since the incident field represents the fundamental mode of the guide.

The field at an arbitrary value of z is given by the series in Eq. (11.110). Now, for all practical waveguides $\Delta \ll 1$ and if we assume the excitation of low order modes only, then Eq. (11.114) becomes

$$\beta_m \approx n_1 k_0 [1 - (m + \tfrac{1}{2})(2\Delta)^{\frac{1}{2}}/ak_0 n_1]$$ (11.121)

Using the above value of β_m and the expression for a_m given by Eq. (11.120), it is possible to sum the infinite series $\sum_m a_m u_m(x)\, e^{-i\beta_m z}$ using Mehler's formula:

$$\sum_m \frac{H_m(\xi)H_m(\xi')}{2^m m!}\, e^{im\alpha} = \frac{1}{(1-\gamma^2)^{\frac{1}{2}}} \exp\left[\frac{2\xi\xi'\gamma}{1-\gamma^2} - \frac{(\xi^2 + \xi'^2)(1+\gamma^2)}{2(1-\gamma^2)}\right]$$

(11.122)

where $\gamma = e^{i\alpha}$. The final expression for the intensity is given by

$$I(x, z) = I_0 \frac{w_i^2}{w^2(z)} \exp\left[-\frac{2x^2}{w^2(z)}\right]$$ (11.123)

where

$$w^2(z) = w_i^2 \left[\left(\frac{w_0}{w_i}\right)^4 \sin^2 \delta z + \cos^2 \delta z\right]$$ (11.124)

represents the changing spot size[†] of the propagating Gaussian beam and

$$\delta = (2\Delta)^{\frac{1}{2}}/a$$

Eq. (11.124) represents the periodic focussing and defocussing of the incident Gaussian beam (cf. Fig. 13.5). Thus the transverse intensity distribution changes with z due to the different phases of various modes. Notice that as $\Delta \to 0$, δ also tends to zero and we get

$$w^2(z) = \frac{w_0^4}{w_i^2}(\delta z)^2 + w_i^2$$

$$= w_i^2 + \lambda_0^2 z^2/n_1^2 \pi^2 w_i^2$$ (11.125)

which represents the diffraction divergence of the Gaussian beam in a homogeneous medium of refractive index n_1 (see Sec. 5.4). For a detailed discussion of the above analysis one may refer to Ghatak and Thyagarajan (1975, 1978).

In general, for any waveguide both discrete and continuum modes are excited and for a beam launched near the axis, the contribution of the continuum radiation modes to the electric field inside the core will be negligible at large values of z.

[†] Notice that when $w_i = w_0$, $w(z) = w_i$ (independent of z) and the beam propagates as the fundamental mode.

The general analysis presented above (Eqs. (11.101)–(11.111)) is rigorously valid for TE modes in planar waveguides. For TM modes and for modes in cylindrical waveguides, the orthonormality condition becomes more involved. Indeed, for guided modes the rigorously correct orthogonality condition is given by (see, e.g., Snyder and Love (1983), Chapter 11)

$$\iint [\mathbf{E}_m(x, y) \times \mathbf{H}_k^*(x, y) \cdot \hat{z}]\, \mathrm{d}x\, \mathrm{d}y = \iint [\mathbf{E}_k(x, y) \times \mathbf{H}_m^*(x, y) \cdot \hat{z}]\, \mathrm{d}x\, \mathrm{d}y = 0$$

$$\text{for } k \neq m \quad (11.126)$$

where the integral is over the *entire* transverse cross section; the refractive index distribution $n^2(x, y)$ can be arbitrary and the modal field is defined by the equation

$$\mathcal{E}_m(x, y, z, t) = \mathbf{E}_m(x, y)\, e^{i(\omega t - \beta_m z)} \qquad (11.127)$$

and a similar equation for $\mathbf{H}$; m represents the mode number. However, in the weakly guiding approximation, which is valid for almost *all* practical waveguides, we can still use the orthonormality condition given by Eq. (11.104) where u_m is assumed to represent the transverse component of the electric (or magnetic) field. In this approximation therefore, the entire analysis for the calculation of the excitation coefficient etc. [Eqs. (11.101)–(11.111)] remains valid.

Problem 11.5: Starting from Eq. (11.66) show that the TM modes satisfy the following orthogonality condition:

$$\int_{-\infty}^{+\infty} (1/n^2) H_y^{(m)}(x) H_y^{(k)*}(x)\, \mathrm{d}x = 0 \qquad m \neq k$$

Show that the above equation is consistent with Eq. (11.126).
(Hint: Use Eq. (11.12))

Problem 11.6: Consider a symmetric planar waveguide with $n_1 = 1.5$, $n_2 = 1.496$ operating at $\lambda_0 = 1\ \mu m$ with the fundamental mode having a propagation constant of $\beta/k_0 = 1.498$. (*a*) Calculate the penetration depth $1/\gamma$ in the cover. (*b*) Obtain the angle at which the rays representing the mode are travelling in the waveguide. (*c*) Calculate the width of the waveguide.
(Answer: (*a*) 2.06 μm, (*b*) 2.96°, (*c*) $\approx 3.2\ \mu m$).

11.9 Maxwell's equations in inhomogeneous media: TE and TM modes in planar waveguides

In this section we will derive the equations which are the starting points for modal analysis. We start with Maxwell's equations, which for an isotropic, linear, non-conducting and nonmagnetic medium take the form

$$\mathbf{\nabla} \times \mathscr{E} = -\partial \mathscr{B}/\partial t = -\mu_0 \partial \mathscr{H}/\partial t \tag{11.128}$$

$$\mathbf{\nabla} \times \mathscr{H} = \partial \mathscr{D}/\partial t = \epsilon_0 n^2 \partial \mathscr{E}/\partial t \tag{11.129}$$

$$\mathbf{\nabla} \cdot \mathscr{D} = 0 \tag{11.130}$$

$$\mathbf{\nabla} \cdot \mathscr{B} = 0 \tag{11.131}$$

where we have used the constitutive relations

$$\mathscr{B} = \mu_0 \mathscr{H} \tag{11.132}$$

$$\mathscr{D} = \epsilon \mathscr{E} = \epsilon_0 n^2 \mathscr{E} \tag{11.133}$$

in which $\mathscr{E}$, $\mathscr{D}$, $\mathscr{B}$ and $\mathscr{H}$ represent the electric field, electric displacement, magnetic induction and magnetic intensity respectively, $\mu_0 (= 4\pi \times 10^{-7}\,\mathrm{N\,s^2/C^2})$ represents the free space magnetic permeability, $\epsilon\,[= \epsilon_0 K = \epsilon_0 n^2]$ represents the dielectric permittivity of the medium, K and n are respectively the dielectric constant and the refractive index and $\epsilon_0\,[= 8.854 \times 10^{-12}\,\mathrm{C^2/N\,m^2}]$ is the permittivity of free space. Now taking the curl of Eq. (11.128) and using Eq. (11.129) we get

$$\mathbf{\nabla} \times (\mathbf{\nabla} \times \mathscr{E}) = -\mu_0 \frac{\partial}{\partial t}(\mathbf{\nabla} \times \mathscr{H}) = -\mu_0 n^2 \frac{\partial^2 \mathscr{E}}{\partial t^2}$$

or

$$\mathbf{\nabla}(\mathbf{\nabla} \cdot \mathscr{E}) - \nabla^2 \mathscr{E} = -\epsilon_0 \mu_0 n^2 \frac{\partial^2 \mathscr{E}}{\partial t^2} \tag{11.134}$$

Further

$$0 = \mathbf{\nabla} \cdot \mathscr{D} = \epsilon_0 \mathbf{\nabla} \cdot (n^2 \mathscr{E}) = \epsilon_0 [\mathbf{\nabla} n^2 \cdot \mathscr{E} + n^2 \mathbf{\nabla} \cdot \mathscr{E}]$$

Thus

$$\mathbf{\nabla} \cdot \mathscr{E} = -(1/n^2)\mathbf{\Delta} n^2 \cdot \mathscr{E} \tag{11.135}$$

Substituting in Eq. (11.134) we obtain

$$\nabla^2 \mathscr{E} + \mathbf{\nabla}\left(\frac{1}{n^2} \mathbf{\nabla} n^2 \cdot \mathscr{E}\right) - \epsilon_0 \mu_0 n^2 \frac{\partial^2 \mathscr{E}}{\partial t^2} = 0 \tag{11.136}$$

The above equation shows that for an inhomogeneous medium the equations for $\mathscr{E}_x$, $\mathscr{E}_y$ and $\mathscr{E}_z$ are coupled. For a homogeneous medium, the second term on the LHS vanishes and each Cartesian component of the electric vector satisfies the scalar wave equation.

In a similar manner taking the curl of Eq. (11.129) and using Eqs. (11.128) and (11.131) we get

$$\nabla^2 \mathcal{H} + \frac{1}{n^2}\nabla n^2 \times (\nabla \times \mathcal{H}) - \epsilon_0\mu_0 n^2 \frac{\partial^2 \mathcal{H}}{\partial t^2} = 0 \qquad (11.137)$$

If the refractive index varies only in the transverse direction, i.e.,

$$n^2 = n^2(x, y) \qquad (11.138)$$

then writing each Cartesian component of Eqs. (11.136) and (11.137) one can easily see that the time and z part can be separated out. Thus, if the refractive index is independent of the z-coordinate then the solutions of Eqs. (11.136) and (11.137) can be written in the form

$$\mathcal{E} = \mathbf{E}(x, y)\,e^{i(\omega t - \beta z)} \qquad (11.139)$$

$$\mathcal{H} = \mathbf{H}(x, y)\,e^{i(\omega t - \beta z)} \qquad (11.140)$$

where β is known as the propagation constant. Eqs. (11.139) and (11.140) define *modes* of the system and were discussed in Sec. 11.2.

We next assume that the refractive index depends only on the x-coordinate, i.e.,

$$n^2 = n^2(x) \qquad (11.141)$$

Then even the y part can be separated out implying that the y and z dependences of the fields will be of the form $e^{-i(\gamma y + \beta z)}$. However, we can *always* choose the z-axis along the direction of propagation of the wave and we may, without any loss of generality put $\gamma = 0$. Thus we may write

$$\mathcal{E}_j = E_j(x)\,e^{i(\omega t - \beta z)}; \quad j = x, y, z \qquad (11.142)$$

$$\mathcal{H}_j = H_j(x)\,e^{i(\omega t - \beta z)}; \quad j = x, y, z \qquad (11.143)$$

Substituting the above expressions for the electric and magnetic fields in Eqs. (11.128) and (11.129) and taking their x, y and z-components we obtain

$$i\beta E_y = -i\omega\mu_0 H_x \qquad (11.144)$$

$$\partial E_y/\partial x = -i\omega\mu_0 H_z \qquad \text{TE modes} \qquad (11.145)$$

$$-i\beta H_x - \partial H_z/\partial x = i\omega\epsilon_0 n^2(x)E_y \qquad (11.146)$$

$$i\beta H_y = i\omega\epsilon_0 n^2(x)E_x \qquad (11.147)$$

$$\partial H_y/\partial x = i\omega\epsilon_0 n^2(x)E_z \qquad \text{TM modes} \qquad (11.148)$$

$$-i\beta E_x - \partial E_z/\partial x = -i\omega\mu_0 H_y \qquad (11.149)$$

These equations were the starting point of Sec. 11.2.

Additional problems

Problem 11.7: We rewrite the wave equation determining the TE modes (Eq. (11.15)) as an eigenvalue equation

$$d^2u_m/dx^2 + k_0^2 n^2(x)u_m = \lambda_m u_m(x) \tag{11.150}$$

where $\lambda_m = \beta_m^2$ represent the eigenvalues of the operator $[(d^2/dx^2 + k_0^2 n^2(x)]$. Prove that all values of λ_m are real and that if $\lambda_m \neq \lambda_k$, the corresponding modal fields are orthogonal.

Solution: We write the complex conjugate of the eigenvalue equation corresponding to the eigenvalue λ_k:

$$d^2u_k^*/dx^2 + k_0^2 n^2(x)u_k^* = \lambda_k^* u_k^*(x) \tag{11.151}$$

We multiply Eq. (11.150) by u_k^* and Eq. (11.151) by u_m and subtract to obtain

$$u_k^* d^2u_m/dx^2 - u_m d^2u_k^*/dx^2 = (\lambda_m - \lambda_k^*)u_k^*(x)u_m(x)$$

The LHS is simply

$$\frac{d}{dx}\left(u_k^* \frac{du_m}{dx} - u_m \frac{du_k^*}{dx} \right)$$

therefore if we integrate from $-\infty$ to $+\infty$ we would obtain

$$(\lambda_m - \lambda_k^*) \int_{-\infty}^{+\infty} u_k^*(x)u_m(x)\,dx = [u_k^*\,du_m/dx - u_m\,du_k^*/dx]_{-\infty}^{+\infty} \tag{11.152}$$

The RHS vanishes because for guided modes the fields vanish at $x = \pm\infty$. Thus we get

$$(\lambda_m - \lambda_k^*) \int_{-\infty}^{+\infty} u_k^*(x)u_m(x)\,dx = 0 \tag{11.153}$$

For $k = m$, the integral $\int_{-\infty}^{+\infty} |u_m(x)|^2\,dx$ is positive definite and therefore we must have

$$\lambda_m = \lambda_m^* \tag{11.154}$$

proving that *all* eigenvalues β_m^2 must be real. Further for $\lambda_m \neq \lambda_k$, we must have

$$\int_{-\infty}^{+\infty} u_k^*(x)u_m(x)\,dx = 0; \qquad (\lambda_m \neq \lambda_k) \tag{11.155}$$

which represents the orthogonality condition.[†]

[†] If $\lambda_m = \lambda_k$, so that u_m and u_k are two independent modal fields belonging to the same value of the propagation constant, then the set of modes are said to be *degenerate* and u_m and u_k are not necessarily orthogonal. It can easily be seen that any linear combination $C_1 u_m + C_2 u_k$ is also a possible mode belonging to the same propagation constant and it is always possible to construct modes which are mutually orthogonal to each other; the details are given in most text books on quantum mechanics, see, e.g., Ghatak and Lokanathan (1984).

Problem 11.8: In the eigenvalue equation for TE modes (Eq. (11.15)) prove that β^2 cannot be greater than the maximum value of $k_0^2 n^2(x)$.

Solution: We rewrite Eq. (11.15) in the form

$$d^2 E_y/dx^2 = \alpha(x)E_y(x)$$

where

$$\alpha(x) = \beta^2 - k_0^2 n^2(x)$$

Now, if β^2 is greater than the maximum value of $k_0^2 n^2(x)$ then $\alpha(x)$ is positive *everywhere* and $d^2 E_y/dx^2$ has the same sign as $E_y(x)$ *everywhere*. Thus if E_y is positive at some value of x then $d^2 E_y/dx^2$ will also be positive. Hence if $E_y' > 0$ then $E_y \to \infty$ as $x \to \infty$ because E_y' will keep on increasing. On the other hand if $E_y' < 0$ then $E_y \to -\infty$ as $x \to -\infty$. Therefore, there must be some region where $\beta^2 < k_0^2 n^2(x)$.

Problem 11.9: In the specific profile shown in Fig. 11.2 show explicitly that for $\beta^2 > k_0^2 n_1^2$, the boundary conditions cannot be satisfied.

Problem 11.10: Show that if $n^2(-x) = n^2(x)$ then the modal field patterns are either symmetric or antisymmetric functions of x.

Solution: We write the wave equation determining the TE modes (Eq. (11.15)) in the form

$$d^2 E_y(x)/dx^2 + k_0^2 n^2(x)E_y(x) = \beta^2 E_y(x) \tag{11.156}$$

Making the transformation $x \to -x$ we get

$$\frac{d^2}{dx^2} E_y(-x) + k_0^2 n^2(x)E_y(-x) = \beta^2 E_y(-x) \tag{11.157}$$

where we have used the fact that $n^2(-x) = n^2(x)$. Comparing Eqs. (11.156) and (11.157) we see that $E_y(x)$ and $E_y(-x)$ satisfy the same equation and therefore they are eigenfunctions belonging to the *same* value of β^2. Thus, if the mode is nondegenerate[†] then $E_y(-x)$ must be a multiple of x, i.e.,

$$E_y(-x) = \lambda E_y(x)$$

Making the transformation $x \to -x$ again, we get

$$E_y(x) = \lambda E_y(-x) = \lambda^2 E_y(x)$$

so that $\lambda^2 = 1$ or $\lambda = \pm 1$. Hence

$$E_y(-x) = \pm E_y(x) \tag{11.158}$$

[†] This theorem is therefore strictly true for nondegenerate modes only (see footnote on p. 342). For degenerate modes the field patterns need not be symmetric or antisymmetric functions of x. However, even for degenerate modes, one can always construct appropriate linear combinations which are either symmetric or antisymmetric functions of x.

proving the theorem. Modal fields belonging to the class $\lambda = +1$ and $\lambda = -1$ are symmetric and antisymmetric functions of x respectively.

Problem 11.11: Show by direct substitution that with n^2 given by Eq. (11.112)

$$e^{-\frac{1}{2}\xi^2}, \; \xi e^{-\frac{1}{2}\xi^2} \quad \text{and} \quad (4\xi^2 - 2)e^{-\frac{1}{2}\xi^2}$$

satisfy Eq. (11.101); the corresponding value of β^2 being given by Eq. (11.114) with $m = 0, 1$ and 2 respectively.

In general show that $H_m(\xi)e^{-\frac{1}{2}\xi^2}$ satisfies Eq. (11.101) the corresponding value of β^2 being given by Eq. (11.114).
(Hint: Hermite functions are polynomial solutions of $\mathrm{d}^2 y/\mathrm{d}\xi^2 - 2\xi\,\mathrm{d}y/\mathrm{d}\xi + 2ny(\xi) = 0$).

Problem 11.12: Consider a waveguide whose bounding surfaces are made of perfect conductors (see Fig. 11.7) so that E_y may be assumed to vanish at $x = \pm d/2$. Consider the TE modes and show that

$$\beta_m^2 = k_0^2 n_1^2 - (m\pi/d)^2, \tag{11.159}$$

where $m = 1, 3, 5, \ldots$ correspond to the symmetric TE modes and $m = 2, 4, 6, \ldots$ correspond to the antisymmetric TE modes (two points should be noted: firstly

Fig. 11.7 A dielectric slab sandwiched between two perfectly conducting surfaces.

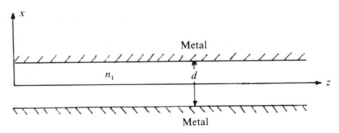

Fig. 11.8 Three dielectric slabs (with $n_1 > n_2$) sandwiched between two perfectly conducting surfaces.

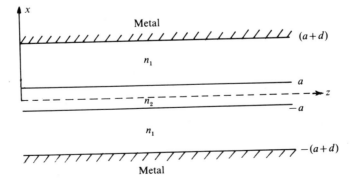

dψ/dx is discontinuous at $x = \pm d/2$, this is because $n^2(x)$ has an infinite discontinuity at $x = \pm d/2$; secondly, for $m > k_0 n_1 d/\pi$, β_m becomes imaginary – these are the attenuating modes, there is, however, no absorption of energy.)

Problem 11.13: We next consider the structure for which the refractive index is n_2 for $|x| < a$ and n_1 for $a < |x| < a + d$ with a metal boundary at $|x| = a + d$ (see Fig. 11.8). For TE modes the field distribution is given by

$$E_y = \begin{cases} A_s \cosh \gamma x; & \text{symmetric} \\ A_a \sinh \gamma x; & \text{antisymmetric} \end{cases} \quad |x| < a$$

$$= \begin{cases} B \sin [\kappa(a + d - |x|)]; & \text{symmetric} \\ \dfrac{x}{|x|} B \sin [\kappa(a + d - |x|)]; & \text{antisymmetric} \end{cases} \quad a < |x| < a + d$$

$$\tag{11.160}$$

where we have assumed $n_2 < \beta/k_0 < n_1$; γ and κ are the same as in Sec. 11.3. Show that the propagation constants are determined from the following transcendental equation

$$-\frac{(V_0^2 - \xi^2)^{\frac{1}{2}}}{\xi} \tan 2\xi = (\coth \gamma a)^{\pm 1}; \qquad V_0 = k_0 d(n_1^2 - n_2^2)^{\frac{1}{2}}/2 \tag{11.161}$$

where the $+$ and $-$ signs correspond to symmetric and antisymmetric modes respectively. Thus for $\gamma a \gg 1$ (i.e., when the two waveguides are well separated) the symmetric and antisymmetric modes are almost degenerate.

Problem 11.14: Fig. 11.8 can be assumed approximately to represent two identical waveguides separated by a certain distance. This is essentially a directional coupler which will be discussed in Sec. 14.6.3 and the previous problem describes the 'supermodes' of the composite structure. Considering the two lowest order modes

Fig. 11.9 The first and second modes of the structure shown in Fig. 11.8. If $\gamma a \gg 1$, the propagation constants of the two modes are nearly equal and the field in the region $|x| < a$ is extremely small.

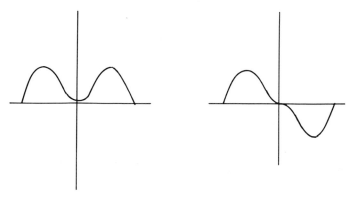

we may write as a solution of the wave equation

$$\Psi = [\psi_s(x)e^{i(\omega t - \beta_s z)} + \psi_a(x)e^{i(\omega t - \beta_a z)}]$$
$$= [\psi_s(x) + \psi_a(x)e^{i(\Delta\beta)z}]e^{i(\omega t - \beta_s z)} \tag{11.162}$$

where $\psi_s(x)$ and $\psi_a(x)$ represent the modal pattern for the symmetric and antisymmetric modes respectively (see Fig. 11.9) and $\Delta\beta = (\beta_s - \beta_a)$. Show that at $z = 0$, $2\pi/\Delta\beta, 4\pi/\Delta\beta,\ldots$ most of the energy is in the first waveguide and at $z = \pi/\Delta\beta$, $3\pi/\Delta\beta,\ldots$ there is almost complete transfer of power to the second waveguide.

Problem 11.15: Consider an interface of a dielectric and a metal, i.e.,

$$n^2 = \begin{cases} n_1^2; & x > 0 \\ n_m^2; & x < 0 \end{cases} \tag{11.163}$$

where n_m^2 is complex having a large negative part. Show that such a structure can support a TM mode whose propagation constant is given by

$$\beta^2/k_0^2 = n_1^2 n_m^2/(n_1^2 + n_m^2) \tag{11.164}$$

This mode is known as the surface plasmon mode and is of considerable interest in integrated optics. Show that the corresponding field decays exponentially in the $x > 0$ and $x < 0$ regions. Show also that for such a structure, TE modes can never exist.

12

Leaky modes in optical waveguides

12.1 Introduction

In this chapter we will give an analysis of leaky modes in optical waveguides which is a subject of considerable importance not only in optical waveguides but also in some laser systems.

We first consider a plane wave incident at the interface of two dielectrics as shown in Fig. 12.1(*a*) and let $n_2 < n_1$. Obviously, if the angle of incidence i is greater than the critical angle $i_c (= \sin^{-1}(n_2/n_1))$ then (although an evanescent wave will exist in the second medium) the wave will undergo total

Fig. 12.1 (*a*) Total internal reflection at an interface; (*b*) frustrated total internal reflection.

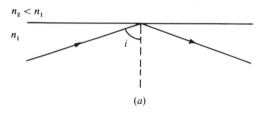

(*a*)

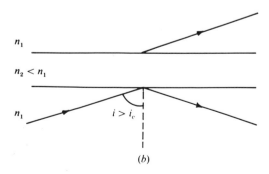

(*b*)

internal reflection at the interface (see Sec. 2.3). Now if we have a third medium of refractive index n_1 (see Fig. 12.1(b)) then even for $i > i_c$, a part of the energy will tunnel through the rarer medium and appear in the last medium. This is known as *frustrated total internal reflection* and the phenomenon is very similar to the quantum mechanical tunnelling of a particle through a potential energy barrier (see, e.g., Ghatak and Lokanathan (1984)).

Because of this phenomenon of frustrated total internal reflection, the structure shown in Fig. 12.2 cannot support a guided mode; however, if the distance d_2 (see Fig. 12.2) is large, then if power is launched into the core of the structure, this energy will gradually leak out of the core. In this chapter we will study the leakage phenomenon which is not only technologically important for our understanding of the losses in waveguides but which also allows us to understand quantitatively the propagation of a packet of radiation modes. We will first carry out an electromagnetic analysis of a planar leaky structure which will bring out all the physics associated with leaky modes in a waveguide. Using the formalism of leaky modes we will, in Sec. 12.4, develop a novel numerical method for obtaining the propagation characteristics of graded/absorbing structures; we will closely follow the

Fig. 12.2 A leaky planar structure.

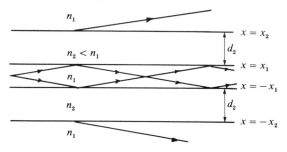

Fig. 12.3 The refractive index profile for (a) a leaky structure and (b) the corresponding guiding structure. We assume a perfectly conducting boundary at $x = 0$ so that the fields vanish there.

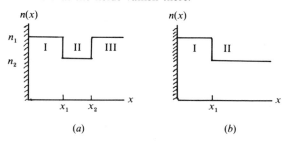

analysis of Ghatak (1985) and Ghatak, Thyagarajan and Shenoy (1987). Finally in Sec. 12.5 we will show that a bent waveguide is also a leaky structure and will suggest a method for the calculation of bending losses.

12.2   Quasi-modes in a planar structure

We consider the TE modes for the refractive index distribution given by (see Fig. 12.3(*a*)):

$$
n(x) = \begin{cases} n_1; & 0 < x < x_1 \quad \text{and} \quad x > x_2 \\ n_2 (< n_1); & x_1 < x < x_2 \end{cases} \tag{12.1}
$$

We assume that at $x = 0$ there is a metallic boundary so that the field vanishes there. By assuming the field to vanish at $x = 0$, we are essentially considering the antisymmetric modes of the structure shown in Fig. 12.2. For such a structure there exists no guided mode, and all values of $\beta^2 < k_0^2 n_1^2$ are allowed; these form the continuum radiation modes of the system. The solution for the TE modes for such a structure for $n_2 < \beta/k_0 < n_1$ is given by

$$
\psi_{\mathrm{I}}(x) = A \sin \alpha x; \quad 0 < x < x_1 \tag{12.2a}
$$

$$
\psi_{\mathrm{II}}(x) = B\,e^{\gamma(x - x_1)} + C\,e^{-\gamma(x - x_1)}; \quad x_1 < x < x_2 \tag{12.2b}
$$

$$
\psi_{\mathrm{III}}(x) = D_+\,e^{i\alpha(x - x_2)} + D_-\,e^{-i\alpha(x - x_2)}; \quad x > x_2 \tag{12.2c}
$$

where

$$
\alpha^2 = k_0^2 n_1^2 - \beta^2; \quad \gamma^2 = \beta^2 - k_0^2 n_2^2 \tag{12.3}
$$

In rejecting the solution $\cos \alpha x$ in the region $0 < x < x_1$ we have used the boundary condition that $\psi(0) = 0$. Continuity of ψ and $d\psi/dx$ at $x = x_1$ and $x = x_2$ gives us

$$
B = \tfrac{1}{2}A[\sin \alpha x_1 + (\alpha/\gamma) \cos \alpha x_1] \tag{12.4a}
$$

$$
C = \tfrac{1}{2}A[\sin \alpha x_1 - (\alpha/\gamma) \cos \alpha x_1] \tag{12.4b}
$$

$$
D_-^* = D_+ = \tfrac{1}{2}B(1 + \gamma/i\alpha)e^{\gamma(x_2 - x_1)} + \tfrac{1}{2}C(1 - \gamma/i\alpha)e^{-\gamma(x_2 - x_1)} \tag{12.4c}
$$

From the above equations it is obvious that there are four continuity conditions and five unknown coefficients. Thus one cannot construct an eigenvalue equation and all values of β^2 (less than $k_0^2 n_1^2$) are allowed and Eqs. (12.2)–(12.4) represent the continuum radiation modes of the structure. In order to get a physical picture of the field we multiply Eq. (12.2c) by $e^{i(\omega t - \beta z)}$ (see Eq. (11.7)) to obtain

$$
\Psi_{\mathrm{III}}(x, z, t) = \psi_{\mathrm{III}}(x)\,e^{i(\omega t - \beta z)}
$$

$$
= D_+ e^{-i\alpha x_2}e^{i(\omega t + \alpha x - \beta z)} + D_- e^{i\alpha x_2}e^{i(\omega t - \alpha x - \beta z)} \tag{12.5}
$$

The two terms on the RHS of the above equation represent plane waves propagating in the downward and upward directions both making the same angle with the z-axis as shown in Fig. 12.4. Thus the solution represented by Eqs. (12.2a)–(12.2c) corresponds to the reflection of a plane wave incident on the structure and different values of β correspond to different angles of incidence. Further, for the specific structure shown in Fig. 12.4, $|D_-| = |D_+|$ implying the reflection is complete – except for a change of phase.

Now, the orthonormality condition for radiation modes (see Appendix E)

$$\int_0^\infty \psi_{\beta'}^*(x)\psi_\beta(x)\,\mathrm{d}x = \delta(\beta - \beta') \tag{12.6}$$

gives us (see Appendix E)

$$|D_\pm| = (\beta/2\pi\alpha)^{\frac{1}{2}} \tag{12.7}$$

If we assume that $\gamma(x_2 - x_1) \gg 1$, then the exponentially decaying term in Eq. (12.4c) can be neglected, and in this approximation

$$D_-^* = D_+ \approx \tfrac{1}{4}LA(1 + \gamma/i\alpha)\,e^{\gamma(x_2 - x_1)} \tag{12.8}$$

where

$$L = \sin \alpha x_1 + (\alpha/\gamma)\cos \alpha x_1 \tag{12.9}$$

It is readily seen that since $\gamma(x_2 - x_1) \gg 1$, unless $L \approx 0$, $|D_+/A|$ and $|D_-/A|$ are very large quantities and the amplitude of the field is large in the region $x > x_2$. However, when $L = 0$, or when

$$\cot \alpha x_1 = -\gamma/\alpha, \tag{12.10}$$

$B = 0$ and therefore we have only the exponentially decaying solution in the region $x_1 < x < x_2$ (see Eq. (12.2b)) and the oscillatory field in the region $x > x_2$ is much weaker (see Fig. 12.5). Thus when $L = 0 = B$ the field has the

Fig. 12.4 Incidence of a plane wave of amplitude D_+ on a structure whose refractive index profile is shown in Fig. 12.3 (a). The angle $\phi_1 = \sin^{-1}(\beta/k_0 n_1)$.

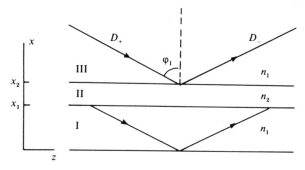

properties of a guided mode in regions I and II although it becomes oscillatory in region III (like a radiating field). These fields with $B = 0$ are referred to as 'quasi-modes'.

Eq. (12.10) represents the transcendental equation corresponding to the guided modes of the structure shown in Fig. 12.3(b) for which

$$n(x) = \begin{cases} n_1; & 0 < x < x_1 \\ n_2; & x > x_1 \end{cases} \tag{12.11}$$

with $\psi(0) = 0$. We write the wave equation in each region and following a method similar to that given in the previous chapter, we obtain the following expressions for the field corresponding to the guided mode:

$$\left. \begin{aligned} \psi_g(x) &= A_g \sin \alpha_g x; & 0 < x < x_1 \\ &= A_g \sin \alpha_g x_1 e^{-\gamma_g(x - x_1)}; & x > x_1 \end{aligned} \right\} \tag{12.12}$$

where the subscript g refers to the guided mode and continuity of $d\psi/dx$ at $x = x_1$ gives us Eq. (12.10) where α_g and γ_g represent those specific values which satisfy Eq. (12.10).

Returning to the profile shown in Fig. 12.3(a), we note that for the radiation modes corresponding to those values of β which satisfy Eq. (12.10), the exponentially amplifying term in Eq. (12.2b) is zero and the field in regions I and II (see Fig. 12.3(a)) is identical to that of the guided mode. This is shown explicitly in Fig. 12.5.

Fig. 12.5 (a) For the quasi-modes we have only the exponentially decaying solution in the region $x_1 < x < x_2$ and the oscillatory field in the region $x > x_2$ is very weak. For β much different from β_g (the subscript g refers to the guided mode) the amplitude of the field is very large in the region $x > x_2$. (b) The guided mode corresponding to the structure shown in Fig. 12.3 (b). (Adapted from Ghatak (1985)).

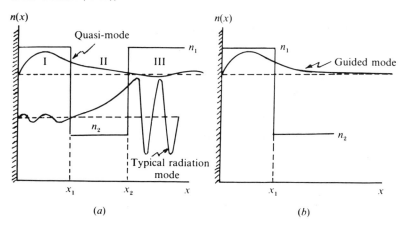

We normalize the guided mode given by Eq. (12.12) such that

$$\int_0^\infty |\psi_g(x)|^2 \, dx = 1 \tag{12.13}$$

and obtain

$$A_g = \left(\frac{2\gamma_g}{1 + \gamma_g x_1} \right)^{\frac{1}{2}} \tag{12.14}$$

12.3   Leakage of power from the core

We next consider the incidence of the fundamental guided mode of the structure shown in Fig. 12.3(*b*) on a leaky structure as shown in Fig. 12.6. Thus

$$\psi(x, z = 0) = \psi_g(x) \tag{12.15}$$

Such an incident field would excite a packet of radiation modes of the leaky structure and we may write

$$\psi(x, z = 0) = \psi_g(x) = \int \phi(\beta)\psi_\beta(x) \, d\beta \tag{12.16}$$

where $\psi_\beta(x)$ represents the normalized radiation modes of the leaky structure shown in Fig. 12.3(*a*). For $z > 0$ the field will be given by

$$\psi(x, z) = \int \phi(\beta)\psi_\beta(x) e^{-i\beta z} \, d\beta \tag{12.17}$$

In order to determine $\phi(\beta)$, we multiply Eq. (12.16) by $\psi_{\beta'}^*(x)$ and integrate

Fig. 12.6 Incidence of the fundamental guided mode of the structure shown in Fig. 12.3 (*b*) on a leaky structure.

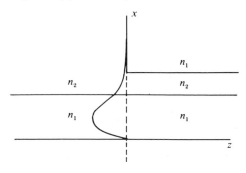

over x to obtain

$$\int \psi_{\beta'}^*(x)\psi_g(x)\,dx = \int d\beta\phi(\beta) \int dx\psi_{\beta'}^*(x)\psi_\beta(x)$$

$$= \int d\beta\phi(\beta)\delta(\beta - \beta') = \phi(\beta')$$

where use has been made of Eq. (12.6). Thus

$$\phi(\beta) = \int_0^\infty \psi_\beta^*(x)\psi_g(x)\,dx \tag{12.18}$$

As discussed earlier, it is only around $\beta \approx \beta_g$, that $\psi_\beta(x)$ has almost the same spatial dependence as $\psi_g(x)$ in regions I and II and therefore in Eq. (12.18), $\phi(\beta)$ will be appreciable only around $\beta \approx \beta_g$; this we will explicitly find later. Thus we may write

$$\psi_\beta(x) \approx (A/A_g)\psi_g(x) \quad \text{regions I and II} \tag{12.19}$$

Since in region III ($x > x_2$) $\psi_g(x)$ has a negligible value, we may neglect the contribution from region III to the integral in Eq. (12.18) to write

$$\phi(\beta) \approx \int_0^{x_2} \psi_\beta^*(x)\psi_g(x)\,dx$$

$$\approx (A/A_g) \int_0^{x_2} |\psi_g(x)|^2\,dx$$

$$\approx (A/A_g) \int_0^\infty |\psi_g(x)|^2\,dx = A/A_g \tag{12.20}$$

In Appendix E we have shown that around $\beta \approx \beta_g$

$$|\phi(\beta)|^2 = \left|\frac{A}{A_g}\right|^2 = \frac{1}{\pi}\frac{\Gamma}{(\beta - \beta_g')^2 + \Gamma^2} \tag{12.21}$$

where

$$\Gamma = \frac{4\alpha_g^3\gamma_g^3}{\beta_g\delta^4}\frac{e^{-2\gamma_g(x_2 - x_1)}}{(1 + \gamma_g x_1)} \tag{12.22}$$

$$\beta_g' = \beta_g + \Delta\beta \tag{12.23}$$

$$\Delta\beta = -\frac{\Gamma(\alpha_g^2 - \gamma_g^2)}{2\alpha_g\gamma_g} \tag{12.24}$$

The fractional power $W(z)$ that remains inside the core at z is approximately given by

$$W(z) \approx \left| \int_0^\infty \psi^*(x, 0)\psi(x, z)\,dx \right|^2 \tag{12.25}$$

which can be evaluated to give (see Appendix E)

$$W(z) = e^{-2\Gamma z} \tag{12.26}$$

The above equation shows how the power inside the core 'leaks' into region III. Eq. (12.22) gives an analytical expression for the attenuation coefficient of the 'quasi-modes'. It should be noted that as $x_2 \to \infty$, $\Gamma \to 0$ and there is no leakage of power.

Now, in Appendix E we have also shown that

$$\left| \frac{A}{D_\pm} \right|^2 \approx \frac{4\gamma_g \alpha_g \Gamma}{\beta_g(1 + \gamma_g x_1)} \frac{1}{(\beta - \beta'_g)^2 + \Gamma^2} \tag{12.27}$$

Thus if we are able to calculate $|A/D_\pm|^2$ as a function of β (i.e., as a function of the angle of incidence in Fig. 12.4), we would get a series of Lorentzians, each Lorentzian corresponding to a quasi-mode of the structure. By fitting each peak to a Lorentzian, we would be able to get β'_g and Γ.

In the next section we will develop a matrix method to calculate $|A/D_\pm|^2$ as a function of β from which we will be able to get the propagation characteristics (including leakage/absorption losses) even in graded/absorbing structures.

12.4   The matrix method for determining the propagation characteristics of planar structures which may be leaky or absorbing.

In this section we will develop a matrix method for calculating $|A/D_\pm|^2$ for arbitrary planar structures which will enable us to determine the propagation characteristics of planar structures which may be leaky or absorbing. In the next section we will discuss the applicability of the method to loss calculations in bent waveguides.

We first consider a layered structure as shown in Fig. 12.7. For a plane wave incident at an angle ϕ_1 as shown in Fig. 12.7, the electric field in each medium may be written in the form

$$\mathscr{E}_i = \hat{\mathbf{e}}_i^+ E_i^+ e^{i\Delta_i}e^{i[\omega t - (k_i\cos\phi_i)x - \beta z]} + \hat{\mathbf{e}}_i^- E_i^- e^{-i\Delta_i}e^{i[\omega t + (k_i\cos\phi_i)x - \beta z]} \tag{12.28}$$

where

$$\Delta_1 = \Delta_2 = 0, \quad \Delta_3 = k_3 d_2 \cos\phi_3 \atop \Delta_4 = k_4(d_2 + d_3)\cos\phi_4 \Bigg\} \tag{12.29}$$

$$k_i = k_0 n_i = (\omega/c)n_i \qquad (12.30)$$

$\hat{\mathbf{e}}_i^+$ and $\hat{\mathbf{e}}_i^-$ represent the unit vectors along the direction of the electric fields. Further,

$$\beta = k_1 \sin \phi_1 = k_2 \sin \phi_2 = \cdots = k_4 \sin \phi_4 \qquad (12.31)$$

is an invariant of the system, E_i^+ and E_i^- represent the electric field amplitudes of waves propagating in the downward and upward directions (see Fig. 12.7) respectively. On applying the appropriate boundary conditions at the interface we obtain (cf. Sec. 2.4).

$$\begin{pmatrix} E_1^+ \\ E_1^- \end{pmatrix} = S_1 \begin{pmatrix} E_2^+ \\ E_2^- \end{pmatrix} = \cdots = S_1 S_2 S_3 \begin{pmatrix} E_4^+ \\ E_4^- \end{pmatrix} \qquad (12.32)$$

where

$$S_i = \frac{1}{t_i} \begin{pmatrix} e^{i\delta_i} & r_i e^{i\delta_i} \\ r_i e^{-i\delta_i} & e^{-i\delta_i} \end{pmatrix} \qquad (12.33)$$

and r_i and t_i represent respectively the amplitude reflection and transmission coefficients at the ith interface; $\delta_i = k_i d_i \cos \phi_i$. For TE polarization (cf. Eqs. (2.58) and (2.59))

$$\left. \begin{aligned} r_i &= \frac{n_i \cos \phi_i - n_{i+1} \cos \phi_{i+1}}{n_i \cos \phi_i + n_{i+1} \cos \phi_{i+1}} \\[2ex] t_i &= \frac{2n_i \cos \phi_i}{n_i \cos \phi_i + n_{i+1} \cos \phi_{i+1}} \end{aligned} \right\} \qquad (12.34)$$

Fig. 12.7 The incidence of a plane wave at an angle ϕ_1 in a stratified medium.

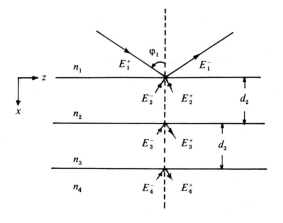

and for TM polarization (cf. Eqs. (2.24) and (2.25))

$$r_i = \left. \frac{n_i \cos \phi_{i+1} - n_{i+1} \cos \phi_i}{n_i \cos \phi_{i+1} + n_{i+1} \cos \phi_i} \right| \qquad (12.35)$$

$$t_i = \frac{2n_i \cos \phi_i}{n_i \cos \phi_{i+1} + n_{i+1} \cos \phi_i} \right|$$

Obviously, for a wave incident from the first medium, there will be no upward propagating wave in the last medium, and thus for the structure shown in Fig. 12.7, $E_4^- = 0$. Using this condition and Eqs. (12.32) and (12.33) one can readily calculate all the fields in terms of E_1^+.

In order to use the above matrix method to determine the propagation characteristics of planar waveguides, the second, third and fourth media (in Fig. 12.7) may correspond to the superstrate, the waveguiding film and the substrate, respectively. If one now evaluates the excitation efficiency of the wave in the film (i.e., $\eta(\beta) = |E_3^+/E_1^+|^2$ or $|E_3^-/E_1^+|^2$ which is similar to $|A/D_+|^2$ of Eq. (12.27)) as a function of β, then one will obtain resonance peaks which are Lorentzian in shape; the value of β at which the peaks appear gives the real part of the propagation constant β_g' and the FWHM represents the power attenuation coefficient 2Γ.

As an example we consider the structure shown in Fig. 12.4(*b*) with

$$n_1 = 1.5, \quad n_2 = 1.4, \quad x_1 = 2\,\mu m$$

Fig. 12.8 The variation of $|E_3^+/E_1^+|^2$ as a function of β/k_0 for the configuration shown in Fig. 12.4 with $n_1 = 1.5$, $n_2 = 1.4$ and $x_1 = 2\,\mu m$. (Adapted from Ghatak *et al* (1987).)

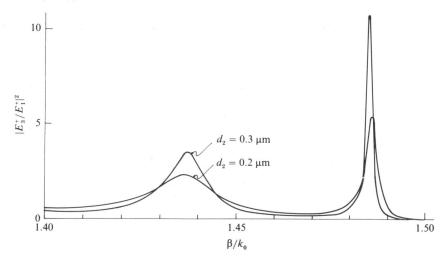

In Fig. 12.8 we have plotted the variation of $|E_3^+/E_1^+|^2$ as a function of β for $d_2 = (x_2 - x_1) = 0.2$ and $0.3\,\mu$m; here E_3^+ and E_1^+ are essentially A and D_+ respectively. The two Lorentzians correspond to the two quasi-modes of the structure.

The above analysis suggests that in order to obtain the propagation characteristics of any planar structure (for example, as shown in Fig. 12.9(a), which may be absorbing or nonabsorbing) we may carry out the following procedure:

Step 1. We first introduce a region, whose refractive index is equal to or greater than the maximum refractive index of the guiding region, above the planar guide as shown in Fig. 12.9(b); the structure is now leaky.

Fig. 12.9 In order to study the propagation characteristics of the structure shown in (a), we should introduce another medium of refractive index $n_1 (> n_2, n_3, \ldots, n_6)$ as shown in (b) and increase d_2.

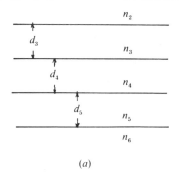

(a)

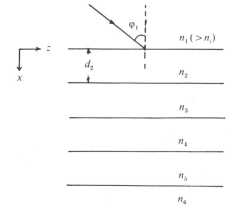

(b)

Step 2. We next consider the incidence of a plane wave on this structure at different angles of incidence and calculate the ratio $|E_{\text{film}}/E_1^+|^2$ as a function of $\beta = k_1 \sin \theta_1$.

Step 3. Each peak is fitted to a Lorentzian which gives the real and imaginary parts of the propagation constant.

If the value of d_2 is increased, then for an absorbing structure β_r and $\beta_i (= \Gamma)$ would converge to particular values giving the complex propagation constant, $\beta = \beta_r - i\beta_i$ of the guiding structure. Usually the limiting values can be obtained to about one part in 10^4 when $\gamma d_2 \sim 4$. On the other hand, for a lossless guide the value of Γ will tend to zero.

The applicability of this technique to absorbing as well as to other structures has recently been demonstrated (Ghatak, *et al.*, 1987).

12.5   Calculation of bending loss in optical waveguides.

Bending loss calculation is a topic of great interest in optical waveguides. We will show that a bent waveguide is essentially a leaky structure and therefore the theory developed in the previous section can be used to calculate the bending loss. We will follow the analysis given by Thyagarajan, Shenoy and Ghatak (1987).

We consider a planar waveguide which is bent along the arc of a circle of radius ρ as shown in Fig. 12.10. We assume the validity of the scalar wave equation

$$\nabla^2\psi + k_0^2 n^2 \psi = 0 \tag{12.36}$$

Fig. 12.10 The coordinate system for a bent planar waveguide.

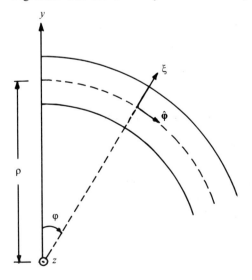

We use a cylindrical system of coordinates (r, ϕ, z) whose origin is at the centre of the arc. The refractive index depends only on r and since the waveguide is of infinite extent in the z-direction we may neglect the z dependence of the fields. Thus we assume a solution of the form

$$\psi = R(r)\,e^{-i\beta\rho\phi} \tag{12.37}$$

and substitute in Eq. (12.36) to obtain

$$\frac{1}{r}\frac{d}{dr}\left(r\frac{dR}{dr}\right) + \left[k_0^2 n^2(r) - \frac{\beta^2\rho^2}{r^2}\right]R(r) = 0 \tag{12.38}$$

Obviously in the limit of $\rho \to \infty$, β will correspond to the propagation constant of the straight guide. If we write

$$R(r) = (1/r^{\frac{1}{2}})u(r) \tag{12.39}$$

Eq. (12.38) becomes

$$d^2u/d\xi^2 + [k_0^2 \tilde{n}^2(\xi) - \beta^2]u(\xi) = 0 \tag{12.40}$$

Fig. 12.11 (a) The solid line gives the effective refractive index profile corresponding to a bent slab waveguide. The dashed line gives the corresponding profile when the waveguide is straight. (b) The effective profile replaced by a large number of homogeneous layers.

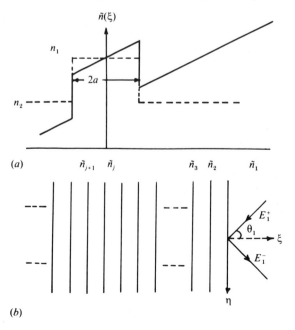

where $\xi = r - \rho$ and

$$\tilde{n}^2(\xi) = n^2(r) + \left\{ \frac{\beta^2}{k_0^2} \left[1 - \frac{\rho^2}{(\rho + \xi)^2} \right] + \frac{1}{4k_0^2(\rho + \xi)^2} \right\} \tag{12.41}$$

We should point out that the only assumption made in obtaining Eq. (12.41) is the validity of the scalar wave equation. Eq. (12.41) represents a one-dimensional wave equation and because of the form of $\tilde{n}^2(\xi)$ (see Fig. 12.11(a)) it cannot support a guided mode. The waveguide is therefore leaky and we may use the matrix method to determine the bending loss. We replace the effective refractive index profile $\tilde{n}(\xi)$ by a series of step variations as shown in Fig. 12.11(b) and consider the incidence of a plane wave from medium 1 (as shown in Fig. 12.11(b)) which is given by $E_1^+ e^{i(k_{1\xi}\xi - \beta\eta)}$ where $k_{1\xi} = k_0\tilde{n}_1 \cos\theta$, $\beta = k_0\tilde{n}_1 \sin\theta$ and $\eta = \rho\phi$. We next calculate the quantity $|E_j^+/E_1^+|^2$ as a function of β; here E_j^+ is the amplitude of the downward travelling plane wave in the jth medium which, in this case, is taken as one of the layers lying inside the core of the waveguide. The quantity $|E_j^+/E_1^+|^2$ will be sharply peaked around each quasi-mode and will be a Lorentzian function given by

$$\left| \frac{E_j^+}{E_1^+} \right|^2 \sim \frac{1}{(\beta - \beta_g')^2 + \Gamma^2} \tag{12.42}$$

where β_g' is the propagation constant of the quasi-mode and 2Γ (which represents the FWHM of the Lorentzian) represents the leakage loss of the mode, i.e., the power inside the core would decrease as $P(\phi) = P(0)e^{-2\Gamma\rho\phi}$. Since β^2 also appears in the expression for $\tilde{n}^2(\xi)$, one can use the following iterative method.

We first substitute the value of β corresponding to the straight waveguide in Eq. (12.41), calculate β_g' and Γ using Eq. (12.42), then replace β in Eq. (12.41) by the obtained value of β_g' and iterate until the value of β_g' converges. The iterated values of β_g' and Γ will give us the propagation constant and loss of the quasi-mode of the bent waveguide.

As an example, we consider a step index slab waveguide with (Thyagarajan, *et al* 1987)

$$n(x) = \begin{cases} n_1 = 1.503; & |x| < d/2 \\ n_2 = 1.500; & |x| > d/2 \end{cases} \tag{12.43}$$

$$d = 4\,\mu m \quad \text{and} \quad \lambda_0 = 1\,\mu m$$

so that $V = (2\pi/\lambda_0)d(n_1^2 - n_2^2)^{\frac{1}{2}} \approx 2.385$ and therefore the straight waveguide will support only one guided mode. Fig. 12.12 shows a typical variation of

Fig. 12.12 Typical variation of $|E_j^+/E_1^+|^2$ as a function of β for $\rho = 1\,\mathrm{cm}$ corresponding to the waveguide defined by Eq. (12.4). The peak corresponds to β_g', and the FWHM, 2Γ gives the loss coefficient of the bent waveguide; j corresponds to a layer inside the core of the waveguide.

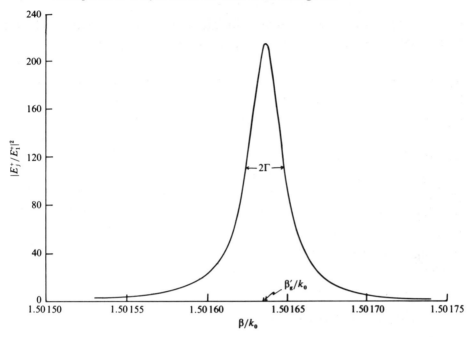

Fig. 12.13 The effective refractive index profile of the planar waveguide defined by Eq. (12.43), and the corresponding field distribution of the quasi-mode for a bend radius of $\rho = 0.5\,\mathrm{cm}$.

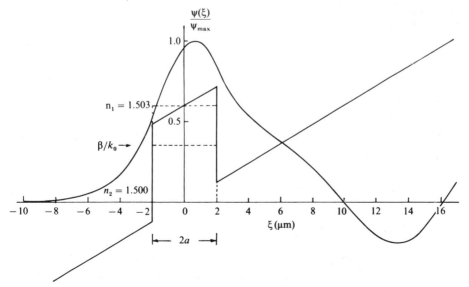

$|E_j^+/E_1^+|^2$ as a function of β for $\rho = 1 \,\text{cm}$. In the calculation $\tilde{n}^2(\xi)$ was replaced by 100 layers which gave a convergence of one part in a million. Fig. 12.13 shows the calculated field distribution for the quasi-mode which clearly shows the shift in the peak of the modal field as well as the oscillatory behaviour from the value of ξ at which $\tilde{n}(\xi) > \beta/k_0$. Fig. 12.14 gives the corresponding bend loss variation with bend radius.

A novel profile for small bend loss Bent waveguides play an important role in the optimization of substrate space and the integration of various integrated optical devices on a single chip, and therefore reduction of bend loss is of considerable interest in integrated optics (Korotky, Marcatili, Vaselka and Bosworth, 1986). We give here an approach for the realization of arbitrarily small bend loss (Thyagarajan, *et al* 1987).

We consider a one-dimensional slab waveguide for which the refractive index profile is given by

$$n(x) = \begin{cases} n_1; & -a < x < a \\ n_2 - \alpha^2 x; & a < x < d \\ n_3; & x > d \end{cases} \tag{12.44}$$

The refractive index in the region $x < -a$ can be arbitrary but less than n_1. For a given value of the parameter α, for ρ greater than a critical radius ρ_c, the term $\alpha^2 \xi$ will be greater than the second term on the RHS of Eq. (12.41); therefore $\tilde{n}^2(\xi)$ beyond $x = a$ will either decrease or remain approximately

Fig. 12.14 Variation of bend loss with radius of curvature for a planar waveguide defined by Eq. (12.43). (Adapted from Thyagarajan, *et al.* (1987)).

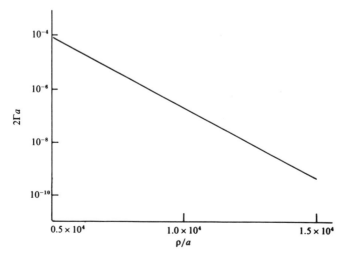

constant till $\xi = d$. If the value of d is sufficiently large, then it is obvious that the loss will be extremely small. Conversely, we can always tailor the profile such that for all radii of curvature greater than a critical value, the bending losses will be extremely small. Such profiles should find considerable applications in the design of integrated optical waveguides.

13

Optical fibre waveguides

13.1 Introduction

Since optical frequencies are extremely large ($\sim 10^{15}$ Hz), as compared to conventional radio waves ($\sim 10^6$ Hz) and microwaves ($\sim 10^{10}$ Hz), a light beam acting as a carrier wave is capable of carrying far more information in comparison to radio waves and microwaves. It is expected that in the not too distant future, the demand for flow of information traffic will be so high that only a light wave will be able to cope with it.

Soon after the discovery of the laser, some preliminary experiments on the propagation of information-carrying light waves through the open atmosphere were carried out, but it was realized that because of the vagaries of the terrestrial atmosphere – e.g., rain, fog etc. – in order to have an efficient and dependable communication system, one would require a guiding medium in which the information-carrying light waves could be transmitted. This guiding medium is the optical fibre which is hair thin and guides the light beam from one place to another (see Fig. 13.1). In addition to the capability of carrying a huge amount of information, fibres fabricated with recently developed technology are characterized by extremely low losses (~ 0.2 dB/km) as a consequence of which the distance between two consecutive repeaters (used for revamping the attenuated signals) can be as large as 250 km. In a recently developed fibre optic system it has been possible to send 140 Mbit/s information through a 220 km link of one optical fibre; this is equivalent to about 450 000 voice channel km! In comparison, the copper cables used today have repeater spacings every few kilometres or so. Typical losses in various guiding media are shown in Fig. 13.2. We should point out here that it has been only recently that the devices used in fibre optic communication systems have become very reliable and the cost of the system has come down considerably so that people throughout the world have seriously begun the switch-over from copper cables to fibre optic systems. In addition to the long distance communication systems, optical fibres are also being extensively used for local area networks (LANS) – networks that wire up telephones, televisions, computers or robots in offices and cities.

In this chapter we will discuss the basic waveguidance properties of various types of optical fibres. We will present detailed modal analyses of step index and parabolic index fibres. Knowledge of the modal field

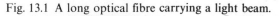

Fig. 13.1 A long optical fibre carrying a light beam.

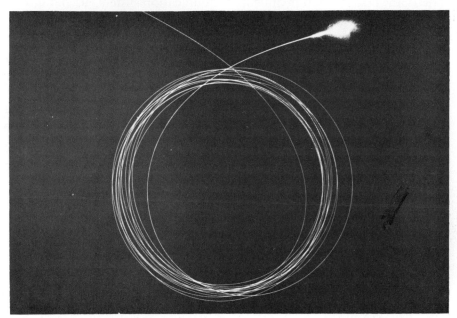

Fig. 13.2 Typical attenuation of various guiding media. The loss curve for the optical fibre appears very sharp because of the logarithmic frequency scale. (Figure adapted from Henry (1985)).

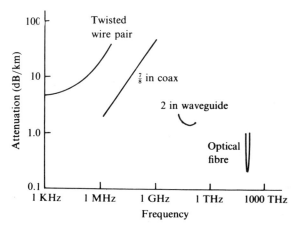

distribution is essential for the calculation of excitation efficiencies, splice losses at joints, bending losses and in development of fibre optic devices. We will also have a detailed discussion on the frequency dependence of the propagation constant. This dependence allows the calculation of pulse dispersion which essentially determines the information-carrying capacity of the optical fibre. We will discuss how the large pulse dispersion in multimode step index fibres can be reduced significantly by using graded core fibres. Indeed, the first generation optical communication systems used multimode graded index fibres at $0.85 \, \mu m$ and as we will show later, the information-carrying capacity was limited by material dispersion. In order to decrease the material dispersion the operating wavelength was shifted to $1.3 \, \mu m$ where the material dispersion is negligibly small which corresponds to the second

Fig. 13.3 (*a*) A glass fibre which consists of a cylindrical central core cladded by a material of slightly lower refractive index. Light rays impinging on the core–cladding interface at an angle greater than the critical angle are trapped inside the core of the waveguide. Rays making larger angles with the axis take a longer time to traverse the length of the fibre. (*b*) Refractive index distribution of a cladded optical fibre which consists of a cylindrical glass structure surrounded by a material of slightly lower refractive index. In a typical (multimode) fibre we may have the core refractive index $n_1 = 1.46$, $\Delta = 0.01$, core radius $a = 25 \, \mu m$, cladding thickness $b = 37.5 \, \mu m$.

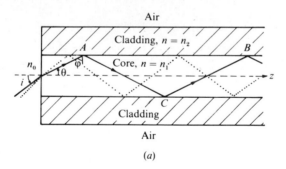

(*a*)

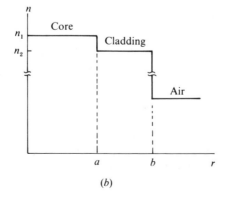

(*b*)

generation optical communication system where the information-carrying capacity is limited by intermodal dispersion. Now, the intermodal dispersion can be eliminated by using single mode fibres. This is indeed the case for third generation optical communication systems which operate at 1.3 μm. We will also discuss the fact that the loss in typical single mode fibres is lowest around 1.55 μm and fourth generation optical communication systems use single mode fibres at 1.55 μm. The refractive index profiles of these single mode fibres are tailored in such a way that the total dispersion is negligible at 1.55 μm.

It is the purpose of this chapter to develop all these concepts step by step and to study the basic physics behind the evolution of different types of fibres.

13.2 The optical fibre

Fig. 13.3(*a*) shows a glass fibre which consists of a (cylindrical) central core cladded by a material of slightly lower refractive index. The corresponding refractive index distribution (in the transverse direction) is shown in Fig. 13.3(*b*) and is given by

$$n(r) = \begin{cases} n_1; & r < a \\ n_2; & r > a \end{cases} \tag{13.1}$$

For a ray entering the fibre, if the angle of incidence (at the core–cladding interface) ϕ is greater than the critical angle $\phi_c (= \sin^{-1}(n_2/n_1))$ then the ray will undergo total internal reflection at that interface. Further, because of the cylindrical symmetry in the fibre structure, this ray will suffer total internal reflection at the lower interface also and will therefore be guided through the core by repeated total internal reflections. Fig. 13.1 shows the actual guidance of a light beam as it propagates through a long optical fibre.

It was felt necessary to use a cladded fibre (as shown in Fig. 13.3) rather than a bare fibre because of the fact that for transmission of light from one place to another, the fibre must be supported, and the supporting structures may considerably distort the fibre thereby affecting the guidance of the light wave. This can be avoided by choosing a sufficiently thick cladding. Further, in a fibre bundle, in the absence of the cladding, light can leak from one fibre to another.[†]

[†] This leakage is due to the fact that when a wave undergoes total internal reflection, it actually penetrates a small region of the rarer medium. The wave in the rarer medium is known as the evanescent wave which can couple light from one fibre to another. Thus in the absence of the cladding, light may leak away to an adjacent fibre.

13.3   The numerical aperture

We return to Fig. 13.3(*a*) and consider a ray which is incident on the entrance aperture of the fibre making an angle *i* with the axis. Let the refracted ray make an angle θ with the axis. Assuming the outside medium to have a refractive index n_0 (which for most practical cases is unity), we get

$$\frac{\sin i}{\sin \theta} = \frac{n_1}{n_0} \tag{13.2}$$

Obviously if this ray has to suffer total internal reflection at the core–cladding interface,

$$\sin \phi (= \cos \theta) > n_2/n_1$$

Thus

$$\sin \theta < [1 - (n_2/n_1)^2]^{\frac{1}{2}}$$

and we must have

$$\sin i < \frac{n_1}{n_0} \left[1 - \left(\frac{n_2}{n_1} \right)^2 \right]^{\frac{1}{2}} = \left(\frac{n_1^2 - n_2^2}{n_0^2} \right)^{\frac{1}{2}}$$

If $(n_1^2 - n_2^2) \geqslant n_0^2$ then for all values of *i*, total internal reflection will occur. Assuming $n_0 = 1$, the maximum value of $\sin i$ for a ray to be guided is given by

$$\sin i_m = \begin{cases} (n_1^2 - n_2^2)^{\frac{1}{2}} & \text{when} \quad n_1^2 < n_2^2 + 1 \\ 1 & \text{when} \quad n_1^2 > n_2^2 + 1 \end{cases} \tag{13.3}$$

Thus, if a cone of light is incident on one end of the fibre, it will be guided through it provided the semiangle of the cone is less than i_m. This angle is a measure of the light gathering power of the fibre and as such, one defines the numerical aperture (NA) of the fibre by the following equation

$$\text{NA} = (n_1^2 - n_2^2)^{\frac{1}{2}} = n_1(2\Delta)^{\frac{1}{2}} \tag{13.4}$$

where

$$\Delta = \frac{n_1^2 - n_2^2}{2n_1^2} \approx \frac{n_1 - n_2}{n_1} \tag{13.5}$$

and in the last step we have assumed $n_1 \approx n_2$ which is indeed the case for all practical optical fibres.

For a typical optical fibre $n_2 = 1.458$, $\Delta = 0.01$ and the corresponding NA is ≈ 0.2. Thus the fibre would accept light incident over a cone with a semiangle $\sin^{-1}(0.2) \approx 11.5°$ about the axis.

13.4 Pulse dispersion in step index fibres

As already shown in Fig. 13.3 the simplest type of optical fibre consists of a thin cylindrical structure of transparent glassy material of uniform refractive index n_1 surrounded by a cladding of another material of uniform but slightly lower refractive index n_2. These fibres are referred to as *step index fibres* due to the step discontinuity of the index profile at the core–cladding interface.

In digital communication systems, information to be sent is first coded in the form of pulses and then these pulses of light are transmitted from the transmitter to the receiver where the information is decoded. The larger the number of pulses that can be sent per unit time and still be resolvable at the receiver end, the larger is the transmission capacity of the system. A pulse of light sent into a fibre broadens in time as it propagates through the fibre; this phenomenon is known as *pulse dispersion* and happens because of the different times taken by different rays to propagate through the fibre; for example, in the fibre shown in Fig. 13.3 the rays making larger angles with the axis have to traverse a longer optical path length and they take a longer time to reach the output end. Consequently, the pulse broadens as it propagates through the fibre (see Fig. 13.4). Hence, even though two pulses may be well resolved at the input end, because of broadening of the pulses they may not be so at the output end. Where the output pulses are not resolvable, no information can be retrieved. *Thus the smaller the pulse dispersion, the greater will be the information-carrying capacity of the system.*

We next calculate the amount of dispersion in a step index fibre. Referring to Fig. 13.3, for a ray making an angle θ with the axis, the distance AB is traversed in time

$$t = \frac{AC + CB}{c/n_1} = \frac{n_1(AB)}{c \cos \theta}$$

where c/n_1 represents the speed of light in a medium of refractive index n_1, c being the speed of light in free space. Since the ray path will repeat itself, the time taken by a ray to traverse a length L of the fibre will be

$$t = n_1 L / c \cos \theta \tag{13.6}$$

The above expression shows that the time taken by a ray is a function of the angle θ made by the ray with the z-axis which leads to pulse dispersion. If we assume that all rays between 0 and θ_c are present then the time taken by rays corresponding respectively to $\theta = 0$ and $\theta = \theta_c = \cos^{-1}(n_2/n_1)$ will be given by

$$t_{max} = n_1 L/c \tag{13.7}$$

$$t_{min} = L n_1^2/c n_2 \tag{13.8}$$

Hence if all the input rays were excited simultaneously, the rays will occupy a time interval at the output end of duration

$$\Delta\tau = t_{max} - t_{min} = \frac{n_1 L}{c}\left(\frac{n_1}{n_2} - 1\right) \tag{13.9}$$

Fig. 13.4 A series of pulses each of width τ_1 (at the input and of the fibre) after transmission through the fibre emerges as a series of pulses of width $\tau_2 (> \tau_1)$. If the broadening of the pulses is large, then adjacent pulses will overlap at the output end and may not be resolvable. Thus, pulse broadening determines the minimum separation between adjacent pulses, which, in turn, determines the maximum information-carrying capacity of the optical fibre. For example, in a 10 Mbits/s system, the pulses would be separated by 10^{-7} s and for the pulses to be resolvable at the end of the fibre, the pulse dispersion should not be greater than 10^{-8} s.

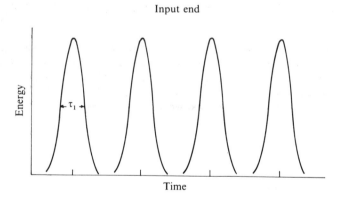

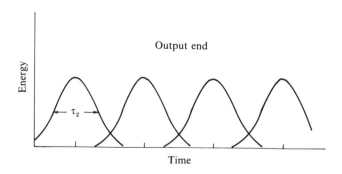

For a typical fibre, if we assume

$$n_1 = 1.46, \quad (n_1 - n_2)/n_1 = 0.01, \quad L = 1 \, \text{km} \tag{13.10}$$

one obtains

$$\Delta\tau \approx 50 \times 10^{-9} \, \text{s/km} = 50 \, \text{ns/km} \tag{13.11}$$

i.e., an impulse after traversing through a fibre of length 1 km broadens to a pulse of duration of about 50 ns. Thus two pulses separated by say 500 ns at the input end would be quite resolvable at the output end of 1 km of the fibre; however, if consecutive pulses are separated by, say 10 ns at the input end, they would be absolutely unresolvable at the output end. Hence in 1 Mbit/s fibre optic system, where we have one pulse every 10^{-6} s, 50 ns/km dispersion would require repeaters to be placed every 3–4 km. On the other hand, in a 1000 Mbit/s fibre optic communication system, where we require the transmission of one pulse every 10^{-9} s, a dispersion of 50 ns/km would result in intolerable broadening even within 50 m or so which would be highly inefficient and uneconomical from a system point of view.

From the above discussion it follows that for a very high information-carrying system, it is necessary to reduce the pulse dispersion; two alternative

Fig. 13.5 (*a*) A parabolic decrease in the refractive index from a maximum at the fibre axis to a constant value at the cladding. (*b*) Typical meridional ray paths in a parabolic index fibre. Because of the periodic focussing and defocussing of the beam, such fibres are often called SELFOC fibres; the word SELFOC is a contraction of the two words self focussing.

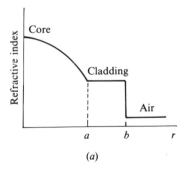

(*a*)

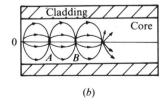

(*b*)

solutions exist: one involving the use of graded index fibres and the other involving single mode fibres.

13.4.1 Multimode graded index fibres

In contrast to a uniform refractive index core, in a graded refractive index fibre, the refractive index in the core decreases continuously in a nearly parabolic manner from a maximum value at the centre of the core to a constant value at the core–cladding interface (see Fig. 13.5). Due to the gradual decrease in refractive index as one moves away from the centre of the core, a ray that enters the fibre is continuously bent towards the axis of the fibre (see Fig. 13.5(*b*)). This follows from Snell's law since as the ray proceeds away from the axis, it continuously encounters a medium of lower refractive index and hence bends away from the normal i.e., it bends towards the axis of the fibre. Such a graded index fibre reduces the transit time of rays travelling obliquely by providing a larger velocity (due to a lower refractive index) which partially compensates the relatively large optical path length that the ray has to traverse. Dispersion due to the differences in transit time delays of different rays in such a fibre can be shown to be extremely low,[†] of the order of 0.05 ns/km, which is almost 1000 times better than an equivalent step index fibre. Thus such graded refractive index profiled fibres would provide an extremely large information carrying capacity. In Sec. 13.9 we will discuss in greater detail the small dispersion of graded index fibres.

13.4.2 Single mode fibres

In the above discussions we have considered the light propagation inside the fibre as a multitude of rays bouncing back and forth at the core–cladding interface. Such fibres which permit a large number of guided optical ray paths are known as multimode fibres. However, it can be shown that if the diameter of the fibre shrinks or if the refractive index difference between the core and the cladding decreases, the number of possible paths for waveguidance reduces. We may recall that while considering planar waveguides, we have explicitly shown that the results obtained by solving Maxwell's equations can be interpreted by assuming that rays can propagate at specific discrete angles from the *z*-axis. Each discrete angle corresponds to

[†] When the refractive index decreases in a parabolic manner (from the axis of the core) the ray paths are sinusoidal and in the paraxial approximation they have the same period as has been shown in Fig. 13.5 (see e.g., Ghatak and Thyagarajan (1978)). Since between the two points *A* and *B* (see Fig. 13.5), an infinitesimal variation of the ray path is also a possible ray path, Fermat's principle tells us that all the (paraxial) rays will take the same time.

a specific mode of the waveguide (see Sec. 11.3.2). For a step index fibre (see Eq. (13.1)); we will show in Sec. 13.6 that if the following quantity (known as the waveguide parameter)

$$V = \frac{2\pi}{\lambda_0} a (n_1^2 - n_2^2)^{\frac{1}{2}} = \frac{2\pi}{\lambda_0} a n_1 (2\Delta)^{\frac{1}{2}} \tag{13.12}$$

is less than 2.4048 then only one guided mode (implying there is only one value of θ) is possible and the fibre is known as a single mode fibre. Practical single mode fibres have Δ varying from 0.2% to 0.5% and typical core diameters in the range 10–6 μm.

It is obvious now that since only a single ray path is possible in a single mode fibre, the dispersion, which is caused by the differences in the velocity of the different rays in a multimode fibre, will be completely absent and hence

Fig. 13.6 Typical dimensions and refractive index profiles of different types of optical fibres.

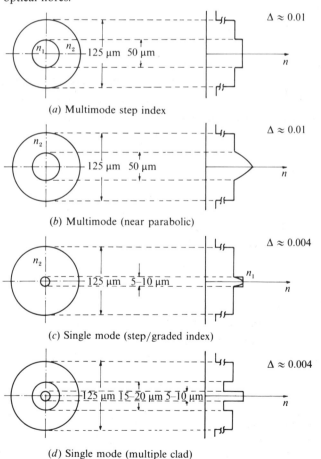

(a) Multimode step index

(b) Multimode (near parabolic)

(c) Single mode (step/graded index)

(d) Single mode (multiple clad)

the information transmission capacity of such a single mode fibre must be much larger than a multimode fibre. However, even in the absence of this source of temporal dispersion, the dispersion in a single mode fibre is still limited due to what are known as material and waveguide dispersions which will be discussed in later sections. Typically in a single mode fibre the dispersion is of the order of a few picoseconds per kilometre and thus single mode fibres are obviously more suitable for uses in which there is a large flow of information.

Fig. 13.6 shows typical dimensions and refractive index profiles of different types of optical fibres which have been and are being used in optical communication systems.

13.5   Scalar wave equation and the modes of a fibre

In Sec. 11.9 we have shown that for an inhomogeneous medium the electric field $\mathscr{E}$ satisfies the equation

$$\nabla^2 \mathscr{E} + \nabla\left(\frac{\nabla n^2}{n^2}\cdot\mathscr{E}\right) - \epsilon_0\mu_0 n^2\frac{\partial^2\mathscr{E}}{\partial t^2} = 0 \tag{13.13}$$

and a similar equation holds for the magnetic field $\mathscr{H}$ (see Eq. (11.137)). The above equation tells us that the different components of the electric field are coupled.

Now, for an infinitely extended homogeneous medium, the second term on the LHS is zero everywhere and each Cartesian component of the electric field satisfies the scalar wave equation:

$$\nabla^2\Psi = \epsilon_0\mu_0 n^2(\partial^2\Psi/\partial t^2) \tag{13.14}$$

The solution of the above equation can be written in the form of plane waves and using Maxwell's equations one can easily show that the waves are transverse (see Chapter 1). For a medium having weak inhomogeneity (i.e., the variation of n is small in a region $\sim \lambda$) the second term on the LHS of Eq. (13.13) can be assumed to be negligible and the waves can be assumed to be nearly transverse with the transverse component of the electric field satisfying Eq. (13.14). The above equation is obtained after neglecting the term depending on ∇n^2 in Eq. (13.13). This is known as the scalar wave approximation and in this approximation the modes have been assumed to be nearly transverse and can have an arbitrary state of polarization. Thus the two independent sets of modes can be assumed to be x-polarized and y-polarized and in the scalar approximation they have the same propagation constants. These linearly polarized modes are usually referred to as LP modes. We may compare this with the discussion in Sec. 11.5 where we mentioned that when $n_1 \approx n_2$, the modes are nearly transverse and the propagation constants of the TE and TM modes are almost equal.

For n^2 depending only on the transverse coordinates (r, ϕ), we may write

$$\Psi(r, \phi, z, t) = \psi(r, \phi)e^{i(\omega t - \beta z)} \qquad (13.15)$$

where ω is the angular frequency and β is known as the propagation constant. The above equation represents the modes of the system. Substituting in Eq. (13.14), we readily obtain

$$\left(\nabla^2 - \frac{\partial^2}{\partial z^2}\right)\psi + \left[\frac{\omega^2}{c^2}n^2(r, \phi) - \beta^2\right]\psi = 0$$

In most practical fibres n^2 depends only on the cylindrical coordinate r and therefore it is convenient to use the cylindrical system of coordinates to obtain[†]

$$\frac{\partial^2 \psi}{\partial r^2} + \frac{1}{r}\frac{\partial \psi}{\partial r} + \frac{1}{r^2}\frac{\partial^2 \psi}{\partial \phi^2} + [k_0^2 n^2(r) - \beta^2]\psi = 0 \qquad (13.16)$$

where

$$k_0 = \omega/c = 2\pi/\lambda_0 \qquad (13.17)$$

is the free space wave number. Since the medium has cylindrical symmetry, we can solve Eq. (13.16) by the method of separation of variables:

$$\psi(r, \phi) = R(r)\Phi(\phi) \qquad (13.18)$$

On substitution and dividing by ψ, we obtain

$$\frac{r^2}{R}\left(\frac{d^2 R}{dr^2} + \frac{1}{r}\frac{dR}{dr}\right) + r^2[n^2(r)k_0^2 - \beta^2] = -\frac{1}{\Phi}\frac{d^2\Phi}{d\phi^2} = +l^2 \qquad (13.19)$$

where l is a constant. The ϕ dependence will be of the form $\cos l\phi$ or $\sin l\phi$ and for the function to be single valued (i.e., $\Phi(\phi + 2\pi) = \Phi(\phi)$) we must have

$$l = 0, 1, 2, \ldots, \text{etc.} \qquad (13.20)$$

(Negative values of l correspond to the same field distribution.) Since for each value of l, there can be two independent states of polarization, modes with $l \geqslant 1$ are four-fold degenerate; modes with $l = 0$ are ϕ independent and have two fold degeneracy. The radial part of the equation gives us

$$r^2\frac{d^2 R}{dr^2} + r\frac{dR}{dr} + \{[n^2(r)k_0^2 - \beta^2]r^2 - l^2\}R = 0 \qquad (13.21)$$

[†] We should mention here that for an infinitely extended parabolic profile $n^2(r) = n_1^2 - \alpha(x^2 + y^2)$ it is equally convenient to use Cartesian coordinates (see Sec. 13.7 and Problem 13.9).

The solution of the above equation for a step index profile will be given in Sec. 13.6. However, we can make some general comments about the solutions of Eq. (13.21) for an arbitrary cylindrically symmetric profile having a refractive index which decreases monotonically from a value n_1 on the axis to a constant value n_2 beyond the core–cladding interface $r = a$. The solutions of Eq. (13.21) can be divided into two distinct classes (cf. Sec. 11.3):

(a) $\quad k_0^2 n_1^2 > \beta^2 > k_0^2 n_2^2$ $\qquad\qquad\qquad\qquad\qquad$ (13.22)

For β^2 lying in the above range, the fields $R(r)$ are oscillatory in the core and decay in the cladding and β^2 assumes only discrete values; these are known as the *guided modes* of the system. For a given value of l, there will be several guided modes which are designated LP_{lm} modes ($m = 1, 2, 3, \ldots$); LP stands for linearly polarized.[†] Further, since the modes are solutions of the scalar wave equation, they can be assumed to satisfy the orthonormality condition

$$\int_0^\infty \int_0^{2\pi} \psi_{lm}^*(r, \phi)\psi_{l'm'}(r, \phi)r\,dr\,d\phi = \delta_{ll'}\delta_{mm'}$$ (13.23)

(b) $\quad \beta^2 < k_0^2 n_2^2$ $\qquad\qquad\qquad\qquad\qquad\qquad\qquad$ (13.24)

For such β values, the fields are oscillatory even in the cladding and β can assume a continuum of values. These are known as the *radiation modes*.

The guided and radiation modes form a complete set of modes in the sense that an arbitrary field distribution can be expanded in terms of these modes, i.e.,

$$\psi(x, y, z) = \sum_\nu a_\nu \psi_\nu(x, y)e^{-i\beta_\nu z} + \int a(\beta)\psi(\beta, x, y)e^{-i\beta z}\,d\beta$$ (13.25)

where the first term represents a sum over discrete modes and the second term an integral over the continuum of modes[‡]. The quantity $|a_\nu|^2$ is proportional to the power carried by the νth mode; the constants a_ν can be determined by knowing the incident field at $z = 0$ and using the orthonormality condition.

[†] If one solves the vector wave equation, the modes are classified as HE_{lm}, EH_{lm}, TE_{0m} and TM_{0m} modes, the correspondence is $\mathrm{LP}_{0m} = \mathrm{HE}_{1m}$; $\mathrm{LP}_{1m} = \mathrm{HE}_{2m}$, TM_{0m} and $\mathrm{LP}_{lm} = \mathrm{HE}_{l+1, m}$, $\mathrm{EH}_{l-1, m}(l \geqslant 2)$. (See Sec. 13.4)

[‡] The radiation modes can be further classified into propagating radiation modes $0 < \beta^2 < k_0^2 n_2^2$ and decaying radiation modes $-\infty < \beta^2 < 0$; the latter correspond to modes which decay exponentially along the z-axis.

The calculation of the modal field distributions and the corresponding propagation constants are of extreme importance in the study of waveguides. For example, knowing the frequency dependence of the propagation constant one can calculate the temporal broadening of a pulse (see Sec. 13.8) which determines the information-carrying capacity. Knowledge of the modal field distribution is essential for the calculation of excitation efficiencies, splice losses at joints and in the development of new fibre optic devices like directional couplers etc. We will now present a detailed modal analysis for step index and parabolic refractive index distributions.

13.6 Modal analysis for a step index fibre

In this section, we will obtain the modal fields and the corresponding propagation constants for a step index fibre for which the refractive index variation is given by Eq. (13.1). For such a fibre it is possible to obtain rigorous solutions of the vector equations (see, e.g., Sodha and Ghatak (1977)). However, most practical fibres used in communication are weakly guiding i.e., relative refractive index difference $(n_1 - n_2)/n_1 \ll 1$ and in such a case the radial part of the transverse component of the electric field satisfies the following equation (see Eq. (13.21)):

$$r^2 \frac{d^2 R}{dr^2} + r \frac{dR}{dr} + \{ [k_0^2 n^2(r) - \beta^2] r^2 - l^2 \} R = 0 \tag{13.26}$$

and the complete transverse field is given by

$$\Psi(r, \phi, z, t) = R(r) e^{i(\omega t - \beta z)} \begin{Bmatrix} \cos l\phi \\ \sin l\phi \end{Bmatrix} \tag{13.27}$$

If we substitute in Eq. (13.26) for $n^2(r)$, we obtain

$$r^2 \frac{d^2 R}{dr^2} + r \frac{dR}{dr} + \left(U^2 \frac{r^2}{a^2} - l^2 \right) R = 0; \quad r < a \tag{13.28}$$

and

$$r^2 \frac{d^2 R}{dr^2} + r \frac{dR}{dr} - \left(W^2 \frac{r^2}{a^2} + l^2 \right) R = 0; \quad r > a \tag{13.29}$$

where

$$U = a(k_0^2 n_1^2 - \beta^2)^{\frac{1}{2}} \tag{13.30}$$

$$W = a(\beta^2 - k_0^2 n_2^2)^{\frac{1}{2}} \tag{13.31}$$

and the normalized waveguide parameter V is defined by

$$V = (U^2 + W^2)^{\frac{1}{2}} = k_0 a(n_1^2 - n_2^2)^{\frac{1}{2}} \tag{13.32}$$

Guided modes correspond to $n_2^2 k_0^2 < \beta^2 < n_1^2 k_0^2$ and therefore for guided modes both U and W are real.

Eq. (13.28) and (13.29) are of the standard form of Bessel's equation (see, e.g., Irving and Mullineux (1959)). The solutions of Eq. (13.28) are $J_l(x)$ and $Y_l(x)$ where $x = Ur/a$. The solution $Y_l(x)$ has to be rejected since it diverges as $x \to 0$. The solutions of Eq. (13.29) are the modified Bessel functions $K_l(\tilde{x})$ and $I_l(\tilde{x})$ with the asymptotic forms

$$K_l(\tilde{x}) \xrightarrow[\tilde{x} \to \infty]{} \left(\frac{\pi}{2\tilde{x}} \right)^{\frac{1}{2}} e^{-\tilde{x}} \tag{13.33}$$

$$I_l(\tilde{x}) \xrightarrow[\tilde{x} \to \infty]{} \frac{1}{(2\pi\tilde{x})^{\frac{1}{2}}} e^{\tilde{x}} \tag{13.34}$$

where $\tilde{x} = Wr/a$. Obviously the solution $I_l(\tilde{x})$ which diverges as $\tilde{x} \to \infty$ has to be rejected. Thus the transverse dependence of the modal field is given by

$$\psi(r, \phi) = \begin{cases} \dfrac{A}{J_l(U)} J_l\left(\dfrac{Ur}{a} \right) \begin{bmatrix} \cos l\phi \\ \sin l\phi \end{bmatrix}; & r < a \\[4mm] \dfrac{A}{K_l(W)} K_l\left(\dfrac{Wr}{a} \right) \begin{bmatrix} \cos l\phi \\ \sin l\phi \end{bmatrix}; & r > a \end{cases} \tag{13.35}$$

where we have assumed the continuity of ψ at the core–cladding interface. Continuity of $\partial\psi/\partial r$ at $r = a$ leads to[†]

$$\frac{U J_l'(U)}{J_l(U)} = \frac{W K_l'(W)}{K_l(W)} \tag{13.36}$$

Using the identities

$$\pm U J_l'(U) = l J_l(U) - U J_{l\pm1}(U) \tag{13.37}$$

$$\pm W K_l'(W) = l K_l(W) \mp W K_{l\pm1}(W) \tag{13.38}$$

$$J_{l+1}(U) = (2l/U)J_l(U) - J_{l-1}(U) \tag{13.39}$$

and

$$K_{l+1}(W) = (2l/W)K_l(W) + K_{l-1}(W) \tag{13.40}$$

Eq. (13.36) can be written in either of the following two forms

$$U \frac{J_{l+1}(U)}{J_l(U)} = W \frac{K_{l+1}(W)}{K_l(W)} \tag{13.41}$$

[†] It should be mentioned that as long as ψ is assumed to satisfy the scalar wave equation (Eq. (13.16)), both ψ and $d\psi/dr$ have to be continuous at any refractive index discontinuity. This follows from the fact that if $d\psi/dr$ does not happen to be continuous then $d^2\psi/dr^2$ will be a Dirac delta function which would therefore be inconsistent with Eq. (13.16) unless, of course, there is an infinite discontinuity in n^2 which indeed happens at the interface between a dielectric and a perfect conductor.

or

$$U\frac{J_{l-1}(U)}{J_l(U)} = -W\frac{K_{l-1}(W)}{K_l(W)} \tag{13.42}$$

However using the proper limiting forms of $K_l(W)$ as $W \to 0$, one can show that[†]

$$\underset{W \to 0}{\mathrm{Lt}}\ W\frac{K_{l-1}(W)}{K_l(W)} \to 0; \quad l = 0, 1, 2, \ldots \tag{13.43}$$

and therefore we use Eq. (13.42) for studying the modes. For $l = 0$ we get

$$U\frac{J_1(U)}{J_0(U)} = W\frac{K_1(W)}{K_0(W)} \tag{13.44}$$

where we have used the relations $J_{-1}(U) = -J_1(U)$ and $K_{-1}(W) = K_1(W)$.

We should mention here that the boundary conditions used in deriving the eigenvalue equation (Eq. (13.42)) are consistent with the approximation involved in using the scalar wave equation. For example, if ψ is assumed to represent E_y then, rigorously speaking, E_y and $\partial E_y/\partial r$ are *not* continuous at $r = a$ for all ϕ; indeed, one must make E_ϕ, E_z and $n^2 E_r$ continuous at the interface $r = a$. However, if $n_1 \approx n_2$ the error involved is negligible (see, e.g., Sodha and Ghatak (1977)).

Since $V^2 = U^2 + W^2$ the solution of the transcendental equation (for given values of l and V) will give us universal curves describing the dependence of U or W on V. It is, however, more convenient to define the normalized propagation constant

$$b = \frac{\beta^2/k_0^2 - n_2^2}{n_1^2 - n_2^2} = \frac{W^2}{V^2} \tag{13.45}$$

using which Eqs. (13.42) and (13.44) reduce to

$$V(1-b)^{\frac{1}{2}}\frac{J_{l-1}(V(1-b)^{\frac{1}{2}})}{J_l(V(1-b)^{\frac{1}{2}})} = -Vb^{\frac{1}{2}}\frac{K_{l-1}(Vb^{\frac{1}{2}})}{K_l(Vb^{\frac{1}{2}})}; \quad l \geq 1 \tag{13.46}$$

and

$$V(1-b)^{\frac{1}{2}}\frac{J_1(V(1-b)^{\frac{1}{2}})}{J_0(V(1-b)^{\frac{1}{2}})} = Vb^{\frac{1}{2}}\frac{K_1(Vb^{\frac{1}{2}})}{K_0(Vb^{\frac{1}{2}})}; \quad l = 0 \tag{13.47}$$

[†] The limiting forms are

$$K_0(W) \xrightarrow[W \to 0]{} -\ln(W/2)$$

and

$$K_l(W) \xrightarrow[W \to 0]{} \tfrac{1}{2}\Gamma(l)(2/W)^l; \quad l > 0$$

Table 13.1. *Cutoff frequencies of various LP_{lm} modes in a step index fibre*

$l = 0$ modes	$(J_1(V_c) = 0)$	$l = 1$ modes	$(J_0(V_c) = 0)$
Mode	V_c	Mode	V_c
LP_{01}	0	LP_{11}	2.4048
LP_{02}	3.8317	LP_{12}	5.5201
LP_{03}	7.0156	LP_{13}	8.6537
LP_{04}	10.1735	LP_{14}	11.7915

$l = 2$ modes	$(J_1(V_c) = 0; \ V_c \neq 0)$	$l = 3$ modes	$(J_2(V_c) = 0; \ V_c \neq 0)$
Mode	V_c	Mode	V_c
LP_{21}	3.8317	LP_{31}	5.1356
LP_{22}	7.0156	LP_{32}	8.4172
LP_{23}	10.1735	LP_{33}	11.6198
LP_{24}	13.3237	LP_{34}	14.7960

respectively. Since for guided modes $k_0^2 n_2^2 < \beta^2 < k_0^2 n_1^2$, we must have $0 < b < 1$. For a given value of l, there will be a finite number of solutions and the mth solution ($m = 1, 2, 3, \ldots$) is referred to as the LP_{lm} mode. In Fig. 13.7 we have plotted the LHS and RHS of the above equations for the $l = 0$ and $l = 1$ modes corresponding to $V = 8$. The points of intersection give the allowed values of b for different modes. By studying the zeroes of Bessel functions (see

Fig. 13.7 Variation of the LHS (solid curves) and the RHS (dashed curves) of Eq. (13.46) for (a) $l = 0$ and (b) $l = 1$ for $V = 8$. The points of intersection represent the discrete modes of the waveguide.

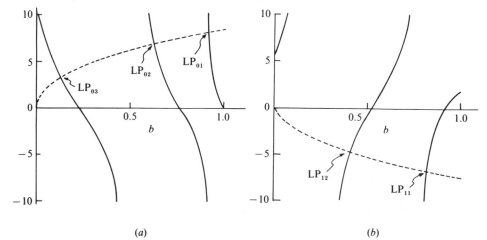

(a)   (b)

Table 13.1) one can immediately see that since $0 < b < 1$, there will be only a finite number of guided modes. The guided modes which are given by the points of intersection in Figs. 13.7(a) and (b) are designated in decreasing values of b as LP_{11}, LP_{12}, LP_{13} etc. The variation of b with V forms a set of universal curves, which are plotted in Fig. 13.8. As can be seen, at a particular V value there are only a finite number of modes.

The condition $b = 0$ (i.e., $\beta^2 = k_0^2 n_2^2$) corresponds to what is known as the *cutoff* of the mode. For $b < 0$, $\beta^2 < k_0^2 n_2^2$ and the fields are oscillatory even in the cladding and we have what are known as radiation modes. Obviously at cut off $\beta = k_0 n_2$ implying

$$b = 0, \quad W = 0, \quad U = V = V_c$$

The cutoffs of various modes are determined from the following equations

$$
\begin{aligned}
l = 0 \quad &\text{modes:} \quad J_1(V_c) = 0 \\
l = 1 \quad &\text{modes:} \quad J_0(V_c) = 0 \\
l \geqslant 2 \quad &\text{modes:} \quad J_{l-1}(V_c) = 0; \quad V_c \neq 0
\end{aligned}
\tag{13.48}
$$

It must be noted that for $l \geqslant 2$, the root $V_c = 0$ must not be included since

$$\underset{V \to 0}{\text{Lt}} \, V \frac{J_{l-1}(V)}{J_l(V)} \neq 0 \quad \text{for } l \geqslant 2 \tag{13.49}$$

Fig. 13.8 Variation of the normalized propagation constant b with normalized frequency V for a step index fibre corresponding to some low order modes. The cutoff frequencies of LP_{2m} and $LP_{0,m+1}$ modes are the same. (Adapted from Gloge (1971).)

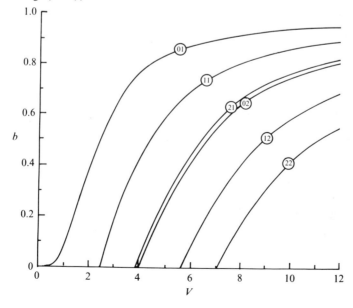

Thus the cutoff V values (also known as normalized cutoff frequencies) occur at the zeroes of Bessel functions and are tabulated in Table 13.1.

As is obvious from the above analysis and also from Fig. 13.8 for a step index fibre with

$$0 < V < 2.4048 \tag{13.50}$$

we will have only one guided mode namely the LP_{01} mode. Such a fibre is referred to as a single mode fibre and is of tremendous importance in optical communication systems. As will be discussed later, the dispersion curves can be used to calculate the group velocity of various modes of a step index fibre.

The radial field distributions and the schematic intensity patterns of some

Fig. 13.9 Radial intensity distributions (normalized to the same power) of some low order modes in a step index fibre for $V = 8$. Notice that the higher order modes have a greater fraction of power in the cladding.

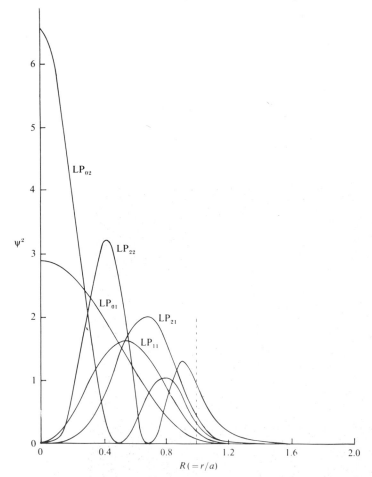

Fig. 13.10 (*a*) Schematic of the modal field patterns for some low order modes in a step index fibre. The arrows represent the direction of the electric field. (*b*) Modal intensity pattern of the $LP_{23,12}$ mode in a multimode fibre. (Photograph courtesy of Dr W. Freude.)

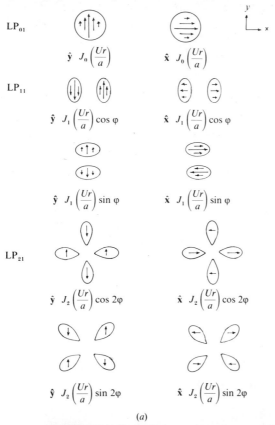

LP_{01}

$\hat{\mathbf{y}} \ J_0\left(\dfrac{Ur}{a}\right)$ $\hat{\mathbf{x}} \ J_0\left(\dfrac{Ur}{a}\right)$

LP_{11}

$\hat{\mathbf{y}} \ J_1\left(\dfrac{Ur}{a}\right)\cos\varphi$ $\hat{\mathbf{x}} \ J_1\left(\dfrac{Ur}{a}\right)\cos\varphi$

$\hat{\mathbf{y}} \ J_1\left(\dfrac{Ur}{a}\right)\sin\varphi$ $\hat{\mathbf{x}} \ J_1\left(\dfrac{Ur}{a}\right)\sin\varphi$

LP_{21}

$\hat{\mathbf{y}} \ J_2\left(\dfrac{Ur}{a}\right)\cos 2\varphi$ $\hat{\mathbf{x}} \ J_2\left(\dfrac{Ur}{a}\right)\cos 2\varphi$

$\hat{\mathbf{y}} \ J_2\left(\dfrac{Ur}{a}\right)\sin 2\varphi$ $\hat{\mathbf{x}} \ J_2\left(\dfrac{Ur}{a}\right)\sin 2\varphi$

(*a*)

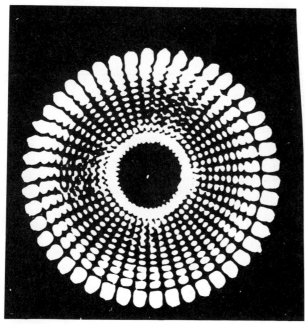

(*b*)

lower order modes are shown in Figs. 13.9 and 13.10(*a*). We may note the following points:

(*a*) The $l = 0$ modes are two fold degenerate corresponding to two independent states of polarization.

(*b*) The $l \geqslant 1$ modes are four fold degenerate because for each polarization, the ϕ dependence could be either $\cos l\phi$ or $\sin l\phi$.

Further

$$\text{number of zeroes in the } \phi \text{ direction} = 2l \qquad (13.51)$$

and

$$\text{number of zeroes in the radial direction (excluding } r = 0) = m - 1$$
$$(13.52)$$

Fig. 13.10(*b*) shows the mode field distribution of a typical higher order mode in a multimode fibre. The figure corresponds to the $LP_{23,12}$ mode; the values of l and m are obtained using Eqs. (13.51) and (13.52). In Appendix F we have shown that when $V \gg 1$, the total number of modes is given by

$$N \approx V^2/2 \qquad (13.53)$$

and such a fibre which supports a large number of guided modes is known as a multimode fibre. For a typical multimode step index fibre

$$n_1 = 1.47, \quad n_2 = 1.46, \quad a = 25\,\mu\text{m} \qquad (13.54)$$

and for $\lambda_0 = 0.8\,\mu\text{m}$, we obtain

$$V \approx 34$$

which would support approximately 580 modes.

Problem 13.1: (*a*) Consider the LP_{0m} modes in a step index fibre. Show that for $m = 1, 2, 3, \ldots$ the eigenvalues will correspond to $U < 2.405$, $U < 5.5201$, $U < 8.654, \ldots$ respectively. Thus show that along the radial direction the LP_{01} mode will have no zeroes, LP_{02} mode will have one zero, LP_{03} mode will have two zeroes etc. (Hint: one may look up the points at which the LHS of Eq. (13.47) becomes infinite – see Fig. 13.7).

(*b*) Similarly consider the LP_{2m} modes and show that (excluding the zero at $r = 0$) along the r direction the LP_{21} mode will have no zeroes, the LP_{22} mode will have one zero etc.

Problem 13.2: Consider a step index fibre with

$$n_1 = 1.461, \quad n_2 = 1.458, \quad a = 5\,\mu\text{m}$$

Show that the fibre is single moded for $\lambda_0 > 1.22\,\mu\text{m}$. Calculate the number of modes

supported by the fibre at $\lambda_0 = 0.6\,\mu m$ and label them. Compare the number of modes obtained with that predicted by Eq. (13.53).
(Answer: Total number of modes = 14).

Problem 13.3: If the fibre in the above problem is to have a cut off at $0.8\,\mu m$, what would be the corresponding core radius.
(Answer: $a \approx 3.3\,\mu m$).

Problem 13.4: Fig. 13.7 corresponds to a fibre with $V = 8$. Consider the fibre of Problem 13.2 and calculate the wavelength of operation at which $V = 8$. Using Fig. 13.7 obtain the corresponding (β/k_0) and plot the transverse field patterns.

13.6.1 *Fractional modal power in the core*

One of the important parameters associated with a fibre optic waveguide is the fractional power carried in the core. Now, the power in the core of the fibre is given by

$$P_{\text{core}} = \text{const.} \int_0^a \int_0^{2\pi} |\psi|^2 r\,dr\,d\phi$$

$$= \frac{2C}{J_l^2(U)} \int_0^a J_l^2\left(\frac{Ur}{a}\right) r\,dr \int_0^{2\pi} \cos^2 l\phi\,d\phi$$

or

$$P_{\text{core}} = C\frac{\pi a^2}{U^2}\frac{2}{J_l^2(U)} \int_0^U J_l^2(x)x\,dx$$

$$= C\pi a^2\left[1 - \frac{J_{l-1}(U)J_{l+1}(U)}{J_l^2(U)}\right] \tag{13.55}$$

where C is a constant and use has been made of the result derived in Problem 13.14. Similarly the power in the cladding is given by

$$P_{\text{clad}} = \text{const.} \int_a^\infty \int_0^{2\pi} |\psi|^2 r\,dr\,d\phi$$

$$= C\pi a^2\left[\frac{K_{l-1}(W)K_{l+1}(W)}{K_l^2(W)} - 1\right] \tag{13.56}$$

The total power is

$$P_{\text{tot}} = P_{\text{core}} + P_{\text{clad}}$$

$$= C\pi a^2\frac{V^2}{U^2}\frac{K_{l+1}(W)K_{l-1}(W)}{K_l^2(W)} \tag{13.57}$$

where use has been made of the equation

$$\frac{U^2 J_{l+1}(U) J_{l-1}(U)}{J_l^2(U)} = - W^2 \frac{K_{l+1}(W) K_{l-1}(W)}{K_l^2(W)} \tag{13.58}$$

which follows from the eigenvalue equations. The fractional power propagating in the core is thus given by

$$\eta = \frac{P_{\text{core}}}{P_{\text{tot}}} = \left[\frac{W^2}{V^2} + \frac{U^2}{V^2} \frac{K_l^2(W)}{K_{l+1}(W) K_{l-1}(W)} \right] \tag{13.59}$$

Thus as the mode approaches cutoff

$$V \to V_{\text{c}}, \quad W \to 0, \quad U \to V_{\text{c}} \tag{13.60}$$

and if we now use the limiting forms of $K_l(W)$ given in the footnote on page 379, we obtain

$$\eta \to \begin{cases} 0 & \text{for } l = 0 \text{ and } 1 \\ (l-1)/l & \text{for } l \geqslant 2 \end{cases} \tag{13.61}$$

In Fig. 13.11 we have plotted the fractional power contained in the core and in the cladding as a function of V for various modes of a step index fibre. Notice that the power associated with a particular mode is concentrated in the core for large values of V, i.e., far from cutoff.

Fig. 13.11 Variation of the fractional power contained in the cladding with V for some low order modes in a step index fibre. (Adapted from Gloge (1971).)

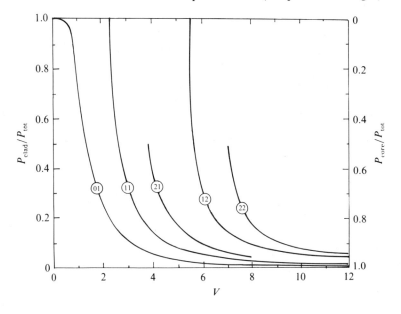

13.7 Modal analysis of a parabolic index medium

In this section we will solve the scalar wave equation (Eq. (13.14)) for a parabolic index medium characterized by the following refractive index distribution:

$$n^2 = n_1^2(1 - 2\Delta(r^2/a^2))$$

$$= n_1^2 \left[1 - \frac{2\Delta}{a^2}(x^2 + y^2) \right] \qquad (13.62)$$

where Δ and a are constants and n_1 represents the axial refractive index. A medium characterized by Eq. (13.62) is usually referred to as an infinitely extended square law medium. We should mention here that the refractive index variation given by Eq. (13.62) cannot extend to infinity and in an actual fibre there is a boundary at $r = a$ beyond which the refractive index is constant. However, the scalar wave equation can be solved exactly for an infinitely extended square law medium and such solutions give a deep physical insight and also bring out most salient aspects of optical waveguides.

If we substitute Eq. (13.62) in the scalar wave equation we obtain

$$\nabla^2 \Psi = \frac{n_1^2}{c^2} \left[1 - 2\Delta \left(\frac{x^2}{a^2} + \frac{y^2}{a^2} \right) \right] \frac{\partial^2 \Psi}{\partial t^2} \qquad (13.63)$$

We assume a solution of the form (cf. Eq. (13.15))

$$\Psi(x, y, z, t) = \psi(x, y)\, e^{i(\omega t - \beta z)} \qquad (13.64)$$

then Eq. (13.63) becomes

$$\frac{\partial^2 \psi}{\partial x^2} + \frac{\partial^2 \psi}{\partial y^2} + \left\{ k_0^2 n_1^2 \left[1 - 2\Delta \left(\frac{x^2}{a^2} + \frac{y^2}{a^2} \right) \right] - \beta^2 \right\} \psi = 0 \qquad (13.65)$$

We use the method of separation of variables and write

$$\psi(x, y) = X(x)Y(y) \qquad (13.66)$$

If we substitute the above solution in Eq. (13.65) and divide by XY we obtain

$$\left(\frac{1}{X} \frac{d^2 X}{dx^2} - k_0^2 n_1^2 \frac{2\Delta}{a^2} x^2 \right) + \left(\frac{1}{Y} \frac{d^2 Y}{dy^2} - k_0^2 n_1^2 \frac{2\Delta}{a^2} y^2 \right)$$

$$+ (k_0^2 n_1^2 - \beta^2) = 0 \qquad (13.67)$$

The variables have indeed separated out and we may write

$$\frac{1}{X} \frac{d^2 X}{dx^2} - k_0^2 n_1^2 \frac{2\Delta}{a^2} x^2 = -\gamma_1 \qquad (13.68)$$

$$\frac{1}{Y}\frac{d^2Y}{dy^2} - k_0^2 n_1^2 \frac{2\Delta}{a^2} y^2 = -\gamma_2 \tag{13.69}$$

with

$$\beta^2 = k_0^2 n_1^2 - \gamma_1 - \gamma_2 \tag{13.70}$$

We now use the variables

$$\xi = \alpha x, \quad \eta = \alpha y \tag{13.71}$$

with

$$\alpha = (k_0^2 n_1^2 2\Delta/a^2)^{\frac{1}{4}} = V^{\frac{1}{2}}/a \tag{13.72}$$

where

$$V = k_0 n_1 a (2\Delta)^{\frac{1}{2}} \tag{13.73}$$

represents the waveguide parameter. Thus

$$d^2 X/d\xi^2 + (\lambda_1 - \xi^2) X(\xi) = 0 \tag{13.74}$$

$$d^2 Y/d\eta^2 + (\lambda_2 - \eta^2) Y(\eta) = 0 \tag{13.75}$$

where

$$\lambda_1 = \gamma_1/\alpha^2 = \gamma_1 (a/k_0 n_1 (2\Delta)^{\frac{1}{2}}) \tag{13.76}$$

and

$$\lambda_2 = \gamma_2/\alpha^2 = \gamma_2 (a/k_0 n_1 (2\Delta)^{\frac{1}{2}}) \tag{13.77}$$

For bounded solutions i.e., for $X(\xi)$ and $Y(\eta)$ to tend to zero as $\xi, \eta \to \pm \infty$ we must have (see Appendix D)

$$\lambda_1 = 2m + 1; \quad m = 0, 1, 2, \dots \tag{13.78}$$

with

$$X(\xi) = X_m(\xi) = N_m H_m(\xi) e^{-\frac{1}{2}\xi^2} \tag{13.79}$$

where $H_m(\xi)$ are the Hermite polynomials,

$$H_0(\xi) = 1, \quad H_1(\xi) = 2\xi, \quad H_2(\xi) = 4\xi^2 - 2 \tag{13.80}$$

and N_m are the normalization constants given by

$$N_m = \alpha^{\frac{1}{2}}/(2^n n! \pi^{\frac{1}{2}})^{\frac{1}{2}} \tag{13.81}$$

such that

$$\int_{-\infty}^{\infty} X_m(x) X_{m'}(x) \, dx = \delta_{mm'} \tag{13.82}$$

Similarly for λ_2 and $Y(\eta)$. Using Eqs. (13.76) and (13.78) (and a similar expression for γ_2) in Eq. (13.70) we obtain the following analytic expression for the propagation constant β_{mn} of the $(m, n)^{\text{th}}$ mode.

$$\beta_{mn} = k_0 n_1 \left[1 - \frac{2(m+n+1)}{k_0 n_1 a} (2\Delta)^{\frac{1}{2}} \right]^{\frac{1}{2}}; \quad m, n = 0, 1, 2, \ldots \quad (13.83)$$

For a typical multimoded parabolic index fibre,

$$n_1 = 1.46, \quad \Delta = 0.01, \quad a = 25 \, \mu m$$

we have for $\lambda_0 \approx 0.8 \, \mu m$

$$2(2\Delta)^{\frac{1}{2}}/k_0 n_1 a \approx 10^{-3}$$

Thus for $m + n \ll 10^3$, one is justified in making a binomial expansion in Eq. (13.83) and if we retain only the first order term we will obtain

$$\beta_{mn} \approx (\omega/c) n_1 - (m + n + 1)(2\Delta)^{\frac{1}{2}}/a \quad (13.84)$$

If we neglect the wavelength dependence of n_1 and Δ, i.e., if we neglect material dispersion we get the very important result

$$d\beta_{mn}/d\omega \approx n_1/c \quad \text{independent of } m \text{ and } n \quad (13.85)$$

In Sec. 13.8 we will show that the group velocity v_g of a particular mode is given by

$$v_g = (d\beta/d\omega)^{-1} \quad (13.86)$$

and therefore Eq. (13.85) implies that different modes in a parabolic index fibre travel with almost the same group velocity. In the language of ray optics, this corresponds to the fact that all rays take approximately the same amount of time to propagate through the fibre.

We should mention here that in an actual fibre the refractive index is given by Eq. (13.62) only for $r < a$ beyond which it has a constant value equal to

$$n = n_1 (1 - 2\Delta)^{\frac{1}{2}} (= n_2) \quad \text{for} \quad (x^2 + y^2) > a^2 \quad (13.87)$$

Thus the guided mode will correspond to only those values of m and n for which

$$k_0 n_2 < \beta_{mn} < k_0 n_1 \quad (13.88)$$

The modal pattern of the $(m, n)^{\text{th}}$ mode is given by

$$\psi_{mn} = N_m N_n H_m(\xi) H_n(\eta) e^{-\frac{1}{2}(\xi^2 + \eta^2)} \quad (13.89)$$

Thus the lowest order mode which corresponds to $m = 0, n = 0$ is

$$\psi_{00} = \frac{\alpha}{\pi^{\frac{1}{2}}} e^{-\frac{1}{2}\alpha^2 r^2} = \frac{V^{\frac{1}{2}}}{a\pi^{\frac{1}{2}}} e^{-\frac{1}{2}V(r/a)^2} \quad (13.90)$$

which has a Gaussian field distribution.

In Fig. 13.12 we have given the schematic field patterns for some of the low order modes. It may be noted that these field patterns can be expressed as linear combinations of the corresponding LP_{lm} modes having the same β value (see also Problem 13.9). Notice that these field patterns are the same as those obtained in a laser (see Fig. 9.5).

13.8   Pulse dispersion

The study of the broadening of an optical pulse as it propagates through the fibre is of great importance as it determines the information-carrying capacity of the communication system.

We first consider a pure harmonic wave of frequency ω_c incident on a fibre at $z = 0$ so that the field distribution is given by

$$\Psi^{\mathrm{p}}(x, y, z = 0, t) = \phi(x, y)\,e^{i\omega_c t} \tag{13.91}$$

where the superscript p refers to the fact that we are considering pure harmonic waves and $\phi(x, y)$ represents the transverse field distribution at $z = 0$. Since the modes of the fibre form a complete set of functions, we expand $\phi(x, y)$ in terms of these modes (cf. Eq. (13.25)):

$$\phi(x, y) = \sum_{v} A_v \psi_v(x, y) \tag{13.92}$$

where $\sum$ denotes a sum over the guided modes and an integration over the continuum of radiation modes; actually the radiation modes may be neglected as they would not contribute to power at large distances from the input end. At an arbitrary value of z, the field will be given by

$$\Psi^{\mathrm{p}}(x, y, z, t) = \sum_{v} A_v \psi_v(x, y)\,e^{i(\omega_c t - \beta_v z)} \tag{13.93}$$

Fig. 13.12 Schematic of the modal field patterns for some low order modes in an infinitely extended square law medium.

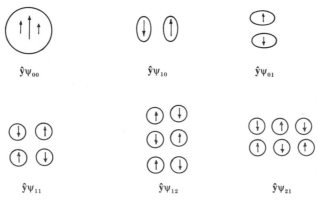

where β_ν represents the propagation constant of the νth mode. The function $e^{-i\beta_\nu z}$ is usually called the amplitude transfer function (ATF). For a temporal pulse, we assume that at $z = 0$ (which is the entrance aperture of the fibre) the field pattern can be written in the form:

$$\Psi(x, y, z = 0, t) = \phi(x, y)f(t) \qquad (13.94)$$

We Fourier analyse $f(t)$:

$$F(\omega) = \frac{1}{(2\pi)^{\frac{1}{2}}} \int_{-\infty}^{\infty} f(t) e^{-i\omega t} \, dt \qquad (13.95)$$

Thus

$$f(t) = \frac{1}{(2\pi)^{\frac{1}{2}}} \int_{-\infty}^{\infty} F(\omega) e^{i\omega t} \, d\omega \qquad (13.96)$$

which implies that $f(t)$ can be written as a superposition of harmonic waves. Hence

$$\Psi(x, y, z = 0, t) = \frac{1}{(2\pi)^{\frac{1}{2}}} \int_{-\infty}^{\infty} \phi(x, y)F(\omega) e^{i\omega t} \, d\omega \qquad (13.97)$$

Using Eq. (13.92), we obtain

$$\Psi(x, y, z = 0, t) = \frac{1}{(2\pi)^{\frac{1}{2}}} \sum_\nu \int_{-\infty}^{\infty} A_\nu \psi_\nu(x, y)F(\omega) e^{i\omega t} \, d\omega \qquad (13.98)$$

Since each frequency component propagates independently, the field at an arbitrary value z is given by

$$\Psi(x, y, z, t) = \frac{1}{(2\pi)^{\frac{1}{2}}} \sum_\nu \int_{-\infty}^{\infty} A_\nu \psi_\nu(x, y)F(\omega) e^{i[\omega t - \beta_\nu(\omega)z]} \, d\omega \qquad (13.99)$$

In all practical cases, the function $F(\omega)$ is extremely sharply peaked around the carrier frequency ω_c. For example, for a 1 ns pulse at $\lambda_0 = 0.8 \, \mu m$ (i.e., $\omega_c \approx 2.4 \times 10^{15} \, s^{-1}$),

$$\Delta\omega/\omega_c \approx 4 \times 10^{-7}$$

where $\Delta\omega$ represents the width of the frequency spectrum $F(\omega)$. Thus the integral in Eq. (13.99) is only over a region $\Delta\omega$ around ω_c and in this narrow frequency domain we may assume A_ν and ψ_ν to be constants and make a Taylor series expansion of β_ν around $\omega = \omega_c$:

$$\beta_\nu(\omega) = \beta_\nu(\omega_c) + \alpha_\nu(\omega - \omega_c) + \tfrac{1}{2}\gamma_\nu(\omega - \omega_c)^2 + \cdots \qquad (13.100)$$

where

$$\alpha_\nu = \frac{d\beta_\nu}{d\omega}\bigg|_{\omega = \omega_c}, \quad \gamma_\nu = \frac{d^2\beta_\nu}{d\omega^2}\bigg|_{\omega = \omega_c} \qquad (13.101)$$

In Eq. (13.100) if we retain only the linear term in $\omega - \omega_c$ and substitute in Eq. (13.99) we will obtain

$$\Psi(x, y, z, t) = \sum_\nu A_\nu \psi_\nu(x, y)\theta_\nu(t) \tag{13.102}$$

where

$$\theta_\nu(t) = \frac{1}{(2\pi)^{\frac{1}{2}}} \int_{-\infty}^{\infty} F(\omega) e^{i[\omega t - \beta_\nu(\omega)z]} d\omega$$

$$\approx \frac{1}{(2\pi)^{\frac{1}{2}}} e^{-i[\beta_\nu(\omega_c) - \omega_c \alpha_\nu]z} \int_{-\infty}^{\infty} F(\omega) e^{i\omega(t - \alpha_\nu z)} d\omega$$

$$= f(t - \alpha_\nu z) e^{-i[\beta_\nu(\omega_c) - \omega_c \alpha_\nu]z} \tag{13.103}$$

represents the time dependence of the output pulse corresponding to the νth mode. From Eq. (13.103) we see (retaining terms up to first order in $\omega - \omega_c$) that except for a phase factor the temporal dependence of the mode remains undistorted and the pulse propagates with the velocity

$$v_\nu = 1/\alpha_\nu = (d\beta_\nu/d\omega)^{-1} \tag{13.104}$$

which is known as the group velocity of the mode. Thus the time taken by the νth mode to traverse a length L of the fibre is given by

$$t_\nu = L/v_\nu = L(d\beta_\nu/d\omega) = -L(\lambda_0^2/2\pi c)(d\beta_\nu/d\lambda_0) \tag{13.105}$$

where $\lambda_0 (= 2\pi c/\omega)$ is the free space wavelength and we have used the relation

$$d\lambda_0/d\omega = -\lambda_0/\omega = -\lambda_0^2/2\pi c \tag{13.106}$$

The group velocities of different modes are, in general, different; thus, if all the guided modes are excited at $z = 0$ (the input end of the fibre) then different modes will take different times to reach the output end of the fibre leading to what is known as *intermodal dispersion*. In the language of ray optics this implies that different rays, in general, take different amounts of time to propagate through a certain length of the fibre.

Now, if the source is characterized by a spectral width $\Delta\lambda_0$, each wavelength component takes, in general, different times to traverse the length of the fibre and thus, even if only one mode is excited, it will undergo broadening and the pulse spread will be given by

$$\Delta t = \left(\frac{dt}{d\lambda_0}\right)\Delta\lambda_0 = -\frac{L\Delta\lambda_0}{2\pi c}\left[2\lambda_0\left(\frac{d\beta}{d\lambda_0}\right) + \lambda_0^2\left(\frac{d^2\beta}{d\lambda_0^2}\right)\right] \tag{13.107}$$

where, for the sake of convenience, we have dropped the subscript ν. The broadening of a particular mode due to the finite spectral width of the source is known as *intramodal dispersion*. For highly multimoded step index fibres the intermodal dispersion is large and intramodal dispersion can be neglected. However, as will be shown in Sec. 13.9, for multimode graded index fibres with nearly parabolic refractive index variation, the intermodal dispersion is small and intramodal dispersion becomes important. Finally, in single mode fibres, the dispersion has only the intramodal component.

Intramodal dispersion is due to material and waveguide dispersion. Material dispersion is due to the dependence of the refractive index of the fibre material on wavelength. In order to evaluate the effect of material dispersion we consider a well guided mode and assume

$$\beta \approx (2\pi/\lambda_0)n_1 \tag{13.108}$$

i.e., we consider a wave propagating in an infinitely extended homogeneous medium of refractive index n_1. Thus

$$\frac{d\beta}{d\lambda_0} = 2\pi\left(-\frac{n_1}{\lambda_0^2} + \frac{1}{\lambda_0}\frac{dn_1}{d\lambda_0}\right) = -\frac{2\pi}{\lambda_0^2}N_1 \tag{13.109}$$

where

$$N_1 = n_1 - \lambda_0\, dn_1/d\lambda_0 \tag{13.110}$$

is called the group index because the group velocity of a wave in an infinitely extended homogeneous medium of refractive index n_1 is

$$v = \left(\frac{d\beta}{d\omega}\right)^{-1} = -\frac{2\pi c}{\lambda_0^2}\left(\frac{d\beta}{d\lambda_0}\right)^{-1} = c/N_1 \tag{13.111}$$

Similarly we may calculate $d^2\beta/d\lambda_0^2$ and on substitution in Eq. (13.107) we obtain

$$\Delta t_m = -\left(\frac{L}{c}\right)\lambda_0^2\left(\frac{\Delta\lambda_0}{\lambda_0}\right)\frac{d^2n_1}{d\lambda_0^2} \tag{13.112}$$

showing that the material dispersion is proportional to $d^2n_1/d\lambda_0^2$ and also to the spectral width $\Delta\lambda_0$ of the source; the subscript m refers to material dispersion. The quantity that is usually calculated is

$$\tau_m = \frac{\Delta t_m}{L\Delta\lambda_0} = -\frac{\lambda_0}{c}\left(\frac{d^2n_1}{d\lambda_0^2}\right)\text{ps/km nm} \tag{13.113}$$

where λ_0 is measured in nanometres and

$$c \approx 3 \times 10^{-7}\,\text{km/ps}$$

As indicated in Eq. (13.113), the dispersion is usually measured in picoseconds per kilometre length of the fibre per nanometre spectral width of the source. Now for pure silica the refractive index variation with wavelength can be very accurately described by the following relation (Paek, Peterson and Carnevale, 1981):

$$n(\lambda) = C_0 + C_1 \lambda_0^2 + C_2 \lambda_0^4 + \frac{C_3}{(\lambda_0^2 - l)} + \frac{C_4}{(\lambda_0^2 - l)^2} + \frac{C_5}{(\lambda_0^2 - l)^3} \qquad (13.114)$$

where

$$C_0 = 1.4508554, \quad C_1 = -0.0031268, \quad C_2 = -0.0000381,$$
$$C_3 = 0.0030270, \quad C_4 = -0.0000779, \quad C_5 = 0.0000018, \quad l = 0.035$$
$$(13.115)$$

and λ_0 is measured in micrometres. Using the above equation one can easily calculate $d^2n/d\lambda_0^2$ which is plotted in Fig. 13.13. One finds that for pure silica

$$\tau_m \begin{cases} \sim 85 \text{ ps/km nm at } \lambda_0 \approx 0.85 \,\mu\text{m} \\ \lesssim 0.1 \text{ ps/km nm at } \lambda_0 \approx 1.3 \,\mu\text{m} \\ \sim 20 \text{ ps/km nm at } \lambda_0 \approx 1.5 \,\mu\text{m} \end{cases} \qquad (13.116)$$

Thus, for pure silica, material dispersion passes through zero around 1.3 μm wavelength. We will use this result in subsequent sections.

Even when the refractive index is assumed to be independent of the wavelength, the group velocity of *each* mode increases with wavelength. Physically this can be understood as follows. If we refer to Fig. 13.11 we find that the fractional power in the core increases with an increase in V, i.e., for a

Fig. 13.13 Variation of $d^2n/d\lambda_0^2$ with λ_0 for pure silica.

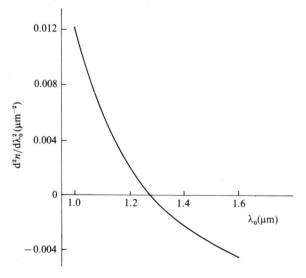

given fibre it will increase with decrease in wavelength. Thus as λ_0 decreases, i.e., as V increases, the mode 'sees' more of the core[†] and therefore the group velocity will decrease to a value closer to c/N_1. Conversely as λ_0 increases, the group velocity increases to a value closer to c/N_2. Thus different wavelength components see different effective indices which, in turn, leads to what is known as waveguide dispersion. However, for multimode fibres, waveguide despersion is much smaller than intermodal dispersion and can be neglected. Waveguide dispersion becomes important only in single mode fibres and therefore we will come back to it in Sec. 13.11.

Problem 13.5: Using Eqs. (13.114) and (13.115) show that $N_1 - n_1$ is approximately 0.014, 0.015 and 0.018 at $\lambda_0 = 0.8\,\mu\text{m}$, $1.3\,\mu\text{m}$ and $1.55\,\mu\text{m}$ respectively and thus that the group index is usually within about 1–2% of the refractive index.

13.9 Multimode fibres with optimum profiles

As discussed in Sec. 13.4.1, the intermodal dispersion in a multimode fibre can be reduced significantly by having a refractive index grading in the core. Indeed, in Sec. 13.7 we have shown that for an infinitely extended square law profile the group velocity is almost independent of the mode number (see Eq. (13.85)). In this section we consider a more general class of refractive index profiles characterized by the following refractive index variation:

$$n^2(r) = \begin{cases} n_1^2[1 - 2\Delta(r/a)^q]; & 0 < r \leqslant a \\ n_1^2(1 - 2\Delta) = n_2^2; & r \geqslant a \end{cases} \tag{13.117}$$

where a represents the radius of the core, Δ is the grading parameter, n_1 and n_2 represent the axial and cladding refractive indices respectively, q represents the exponent of the power law profile. In Fig. 13.14 we have plotted the refractive index variation as given by Eq. (13.117) for different values of q; $q = 2$ corresponds to a parabolic profile and $q = \infty$ corresponds to a step index profile. In Appendix F we have given a WKB analysis corresponding to the profile given by Eq. (13.117) and if we label the propagation constants of the various modes as $\beta_1, \beta_2, \ldots$ (β_1 corresponding to the maximum value of β) we obtain (see Appendix F).

$$\beta_v^2 \approx k_0^2 n_1^2[1 - 2\Delta(v/N)^{q/(q+2)}]; \quad v = 1, 2, \ldots, N \tag{13.118}$$

where

$$N = \frac{1}{2}\frac{q}{q+2}V^2 = \frac{q}{q+2}k_0^2 a^2 n_1^2 \Delta \tag{13.119}$$

[†] This can also be seen from Eq. (13.90) which shows that the spot size of the fundamental mode decreases with decrease in wavelength (see also Eq. (13.155)).

represents the total number of guided modes and

$$V = k_0 a (n_1^2 - n_2^2)^{\frac{1}{2}} = k_0 a n_1 (2\Delta)^{\frac{1}{2}} \qquad (13.120)$$

represents the normalized waveguide parameter. The WKB analysis is valid for highly multimode fibres with $V \gg 1$.

For a typical graded index fibre with $q = 2$

$$n_1 = 1.47, \qquad \Delta = 0.01, \qquad a = 25\,\mu m$$

so that at $\lambda_0 = 0.8\,\mu m$,

$$V \approx 41, \qquad N \approx 420$$

The same fibre would support about 160 modes at 1.3 μm. In either case the fibre is highly multimoded and the WKB analysis is valid.

Problem 13.6: In Eq. (13.118), v actually represents the number of modes having a propagation constant greater than β_v. Consider a $q = 2$ fibre and show that Eq. (13.118) is consistent with Eq. (13.83).

Solution: For $q = 2$, Eq. (13.118) becomes

$$\beta_v^2 = k_0^2 n_1^2 - \frac{k_0 n_1 2 (2\Delta)^{\frac{1}{2}}}{a} v^{\frac{1}{2}} \qquad (13.121)$$

We write Eq. (13.83) as

$$\beta_{mn}^2 = k_0^2 n_1^2 - \frac{k_0 n_1 2 (2\Delta)^{\frac{1}{2}}}{a} (m + n + 1) \qquad (13.122)$$

Fig. 13.14 The refractive index variation for a power law profile (see Eq. (13.117)) for different values of q.

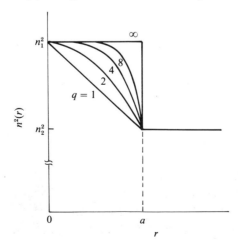

which may be rewritten in the form

$$s = m + n + 1 = \frac{a}{k_0 n_1 2(2\Delta)^{\frac{1}{2}}} (k_0^2 n_1^2 - \beta_{mn}^2) \qquad (13.123)$$

Thus

v = total number of modes above the propagation constant $\beta_v = \beta_{mn}$
  = 2 × total number of sets of integers (m, n) such that $m + n + 1 < s$
  = $2 \times [s + (s - 1) + \cdots + 1]$
  $\approx (m + n + 1)^2 \qquad (13.124)$

where the factor 2 is due to the two independent states of polarization and we have used the fact that when $m = 0, 1, 2, \ldots$, etc., n can take $s - 1$, $s - 2, \ldots$ values. Thus Eq. (13.118) can be considered to be an approximate representation of a large number of modes.

In Appendix F the group velocity for different modes has been calculated and using this the time taken for the vth mode to propagate through a distance L of the fibre can be expressed by

$$t_v = \frac{L}{v_v} = L \frac{d\beta_v}{d\omega}$$

$$\approx \frac{N_1 L}{c} \left[1 + \frac{q - 2 - \epsilon}{q + 2} \delta + \frac{3q - 2 - 2\epsilon}{2(q + 2)} \delta^2 + O(\delta^3) \right] \qquad (13.125)$$

where N_1 is the group index as given by Eq. (13.110) and

$$\epsilon = \frac{2k}{\Delta} \frac{d\Delta}{dk} = -\frac{2n_1}{N_1} \frac{\lambda_0}{\Delta} \Delta' \qquad (13.126)$$

leads to what is known as profile dispersion and

$$\delta = \Delta (v/N)^{q/(q+2)} \qquad (13.127)$$

Obviously $0 < \delta < \Delta$. In Eq. (13.126) primes denote differentiation with respect to λ_0.

We first neglect material dispersion i.e., we put $n_1' = \Delta' = 0$. We then obtain

$$t_v = \frac{n_1 L}{c} \left(1 + \frac{q - 2}{q + 2} \delta + \frac{3q - 2}{q + 2} \frac{\delta^2}{2} + \cdots \right) \qquad (13.128)$$

which was first derived by Gloge and Marcatili (1973). Since $0 < \delta < \Delta$ and Δ is usually of the order of 0.01, unless q is very close to 2, we may neglect terms which are proportional to δ^2, δ^3 etc. Thus for q not very close to 2 we retain only the term linear in δ and immediately find that the time taken increases monotonically as δ increases from 0 to Δ, i.e., higher order modes take a longer time to reach the output end of the fibre. Consequently, t_{max} and t_{min} correspond to $\delta = \Delta$ and $\delta = 0$ respectively, giving

$$t_{max} - t_{min} \approx \frac{n_1 L}{c} \frac{q-2}{q+2} \Delta; \quad \text{except for } q \approx 2 \qquad (13.129)$$

For a step index fibre, $q = \infty$ and

$$\Delta \tau = t_{max} - t_{min} \approx (n_1 L/c) \Delta \qquad (13.130)$$

which is the same as Eq. (13.9) when $n_1 \approx n_2$. For $q = 2$, Eq. (13.128) gives

$$t_v \approx (n_1 L/2c) \delta^2 \qquad (13.131)$$

and t_v again increases monotonically with δ (see Fig. 13.15). Thus since $0 < \delta < \Delta$, we obtain

$$\Delta \tau = t_{max} - t_{min} \approx (n_1 L/2c) \Delta^2 \qquad (13.132)$$

Fig. 13.15 Variation of $(ct_v/n_1 z - 1)$ versus δ for various values of q. The optimum profile corresponds to $q = 2 - 2\Delta$ when the modes corresponding to $\delta = 0$ and $\delta = \Delta$ take the same amount of time.

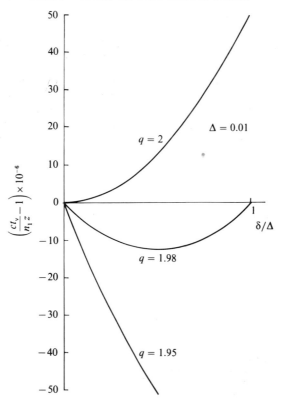

Notice from Eqs. (13.130) and (13.132) that for $\Delta = 0.01$, $\Delta\tau$ for a $q = 2$ fibre is 200 times smaller than the corresponding $\Delta\tau$ for a $q = \infty$ fibre.

For $q \lesssim 2$ the delay time first decreases and then increases with δ (see Fig. 13.15). The minimum value of t_v occurs for

$$\delta = \delta_0 = (2 - q)/(3q - 2) \tag{13.133}$$

which corresponds to $dt_v/d\delta = 0$. Minimum pulse dispersion occurs when the modes corresponding to $\delta = 0$ and $\delta = \Delta$ take the same amount of time and this happens when

$$q = q_0 \approx 2 - 2\Delta \tag{13.134}$$

which represents the optimum profile for minimum intermodal dispersion. The corresponding pulse dispersion is given by

$$\Delta\tau = t_{max} - t_{min} \approx \frac{n_1 L}{c} \frac{\Delta^2}{8} \tag{13.135}$$

We may summarize that

$$\Delta\tau(q = \infty) = \frac{n_1}{c}\left(\frac{n_1}{n_2} - 1\right)L \approx \frac{n_1\Delta}{c}L \quad \text{step profile} \tag{13.136}$$

$$\Delta\tau(q = 2) = \frac{n_1}{2c}\Delta^2 L \quad \text{parabolic profile} \tag{13.137}$$

Fig. 13.16 Variation of the improvement of maximum time delay difference of a graded core fibre over a step index fibre as a function of q. Observe the very sharp variation of the improvement factor with q near the optimum value. Thus to obtain good performance graded index fibres the profile has to be controlled very well.

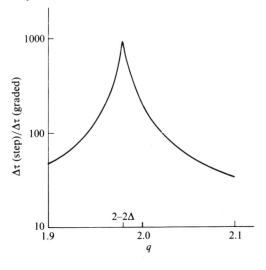

$$\Delta\tau(q = 2 - 2\Delta) \approx \frac{n_1}{8c}\Delta^2 L \quad \text{optimum profile} \tag{13.138}$$

For $n_1 \approx 1.47$, $\Delta \approx 0.01$, we readily obtain

$$\text{Pulse dispersion} \approx \begin{cases} 50\,\text{ns/km}; \ q = \infty, \ \text{step profile} \\ 0.25\,\text{ns/km}; \ q = 2, \ \text{parabolic profile} \\ 0.06\,\text{ns/km}; \ q = 1.98, \ \text{optimum profile} \end{cases} \tag{13.139}$$

From above we may note the following:

(a) For the optimum profile, the pulse dispersion is reduced by a factor of about 1000 in comparison to the step index profile. Because of this, first and second generation optical communication systems used near parabolic index multimode fibres.

(b) The pulse dispersion is very sensitive to the q value – indeed, a 1% deviation from the optimum q value results in an increase of pulse width by a factor of about 4. (see Fig. 13.16).

If we also take ϵ into account, one can show that minimum pulse dispersion occurs when

$$q = q_0 \approx 2 - 2\Delta + \epsilon(1 - \tfrac{1}{2}\Delta) \tag{13.140}$$

which shows the effect of material dispersion on the optimum profile. It should be noted that, in general, ϵ and Δ are functions of wavelength and therefore the optimum profile is also a function of wavelength. Indeed, q_0 is very sensitive to λ_0 for the most common germanium dioxide doped silica glass (Olshansky and Keck, 1976).

13.10 First and second generation fibre optic communication systems

Eq. (13.128) and the subsequent discussions considered only the intermodal component of pulse dispersion. We will now show that material dispersion is almost negligible for step index multimode fibres while it becomes dominant for graded index multimode fibres with near optimum refractive index profiles. Now, the first generation optical communication systems used GaAlAs LED sources around $\lambda_0 \approx 0.85\,\mu\text{m}$. For such LED sources, $\Delta\lambda_0 \sim 25\,\text{nm}$ and therefore from Eq. (13.112) and (13.116) we find

$$\Delta\tau_m \sim 2.1\,\text{ns/km} \quad \text{at } \lambda_0 = 0.85\,\mu\text{m} \tag{13.141}$$

Thus if we compare with Eq. (13.139) we can conclude that:

Table 13.2. *Typical characteristics of fibre optic communication systems at different stages*

		Bit rate	Type of fibre	Loss (dB/km)	Repeater spacing (km)
I	Generation (0.8–0.9 μm)	1977	~45 Mbit/s	Mutimode (graded-index)	~3	~10
II	Generation (1.3 μm)	1981	~45 Mbit/s	Multimode (graded-index)	~1	~30
III	Generation (1.3 μm)	At present	~500 Mbit/s Upgradable to ~1 Gb/s	Single mode	≤1	~40
IV	Generation (1.55 μm)	Immediate future	≥10 Gbit/s(≡150 000 telephone channels	Single mode	<0.3	≥100

Futuristic system: Infrared fibres ($\lambda_0 > 2$ μm)
Extremely low loss ($< 10^{-2}$ dB/km)
⇒ Repeater Spacing > 1000 km.

(a) For step index multimode fibres the material dispersion can be neglected and the information-carrying capacity is limited by intermodal dispersion. For a 1 km length of such a fibre (with $n_1 \approx 1.5$ and $\Delta \approx 0.01$), the interpulse separation should be at least 50 ns. The corresponding bit rate would be given by

$$\text{bit rate} \approx 1/(50 \times 10^{-9}) \approx 20 \,\text{Mbits/s}$$

for a 1 km length of the (multimode step index) fibre.

(b) On the other hand for near parabolic index multimode fibres, the intermodal dispersion becomes very small (see Eq. (13.139)) and the information-carrying capacity is limited by material dispersion. With material dispersion of ~ 2.1 ns/km, the corresponding bit rate would be ~ 400 Mbits/s for a 1 km length of the optical fibre and about 40 Mbit/s for a 10 km length of the fibre. Indeed, first generation optical communication systems (prior to 1980) used graded index (near-parabolic) fibres with (typically) GaAlAs LED sources and Si PIN/APD receivers. In a typical system, the spacing between two repeaters was ~ 12 km, the loss was about 3 dB/km and the bit rate was ~ 45 Mbit/s (see Table 13.2).

In order to increase the information-carrying capacity, the pulse dispersion has to be decreased and therefore the operating wavelength has to be shifted to a region where $d^2 n/d\lambda_0^2$ is very small. Fig. 13.13 shows the variation of $d^2 n/d\lambda_0^2$ with λ_0 for pure silica. It can be seen that the material dispersion of pure silica passes through zero around 1.274 μm. Thus operating at

Fig. 13.17 The solid curve represents a typical loss spectrum of a silica fibre fabricated around 1979 and has been adapted from Miya, Terunama, Hosaka and Miyashita (1979). The dotted curve represents the Rayleigh scattering loss which varies as $1/\lambda_0^4$ and the dashed curve represents the infrared absorption tail.

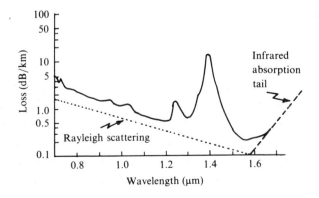

$\lambda_0 \sim 1.3\,\mu$m with multimode fibres which have optimum refractive index profiles, the pulse dispersion would be limited mainly by intermodal dispersion since material dispersion becomes very small.

The shifting of the operating wavelength to the $1.3\,\mu$m range has another important advantage. The corresponding loss is much smaller. Fig. 13.17 shows the loss spectrum for a typical optical fibre and as can be seen the losses at $\lambda_0 \sim 1.3\,\mu$m are $\sim 1\,$dB/km which is smaller by a factor of about 3 when compared with losses at $\lambda_0 \sim 0.8\,\mu$m. Optical communication systems around 1981 used graded index multimode fibres with $1.3\,\mu$m sources; typical sources used were InGaAsP/InP LEDs ($\Delta\lambda_0 \approx 25\,$nm) and laser diodes ($\Delta\lambda_0 \approx 2\,$nm) with Ge APDs as detectors. The losses in the fibres were $\sim 1\,$dB/km. These systems are usually said to belong to second generation optical communication systems. In a typical operating system the repeater spacing was 30 km and the bit rate was $\sim 45\,$Mbit/s (see Table 13.2). Thus, in comparison to first generation systems, the repeater spacing was increased by a factor of about 2.5 which was primarily due to the decrease in pulse dispersion as well as in the losses in optical fibres.

As shown above by operating at $1.3\,\mu$m wavelength, material dispersion has become extremely small and therefore the information-carrying capacity is limited by intermodal dispersion. This intermodal dispersion can be totally eliminated by using single mode fibres where only one guided mode is possible. This will be discussed in the following section.

13.11 Single mode fibres

In Sec. 13.6 we have shown that for a step index fibre with a normalized V parameter such that[†] (see Eq. (13.50))

$$0 < V < 2.4048 \tag{13.142}$$

there is only one guided mode which can propagate through the fibre. Since $V = (2\pi/\lambda_0)a(n_1^2 - n_2^2)^{\frac{1}{2}}$ the fibre will be single moded for

$$\lambda_0 > \lambda_{co} = \frac{2\pi a(n_1^2 - n_2^2)^{\frac{1}{2}}}{2.4048}$$

The parameter λ_{co} is called the cutoff wavelength of the given step index fibre. Such fibres are referred to as single mode fibres and play a very important role in high bandwidth optical fibre communication systems. Due to the presence of only a single mode, such fibres are free from intermodal

[†] The largest V value for single mode operation depends on the refractive index profile. For a parabolic index fibre, single mode operation requires $0 < V < 3.518$. For other q values one may refer to Ghatak and Thyagarajan (1980).

dispersion, the major factor which limits the information capacity of multimode fibres. The only form of dispersion in such fibres is the intramodal dispersion caused by the dependence of the refractive index of the fibre material on wavelength and waveguide dispersion, a qualitative discussion of which was given in Sec. 13.8. We may recall from Sec. 13.8 that the material dispersion is given by

$$\Delta\tau_m = -\frac{z}{c}\lambda_0^2\frac{d^2 n_1}{d\lambda_0^2}\left(\frac{\Delta\lambda_0}{\lambda_0}\right) \tag{13.143}$$

which is thus proportional to $d^2 n_1/d\lambda_0^2$ and to the spectral width of the source $(\Delta\lambda_0)$.

As discussed in Sec. 13.8, even if the material dispersion is zero, we have waveguide dispersion. In order to calculate this we write

$$\frac{d\beta}{d\lambda_0} = \frac{d\beta}{dV}\frac{dV}{d\lambda_0} = -\frac{V}{\lambda_0}\frac{d\beta}{dV} \tag{13.144}$$

where we have neglected the wavelength dependences of n_1 and Δ. Now, for weakly guiding fibres $(n_1 \approx n_2)$

$$b = \frac{(\beta/k_0)^2 - n_2^2}{n_1^2 - n_2^2} \approx \frac{\beta/k_0 - n_2}{n_1 - n_2} \approx \frac{\beta/k_0 - n_2}{n_2\Delta} \tag{13.145}$$

Thus

$$\beta \approx \frac{2\pi n_2}{\lambda_0}(1 + b\Delta) \tag{13.146}$$

and

$$\frac{d\beta}{dV} = -\frac{2\pi n_2}{\lambda_0^2}\frac{d\lambda_0}{dV}(1 + b\Delta) + \frac{2\pi n_2}{\lambda_0}\Delta\frac{db}{dV} \tag{13.147}$$

Simple manipulations give us the following expression for the group delay (see Eq. (13.105)):

$$\tau = L(d\beta/d\omega) \approx \frac{Ln_2}{c}\left[1 + b\Delta + \Delta\left(V\frac{db}{dV}\right)\right] \tag{13.148}$$

The broadening of the pulse due to waveguide dispersion is given by (cf. Eq. (13.107))

$$\Delta\tau_w = \frac{d\tau}{d\lambda_0}\Delta\lambda_0$$

$$\approx -L\frac{n_2\Delta}{c}\left(\frac{\Delta\lambda_0}{\lambda_0}\right)V\frac{d^2(bV)}{dV^2} \tag{13.149}$$

In Sec. 13.6 we have shown that for a step index fibre one obtains universal curves for b as a function of V. Thus for such a fibre, the quantity $V d^2(bV)/dV^2$ depends only on the V value; this is, however, true only when $n_1 \approx n_2$ which is indeed the case for all practical fibres. Now, in order to get a numerical appreciation we use the following empirical relation (Rudolph and Neumann, 1976):

$$b = (A - B/V)^2; \ 1.5 < V < 2.5, \ A = 1.428, \ B = 0.996 \qquad (13.150)$$

The above relation is accurate to within 0.2% of the exact values. Although the above empirical formula is not very accurate for the calculation of derivatives, nevertheless we can get an idea of the interplay between material and waveguide dispersions. Substituting in Eq. (13.149), we obtain

$$\frac{\Delta\tau_w}{L\Delta\lambda_0} = -\frac{2n_2\Delta}{c\lambda_0}\frac{B^2}{V^2}$$

$$\approx -\frac{\lambda_0}{a^2}\frac{B^2}{4\pi^2 n_2 c} \qquad (13.151)$$

where we have used the relation $V = k_0 a n_1 (2\Delta)^{\frac{1}{2}} \approx k_0 a n_2 (2\Delta)^{\frac{1}{2}}$.

The total dispersion in a step index fibre is approximately the sum of the material and waveguide dispersions:

$$\Delta\tau_{tot} \approx \Delta\tau_m + \Delta\tau_w \qquad (13.152)$$

We next study the relative contributions between the material and waveguide dispersions. Fig. 13.18 shows a typical variation of

$$\Delta\tau_m/(L\Delta\lambda_0)$$

Fig. 13.18 Variation of $\Delta\tau_m/L\Delta\lambda_0$ and $\Delta\tau_w/L\Delta\lambda_0$ and $\Delta\tau_{tot}/L\Delta\lambda_0$ as a function of λ_0 for a typical step index fibre. The dots denote experimental data for total dispersion (Kimura, 1979).

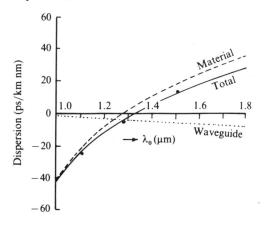

as a function of λ_0 for pure silica. For

$$\lambda_0 = 0.8\,\mu\text{m}, \quad \text{d}^2 n/\text{d}\lambda_0^2 \approx 4 \times 10^{-2}\,\mu\text{m}^{-2}$$

implying

$$\Delta\tau_m/L\Delta\lambda_0 \approx 100\,\text{ps/km nm}$$

For a typical step index single mode fibre,

$$a = 3\,\mu\text{m}, \qquad \Delta = 0.00154, \qquad n_2 = 1.45$$

The values have been so chosen that $V \simeq 1.9$ at $\lambda_0 = 0.8\,\mu\text{m}$. Using Eq. (13.151) with $B = 0.996$, we obtain

$$\Delta\tau_w/L\Delta\lambda_0 \approx -5\,\text{ps/km nm}$$

Thus in the wavelength region around $0.8\,\mu\text{m}$, the contribution due to material dispersion is much greater than that due to waveguide dispersion and therefore the dispersion is mainly due to material dispersion which is large.

As evident from Fig. 13.18 the material dispersion becomes extremely small around $\lambda_0 \approx 1.3\,\mu\text{m}$ and changes sign; this can be used to cancel the effect due to waveguide dispersion. Thus one may expect zero total dispersion around $\lambda_0 \approx 1.3\,\mu\text{m}$. In order to illustrate this we first calculate the material dispersion. Now at

$$\lambda_0 = 1.3\,\mu\text{m}, \text{d}^2 n/\text{d}\lambda_0^2 = -5.5 \times 10^{-4}\,\mu\text{m}^{-2}$$

so that

$$\Delta\tau_m/L\Delta\lambda_0 \approx 2.4\,\text{ps/km nm}$$

We now consider a step index single mode fibre with

$$a = 5.6\,\mu\text{m}, \Delta = 0.00117, n_2 = 1.45$$

The fibre parameters have been so chosen that

$$V \approx 1.9 \quad \text{at } \lambda_0 = 1.3\,\mu\text{m}$$

and (using Eq. (13.151))

$$\Delta\tau_w/L\Delta\lambda_0 \approx -2.4\,\text{ps/km nm}$$

Thus the waveguide and material dispersion would cancel one another giving rise to zero total dispersion and therefore a very large bandwidth. We must mention here that although the material dispersion given by Eq. (13.143) is quite accurate, the corresponding calculation of waveguide dispersion is not very accurate because of the use of the empirical formula[†] given by Eq. (13.150). Nevertheless the above procedure tells us how one may obtain zero dispersion; for a more accurate estimation one must use a more accurate value of $V\mathrm{d}^2(bV)/\mathrm{d}V^2$ in Eq. (13.149).

Fig. 13.17 gives the loss spectrum corresponding to an extremely low loss fibre and as can be seen the loss attains a minimum value of 0.2 dB/km at 1.55 μm. Thus operating at $\lambda_0 = 1.3$ μm, the system would be limited by the loss in the fibre. Since the lowest loss lies at $\lambda_0 \approx 1.55$ μm, if the zero dispersion wavelength could be shifted to the $\lambda_0 \approx 1.55$ μm region, one could have both minimum loss and very low dispersion. This would lead to very high bandwidth systems with very long (~ 100 km) repeaterless transmission. The shift of the zero total dispersion wavelength to a region around 1.5 μm can indeed be accomplished by changing the fibre parameters.

In order to illustrate this we first calculate the material dispersion at $\lambda_0 = 1.55$ μm. Now

$$\mathrm{d}^2n_1/\mathrm{d}\lambda_0^2 = -4.2 \times 10^{-3}\,\mu\mathrm{m}^{-2} \quad \text{at } \lambda_0 = 1.55\,\mu\mathrm{m}$$

so that

$$\Delta\tau_{\mathrm{m}}/L\Delta\lambda_0 \approx 22\,\mathrm{ps/km\,nm}$$

In order to cancel this large material dispersion, we see from Eq. (13.151) that a must be reduced. Indeed if we assume the validity of Eq. (13.150) we must have

$$a = 2.02\,\mu\mathrm{m}, \qquad \Delta = 0.013, \qquad n_2 = 1.45$$

The value of Δ has been so chosen that $V = 1.9$ at $\lambda_0 = 1.55$ μm. For the above fibre

$$\Lambda\tau_{\mathrm{w}}/L\Delta\lambda_0 \approx -22\,\mathrm{ps/km\,nm}$$

Thus the fibre will have a zero dispersion wavelength around $\lambda_0 = 1.55$ μm. Such a fibre in which the zero dispersion wavelength occurs around $\lambda_0 = 1.55$ μm is referred to as a dispersion shifted fibre.

Fig. 13.19 shows a typical variation of total dispersion of two single mode step index fibres having zero dispersion around $\lambda_0 \approx 1.3$ μm and $\lambda_0 \approx 1.55$ μm. For a typical single mode fibre fabricated by Corning Glass Works, USA (quoted by Blank, Bickers and Walker (1985))

[†] We should mention here that we are consistently working at $V = 1.9$ where the value of $V\mathrm{d}^2(bV)/\mathrm{d}V^2 = 2B^2/V^2$ (obtained from the empirical formula– Eq (13.150)) is accurate (Sammut, 1979). A more accurate empirical expression for $1.3 < V < 2.6$, is given by (Marcuse, 1979) $V(bV)'' \approx 0.080 + 0.549(2.834 - V)^2$.

Attenuation: 0.215 dB/km at $\lambda_0 = 1.55\,\mu$m
0.23 dB/km at $\lambda_0 = 1.52\,\mu$m

Dispersion: Zero dispersion wavelength: $1.55\,\mu$m

Dispersion slope $= 0.075$ ps/nm^2 km (maximum)
$\Rightarrow$ dispersion < 0.75 ps/nm km for $1.54 < \lambda_0 < 1.56\,\mu$m

Using multimode lasers, Blank, *et al.* (1985) could achieve a 140 Mbit/s system for a *repeaterless link* of 222.8 km; the 222.8 km span comprised 37 spliced fibre lengths with a total span loss of 50.6 dB. We should mention here that repeater spacings $\gtrsim 250$ km are of tremendous interest because about 40% of undersea systems are less than 250 km in length and hence use of such fibre optic communication systems would not require any repeaters.

In the above we have shown that by reducing the core radius the zero dispersion wavelength can be shifted to the minimum loss wavelength window. The same can be achieved by suitably grading the refractive index profile in the core of the fibre. For example curve (ii) in Fig. 13.20 can also be achieved by a linear variation of refractive index inside the core. One can also tailor the refractive index profile in such a way that the waveguide dispersion almost balances the material dispersion over a range of wavelengths. Such fibres which have a low dispersion over a range of wavelengths are known as *dispersion flattened* fibers; curves (iii) and (iv) in Fig. 13.20 give the dispersion characteristic of such fibres and the corresponding refractive index variations are shown in the inset of Fig. 13.20. Using dispersion flattened fibres one can enormously increase the information-carrying capacity by employing wavelength division multiplexing. Olsson *et al.* (1985) have used ten

Fig. 13.19 Variation of total dispersion as a function of λ_0 in a step index fibre. The solid curve corresponds to a fibre with $\Delta = 0.0027$, $a = 4.1\,\mu$m, $\lambda_{co} = 1.13\,\mu$m while the dashed curve corresponds to a fibre with $\Delta = 0.0075$, $a = 2.3\,\mu$m and $\lambda_{co} = 1.06\,\mu$m; here λ_{co} is the cutoff wavelength. (Adapted from Kimura (1980).)

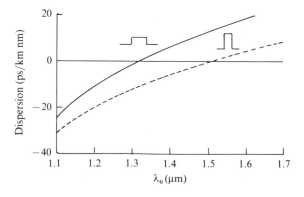

Fig. 13.20 The variation of total dispersion for four classes of fibres. Curve (i) corresponds to conventional fibres. Curve (ii) corresponds to dispersion shifted fibres. Curves (iii) and (iv) correspond to dispersion flattened fibres; the corresponding refractive index profiles are shown in the inset of the figure (Adapted from Cohen (1985)).

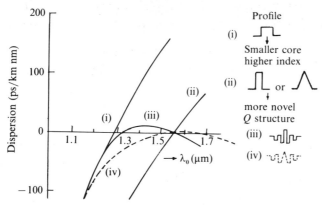

Fig. 13.21 Typical fundamental modal field shapes $\psi(r)$ for fibres with different profiles. The solid curves represent exact field variations while the dashed curves represent the best fitted Gaussian function (Adapted from Ghatak and Sharma (1986).)

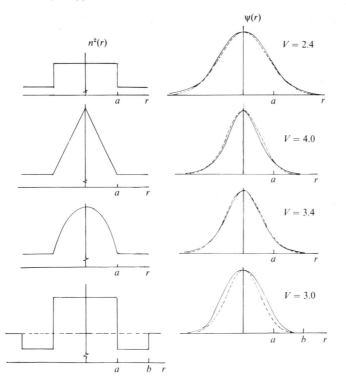

wavelengths between 1.529 μm and 1.561 μm with a link length of 68.5 km; each channel has an information-carrying capacity of 2 Gbit/s which implies

Total information carrying capacity for one fibre $= 10 \times 2 \times 68.5$ Gbit km/s
$$= 1.37 \, \text{Tbit km/s}.$$

which is equivalent to about 20 million voice channel km through a single fibre – an enormous amount of information indeed.

13.12   The Gaussian approximation

The fundamental mode field distribution for a single mode fibre is a very important characteristic which determines various important parameters such as splice loss at joints, launching efficiencies, bending loss etc. In Fig. 13.21 we have given some typical refractive index profiles of single mode fibers and their corresponding fundamental mode field distributions. Also plotted in the figure are best fitted Gaussian functions and it can be seen that irrespective of the refractive index profile, the fundamental mode field distribution can be well approximated by a Gaussian function[†] which may be written in the form

$$\psi(r) = A \, e^{-r^2/w^2} \tag{13.153}$$

where w is referred to as the spot size of the mode field pattern. One of the criterion widely used to choose the value of w is that which leads to the maximum launching efficiency of the exact fundamental mode field by an incident Gaussian field, i.e., one maximizes the quantity

$$\eta = \frac{\displaystyle\int_0^\infty e^{-r^2/w^2} R(r) r \, dr}{\left[\displaystyle\int_0^\infty e^{-2r^2/w^2} r \, dr \int_0^\infty R^2(r) r \, dr \right]^{\frac{1}{2}}} \tag{13.154}$$

where $R(r)$ represents the exact modal field. For example, for a step index fibre $R(r)$ can be expressed in terms of Bessel functions and one has the following empirical expression for w (Marcuse, 1977):

$$w/a \approx (0.65 + 1.619/V^{3/2} + 2.879/V^6); \; 0.8 \lesssim V \lesssim 2.5 \tag{13.155}$$

where a is the core radius. The above empirical formula gives a value of w (as obtained by maximizing η) to within about 1%.

We should mention here that the different refractive index profiles shown in Fig. 13.21 are of considerable technological importance; for example, as we have seen in the previous section, one can tailor the dispersion

[†] Indeed for an infinitely extended parabolic index medium, the fundamental modal field is exactly Gaussian (see Eq. (13.90)).

characteristics by a proper choice of refractive index variation inside the core.

13.13 Splice loss

One of the great advantages of the Gaussian approximation is the fact that it gives us simple analytical expressions for losses at fibre splices. At the joint there could be three types of misalignments: (*a*) longitudinal offset (*b*) transverse offset, (*c*) angular misalignment (see Fig. 13.22). As a specific example, we consider transverse misalignment (Fig. 13.22(*b*)). The two single mode fibres are represented by Gaussian fundamental modes with spot sizes w_1 and w_2. Let us consider the direction of misalignment to be along the x-direction. With respect to the coordinate axes fixed on the fibre the normalized Gaussian modes can be represented by

$$\psi_1(x, y) = \left(\frac{2}{\pi}\right)^{\frac{1}{2}} \frac{1}{w_1} e^{-(x^2 + y^2)/w_1^2} \tag{13.156}$$

$$\psi_2(x, y) = \left(\frac{2}{\pi}\right)^{\frac{1}{2}} \frac{1}{w_2} e^{-[(x - \xi)^2 + y^2]/w_2^2} \tag{13.157}$$

where ξ is the transverse misalignment and the factors are such that

$$\int\int_{-\infty}^{+\infty} \psi_{1,2}^2 \, dx \, dy = 1$$

The fractional power that is coupled to the fundamental mode of the second

Fig. 13.22 Various misalignments at a splice between two fibres: (*a*) Longitudinal misalignment, (*b*) transverse misalignment, (*c*) angular misalignment.

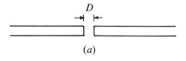

(*a*)

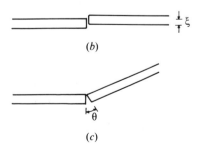

(*b*)

(*c*)

fibre is given by

$$T = \left| \int\int_{-\infty}^{+\infty} \psi_1 \psi_2^* \, dx \, dy \right|^2 \tag{13.158}$$

Thus

$$T = \left| \frac{2}{\pi w_1 w_2} \int\int_{-\infty}^{+\infty} \exp\left[-x^2 \left(\frac{1}{w_1^2} + \frac{1}{w_2^2} \right) + \frac{2x\xi}{w_2^2} \right. \right.$$
$$\left. \left. - y^2 \left(\frac{1}{w_1^2} + \frac{1}{w_2^2} \right) - \frac{\xi^2}{w_2^2} \right] dx \, dy \right|^2$$
$$= \left(\frac{2w_1 w_2}{w_1^2 + w_2^2} \right)^2 \exp\left[-\frac{2\xi^2}{(w_1^2 + w_2^2)} \right] \tag{13.159}$$

The maximum power coupling appears at $\xi = 0$ and the transmitted fractional power is

$$T_{\text{max}} = \left(\frac{2w_1 w_2}{w_1^2 + w_2^2} \right)^2 \tag{13.160}$$

which is unity for two fibres having identical Gaussian fundamental modes. Also the transmitted energy decreases by a factor of $1/e$ for a misalignment of

$$\xi_e = (w_1^2 + w_2^2)^{\frac{1}{2}}/\sqrt{2} \tag{13.161}$$

For two identical fibres, the transverse offset loss varies as

$$T = e^{-\xi^2/w_1^2} \tag{13.162}$$

Hence by measuring the transmission loss across a splice as a function of ξ one can experimentally obtain the spot size w_1.

The corresponding expressions for tilt loss and longitudinal separations for a pair of identical fibres are given by (see Problems 13.11 and 13.13)

$$T_t = e^{-(k_0 n_2 \theta w)^2/4} \tag{13.163}$$

$$T_1 = \frac{1 + 4\tilde{D}^2}{\tilde{D}^2 + (1 + 2\tilde{D}^2)^2}; \quad \tilde{D} = \frac{4D}{n_2 k_0 w^2} \tag{13.164}$$

where θ is the angle between the two fibre axes and D the separation between the two fibre end faces; n_2 is the refractive index of the medium between the fibre ends.

Problem 13.7: Consider a Gaussian temporal pulse incident at $z = 0$:

$$f(t) = e^{-t^2/2\tau^2} e^{i\omega_c t} \tag{13.165}$$

Show that

$$F(\omega) = \tau e^{-(\omega - \omega_c)^2 \tau^2/2} \tag{13.166}$$

In Eq. (13.100) retain terms up to $(\omega - \omega_c)^2$ and show that the intensity associated with the νth mode at the output at $z = L$ is given by

$$|\theta_\nu(t)|^2 = \left[1 + \left(\frac{L\gamma_\nu}{\tau^2}\right)^2\right]^{-\frac{1}{2}} \exp\left\{-\frac{(t - \alpha_\nu L)^2}{\tau^2[1 + (\gamma_\nu L/\tau^2)^2]}\right\} \tag{13.167}$$

Discuss the broadening of the pulse. (cf. Sec. 5.2)

Problem 13.8: In the previous problem the pulse full width for large $L (\gg \tau^2/\gamma_\nu)$ for the νth mode is given by $2\gamma_\nu L/\tau$. Show that this is consistent with Eq. (13.107).

Solution: The pulse width at large L satisfying $L \gg \tau^2/\gamma_\nu$ is given by $2\gamma_\nu L/\tau$. Thus the pulse spread is given by

$$\Delta t = 2\gamma_\nu L/\tau - 2\tau \approx 2\gamma_\nu L/\tau \tag{13.168}$$

since $\gamma_\nu L/\tau \gg \tau$. The frequency spectrum of a Gaussian pulsed monochromatic source is given by Eq. (13.166). Thus the spread in frequency is given by

$$\Delta\omega \approx 2/\tau \tag{13.169}$$

where $\Delta\omega$ is the width of the power spectrum defined at the e^{-1} points. Thus the dispersion is given by

$$\Delta t = L\Delta\omega (d^2\beta/d\omega^2)_{\omega_c} \tag{13.170}$$

where we have used Eq. (13.101). The above equation is consistent with Eq. (13.107).

Problem 13.9: Consider an infinitely extended parabolic index fibre characterized by a refractive index variation of the form given by Eq. (13.62). Solve the scalar wave equation in cylindrical coordinates and obtain the LP_{lm} modes of the fibre and show that the results are consistent with those obtained in Sec. 13.7.

Solution: Substituting for the refractive index variation in Eq. (13.26), we obtain

$$\frac{d^2R}{dr^2} + \frac{1}{r}\frac{dR}{dr} + \left(k_0^2 n_1^2 - \beta^2 - k_0^2 n_1^2 \frac{2\Delta}{a^2} r^2 - \frac{l^2}{r^2}\right) R(r) = 0 \tag{13.171}$$

Substituting

$$\xi = \alpha^2 r^2; \quad \alpha^2 = \frac{k_0 n_1}{a} (2\Delta)^{\frac{1}{2}}; \quad \chi = e^{\xi/2} \alpha^l \xi^{-1/2} R(r) \tag{13.172}$$

in Eq. (13.171) we obtain

$$\xi \frac{d^2\chi}{d\xi^2} + (l + 1 - \xi)\frac{d\chi}{d\xi} + \frac{1}{4}(\gamma - 2l - 2)\chi(\xi) = 0 \tag{13.173}$$

where

$$\gamma = \frac{k_0^2 n_1^2 - \beta^2}{k_0 n_1 (2\Delta)^{\frac{1}{2}}} a \tag{13.174}$$

The solutions of the above equation are associated Laguerre functions. The solution behaves as e^ξ for large ξ (i.e., R behaves as $e^{\xi/2}\xi^{l/2}$) unless

$$\tfrac{1}{4}(\gamma - 2l - 2) = m - 1; \qquad m = 1, 2, 3, \dots \tag{13.175}$$

For such a case $R(r) \to 0$ as $r \to \infty$ which corresponds to guided modes. Using the value of γ we get

$$\beta_{lm}^2 = k_0^2 n_1^2 - 2(2m + l - 1)k_0 n_1 (2\Delta)^{\frac{1}{2}}/a \tag{13.176}$$

β_{lm} is the propagation constant of the LP_{lm} mode. The corresponding field distribution is

$$\psi_{lm}(r, \phi, z, t) = N_{lm} r^l e^{-\alpha^2 r^2/2} L_{m-1}^l(\alpha^2 r^2) \begin{Bmatrix} \cos l\phi \\ \sin l\phi \end{Bmatrix} e^{i(\omega t - \beta_{lm} z)} \tag{13.177}$$

where $L_{m-1}^l(\alpha^2 r^2)$ is the associated Laguerre polynomial defined by

$$L_{m-1}^l(x) = \frac{e^x x^{-l}}{(m-1)!} \frac{d^{m-1}}{dx^{m-1}} (e^{-x} x^{m+l-1}) \tag{13.178}$$

For example

$$L_0^l(x) = 1, \quad L_1^0(x) = 1 - x, \quad L_2^0(x) = (1 - 2x - x^2/2), \quad L_1^1(x) = 2 - x \text{ etc.} \tag{13.179}$$

The fundamental mode corresponds to $l = 0, m = 1$ and can easily be shown to have the same distribution as given by Eq. (13.90).

Problem 13.10: For any incident field distribution that is symmetric about the z-axis, the modes described by Eq. (13.177) with $l = 0$ can be used to study the propagation of a beam. Consider an incident Gaussian beam of the form

$$\psi(x, y, z = 0) = (\gamma/\pi^{\frac{1}{2}}) e^{-\gamma^2 r^2/2} \tag{13.180}$$

show that

$$\psi(x, y, z) = \sum_{m=1}^{\infty} A_m [(\alpha/\pi^{\frac{1}{2}}) L_{m-1}(\alpha^2 r^2) e^{-\alpha^2 r^2/2}] e^{-i\beta_m z} \tag{13.181}$$

where

$$A_m = \frac{2w}{1 + w^2} \left(\frac{w^2 - 1}{w^2 + 1} \right)^m; \quad w = \gamma/\alpha \tag{13.182}$$

Problem 13.11: Consider two single mode fibres and assume that the fundamental modes of the two fibres can be represented by Gaussian distributions. Calculate the power transmission loss as a function of angular misalignment between the two fibres. Assume that the fibre ends are placed in a liquid of refractive index n_2.

Solution: We describe the fundamental modes of the two fibres by Eq. (13.156) and a similar equation with w_1 replaced by w_2. Let us assume that there is a small angular misalignment of θ between the two fibres (see Fig. 13.22(c)). In order to calculate the power transmission loss, we must transform the Gaussian beam of the first fibre into the coordinate system of the second fibre. If (x, y, z) and (x', y', z') represent the coordinate systems of the first and second fibres respectively then we have

$$\left.\begin{array}{l} x = x' \cos \theta + z' \sin \theta \\ y = y' \\ z = - x' \sin \theta + z' \cos \theta \end{array}\right\} \tag{13.183}$$

where we assume an angular misalignment in the x–z plane. Since the medium between the two fibres is of refractive index n_2, the Gaussian mode of the first fibre, as it emerges, will propagate aproximately as

$$\psi_1(x, y, z) \approx \left(\frac{2}{\pi}\right)^{\frac{1}{2}} \frac{1}{w_1} e^{-(x^2 + y^2)/w_1^2} e^{-ik_0 n_2 z} \tag{13.184}$$

for small values of z so that we can neglect diffraction effects. We transform Eq. (13.184) into the (x', y', z') coordinate system and obtain for the incident beam at the input plane $z' = 0$ of the second fibre

$$\psi_1(x', y, z') = \left(\frac{2}{\pi}\right)^{\frac{1}{2}} \frac{1}{w_1} e^{-(x'^2 + y^2)/w_1^2} e^{ik_0 n_2 x' \theta} \tag{13.185}$$

where we have assumed $\sin \theta \approx \theta$ and $\cos \theta \approx 1$. Calculation of the overlap of this field with the field of the second fibre and taking modulus squared gives us the power transmission coefficient as

$$T(\theta) = \left(\frac{2w_1 w_2}{w_1^2 + w_2^2}\right)^2 \exp\left[-\frac{k_0^2 n_2^2 \theta^2 w_1^2 w_2^2}{2(w_1^2 + w_2^2)}\right] \tag{13.186}$$

Thus the coupled power decreases to e^{-1} of its maximum value (corresponding to $\theta = 0$) in an angle

$$\theta_e = \sqrt{2(w_1^2 + w_2^2)^{\frac{1}{2}}}/w_1 w_2 k_0 n_2 \tag{13.187}$$

It is interesting to note that for two identical fibres with $w_1 = w_2 = w$ and using Eqs. (13.161) and (13.187) we get

$$\xi_e \theta_e = 2/k_0 n_2 = \lambda_0/\pi n_2 \tag{13.188}$$

which is independent of w_1. Eq. (13.188) is similar to the uncertainty principle in quantum mechanics (see e.g., Ghatak and Lokanathan (1984)). Eq. (13.188) implies that increasing the mode spot size w_1 would increase ξ_e, i.e., tolerance towards transverse misalignment but this would correspondingly decrease θ_e, i.e., the tolerance towards angular misalignment.

Problem 13.12: At the wavelength at which $d^2 n/d\lambda_0^2 = 0$, estimate the broadening caused due to the term $d^3 n/d\lambda_0^3$.

Solution: Let us consider two close wavelengths λ_0 and $\lambda_0 + \Delta\lambda_0$. If $\tau(\lambda_0)$ and $\tau(\lambda_0 + \Delta\lambda_0)$ represent the respective time delays, then we may write

$$\tau(\lambda_0 + \Delta\lambda_0) = \tau(\lambda_0) + \Delta\lambda_0 \, d\tau/d\lambda_0 + \tfrac{1}{2}(\Delta\lambda_0)^2 d^2\tau/d\lambda_0^2 \tag{13.189}$$

Since $d\tau/d\lambda_0 = -(L\lambda_0/c)d^2n/d\lambda_0^2$ which is zero at the zero material dispersion point, we have

$$\Delta\tau = \tau(\lambda_0 + \Delta\lambda_0) - \tau(\lambda_0) = \tfrac{1}{2}(\Delta\lambda_0)^2 \, d^2\tau/d\lambda_0^2 \qquad (13.190)$$

Also

$$\frac{d^2\tau}{d\lambda_0^2} = \frac{d}{d\lambda_0}\left(\frac{d\tau}{d\lambda_0}\right) = \frac{d}{d\lambda_0}\left(-\frac{L\lambda_0}{c}\frac{d^2n}{d\lambda_0^2}\right)$$

$$= -\frac{L\lambda_0}{c}\frac{d^3n}{d\lambda_0^3}$$

since $d^2n/d\lambda_0^2 = 0$. Thus

$$\Delta\tau = -L(\Delta\lambda_0)^2\left(\frac{\lambda_0}{2c}\frac{d^3n}{d\lambda_0^3}\right) \qquad (13.191)$$

which gives the broadening due to material dispersion. Observe that the broadening is proportional to $(\Delta\lambda_0)^2$ and to L.

Problem 13.13: Derive Eq. (13.164). (Hint: use Eqs. (5.10) and (13.158).)

13.14   The vector modes

The entire analysis in this chapter is based on the scalar wave approximation in which the modes are assumed to be linearly polarized. Since the actual vector wave equation involves a term depending on ∇n^2 (see Eq. (13.13)) the linearly polarized modes are not the vector modes of the fibre. This can also be seen from the fact that if we consider an x-polarized LP mode then according to our assumption of continuity of fields E_x must be continuous at $r = a$. However, since E_x is a tangential component only at $y = \pm a$ rigorously speaking it would not be continuous for $r = a$ for all values of ϕ. Hence the LP modes cannot represent the polarization of the actual modes even in the weakly guiding approximation. Since the LP modes are degenerate (i.e., they can be x or y-polarized) one can choose appropriate linear combinations of these (LP) modes which satisfy the necessary continuity conditions; these are referred to as the zero order vector modes. We follow the analysis given by Sharma (1988) which is essentially based on the fact that for cylindrically symmetric systems (i.e., $n^2(r, \phi) = n^2(r)$) the boundary conditions that must be rigorously satisfied are the continuity of E_z and E_ϕ (which represent tangential components) and n^2E_r, at each refractive index discontinuity. Thus to obtain the vector modes it is the ϕ dependence of the fields E_z, E_r and E_ϕ (and not E_x, E_y, E_z) that is assumed to be of the form $\cos l\phi$ and $\sin l\phi$. Hence the zero order vector modes can be obtained from the LP modes by choosing such combinations that give the ϕ dependence of E_r or E_ϕ to be of the form $\cos l\phi$ or $\sin l\phi$.

Now

$$E_r = E_x \cos \phi + E_y \sin \phi \atop E_\phi = - E_x \sin \phi + E_y \cos \phi \Bigg\}$$ (13.192)

As discussed earlier for a given (l, m) we have the following four degenerate LP_{lm} modes:

$$\mathbf{E}_1 = \hat{\mathbf{x}} \psi_{lm} \cos l\phi, \quad \mathbf{E}_2 = \hat{\mathbf{y}} \psi_{lm} \cos l\phi \atop \mathbf{E}_3 = \hat{\mathbf{x}} \psi_{lm} \sin l\phi, \quad \mathbf{E}_4 = \hat{\mathbf{y}} \psi_{lm} \sin l\phi \Bigg\}$$ (13.193)

The corresponding E_ϕ components in each of the above LP modes are

$$\begin{aligned} E_{1\phi} &= - \psi_{lm} \cos l\phi \sin \phi \\ E_{2\phi} &= \psi_{lm} \cos l\phi \cos \phi \\ E_{3\phi} &= - \psi_{lm} \sin l\phi \sin \phi \\ E_{4\phi} &= \psi_{lm} \sin l\phi \cos \phi \end{aligned} \Bigg\}$$ (13.194)

Thus the LP modes by themselves do not have the correct ϕ dependence. As shown by Sharma (1988), by using the simple combination properties of trigonometric functions, the only combinations of the four functions in Eq. (13.194) which will reduce to single sine and cosine functions are

$$\begin{aligned} \tilde{E}_{1\phi} &= E_{1\phi} + E_{4\phi} = \psi_{lm} \sin [(l - 1)\phi] \\ \tilde{E}_{2\phi} &= E_{1\phi} - E_{4\phi} = - \psi_{lm} \sin [(l + 1)\phi] \\ \tilde{E}_{3\phi} &= E_{2\phi} + E_{3\phi} = \psi_{lm} \cos [(l + 1)\phi] \\ \tilde{E}_{4\phi} &= E_{2\phi} - E_{3\phi} = \psi_{lm} \cos [(l - 1)\phi] \end{aligned} \Bigg\}$$ (13.195)

Thus the zeroeth order vector modes are

$$\tilde{\mathbf{E}}_1 = \mathbf{E}_1 + \mathbf{E}_4 = (\hat{\mathbf{x}} \cos l\phi + \hat{\mathbf{y}} \sin l\phi) \psi_{lm}$$ (13.196)

$$\tilde{\mathbf{E}}_2 = \mathbf{E}_1 - \mathbf{E}_4 = (\hat{\mathbf{x}} \cos l\phi - \hat{\mathbf{y}} \sin l\phi) \psi_{lm}$$ (13.197)

$$\tilde{\mathbf{E}}_3 = \mathbf{E}_2 + \mathbf{E}_3 = (\hat{\mathbf{x}} \sin l\phi + \hat{\mathbf{y}} \cos l\phi) \psi_{lm}$$ (13.198)

$$\tilde{\mathbf{E}}_4 = \mathbf{E}_2 - \mathbf{E}_3 = (- \hat{\mathbf{x}} \sin l\phi + \hat{\mathbf{y}} \cos l\phi) \psi_{lm}$$ (13.199)

Notice from the above equations that for $l = 0$ modes, the zero order vector modes are also linearly polarized. If we now use

$$\hat{\mathbf{x}} = \hat{\mathbf{r}} \cos \phi - \hat{\boldsymbol{\phi}} \sin \phi \atop \hat{\mathbf{y}} = \hat{\mathbf{r}} \sin \phi + \hat{\boldsymbol{\phi}} \cos \phi$$ (13.200)

then Eq. (13.196) may be written in the form

$$\begin{aligned} \tilde{\mathbf{E}}_1 &= [(\hat{\mathbf{r}} \cos \phi - \hat{\boldsymbol{\phi}} \sin \phi) \cos l\phi + (\hat{\mathbf{r}} \sin \phi + \hat{\boldsymbol{\phi}} \cos \phi) \sin l\phi] \psi_{lm} \\ &= \{\hat{\mathbf{r}} \cos [(l - 1)\phi] + \hat{\boldsymbol{\phi}} \sin [(l - 1)\phi]\} \psi_{lm} \end{aligned}$$ (13.201)

The above equation implies

$$E_{1r} = \psi_{lm} \cos[(l-1)\phi] \atop E_{1\phi} = \psi_{lm} \sin[(l-1)\phi] \Big\} \text{even EH}_{(l-1)m} \text{ mode} \tag{13.202}$$

The mode characterized by $\tilde{E}_1$ is called an even EH$_{(l-1)m}$ mode; it is said to be even since the ϕ dependence of E_r is a cosine function. Similarly we can obtain the r and ϕ components of $\tilde{E}_2$, $\tilde{E}_3$ and $\tilde{E}_4$; the final results are

$$E_{2r} = \psi_{lm} \cos[(l+1)\phi] \atop E_{2\phi} = -\psi_{lm} \sin[(l+1)\phi] \Big\} \text{even HE}_{(l+1)m} \tag{13.203}$$

$$E_{3r} = \psi_{lm} \sin[(l+1)\phi] \atop E_{3\phi} = \psi_{lm} \cos[(l+1)\phi] \Big\} \text{odd HE}_{(l+1)m} \tag{13.204}$$

$$E_{4r} = -\psi_{lm} \sin[(l-1)\phi] \atop E_{4\phi} = \psi_{lm} \cos[(l-1)\phi] \Big\} \text{odd EH}_{(l-1)m} \tag{13.205}$$

The classification of the modes can also be carried out through the z-component of the fields. In order to do this, we first write Maxwell's curl equation:

$$\nabla \times \mathcal{E} = -\partial \mathcal{B}/\partial t = -i\omega\mu_0 \mathcal{H} \tag{13.206}$$

In the cylindrical system of coordinates, we have

$$-i\omega\mu_0 \mathcal{H} = \begin{vmatrix} \hat{\mathbf{r}} & \hat{\boldsymbol{\phi}} & \hat{\mathbf{z}} \\ \dfrac{\partial}{\partial r} & \dfrac{1}{r}\dfrac{\partial}{\partial \phi} & \dfrac{\partial}{\partial z} \\ \mathcal{E}_r & \mathcal{E}_\phi & \mathcal{E}_z \end{vmatrix} \tag{13.207}$$

Thus

$$-i\omega\mu_0 H_r = \frac{1}{r}\frac{\partial E_z}{\partial \phi} + i\beta E_\phi \approx i\beta E_\phi \tag{13.208}$$

$$-i\omega\mu_0 H_\phi = -i\beta E_r - \frac{\partial E_z}{\partial r} \approx -i\beta E_r \tag{13.209}$$

$$-i\omega\mu_0 H_z = \frac{\partial E_\phi}{\partial r} - \frac{1}{r}\frac{\partial E_r}{\partial \phi} \tag{13.210}$$

where

$$\mathcal{H} = \mathbf{H}\, e^{i(\omega t - \beta z)} \tag{13.211}$$

$$\mathcal{E} = \mathbf{E}\, e^{i(\omega t - \beta z)} \tag{13.212}$$

and we have assumed the fields to be nearly transverse which is valid in the weakly guiding approximation (i.e., $n_1 \approx n_2$). Similarly from the equation

$$\nabla \times \mathcal{H} = \partial \mathcal{D}/\partial t = i\omega\epsilon\mathcal{E} \tag{13.213}$$

we have

$$i\omega\epsilon E_r \approx i\beta H_\phi \tag{13.214}$$

$$i\omega\epsilon E_\phi \approx -i\beta H_r \tag{13.215}$$

$$i\omega\epsilon E_z = \frac{\partial H_\phi}{\partial r} - \frac{1}{r}\frac{\partial H_r}{\partial \phi} \tag{13.216}$$

Now using Eqs. (13.208) and (13.209), Eq. (13.216) becomes

$$E_z = -\frac{i\beta}{\omega^2\epsilon\mu_0}\left(\frac{\partial E_r}{\partial r} + \frac{1}{r}\frac{\partial E_\phi}{\partial \phi}\right) \tag{13.217}$$

Also from Eq. (13.210),

$$H_z = \frac{i}{\omega\mu_0}\left(\frac{\partial E_\phi}{\partial r} - \frac{1}{r}\frac{\partial E_r}{\partial \phi}\right) \tag{13.218}$$

We can now calculate the z-component of each mode.
Even $EH_{(l-1)m}$:

$$E_{1z} = -\frac{i\beta}{\omega^2\epsilon\mu_0}\left\{\frac{d\psi_{lm}}{dr}\cos\left[(l-1)\phi\right] + \frac{(l-1)}{r}\psi_{lm}\cos\left[(l-1)\phi\right]\right\}$$

$$= -\frac{i\beta}{\omega^2\epsilon\mu_0}\cos\left[(l-1)\phi\right]\left[\frac{d\psi_{lm}}{dr} + \frac{(l-1)}{r}\psi_{lm}\right] \tag{13.219}$$

$$H_{1z} = \frac{i}{\omega\mu_0}\sin\left[(l-1)\phi\right]\left[\frac{d\psi_{lm}}{dr} + \frac{(l-1)}{r}\psi_{lm}\right] \tag{13.220}$$

Even $HE_{(l+1)m}$:

$$E_{2z} = -\frac{i\beta}{\omega^2\epsilon\mu_0}\cos\left[(l+1)\phi\right]\left[\frac{d\psi_{lm}}{dr} - \frac{(l+1)}{r}\psi_{lm}\right] \tag{13.221}$$

$$H_{2z} = -\frac{i}{\omega\mu_0}\sin\left[(l+1)\phi\right]\left[\frac{d\psi_{lm}}{dr} - \frac{(l+1)}{r}\psi_{lm}\right] \tag{13.222}$$

Odd $HE_{(l+1)m}$:

$$E_{3z} = -\frac{i\beta}{\omega^2\epsilon\mu_0}\sin\left[(l+1)\phi\right]\left[\frac{d\psi_{lm}}{dr} - \frac{(l+1)}{r}\psi_{lm}\right] \tag{13.223}$$

$$H_{3z} = \frac{i}{\omega\mu_0}\cos\left[(l+1)\phi\right]\left[\frac{d\psi_{lm}}{dr} - \frac{(l+1)}{r}\psi_{lm}\right] \tag{13.224}$$

Odd $EH_{(l-1)m}$:

$$E_{4z} = \frac{i\beta}{\omega^2 \epsilon \mu_0} \sin\left[(l-1)\phi\right] \left[\frac{d\psi_{lm}}{dr} + \frac{(l-1)}{r}\psi_{lm}\right] \tag{13.225}$$

$$H_{4z} = \frac{i}{\omega\mu_0} \cos\left[(l-1)\phi\right] \left[\frac{d\psi_{lm}}{dr} + \frac{(l-1)}{r}\psi_{lm}\right] \tag{13.226}$$

From above we see that the LP_{lm} mode is composed of $HE_{(l+1)m}$ and $EH_{(l-1)m}$ modes. We also note that for the even EH_{0m} modes, $H_z = 0$ and therefore they correspond to TM_m modes. Similarly for the odd EH_{0m} modes, $E_z = 0$ and therefore they correspond to TE_m modes. Since E_z and H_z do not have the same ϕ dependence, we define a quantity

$$P = \frac{H_z}{(dE_z/d\phi)} \tag{13.227}$$

Simple calculations show that

$$P = \omega\epsilon/\beta(l-1) \qquad \text{for even and odd } EH_{(l-1)m} \text{ modes} \tag{13.228}$$
$$P = -\omega\epsilon/\beta(l+1) \quad \text{for even and odd } HE_{(l+1)m} \text{ modes}$$

Indeed one can use the above definition for the classification of the various modes.

Problem 13.14: Derive Eq. (13.55)

Solution: The function $y = J_l(x)$ satisfies the following differential equation

$$x^2 y'' + xy' + (x^2 - l^2)y(x) = 0$$

Multiplying by $2y'$ we get

$$\frac{d}{dx}[x^2 y'^2 - l^2 y^2 + x^2 y^2] = 2xy^2$$

Integrating we get

$$2\int xy^2 \, dx = x^2 y'^2 + (x^2 - l^2)y^2 + \text{constant}$$

Using $J_l' = (J_{l-1} - J_{l+1})/2$ and Eq. (13.39) we obtain

$$x^2 J_l'^2 = (l^2 J_l^2 - x^2 J_{l-1} J_{l+1})$$

from which Eq. (13.55) follows.

14

Integrated optics

14.1 Introduction

Integrated optics is a new and exciting field of activity which is primarily based on the fact that light can be guided and confined in very thin films (with dimensions $\sim$ wavelength of light) of transparent materials on suitable substrates. By a proper choice of substrates and films and a proper configuration of the waveguides, one can perform a wide range of operations such as modulation, switching, multiplexing, filtering or generation of optical waves. Due to the miniature size of these components, it is possible to obtain a high density of optical components in space unlike the case in bulk optics. These devices are expected to be rugged in construction, have good mechanical and thermal stability, be mass producible with high precision and reproducibility, and have a small power consumption.

One of the most promising applications of integrated optics is expected to be in the field of optical fibre communications. As discussed in Chapter 13, the field of optical fibre communication has assumed tremendous importance because of its high information-carrying capacity; it is here that integrated optics is expected to play an important role in optical signal processing at the transmitting and receiving ends and on regeneration at the repeaters. Other important applications of integrated optics are envisaged to be in spectrum analysis (see Chapter 19) and optical signal processing.

In addition to the above, use of integrated optic techniques may lead to the realization of new devices which may be too cumbersome to be fabricated in bulk optics. The very high concentration of energy in very small regions in the optical waveguide leads to enormous intensities leading to the realization of nonlinear devices employing second harmonic generation etc. The confinement of light energy in small regions of space also leads to an efficient interaction of the optical energy with an applied electric field or an acoustic wave, thus leading to much more efficient electrooptic and acoustooptic modulators and deflectors requiring very low drive powers.

In integrated optics, two main kinds of waveguides are used; these are the planar and strip guides (see Figs. 14.1 and 14.2). In planar waveguides, the confinement of the light energy is only along one transverse dimension and the light energy can diffract in the other transverse dimension. In contrast to planar waveguides, strip waveguides confine the light energy in both transverse dimensions; this confinement is a desirable feature for the fabrication of devices such as amplitude or intensity modulators, directional couplers, optical switches etc. In order to understand the functioning of the devices, it is necessary to understand the guiding properties of waveguides quantitatively; these are discussed in Section 14.2. In Section 14.3 we will discuss some important devices such as the phase modulator, the Mach–Zehnder interferometer modulator and the directional coupler switch.

14.2   Modes in an asymmetric planar waveguide

In Chapter 11 we have shown that for the one-dimensional waveguide with

$$n^2 = n^2(x) \tag{14.1}$$

Fig. 14.1 (*a*) A planar waveguide consisting of a film of refractive index n_f deposited on a substrate of refractive index n_s and with a cover of refractive index n_c. The refractive index varies only along the *x*-direction and is independent of *y* and *z*. Light confinement is only along the *x*-direction; light can diffract in the *y–z* plane. (*b*) Transverse cross section of the wave guide.

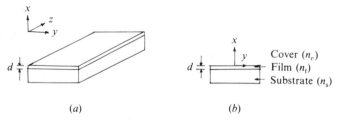

(*a*)   (*b*)

Fig. 14.2. (*a*) A strip waveguide consisting of a high index region of refractive index n_f surrounded by media of lower refractive indices. The light confinement is along both the *x* and *y*-directions; the direction of propagation is the *z*-direction. (*b*) Transverse cross section of the wave guide.

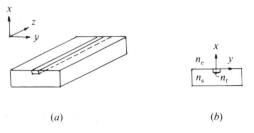

(*a*)   (*b*)

Maxwell's equations reduce to two independent sets of equations. The first set (see Eqs. (11.9)–(11.11)) corresponds to non-vanishing values of E_y, H_x and H_z (with E_x, E_z and H_y vanishing) giving rise to TE modes. The second set (see Eqs. (11.12)–(11.14)) corresponds to nonvanishing values of E_x, E_z and H_y (with E_y, H_x and H_z vanishing) giving rise to TM modes. The propagation of waves in such planar waveguides (characterized by Eq. (14.1)) may thus be described in terms of TE and TM modes.

In Chapter 11, we gave a detailed analysis of the symmetrical step index planar waveguide bringing out most of the salient aspects of the propagation characteristics of optical waveguides. However, most thin film waveguides used in integrated optics are not symmetrical in nature and can usually be characterized by a refractive index variation of the form

$$n(x) = \begin{cases} n_c; & x > 0, \text{ cover} \\ n_f; & -d < x < 0, \text{ film} \\ n_s; & x < -d, \text{ substrate} \end{cases} \qquad (14.2)$$

The above refractive index variation corresponds to a thin film of refractive index n_f and thickness d (which is usually of the order of a few microns) deposited on a transparent substrate of refractive index n_s ($n_s < n_f$). The region above the film is referred to as the cover (or superstrate) and has a refractive index n_c (which is also less than n_f). In most cases, the cover is air ($n_c = 1$).

We will assume without any loss of generality that $n_f > n_s \geqslant n_c$. Light guidance in such planar guides can easily be understood on the basis of total internal reflection. Consider a light wave incident on the film–cover interface subtending an angle ϕ with the normal to the interface. (see Fig. 14.3). For small values of ϕ which are less than the critical angle for total internal reflection between film and cover, i.e., for $\phi < \phi_c$ where $\phi_c = \sin^{-1} (n_c/n_f)$, only a part of the energy is reflected, the remaining portion is transmitted. Since we have assumed $n_c < n_s$, the reflected portion is incident on the film–substrate interface at an angle less than the corresponding critical angle ϕ_s [$= \sin^{-1}(n_s/n_f)$] and again the wave undergoes only a partial reflection at the film–substrate boundary. Thus the wave is not guided in the film but loses its energy to the substrate and cover; these correspond to what are referred to as *cover modes*. If we start increasing ϕ, we will reach a condition when ϕ will be just greater than ϕ_c but still less than ϕ_s (see Fig. 14.3(*b*)). For such a case the wave is total internally reflected at the film–cover boundary but is transmitted at the film–substrate boundary; these correspond to what are known as *substrate modes*. Both cover and substrate modes belong to a class of modes referred to as radiation modes which refers to the fact that the waves associated with these modes are not guided in the film. On further increase of ϕ when $\phi > \phi_s$, the light wave is total internally reflected at both the film–cover and film–substrate interfaces and the wave is said to be guided in the film. Such a wave corresponds to what is known as a *guided mode*. Thus

$$\phi > \phi_s = \sin^{-1}(n_s/n_f) \Leftrightarrow \text{guided modes} \tag{14.3}$$

We will now carry out a modal analysis of waveguides characterized by Eq. (14.2) and highlight some interesting features associated with them. We discuss first TE modes and then TM modes.

For TE modes, E_y satisfies the equation (see Eq. (11.15))

$$d^2 E_y/dx^2 + [k_0^2 n^2(x) - \beta^2] E_y(x) = 0 \tag{14.4}$$

Fig. 14.3 (*a*) A ray incident on the film–substrate interface at an angle ϕ less than the critical angle between the film and cover ($\phi < \phi_c = \sin^{-1}(n_c/n_f)$) undergoes only a partial reflection. Since we are assuming $n_c < n_s$, the ray is also incident on the film–substrate interface at an angle less than the critical angle $\phi_s (= \sin^{-1}(n_s/n_f))$ and hence is partially reflected. Such rays correspond to cover radiation modes. (*b*) As ϕ increases, a stage is reached when the ray undergoes total internal reflection at the film–cover interface but undergoes only a partial reflection at the film–substrate interface ($\phi_c < \phi < \phi_s$). Such rays correspond to substrate radiation modes. (*c*) If ϕ is increased to such a value that it is greater than ϕ_s, then the ray undergoes total internal reflection at both the film–substrate and film–cover boundaries. Such rays correspond to the guided modes of the waveguide. The dashed lines perpendicular to the ray represent the phase front of the upgoing plane wave.

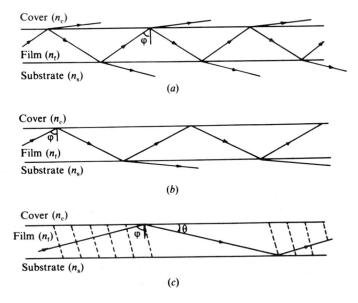

Thus the field component E_y satisfies the following equations in the three media:

$$d^2 E_y/dx^2 - \gamma_c^2 E_y = 0; \quad x > 0, \text{ cover} \tag{14.5a}$$

$$d^2 E_y/dx^2 + \kappa_f^2 E_y = 0; \quad 0 > x > -d, \text{ film} \tag{14.5b}$$

$$d^2 E_y/dx^2 - \gamma_s^2 E_y = 0; \quad x < -d, \text{ substrate} \tag{14.5c}$$

where

$$\gamma_c^2 = \beta^2 - k_0^2 n_c^2 \tag{14.6}$$

$$\kappa_f^2 = k_0^2 n_f^2 - \beta^2 \tag{14.7}$$

$$\gamma_s^2 = \beta^2 - k_0^2 n_s^2 \tag{14.8}$$

For guided modes, we require that the field should decay in the cover and substrate so that most of the energy associated with the mode lies inside the film. Thus, for guided modes γ_c^2 and γ_s^2 must be positive, so that Eqs. (14.5a) and (14.5c) may yield exponentially decaying solutions rather than oscillatory solutions. Thus for guided modes (since $n_s \geqslant n_c$)

$$k_0^2 n_s^2 < \beta^2$$

Also, since the fields have to be oscillatory inside the film, κ_f^2 must be positive[†] and hence from Eq. (14.7) we get

$$\beta^2 < k_0^2 n_f^2$$

Hence for guided modes

$$k_0^2 n_s^2 < \beta^2 < k_0^2 n_f^2 \tag{14.9}$$

and we obtain for the solution of the scalar wave equation in the three regions

$$E_y = \begin{cases} A e^{-\gamma_c x}; & x > 0 \\ B e^{i\kappa_f x} + C e^{-i\kappa_f x}; & 0 > x > -d \\ D e^{\gamma_s x}; & x < -d \end{cases} \tag{14.10}$$

where A, B, C and D are constants (to be determined from the boundary conditions) and in writing the solutions in the regions $x > 0$ and $x < -d$ we have neglected the exponentially amplifying solution. We now use the boundary conditions, namely the continuity of E_y and dE_y/dx *at* $x = 0$ and

---

[†] If κ_f^2 is allowed to take negative values, then the solutions will be such that the boundary conditions can never be satisfied (see Problem 11.8).

$x = -d$ to obtain

$$A = B + C \tag{14.11}$$

$$B e^{-i\kappa_f d} + C e^{i\kappa_f d} = D e^{-\gamma_s d} \tag{14.12}$$

$$-\gamma_c A = i\kappa_f B - i\kappa_f C \tag{14.13}$$

$$i\kappa_f B e^{-i\kappa_f d} - i\kappa_f C e^{i\kappa_f d} = \gamma_s D e^{-\gamma_s d} \tag{14.14}$$

Eliminating A from Eqs. (14.11) and (14.13) we get

$$(\gamma_c + i\kappa_f)B + (\gamma_c - i\kappa_f)C = 0 \tag{14.15}$$

Similarly eliminating D from Eqs. (14.12) and (14.14) we obtain

$$(\gamma_s - i\kappa_f)B e^{-i\kappa_f d} + (\gamma_s + i\kappa_f)C e^{i\kappa_f d} = 0 \tag{14.16}$$

From Eqs. (14.15) and (14.16) we get

$$\tan \kappa_f d = \frac{\gamma_c/\kappa_f + \gamma_s/\kappa_f}{1 - (\gamma_c/\kappa_f)(\gamma_s/\kappa_f)} \qquad \text{TE modes} \tag{14.17}$$

which represents the transcendental equation determining the propagation constant of the TE mode. Eq. (14.17) can be written in the following convenient form

$$\tan\left[V(1-b)^{\frac{1}{2}}\right] = \frac{\left(\dfrac{b}{1-b}\right)^{\frac{1}{2}} + \left(\dfrac{b+a}{1-b}\right)^{\frac{1}{2}}}{1 - \dfrac{[b(b+a)]^{\frac{1}{2}}}{(1-b)}} \qquad \text{TE modes} \tag{14.18}$$

where

$$\left.\begin{array}{l} b = \dfrac{\beta^2/k_0^2 - n_s^2}{n_f^2 - n_s^2} = \dfrac{n_{\text{eff}}^2 - n_s^2}{n_f^2 - n_s^2}; \quad n_{\text{eff}} = \dfrac{\beta}{k_0} \\[3mm] V = k_0 d(n_f^2 - n_s^2)^{\frac{1}{2}}, \quad a = \dfrac{n_s^2 - n_c^2}{n_f^2 - n_s^2} \end{array}\right\} \tag{14.19}$$

The parameters b, V and a are known as the normalized propagation constant, the normalized waveguide parameter and the asymmetry parameter respectively; n_{eff} is called the effective index of the mode. Observe that since for guided modes $n_s < \beta/k_0 < n_f$, b lies between 0 and 1. The asymmetry parameter a may vary from zero (for symmetric waveguides, $n_s = n_c$) to very large values for highly asymmetric waveguides (with $n_s \neq n_c$ and $n_s \to n_f$).

Using Eqs. (14.15) and (14.16) we obtain the following expression for the field pattern

$$E_y = \begin{cases} A\,e^{-\gamma_c x}; & x \geqslant 0 \\ A[\cos \kappa_f x - (\gamma_c/\kappa_f)\sin \kappa_f x]; & 0 \geqslant x \geqslant -d \\ A[\cos \kappa_f d + (\gamma_c/\kappa_f)\sin \kappa_f d]\,e^{\gamma_s(x+d)}; & x \leqslant -d \end{cases} \qquad (14.20)$$

Fig. 14.4 shows the normalized plot of the variation of b with V for different modes of waveguides which have different values of the asymmetry parameter a. As can be seen, $b \to 0$ as a mode nears cutoff and far from cutoff $b \to 1$ for all modes. It can also be seen that a given guide configuration and a given frequency of operation specifies the value of V and for a particular value of V only a definite number of guided modes may propagate. Thus, for example, for $V = 5$, only two TE modes exist. Fig. 14.5 shows schematic plots of the field variation of TE modes with $m = 0, 1, 2$ and 3 across the cross section of the waveguide. The mode number m corresponds to the number of times the field crosses the axis (see also Section 11.3.1). It can be seen that higher order modes penetrate deeper into the substrate and cover than lower order modes. Fig. 14.6(a) and (b) show qualitatively the field distribution of the fundamental mode for two different wavelengths and as can be seen the

Fig. 14.4 Variation of the normalized propagation constant b versus the waveguide parameter V for different TE modes characterized by $m = 0, 1$, and 2 and for different asymmetry measures. It can be seen that as V becomes large the normalized propagation constant b tends to unity for all values of m and a (i.e., $\beta \to k_0 n_f$). Also for every mode at a specific value of V, $b = 0$; this value of V corresponds to the cutoff V value for this mode. At a given value of V only a definite number of guided TE modes may propagate.

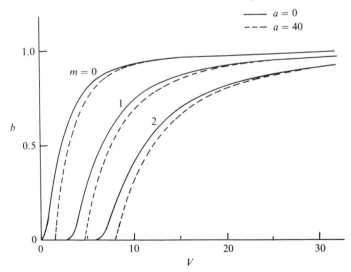

mode penetrates deeper into the substrate as λ_0 increases i.e., as the mode nears cutoff.

In a similar manner we can obtain the transcendental equation which determines the propagation constant for TM modes. The procedure is similar to the one used in Sec. 11.4. The final result is

$$\tan\left[V(1-b)^{\frac{1}{2}}\right] = \frac{\dfrac{1}{\gamma_1}\left(\dfrac{b}{1-b}\right)^{\frac{1}{2}} + \dfrac{1}{\gamma_2}\left(\dfrac{b+a}{1-b}\right)^{\frac{1}{2}}}{1 - \dfrac{1}{\gamma_1\gamma_2}\dfrac{[b(b+a)]^{\frac{1}{2}}}{(1-b)}} \qquad \text{TM modes}$$

$$(14.21)$$

where

$$\gamma_1 = (n_s/n_f)^2, \qquad \gamma_2 = (n_c/n_f)^2 = \gamma_1 - a(1-\gamma_1) \tag{14.22}$$

Fig. 14.5 Schematic of the field distribution of the TE modes corresponding to $m = 0, 1, 2$ and 3 in a planar waveguide. Observe that higher order modes (corresponding to higher values of m) penetrate deeper into the substrate and cover. Also the number of intersections of the field with the axis is just m.

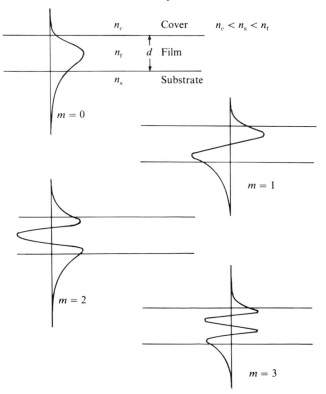

The corresponding field pattern is given by

$$H_y = \begin{cases} A\,e^{-\gamma_c x}; & x \geqslant 0 \\ A\left[\cos \kappa_f x - \dfrac{n_f^2}{n_c^2}\dfrac{\gamma_c}{\kappa_f}\sin \kappa_f x\right]; & 0 \geqslant x \geqslant -d \\ A\left[\cos \kappa_f d + \dfrac{n_f^2}{n_c^2}\dfrac{\gamma_c}{\kappa_f}\sin \kappa_f d\right]e^{\gamma_s(x+d)}; & x \leqslant -d \end{cases} \qquad (14.23)$$

If we substitute the cutoff condition

$$\beta/k_0 = n_s \Rightarrow b = 0 \qquad (14.24)$$

in Eqs. (14.18) and (14.21), we will get the following expressions for the cutoff frequencies[†]

$$V_c^{TE} = \tan^{-1} a^{\frac{1}{2}} + m\pi \qquad \text{TE modes} \qquad (14.25)$$

$$V_c^{TM} = \tan^{-1} (a^{\frac{1}{2}}/\gamma_2) + m\pi \qquad \text{TM modes} \qquad (14.26)$$

Fig. 14.6 Qualitative variation of the fundamental mode field pattern for (*a*) far from cutoff and (*b*) near cutoff. As can be seen, the mode penetrates deeper into the substrate as it nears cutoff.

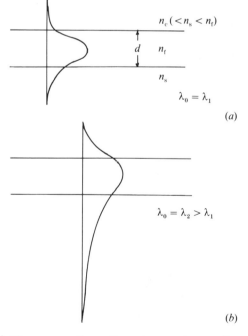

(*a*)

(*b*)

[†] We may recall our discussion on the cutoff condition in Sec. 11.7; for $\beta < k_0 n_s$ the mode is no longer decaying exponentially in the substrate and it does not correspond to a guided mode.

where $m = 0, 1, 2, \ldots$. Since $n_f > n_s > n_c$, we have

$$\gamma_2 < 1$$

and therefore

$$V_c^{TM} > V_c^{TE} \tag{14.27}$$

i.e., the cutoff frequencies of TM modes are greater than the corresponding cutoff frequencies of TE modes (see Fig. 14.7). Considering the fundamental mode ($m = 0$) when

$$\tan^{-1} a^{\frac{1}{2}} < V < \tan^{-1}(a^{\frac{1}{2}}/\gamma_2) \tag{14.28}$$

we will only have the TE mode and the waveguide is usually referred to as absolutely single polarization single moded. This is an example of an SPSM (single polarization single moded) waveguide. It may be noted that for $V < \tan^{-1} a^{\frac{1}{2}}$ there are *no* guided modes of the system. Notice that for a symmetric waveguide $a = 0$ and both TE_0 and TM_0 modes have zero cutoff.

We should mention that in most practical single mode waveguides, there is

Fig. 14.7 Qualitative variation of the normalized propagation constant b with V for the TE and TM modes of symmetric ($a = 0$) and asymmetric ($a \neq 0$) planar waveguides. The region indicated by the horizontal arrow corresponds to a SPSM waveguide.

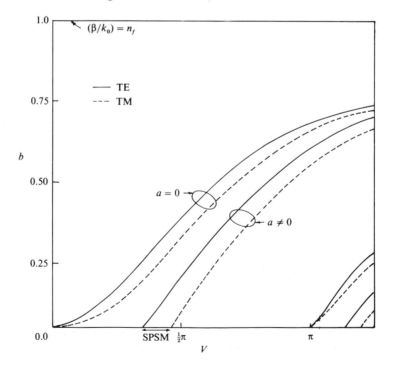

one TE mode and one TM mode; however, the incident field is usually linearly polarized and if $\mathbf{E}$ is along the y-axis, the TE mode is excited and if $\mathbf{E}$ is along the x-axis, the TM mode is excited. This immediately follows from the fact that the nonvanishing field components for TE and TM modes are (E_y, H_x, H_z) and (H_y, E_x, E_z) respectively (see also Sec. 11.4).

Problem 14.1: Consider a waveguide with

$$n_f = 1.5, \qquad n_s = 1.49, \qquad n_c = 1.0$$

Show that the waveguide will be SPSM for

$$1.416 < V < 1.502$$

implying that for $\lambda_0 = 1\,\mu\text{m}$, one must have

$$1.30 < d < 1.38\,\mu\text{m}$$

for SPSM operation.

Problem 14.2: Show that if the value of γ_2 is kept constant and the asymmetry parameter a is increased from zero (i.e., as n_s increases from a value n_c) the range of V values for which we have SPSM operation first increases and then decreases back to zero. In particular show that for $a = 0.5$ and $a = 5$, the range of V values for SPSM operation is given by

$$0.62 < V < 1.01 \qquad \text{and} \qquad 1.15 < V < 1.38$$

respectively. (Assume $n_f = 1.5$, $n_c = 1.0$.)

Problem 14.3: (a) Show that Eq. (14.17) can be written in the form

$$\kappa_f d = \Phi_s + \Phi_c + m\pi; \qquad m = 0, 1, 2, \ldots \tag{14.29}$$

where

$$\tan\Phi_s = \frac{\gamma_s}{\kappa_f} = \left(\frac{\beta^2 - k_0^2 n_s^2}{k_0^2 n_f^2 - \beta^2}\right)^{\frac{1}{2}} \tag{14.30}$$

$$\tan\Phi_c = \frac{\gamma_c}{\kappa_f} = \left(\frac{\beta^2 - k_0^2 n_c^2}{k_0^2 n_f^2 - \beta^2}\right)^{\frac{1}{2}} \tag{14.31}$$

(b) Using Eqs. (2.58) and (2.82), show that $-2\Phi_s$ and $-2\Phi_c$ are the phase shifts at the film–substrate and film–cover interface respectively. (Hint: see footnote on page 44].

Problem 14.4: Show that the field inside the film (see Eq. (14.20)) can be written in the form (cf. see Sec. 11.3.2).

$$\mathscr{E}_y = B' e^{-i\Phi_c} e^{i(\omega t - \kappa_f x - \beta z)} + B' e^{i\Phi_c} e^{i(\omega t + \kappa_f x - \beta z)} \tag{14.32}$$

where Φ_c is given by Eq. (14.31) and $B' = A/2\cos\Phi_c$. Thus show that a guided mode can be considered to be a superposition of plane waves which are propagating at angles $\pm\theta$ with the z-axis where

$$\cos\theta = \beta/k_0 n_f \tag{14.33}$$

Problem 14.5: Show that Eq. (14.21) for the TM modes can be written in the form

$$V(1-b)^{\frac{1}{2}} = \tan^{-1}\left[\frac{1}{\gamma_1}\left(\frac{b}{1-b}\right)^{\frac{1}{2}}\right] + \tan^{-1}\left[\frac{1}{\gamma_2}\left(\frac{b+a}{1-b}\right)^{\frac{1}{2}}\right] + m\pi \tag{14.34}$$

(cf. Eq. (14.29)).

Problem 14.6: The analysis in this section will correspond to that of a symmetric waveguide if we substitute

$$n_s = n_c \qquad \text{(giving } a = 0\text{)}$$

Show that Eq. (14.18) directly leads to Eq. (11.45) and Eq. (11.46) and Eq. (14.21) leads to Eq. (11.76) and (11.77). (Hint: use the equation $\tan 2\theta = 2\tan\theta/(1 - \tan^2\theta)$ and the fact that the solutions of the quadratic equation

$$\frac{2x}{1-x^2} = \frac{2\alpha}{1-\alpha^2}$$

are $x = +\alpha$ and $x = -1/\alpha$.)

14.3 Ray analysis of planar waveguides

In this section we will discuss the ray optics model for analysing the characteristics of planar waveguides. As we discussed in Sec. 14.2, we can consider a guided mode as a set of waves undergoing total internal reflection at the film–substrate and film–cover boundaries (see Fig. 14.3) and the resultant field at any point is the sum of the incident and repeatedly reflected fields. Since a mode corresponds to a specific field distribution across the cross section of the guide which propagates down the guide unchanged in form, we require that for a mode, the total transverse phase shift suffered by the wave in one complete round trip (consisting of two traverses across the film and a reflection at each interface) must be an integral multiple of 2π. It is the imposition of this condition which leads to the set of discrete angles of propagation corresponding to different guided modes and corresponds to the discrete spectrum of the propagation constant for the guided modes (see Section 14.2).

In Fig. 14.3(c) we have drawn a set of plane waves whose wave normals follow a zigzag path undergoing total internal reflection at the two boundaries of the guide. The plane waves have a propagation vector $\mathbf{k}_0 n_f$ which is directed along the wave normals. The z-component of this represents the propagation constant β of the mode. Hence (see Eq. (14.33))

$$\beta = k_0 n_f \cos\theta \tag{14.35}$$

The transverse component of the wave vector is $k_0 n_f \sin \theta$. In one complete round trip the wave undergoes a phase shift of $2 d k_0 n_f \sin \theta$ on passage twice through the thickness d of the guide. In addition, the wave suffers phase shifts at total internal reflections from the cover and substrate boundaries. These phase shifts for the TE case can be shown to be $-2\Phi_c$ and $-2\Phi_s$ where (see Problem 14.3(b))

$$\tan \Phi_c = (n_f^2 \cos^2 \theta - n_c^2)^{\frac{1}{2}}/n_f \sin \theta \tag{14.36}$$

and

$$\tan \Phi_s = (n_f^2 \cos^2 \theta - n_s^2)^{\frac{1}{2}}/n_f \sin \theta \tag{14.37}$$

For the TM case

$$\tan \Phi_c = \frac{n_f^2}{n_c^2} \frac{(n_f^2 \cos^2 \theta - n_c^2)^{\frac{1}{2}}}{n_f \sin \theta} \tag{14.38}$$

$$\tan \Phi_s = \frac{n_f^2}{n_s^2} \frac{(n_f^2 \cos^2 \theta - n_s^2)^{\frac{1}{2}}}{n_f \sin \theta} \tag{14.39}$$

Hence for guided modes one must have

$$2 k_0 n_f d \sin \theta - 2\Phi_c - 2\Phi_s = 2 m \pi; \qquad m = 0, 1, 2, \dots$$

or

$$k_0 n_f d \sin \theta = \Phi_c + \Phi_s + m \pi \tag{14.40}$$

Eq. (14.40) is identical to the one obtained using a more rigorous wave analysis (Eqs. (14.29) and (14.34)).

14.4   WKB analysis of inhomogeneous planar waveguides

In Sec. 14.2 we have carried out a detailed modal analysis of planar waveguides consisting of homogeneous materials. However, in many integrated optical devices (which are fabricated using diffusion techniques) there is a continuous variation of refractive index and it is necessary to understand the methods used to determine the propagation characteristics of these waveguides.

Referring to Eq. (14.4) we see that for TE modes in such a waveguide, the component E_y rigorously satisfies the scalar wave equation. There are a number of refractive index profiles, for which exact solutions for TE modes can be obtained. For example, exact solutions have been obtained for the following refractive index profiles (see, e.g., Adams (1981), Sodha and Ghatak (1977))

$$\left. \begin{aligned} n^2(x) &= n_0^2 - \Delta n x^2/a^2 \\ n^2(x) &= n_0^2 + \Delta n \operatorname{sech}^2(x/d) \\ n^2(x) &= n_0^2 + \Delta n \, e^{-|x|/d} \end{aligned} \right\} \tag{14.41}$$

On the other hand for TM modes (having nonvanishing E_x, E_z and H_y) none of the field components satisfy a scalar wave equation. Even for TM modes, it has been possible to obtain exact solutions for the following profiles (Love and Ghatak, 1979)

$$
\left.
\begin{aligned}
n^2(x) &= n^2(0)\,e^{-|x|/a} \\
n^2(x) &= n^2(0)\,\mathrm{sech}^2\,(2x/a) \\
n^2(x) &= n^2(0)(1 + 2|x|/a)^{-2}
\end{aligned}
\right\}
\tag{14.42}
$$

In this section we will use the WKB method to study the propagation characteristics of inhomogeneous waveguides. This method has been shown to give accurate results for most practical waveguides.

The theory behind the WKB method is fairly involved and is discussed in most books on quantum mechanics (see, e.g., Ghatak and Lokanathan (1984)). The WKB method is applicable when the fractional variation of the refractive index is very small in distances of the order of λ; however, it is often possible to incorporate necessary corrections to take into account abrupt variation of refractive index.

As we have already seen when the refractive index depends only on the x-coordinate, the field component $E_y(x)$ satisfies the scalar wave equation for TE modes:

$$
\mathrm{d}^2\psi/\mathrm{d}x^2 + \kappa^2(x)\psi = 0
\tag{14.43}
$$

where

$$
\kappa(x) = (k_0^2 n^2(x) - \beta^2)^{\frac{1}{2}}
\tag{14.44}
$$

represents the x-component of the propagation vector of the local plane wave (see Fig. 14.8). Since a mode is a standing wave pattern in the transverse direction, the total transverse phase change of the propagating wave in going

Fig. 14.8 According to Snell's law a ray propagating in an inhomogeneous medium bends towards the region of higher refractive index. At any point x, the ray will correspond to a local plane wave with the wave vector $\mathbf{k}_0 n(x)$ directed along the tangent to the ray at that point. If θ is the angle made by the tangent with the z-axis, then β and $\kappa(x)$ will represent respectively the z and x-components of the wave vector. To the right of the figure is a typical refractive index variation.

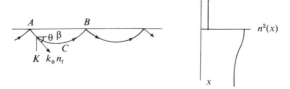

from point A to point B (one cycle away) must be an integral multiple of 2π (see Section 14.3). Thus, the condition which will determine the β-values will be:

$$\int_A^B \kappa(x)\,dx - 2\Phi_1 - 2\Phi_2 = 2m\pi \tag{14.45}$$

where $-2\Phi_1$ and $-2\Phi_2$ are the abrupt phase changes suffered by the wave at the points A and C respectively (see Figs. 14.8 and 14.9). Eq. (14.45) is known as the WKB quantization condition. We now consider two cases.

Case 1 The first corresponds to an $n^2(x)$ variation such that there is no discontinuity in $n^2(x)$ at the turning points (see Fig. 14.9). For such a case $\Phi_1 = \Phi_2 = \frac{1}{4}\pi$, (see Problem 14.11) and

$$\int_A^B \kappa(x)\,dx = 2\int_A^C \kappa(x)\,dx = (2m+1)\pi$$

or

$$\int_A^C [k_0^2 n^2(x) - \beta^2]^{\frac{1}{2}}\,dx = (m+\tfrac{1}{2})\pi \tag{14.46}$$

The points A and C are the points at which $\kappa(x)$ vanishes; these points are known as the turning points or the caustics. At the turning points $\beta = k_0 n(x)$ and the ray turns back.

We illustrate the use of Eq. (14.46) for an infinitely extended parabolic profile for which

$$n^2(x) = n_1^2[1 - 2\Delta(x/a)^2] \tag{14.47}$$

The turning points are

$$x_{1,2} = \mp \left[\frac{n_1^2 - \beta_m^2/k_0^2}{2\Delta n_1^2/a^2} \right]^{\frac{1}{2}} = \mp\alpha \tag{14.48}$$

Fig. 14.9 A ray propagating in a graded index waveguide. The points C and A are caustics where phase changes of $-\frac{1}{2}\pi$ occur.

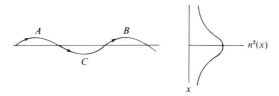

Eq. (14.46) takes the form

$$(m + \tfrac{1}{2})\pi = \int_{-\alpha}^{\alpha} [(k_0^2 n_1^2 - \beta_m^2) - 2n_1^2 k_0^2 \Delta(x/a)^2]^{\frac{1}{2}} \, dx$$

$$= n_1 k_0 (2\Delta/a^2)^{\frac{1}{2}} \int_{-\alpha}^{\alpha} (\alpha^2 - x^2)^{\frac{1}{2}} \, dx$$

$$= \pi n_1 k_0 (2\Delta/a^2)^{\frac{1}{2}} \alpha^2 / 2$$

On simplification we obtain

$$\beta_m = n_1 k_0 [1 - (2m + 1)(2\Delta/k_0^2 n_1^2 a^2)^{\frac{1}{2}}]^{\frac{1}{2}} \tag{14.49}$$

which happens to be identical to the result obtained from the exact solution of the scalar wave equation (see Appendix D).

Case 2 When there is an abrupt change in the refractive index at the turning point, the phase change on reflection would be

$$-2\Phi = -2\tan^{-1} \left(\frac{\beta^2 - k_0^2 n_2^2}{k_0^2 n_t^2 - \beta^2} \right)^{\frac{1}{2}} \tag{14.50}$$

where n_t and n_2 represent the refractive index at the turning point as we approach the turning point from the two directions.

As an example we consider the step index symmetric profile having a film of refractive index n_f surrounded by a medium of refractive index n_s:

$$\Phi_1 = \Phi_2 = \tan^{-1} \left(\frac{\beta^2 - k_0^2 n_s^2}{k_0^2 n_f^2 - \beta^2} \right)^{\frac{1}{2}} \tag{14.51}$$

Thus Eq. (14.45) becomes

$$d(k_0^2 n_f^2 - \beta^2)^{\frac{1}{2}} = 2\tan^{-1} \left(\frac{\beta^2 - k_0^2 n_s^2}{k_0^2 n_f^2 - \beta^2} \right)^{\frac{1}{2}} + m\pi; \qquad m = 0, 1, 2, \ldots \tag{14.52}$$

which is again identical to the exact result (see Eq. (14.29)).

For general refractive index profiles, it may not be possible to evaluate the integral appearing in Eq. (14.45) analytically, and one has then to resort to numerical methods; this procedure obviously involves much less computational effort as compared to solving numerically the scalar wave equation.

As an example, we consider the exponential profile for which

$$n^2(x) = K(x) = \begin{cases} K_0 + \Delta K \, e^{-x/d}; & x > 0 \\ K_1; & x < 0 \end{cases} \tag{14.53}$$

where the region $x < 0$ corresponds to the cover, $(K_0 + \Delta K)^{\frac{1}{2}}$ is the refractive index on the surface of the waveguide, $K_0^{\frac{1}{2}}$ is the substrate refractive index and d represents the depth of the waveguide. One can obtain exact solutions for the TE modes of such a waveguide (see Problem 14.8). Hocker and Burns (1975) have used simple numerical techniques for solving Eq. (14.45) for an exponential profile and have shown that the WKB method gives sufficiently accurate results. For further details on using the WKB method readers may look up Gedeon (1974), Hocker and Burns (1975), Janta and Ctyroky (1978).

14.5 Strip waveguides

In the planar waveguides discussed above, no lateral confinement of the light wave exists, i.e., the light wave is confined near the surface of the substrate but can diffract in the plane parallel to the surface. One can provide a lateral confinement also and one then obtains what is referred to as a strip waveguide (Fig. 14.2); lateral confinement is essential for making efficient modulators, directional couplers etc. (see Sec. 14.6). Indeed there are a large class of waveguides used in integrated optics which consist of rectangular or near rectangular cores surrounded by media of lower refractive indices. Obviously in order to be able to design efficient integrated optical devices such as directional couplers, filters, etc., it is very important to understand the modal properties of such rectangular core waveguides. However, an accurate analytical solution of such waveguides is extremely difficult due to the presence of corners. In this section we will give exact solutions of the scalar wave equation for profiles which can be written in the following form:

$$n^2(x, y) = n'^2(x) + n''^2(y) \tag{14.54}$$

where the functions $n'^2(x)$ and $n''^2(y)$ are judiciously chosen so that $n^2(x, y)$ closely resembles the actual waveguide. We illustrate the use of the method by considering a specific example.

We follow Kumar, Thyagarajan and Ghatak (1983) and assume

$$n'^2(x) = \begin{cases} \frac{1}{2}n_1^2; & |x| < a/2 \\ n_2^2 - \frac{1}{2}n_1^2; & |x| > a/2 \end{cases} \tag{14.55}$$

$$n''^2(y) = \begin{cases} \frac{1}{2}n_1^2; & |y| < b/2 \\ n_2^2 - \frac{1}{2}n_1^2; & |y| > b/2 \end{cases} \tag{14.56}$$

The corresponding refractive index distribution is shown in Fig. 14.10(a), such a waveguide closely resembles the rectangular core waveguide shown in Fig. 14.10(b). Since in most practical cases, $n_1 \approx n_2$, the effect of the difference in the refractive index can be accurately incorporated by using first order perturbation theory (see Problem 14.12). Also for well guided modes (i.e., for large V numbers of the waveguide) since the fractional energy contained in the corner regions (where $n^2 = 2n_2^2 - n_1^2$) will be small, the propagation constant of the modes of the waveguide shown in Fig. 14.10(a) are expected to be very close to those shown in Fig. 14.10(b).

We next solve the scalar wave equation

$$\partial^2\psi/\partial x^2 + \partial^2\psi/\partial y^2 + [k_0^2 n^2(x, y) - \beta^2]\psi = 0 \tag{14.57}$$

where k_0 is the free space wavenumber and β is the propagation constant. For $n^2(x, y)$ given by Eq. (14.54), Eq. (14.57) can be solved by using the method of separation of variables, i.e., assuming

$$\psi(x, y) = X(x)Y(y) \tag{14.58}$$

where $X(x)$ and $Y(y)$ satisfy the one-dimensional equations

$$d^2 X/dx^2 + [n'^2(x)k_0^2 - \beta_1^2]X = 0 \tag{14.59}$$

$$d^2 Y/dy^2 + [n''^2(y)k_0^2 - \beta_2^2]Y = 0 \tag{14.60}$$

where

$$\beta^2 = \beta_1^2 + \beta_2^2 \tag{14.61}$$

If we now substitute for $n'^2(x)$ from Eq. (14.55) in Eq. (14.59) we get

$$\begin{aligned} d^2 X/d\xi^2 + \mu_1^2 X = 0; & \qquad |\xi| \leqslant 1 \\ d^2 X/d\xi^2 - (V_1^2 - \mu_1^2)X = 0; & \qquad |\xi| \geqslant 1 \end{aligned} \tag{14.62}$$

where

$$\xi = 2x/a, \quad V_1 = k_0(a/2)(n_1^2 - n_2^2)^{\frac{1}{2}} \tag{14.63}$$

Fig. 14.10 (a) The rectangular core waveguide described by Eqs. (14.54)–(14.56). This waveguide closely resembles the waveguide shown in (b) when $n_2 \approx n_1$. Also for well guided modes the fraction of the energy present in the corner regions (where $n^2 = 2n_2^2 - n_1^2$) will be very small and the propagation constants of the waveguide shown in (a) are expected to closely match those shown in (b).

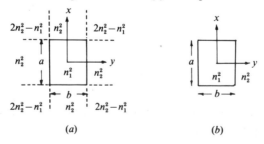

(a)   (b)

and

$$\mu_1 = (a/2)(k_0^2 n_1^2/2 - \beta_1^2)^{\frac{1}{2}} \tag{14.64}$$

since $n'^2(x)$ is symmetric about $x = 0$, the function $X(x)$ will be either a symmetric or an antisymmetric function of x. For the symmetric case the solution will be (see Chapter 11)

$$X(x) = \begin{cases} A \cos \mu_1 \xi & |\xi| \leqslant 1 \\ B e^{-(V_1^2 - \mu_1^2)^{\frac{1}{2}} |\xi|} & |\xi| \geqslant 1 \end{cases} \tag{14.65}$$

Continuity of $X(x)$ and its derivative at $\xi = 1$ will lead to the following transcendental equation

$$\mu_1 \tan \mu_1 = (V_1^2 - \mu_1^2)^{\frac{1}{2}} \tag{14.66}$$

which will determine the allowed values of μ_1. Similarly for the antisymmetric mode we will have

$$\mu_1 \cot \mu_1 = -(V_1^2 - \mu_1^2)^{\frac{1}{2}} \tag{14.67}$$

In a similar manner, we can consider the y dependent solutions which will be either symmetric or antisymmetric in the y-coordinate. Continuity conditions of $Y(y)$ at $y = b/2$ will give us

$$\mu_2 \tan \mu_2 = (V_2^2 - \mu_2^2)^{\frac{1}{2}} \quad \text{symmetric in } y \tag{14.68}$$

$$\mu_2 \cot \mu_2 = -(V_2^2 - \mu_2^2)^{\frac{1}{2}} \quad \text{antisymmetric in } y \tag{14.69}$$

where

$$\left. \begin{aligned} V_2 &= k_0(b/2)(n_1^2 - n_2^2)^{\frac{1}{2}} \\ \mu_2 &= (b/2)(k_0^2 n_1^2/2 - \beta_2^2)^{\frac{1}{2}} \end{aligned} \right\} \tag{14.70}$$

Thus

$$\beta^2 = \beta_1^2 + \beta_2^2 = n_1^2 k_0^2 - 4\mu_1^2/a^2 - 4\mu_2^2/b^2 \tag{14.71}$$

It is convenient to define the normalized propagation constant

$$\mathscr{P}^2 = \frac{\beta^2 - k_0^2 n_2^2}{k_0^2(n_1^2 - n_2^2)} = 1 - \frac{\mu_1^2}{V_1^2} - \frac{\mu_2^2}{V_2^2} \tag{14.72}$$

where μ_1 and μ_2 are obtained by solving the transcendental equations given by Eqs. (14.66)–(14.69) depending on whether the mode is symmetric or antisymmetric along x and y.

The field patterns can easily be written down. For example, for the mode which is symmetric both along x and y,

$$\psi(x,y) = \begin{cases} A \cos \mu_1 \xi \cos \mu_2 \eta; \quad |\xi| \leqslant 1, |\eta| \leqslant 1 \\[1em] \dfrac{A \cos \mu_2}{\exp\left[-(V_2^2 - \mu_2^2)^{\frac{1}{2}}\right]} \cos \mu_1 \xi \exp\left[-(V_2^2 - \mu_2^2)^{\frac{1}{2}}\eta\right]; \\[1em] \hspace{6cm} |\xi| \leqslant 1, |\eta| \geqslant 1 \\[1em] \dfrac{A \cos \mu_1}{\exp\left[-(V_1^2 - \mu_1^2)^{\frac{1}{2}}\right]} \cos \mu_2 \eta \exp\left[-(V_1^2 - \mu_1^2)^{\frac{1}{2}}\xi\right]; \\[1em] \hspace{6cm} |\xi| \geqslant 1, |\eta| \leqslant 1 \\[1em] \dfrac{A}{\exp\left[-(V_1^2 - \mu_1^2)^{\frac{1}{2}}\right]\exp\left[-(V_2^2 - \mu_2^2)^{\frac{1}{2}}\right]} \\[1em] \quad \times \left(\exp\left[-(V_1^2 - \mu_1^2)^{\frac{1}{2}}\xi\right]\exp\left[-(V_2^2 - \mu_2^2)^{\frac{1}{2}}\eta\right]; \\[1em] \hspace{5cm} |\xi| \geqslant 1, |\eta| \geqslant 1 \quad (14.73) \end{cases}$$

where $\eta = 2y/b$. In a similar manner, we can construct fields which are symmetric along x and antisymmetric along y, antisymmetric along x and symmetric along y and antisymmetric along both x and y.

It is of interest to mention that using the modal fields corresponding to the configuration shown in Fig. 14.10(a), it is possible to get an accurate expression for the propagation constant for the profile given in Fig. 14.10(b) by using the perturbation method (see Problem 14.12). The results are shown in Fig. 14.11. For a comparison, we have also plotted the propagation constant calculated using more accurate numerical calculations (Goell,

Fig. 14.11 Variation of the normalized propagation constant $\mathscr{P}^2$ versus the waveguide parameter V_1 for a rectangular cross section guide with $b = 2a$. The solid curves correspond to the calculations using Eq. (14.72), the dotted curves to the results after applying first order perturbation theory (see Problem 14.12) and the dashed curves to the numerical calculations of Goell (1969). (Adapted from Kumar, *et al.* (1983).)

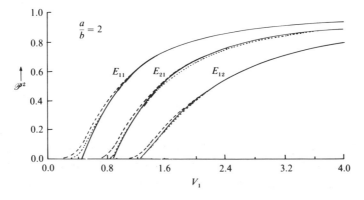

1969). It can be seen that the approximate analysis becomes more and more accurate as the V value becomes farther and farther away from the cutoff V value of the mode.

In the above analysis we have considered the scalar modes of a rectangular core waveguide structure. The exact vector modes are extremely difficult to obtain; however, the above analysis can be modified to obtain the propagation constant of the predominantly x-polarized (E^x_{pq}) mode and the predominantly y-polarized (E^y_{pq}) mode. For example, the E^x_{pq} mode will approximately correspond to a TM mode along the x-direction and a TE mode along the y-direction. In order to get an eigenvalue equation, we make E_x and $\partial E_x/\partial y$ continuous at $y = \pm b/2$ and $n^2 E_x$ and $(1/n^2)\,\partial(n^2 E_x)/\partial x$ continuous at $x \pm a/2$. The continuity conditions at $y = \pm b/2$ will lead to the same transcendental equations (Eqs. (14.68) and (14.69)): however, the transcendental equation corresponding to the continuity conditions at $x = \pm a/2$ will be

$$\mu_1 \tan \mu_1 = \frac{1}{\gamma_1}(V_1^2 - \mu_1^2)^{\frac{1}{2}} \qquad \text{symmetric along } x \qquad (14.74)$$

$$\mu_1 \cot \mu_1 = -\frac{1}{\gamma_1}(V_1^2 - \mu_1^2)^{\frac{1}{2}} \qquad \text{antisymmetric along } x \qquad (14.75)$$

where $\gamma_1 = n_2^2/n_1^2$. The propagation constant will again be given by Eq. (14.72).

We conclude this section by noting that the above method can easily be extended to profiles where $n'^2(x)$ and $n''^2(y)$ are asymmetric functions. For example, by appropriate choice of these functions one can carry out analysis for a profile which is very close to an embedded strip waveguide (see Problem 14.13).

14.6 Some guided wave devices

In the earlier sections we have discussed in detail the propagation of light in planar and strip waveguides. By using various effects such as the electrooptic effect, the acoustooptic effect or the magnetooptic effect one can modulate, switch or deflect the light beam propagating through these waveguides. The electrooptic effect refers to the change in the refractive index of a material on being acted upon by an external electric field; this is discussed in considerable detail in Chapter 15. The acoustooptic effect refers to the change in the refractive index of the material on applying an acoustic field (see Chapters 16–19) and the magnetooptic effect refers to the effect of an external magnetic field on the material. Interaction of the propagating light beam with the various externally applied fields leads to their modulation.

The electrooptic effect is one of the major effects used in fabricating active devices in integrated optics and we will restrict our discussion to just this effect. In Chapter 19 we will discuss the integrated optic spectrum analyser which uses the acoustooptic effect. Details of various other integrated optical devices can be found in, e.g., Tamir (1979), Tien (1977), Alferness (1981).

In the following we will discuss a phase modulator, an amplitude modulator using the Mach–Zehnder configuration and a directional coupler switch.

14.6.1 *Phase modulator*

Let us consider a strip waveguide formed in an electrooptic material such as lithium niobate or lithium tantalate. If an electric field is applied to the strip waveguide then the refractive index of the waveguide will change due to the electrooptic effect. This would lead to a corresponding change in the effective index (β/k_0) of the mode propagating through the waveguide.

Fig. 14.12 A typical strip waveguide phase modulator as fabricated by Kaminow, *et al.* (1975) in Lithium niobate. The *c*-axis of the crystal is along the *y*-direction perpendicular to the propagation direction and to the normal to the waveguide air interface. Electric fields are applied through planar aluminium electrodes deposited on either side of the strip waveguide. Voltage applied between the electrodes generates an electric field parallel to the *c*-axis. Light propagating through the strip waveguide becomes phase modulated because of a change in β of the propagating mode because of the electrooptic effect (Kaminow, *et al.*, 1975).

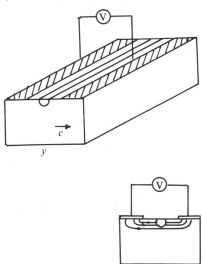

This change in turn modulates the phase of the propagating light beam. This is the basic principle behind a phase modulator.

Fig. 14.12 shows the strip waveguide phase modulator reported by Kaminow, Stulz and Turner (1975). The strip waveguide is formed in lithium niobate by in-diffusion of titanium; the strip waveguide is about 4.6 μm in width and 1 μm in depth. In order to apply the electric field efficiently, two aluminium electrodes 30 mm long, 25 μm wide and 9.3 μm apart are evaporated on either side of the channel waveguide. For a pair of such electrodes whose separation w_1 is much greater than the optical waveguide depth, the electric field is nearly uniform over the waveguide cross section and is approximately given by (Kaminow, *et al.* 1975)

$$E = 2V/\pi w_1 \tag{14.76}$$

This field is directed nearly parallel to the surface of the guide, i.e., along the y-direction (see Fig. 14.12). The configuration chosen by Kaminow, *et al.* (1975) is such that the c-axis of the lithium niobate crystal is along the y-direction. Thus for a mode propagating in the waveguide and predominantly polarized along the y-direction, the electrooptic change in effective index will be (see Sec. 15.4)

$$-\frac{n_e^3}{2} r_{33} \frac{2V}{\pi w_1} \tag{14.77}$$

where it is assumed that the effective index of the mode is approximately equal to n_e. If the length of the electrode is l, then the electrooptically induced phase change will be

$$\Delta\phi = -\frac{\omega}{c} \frac{n_e^3}{2} r_{33} \frac{2V}{\pi w_1} l \tag{14.78}$$

Hence to produce 1 rad phase shift with $l = 3$ cm, the required voltage will be

$$V = \frac{c}{\omega} \frac{2}{n_e^3} \frac{\pi w_1}{2l} \frac{1}{r_{33}} = \frac{\lambda_0 w_1}{2 n_e^3 r_{33} l} \tag{14.79}$$
$$\approx 0.3 \text{ V}$$

at $\lambda_0 = 6328$ Å (see Table 15.1 for values of n_e and r_{33}); this has also been confirmed experimentally. The calculated value of the capacitance of the modulator is 10 pF while the measured value is 12 pF.

For the modulator drive configuration used by Kaminow, *et al.* (1975) the bandwidth of the modulator is

$$\Delta f = 1/\pi RC = 530 \text{ MHz}$$

Fig. 14.13 (*a*) An integrated optic amplitude modulator using the Mach–Zehnder interferometer configuration. A single mode waveguide splits into two single mode waveguides which after a certain length rejoin to form a single mode waveguide. Oppositely directed electric fields are applied on the two arms of the interferometer by the electrode pattern as shown. A voltage V applied at the electrodes creates a phase shift of $+\frac{1}{2}\Delta\phi$ and $-\frac{1}{2}\Delta\phi$ on the light propagating through the two arms, leading to a phase difference of $\Delta\phi$. The beams of light from each arm interfere at the output leading to amplitude modulation in accordance with the applied voltage. The branching angle is typically 1° and the length of arms are $\sim 5\,\text{mm}$. The c-axis of the Lithium niobate crystal is as shown; thus light polarized parallel to the c-axis will be modulated by the r_{33} coefficient. (*b*) An oscilloscope trace of the variation of the output light intensity versus the applied voltage in a Mach–Zehnder interferometer modulator. The upper trace corresponds to the applied voltage which rises linearly from 0 to 8 V. The lower trace shows the output light intensity and shows the sinusoidal response with applied voltage. One half cycle of intensity modulation corresponds to about 1.6 V. The light used was at $\lambda = 5145\,\text{Å}$. (From Papuchon and Puech (1977).)

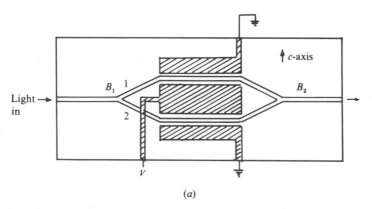

(*a*)

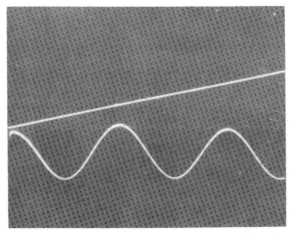

(*b*)

where we have taken $R = 50\,\Omega$ and $C = 12\,\text{pF}$. The drive power required per unit bandwidth will be

$$\frac{P}{\Delta f} = \frac{V^2/2R}{1/\pi RC} = \frac{\pi C V^2}{2} \tag{14.80}$$

Thus for 1 rad phase shift the power per unit bandwidth will be

$$P/\Delta f = \pi \times 12 \times 10^{-12} \times 0.09/2 \approx 1.7\,\mu\text{W/MHz}$$

Comparing the drive voltages and the power per unit bandwidth of the integrated optic modulator with those of bulk modulators (see Chapter 15), one can see a clear advantage in the integrated optical modulator devices.

14.6.2 *Mach–Zehnder interferometer modulator and switch*

We just saw how one can obtain phase modulation of a light beam propagating in a strip waveguide through the electrooptic effect. This phase modulation can easily be converted into amplitude or intensity modulation by using simple interferometric methods. Fig. 14.13(a) shows a Mach–Zehnder arrangement of waveguides in which a single mode strip waveguide divides into two single mode waveguides, which rejoin to form another single mode waveguide. This arrangement is similar to the Mach–Zehnder interferometer used in bulk optics. Electric fields are applied on either waveguide through planar electrodes as shown in Fig. 14.13(a). Opposite phase changes are induced in the light propagating in each arm by applying oppositely directed electric fields on the two waveguides.

Light entering through the input single mode waveguide splits into two parts, propagating through the two single mode waveguides. Electric fields applied to the two waveguides introduce a phase difference of $\Delta\phi$ between the light propagating through the two waveguides. Interference between these beams at the exit leads to a modulation of the amplitude. A phase difference of π introduced between the arms, results in a field distribution at the output branch which corresponds roughly to the second mode of the output single mode waveguide and hence is radiated away (see Fig. 14.14).

Let us assume a mode of amplitude E incident on the branch B_1 of the interferometer (see Fig. 14.13(a)). Let the amplitudes of the modes excited in the arms 1 and 2 of the interferometer be E_1 and E_2. If $\Delta\phi$ is the phase difference introduced between the two arms due to the electrooptic effect, then the light intensity emerging at the output of the interferometer will be proportional to

$$E_1^2 + E_2^2 + 2E_1 E_2 \cos \Delta\phi$$

If the branching at B_1 and B_2 are perfect, then $E_1 = E_2$ and the light emerging from the output would vary between 0 and maximum corresponding to high contrast.

In the experiment reported by Papuchon (1978), the strip waveguides are formed by in-diffusion of titanium in lithium niobate and are 2 μm wide. The length of each arm is 5 mm and the branching angle between the two waveguides is 1°. The electrodes are also 5 mm long and are separated by 5 μm. The c-axis of the lithium niobate crystal is as shown in Fig. 14.13(a).

Fig. 14.13(b) shows the variation of the output intensity at the output waveguide as a function of the applied electric field. The upper trace corresponds to the applied voltage and varies from 0 to 8 V. The lower trace is the optical output. One can clearly see a sinusoidal variation in output intensity due to the $\cos \Delta \phi$ term where $\Delta \phi$ is proportional to the applied voltage. The voltage required for a phase difference of π was only 1.6 V at $\lambda_0 = 5145$ Å. The rise time of the circuit has been shown to be less than 700 ps.

It should be mentioned that a similar arrangement is also well known in bulk optics but in the integrated optical format, the device requires only low voltages for operation, is compact and easily reproducible, has a fast response and is also much more stable.

Fig. 14.14 (a) Modes arriving in phase at the output branch add up approximately to form the fundamental mode of the output single mode waveguide and thus propagate down the output guide. (b) Modes arriving π out of phase at the output branch add up approximately to form the second mode and thus are not guided by the output single mode waveguide and are radiated away.

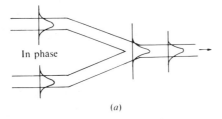

(a)

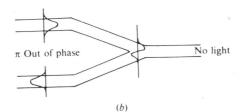

(b)

The Mach–Zehnder interferometer arrangement can be used to design integrated optical analog to digital converters (Leonberger, Woodward and Spears, 1979; King and Jackson, 1980), optically bistable devices (Ito, Ogawa and Inaba, 1979; Schnapper, Papuchon and Puech, 1979).

14.6.3 The optical directional coupler

We have seen in Sec. 14.2 that a guided mode has fields extending beyond the guiding region into the surrounding medium, decaying exponentially as one moves far from the waveguide boundary. If one brings two such waveguides close together so that the evanescent fields of the propagating modes in the two waveguides overlap, then there is a coupling between the two waveguides (see Problem 11.13). Such a device is called a directional coupler and is shown in Fig. 14.15. If the separation between the two waveguides is so large that the evanescent field of one waveguide becomes insignificant at the other waveguide, then the interaction is very small and the energy in either waveguide propagates independently of the other waveguide. On the other hand, when the separation between the guides becomes so small that the evanescent fields of the two waveguides overlap to a considerable extent, then, as we will show, there is an energy exchange between the two guides. This energy exchange can be used in building an optical modulator or an optical switch. The power transfer due to evanescent field coupling may also lead, in some cases, to undesirable cross talk between two waveguides.

Fig. 14.15 shows a directional coupler consisting of two single mode strip waveguides which are in close proximity over a length L. We will show below that if energy is incident on one of the waveguides, then there is a periodic exchange of energy between the two waveguides. If the propagation constants of the modes in the two waveguides are identical then there is a complete transfer of energy from one waveguide to the other and if the

Fig. 14.15 An optical directional coupler in which two waveguides are at close proximity over a length L. The evanescent fields associated with the propagating modes in the two waveguides interact and lead to a periodic exchange of energy between the two waveguides. For two identical waveguides, the energy incident in waveguide 1 may appear in the output 3 or 4 depending on the length L.

propagation constants are not identical then there is only an incomplete transfer of energy.[†]

We will now briefly discuss the theory behind the operation of a directional coupler switch. As discussed earlier the modal field of a single waveguide varies with z in the form $e^{-i\beta_1 z}$ and hence if we write the amplitude of a mode at z as $a(z)$, then we may write

$$da/dz = -i\beta_1 a \tag{14.81}$$

Here β_1 is the propagation constant of the mode in the waveguide 1. Similarly if $b(z)$ represents the amplitude at z of a mode in waveguide 2 with propagation constant β_2, one may write

$$db/dz = -i\beta_2 b \tag{14.82}$$

Eqs. (14.81) and (14.82) are valid as long as waveguides 1 and 2 are noninteracting i.e., the modal fields in waveguides 1 and 2 do not overlap at all. When the two waveguides are close together, then modes in the two waveguides can interact through the evanescent field. Hence in the presence of such an interaction, one may write for the variation of amplitudes of the modes in the two waveguides as

$$da/dz = -i\beta_1 a - i\kappa_{12} b(z) \tag{14.83}$$

$$db/dz = -i\beta_2 b - i\kappa_{21} a(z) \tag{14.84}$$

where the constants κ_{12} and κ_{21} represent the strength of interaction between the two modes and are referred to as coupling constants; in the absence of any interaction between the two modes $\kappa_{12} = \kappa_{21} = 0$. The above equations show that in the presence of interaction, the amplitude of the mode in a waveguide depends on the amplitude of the mode in the other waveguide. The coupling constants depend upon the waveguide parameters, the separation between the waveguides and the wavelength of operation. The determination of the coupling constant for a given geometry of interacting waveguides is an important problem and can be carried out using the coupled mode theory which is briefly discussed in Appendix G. Also for identical waveguides one can show that

$$\kappa_{21} = \kappa_{12} \tag{14.85}$$

Eqs. (14.83) and (14.84) now describe the evolution of $a(z)$ and $b(z)$ with z.

[†] This problem is very similar to a pair of coupled simple pendulums. If the time periods of both the pendulums are same then there will be a periodic and complete exchange of energy from one pendulum to the other; on the other hand for nonidentical pendulums, there is only an incomplete transfer of energy.

In order to solve this set of equations, we write

$$a(z) = a_0 e^{-i\beta z} \tag{14.86}$$

$$b(z) = b_0 e^{-i\beta z} \tag{14.87}$$

i.e., we postulate the existence of a wave in the system consisting of the two coupled waveguides, which propagates with a phase constant β and which is a superposition of the modes of waveguides 1 and 2 with amplitudes a_0 and b_0. Substituting from Eqs. (14.86) and (14.87) in Eqs. (14.83) and (14.84) we get

$$-i\beta a_0 = -i\beta_1 a_0 - i\kappa_{12} b_0$$

or

$$a_0(\beta - \beta_1) - \kappa_{12} b_0 = 0 \tag{14.88}$$

and

$$b_0(\beta - \beta_2) - \kappa_{21} a_0 = 0 \tag{14.89}$$

In order that Eqs. (14.88) and (14.89) have nontrivial solutions, one must have

$$\beta^2 - \beta(\beta_1 + \beta_2) + (\beta_1 \beta_2 - \kappa^2) = 0$$

where $\kappa = (\kappa_{12}\kappa_{21})^{\frac{1}{2}}$; thus

$$\beta_{s,a} = \tfrac{1}{2}(\beta_1 + \beta_2) \pm [\tfrac{1}{4}(\beta_1 - \beta_2)^2 + \kappa^2]^{\frac{1}{2}} \tag{14.90}$$

Thus, in the coupled waveguides, one has two independent sets of modes, one propagating with a propagation constant β_s and the other with β_a. Therefore, the general solution of Eqs. (14.83) and (14.84) can be written as

$$a(z) = a_s e^{-i\beta_s z} + a_a e^{-i\beta_a z} \tag{14.91}$$

$$b(z) = \frac{\beta_s - \beta_1}{\kappa_{12}} a_s e^{-i\beta_s z} + \frac{(\beta_a - \beta_1)}{\kappa_{12}} a_a e^{-i\beta_a z} \tag{14.92}$$

It is interesting to note that when the two waveguides are identical, i.e., when $\beta_1 = \beta_2 = \beta_0$, then

$$\beta_s = \beta_0 + \kappa \tag{14.93}$$

$$\beta_a = \beta_0 - \kappa \tag{14.94}$$

and we have for the mode with propagation constant $\beta = \beta_s$ using Eq. (14.88)

$$b_0 = a_0 \tag{14.95}$$

and for the mode with propagation constant $\beta = \beta_a$, using Eq. (14.88)

$$b_0 = -a_0 \tag{14.96}$$

Hence for this case the normal modes of the coupled waveguide are as shown in Fig. 14.16. The subscripts s and a stand for symmetric and antisymmetric modes.

We now assume that at $z = 0$, the mode in waveguide 1 is launched with unit power and that there is no power in waveguide 2. Then, we have the initial conditions, at $z = 0$,

$$a_s + a_a = 1 \tag{14.97}$$

$$\frac{\beta_s - \beta_1}{\kappa_{12}} a_s + \frac{\beta_a - \beta_1}{\kappa_{12}} a_a = 0 \tag{14.98}$$

Fig. 14.16 (*a*) and (*b*) correspond to the two lowest order modes of the pair of waveguides in a directional coupler considered as a single waveguide structure. Energy incident on one waveguide can be considered to be a linear combination of the two modes (*c*). As the modes propagate down the waveguides, they develop a phase difference due to the difference in β values of the two modes. Thus at some point a phase difference of π is developed which corresponds to the total energy in the other waveguide.

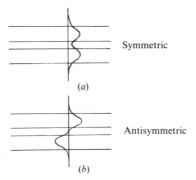

Symmetric

(*a*)

Antisymmetric

(*b*)

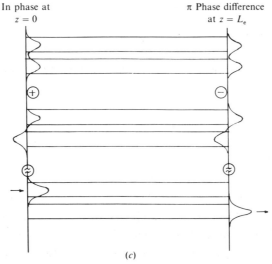

In phase at
$z = 0$

π Phase difference
at $z = L_e$

(*c*)

which give us

$$a_\mathrm{s} = \frac{\beta_1 - \beta_\mathrm{a}}{\beta_\mathrm{s} - \beta_\mathrm{a}}; \qquad a_\mathrm{a} = -\frac{\beta_1 - \beta_\mathrm{s}}{\beta_\mathrm{s} - \beta_\mathrm{a}} \qquad (14.99)$$

The power in guides 1 and 2 is proportional to $|a(z)|^2$ and $|b(z)|^2$ respectively. On substituting from Eq. (14.99) in Eqs. (14.91) and (14.92), one can show that

$$|a(z)|^2 = 1 - \frac{\kappa^2}{\frac{1}{4}\Delta\beta^2 + \kappa^2} \sin^2\left[\left(\tfrac{1}{4}\Delta\beta^2 + \kappa^2\right)^{\frac{1}{2}} z\right] \qquad (14.100)$$

$$|b(z)|^2 = \frac{\kappa^2}{\frac{1}{4}\Delta\beta^2 + \kappa^2} \sin^2\left[\left(\tfrac{1}{4}\Delta\beta^2 + \kappa^2\right)^{\frac{1}{2}} z\right] \qquad (14.101)$$

where

$$\Delta\beta = \beta_1 - \beta_2.$$

As can be seen from Eqs. (14.100) and (14.101), there is a periodic exchange of energy between the two waveguides with a period

$$h = \pi/\left(\tfrac{1}{4}\Delta\beta^2 + \kappa^2\right)^{\frac{1}{2}} \qquad (14.102)$$

The length

$$L_\mathrm{c} = h/2 = \pi/2\left(\tfrac{1}{4}\Delta\beta^2 + \kappa^2\right)^{\frac{1}{2}} \qquad (14.103)$$

is called the coupling length of the directional coupler and corresponds to the minimum interaction length required for the maximum energy transfer. It can also be seen from Eq. (14.101) that the maximum energy coupled is

$$|b|^2_\mathrm{max} = \kappa^2/\left(\tfrac{1}{4}\Delta\beta^2 + \kappa^2\right) \qquad (14.104)$$

and thus depends on $\Delta\beta$ which is the difference in the propagation constant between the two single mode waveguides. Complete energy transfer can take place only if $\Delta\beta = 0$, i.e., the propagation constants of the mode in each waveguide are the same. For such a case the coupling length becomes (see Fig. 14.16(c))

$$L_\mathrm{c0} = \pi/2\kappa = \pi/(\beta_\mathrm{s} - \beta_\mathrm{a}) \qquad (14.105)$$

For waveguides fabricated by titanium in-diffusion in lithium niobate coupling lengths may range from 200 μm to 1 cm (Alferness, Schmidt and Turner, 1979).

Fig. 14.17(a) shows the variation of $|a(z)|^2$ in the waveguide of a directional coupler for $\Delta\beta = 0$ and $\Delta\beta = \sqrt{(12)}\kappa$; for the latter value of $\Delta\beta$, the coupling length is exactly half of the coupling length for $\Delta\beta = 0$. Fig. 14.17(b) shows the periodic exchange of energy between the two waveguides with $\Delta\beta \approx 0$.

The fact that with a finite $\Delta\beta$ the coupling length decreases can be used when making an optical switch. Let us suppose that we have two identical waveguides ($\Delta\beta = 0$) and we make a directional coupler of length equal to L_{c0}. Thus the energy incident at the input of one waveguide comes out of the other waveguide at the exit. If now by some means we can introduce a finite $\Delta\beta$ between the two waveguides such that the coupling length is $\frac{1}{2}L_{c0}$, then at the exit the energy will again emerge from the first waveguide (see Fig. 14.17(a)). Thus one can switch the light energy from one waveguide to the other. This is the principle behind the optical directional coupler switch. The first state where light entering waveguide 1 emerges from waveguide 2 is referred to as the cross state ($\otimes$) and the latter state when the light entering waveguide 1 emerges from the same waveguide is called the parallel state

Fig. 14.17 (a) The variation of power in the two waveguides as a function of z for $\Delta\beta = 0$ and $\Delta\beta = \sqrt{(12)}\kappa$. For $\Delta\beta = 0$, the energy exchange is complete between the two waveguides and the coupling length is $\pi/2\kappa$. For $\Delta\beta \neq 0$, the energy exchange is incomplete and the coupling length is less than $\pi/2\kappa$. For a coupler of length $\pi/2\kappa$, energy incident in one waveguide emerges from the other waveguide ($\otimes$ state). If one introduces a $\Delta\beta = \sqrt{(12)}\kappa$ between the two waveguides, at the length $L = \pi/2\kappa$, the energy emerges from the first waveguide itself ($\ominus$ state). (b) Photograph showing energy exchange between two identical waveguides of a directional coupler. The propagating light is made visible by covering the waveguides by a polymer film (e.g., polyurethane or polymethylmethacrylate) doped with an organic dye (Rhodamine B or Rhodamine 6G) which fluoresces due to absorption of the energy in the evanescent field of the propagating mode (from Papuchon, (1978)).

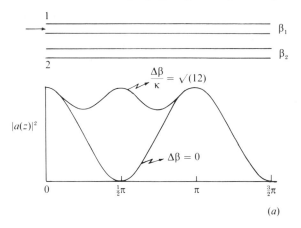

(a)

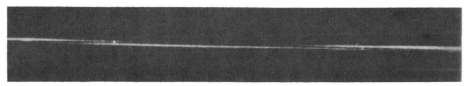

(b)

($\ominus$). The minimum $\Delta\beta$ required for going over from the cross state to the parallel state is

$$\Delta\beta = \sqrt{(12)}\kappa = \sqrt{3}(\pi/L_{\mathrm{co}}) \tag{14.106}$$

The propagation constants of the two waveguides can be changed through the electro optic effect. Thus by applying a voltage, light can be switched back and forth between the waveguides. In addition, applying an oscillating voltage will generate an oscillating $\Delta\beta$ and thus the light coming out of waveguides 1 and 2 will be modulated.

Fig. 14.18 shows a typical directional coupler switch configuration in lithium niobate (Papuchon, *et al.*, 1975). The substrate crystal has its c-axis perpendicular, to the surface of the substrate and the light propagating in the waveguide is predominantly polarized parallel to the c-axis. Oppositely directed electric fields are applied to the two waveguides by metal electrodes deposited on the waveguides so that the effective index of one waveguide mode increases while that of the other decreases. Since the light is polarized along the c-axis and the electric field is also applied along the c-axis, the refractive index change in the waveguide is determined by the r_{33} coefficient (see Chapter 15)

$$\Delta\beta \approx (2\pi/\lambda_0)\Delta n = (2\pi/\lambda_0)n_e^3 r_{33} V/d \tag{14.107}$$

The change in κ due to the applied electric field can be neglected to first

Fig. 14.18 An optical directional coupler switch in Lithium niobate. The c-axis points along the normal to the substrate surface and the electrodes are deposited on the waveguides. The electric fields point in opposite directions in the two waveguides and are approximately parallel to the c-axis, thus creating opposite refractive index changes in the two waveguides. The propagating mode is predominantly polarized parallel to the c-axis and thus the changes in the index produced correspond to the largest r_{33} coefficient (Papuchon, *et al* 1975).

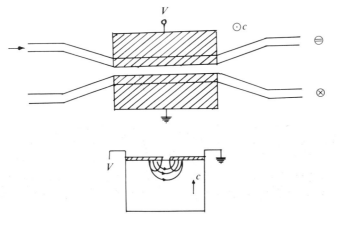

order. Since Eq. (14.106) gives the required $\Delta\beta$ to switch from the $\otimes$ state to $\ominus$ state it can be seen that the switch voltage can be reduced either by increasing the length or by reducing the electrode gap.

In the above discussion of the optical directional coupler switch, we have assumed that the coupler is of such a length that when no electric fields are applied the switch is in the cross state. This requires accurate control of the fabrication since one cannot adjust the $\otimes$ state electrically while the $\ominus$ state can be adjusted electrically. This limitation can be overcome if one uses the split electrode configuration as shown in Fig. 14.19; this configuration is referred to as a stepped $\Delta\beta$ reversed directional coupler switch. It can be shown (see e.g., Kogelnik and Schmidt, 1976) that with such a configuration where in the first half one introduces a propagation constant difference $+\Delta\beta$ and in the second half a difference of $-\Delta\beta$, both the $\otimes$ and $\ominus$ states can be adjusted electrically. This considerably eases the fabrication tolerance. Indeed one can have multiple sections of electrodes with alternating $\Delta\beta$. Typical switching voltages of 5 V for couplers 1 cm long have been demonstrated experimentally (Schmidt and Cross, 1978).

The directional coupler switch that we have discussed should find application in signal routing, in time division multiplexing of lower bit rate signals into one channel to be transmitted over a single fibre etc. The directional coupler also forms a basic configuration which can be used for modulation, wavelength filtering (Taylor, 1973, Alferness and Schmidt, 1978), bistable devices etc.

14.6.4 *Comparison of bulk and integrated optic modulators*

In Chapter 15, we will show that the power required per unit bandwidth (also called the specific energy) for driving a modulator is proportional to d^2/l, where d is the transverse dimension of the modulator

Fig. 14.19 An optical directional coupler switch with stepped $\Delta\beta$ reversal. In the first half of the waveguide, one introduces a propagation constant difference $+\Delta\beta$ between the two waveguides while in the second half the propagation constant difference is reversed to $-\Delta\beta$. In this configuration both the $\ominus$ and $\otimes$ states can be adjusted electrically.

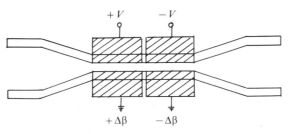

(assumed to be a square cross section) and l is the length of the modulator.

When one uses a bulk electrooptic modulator, the factor d^2/l is restricted to a minimum value of $4\lambda_0/n_0\pi$ because of diffraction effects (see Sec. 15.5). Thus because of diffraction of the beam to be modulated, for a given length of the modulator device, one requires a minimum value of the cross sectional dimension. This restriction is overcome by allowing the light to be guided in an optical waveguide of transverse dimensions comparable to the wavelength of light. This guiding phenomenon allows us to have cross sectional dimensions of the beam of the order of λ for essentially unrestricted lengths. For a strip waveguide whose transverse dimensions are of the order of λ, $d^2/l \sim \lambda^2/l$. Thus we get

$$\frac{(P/\Delta f)_{\text{waveguide}}}{(P/\Delta f)_{\text{bulk}}} \sim \frac{\lambda^2/l}{\lambda} = \frac{\lambda}{l} \tag{14.108}$$

For a modulator operating at $\lambda = 1\,\mu$m and a length of 1 mm, the specific energy required in the waveguide configuration is about a thousand times less than in the corresponding bulk case. We have assumed above that there is complete overlap between the optical field of the mode and the applied electric field. Thus by using a guided wave configuration reductions by a factor of a thousand are possible in the drive power requirements. Considerable reductions in drive power have indeed been realized experimentally.

In addition to the above advantage of the reduction in specific energy while using a waveguide modulator, the voltage required for a given depth of modulation is also very much reduced. This is because in the waveguide configuration, one can place the electrodes for applying the electric field across a gap of the order of the wavelength of light. Thus even with only a few volts applied on the electrodes one can generate electric fields of 10^6 V/m.

The integrated optical modulators that we have discussed are expected to find applications in very high bandwidth systems (at frequencies of modulation much above 1 GHz) where direct modulation of the laser transmitter itself may not be feasible, in realizing practical devices which may otherwise be too cumbersome in bulk optics, in wavelength division multiplexing etc.

Additional problems

Problem 14.7: Using an argument similar to the one used for obtaining Eq. (14.108), show that in a planar waveguide the reduction in specific energy as compared to the bulk case is $\sim (\lambda/l)^{\frac{1}{2}}$.

Problem 14.8: Consider a planar waveguide described by a dielectric constant variation of the form

$$K(x) = n^2(x) = \begin{cases} K_1; & x < 0 \\ K_0 + \Delta K e^{-x/d}; & x > 0 \end{cases} \tag{14.109}$$

$(K_1 < K_0 + \Delta K)$. Such waveguides are formed while fabricating waveguides using diffusion techniques. Obtain the TE modes of this waveguide.

Solution: For TE modes, the field component E_y satisfies the scalar wave equation

$$d^2 E_y/dx^2 + (k_0^2 n^2(x) - \beta^2)E_y = 0$$

Guided modes in the waveguide will have exponentially decaying solutions for $x < 0$ and for $x \to \infty$. Since the dielectric constant as $x \to \infty$ tends to K_0 and the maximum value of $K(x)$ is $K_0 + \Delta K$, the range of β values for guided modes will be

$$K_0 < \beta^2/k_0^2 < K_0 + \Delta K \tag{14.110}$$

assuming $K_0 > K_1$; otherwise the lower limit on β^2/k_0^2 would be K_1.
  For the region $x < 0$, the solution will be

$$E_y = A e^{\gamma_1 x}; \quad x < 0 \tag{14.111}$$

where

$$\gamma_1 = (\beta^2 - k_0^2 K_1)^{\frac{1}{2}} \tag{14.112}$$

The equation for E_y for $x > 0$ is

$$d^2 E_y/dx^2 + (k_0^2 K_0 + k_0^2 \Delta K e^{-x/d} - \beta^2)E_y = 0 \tag{14.113}$$

We now make the following substitution

$$\xi = 2k_0 d(\Delta K)^{\frac{1}{2}} e^{-x/2d} \tag{14.114}$$

and Eq. (14.115) reduces to

$$\xi^2 d^2 E_y/d\xi^2 + \xi dE_y/d\xi + (\xi^2 - \alpha^2)E_y = 0 \tag{14.115}$$

where

$$\alpha = 2d(\beta^2 - k_0^2 K_0)^{\frac{1}{2}} \tag{14.116}$$

Eq. (14.115) is nothing but the Bessel equation of order α. The solutions are $J_\alpha(\xi)$ and $J_{-\alpha}(\xi)$; the second solution tends to ∞ as $\xi \to 0$, i.e., $x \to \infty$. Thus the solution for guided modes is

$$E_y = B J_\alpha(2k_0 d(\Delta K)^{\frac{1}{2}} e^{-x/2d}); \quad x > 0 \tag{14.117}$$

Applying the boundary conditions of continuity of E_y and dE_y/dx at $x = 0$ and using the recurrence relations for Bessel functions, one finally obtains the following transcendental equation which determines the propagation constant of the modes:

$$\frac{J_{\alpha+1}(\xi_0) - J_{\alpha-1}(\xi_0)}{J_\alpha(\xi_0)} = \frac{\gamma_1 \lambda_0}{\pi(\Delta K)^{\frac{1}{2}}} \tag{14.118}$$

where

$$\xi_0 = 2k_0 d(\Delta K)^{\frac{1}{2}} \tag{14.119}$$

Problem 14.9: The energy flow is an electromagnetic field is given by the Poynting vector **S** defined by (see Eq. (11.86))

$$S = \tfrac{1}{2}\mathrm{Re}(E \times H^*) \tag{14.120}$$

Thus the total power in a mode (per unit length along y) propagating along the z direction is given by Eq. (14.45)

$$P_z = \int_{-\infty}^{\infty} S_z \, dx \tag{14.121}$$

Show that for the TE and TM modes $S_x = S_y = 0$ and

$$P_z = \frac{\beta}{2\omega\mu_0} \int_{-\infty}^{\infty} |E_y|^2 \, dx \qquad \text{TE modes} \tag{14.122}$$

$$P_z = \frac{\beta}{2\omega\epsilon_0} \int_{-\infty}^{\infty} \frac{1}{n^2} |H_y|^2 \, dx \qquad \text{TM modes} \tag{14.123}$$

Using the field expressions given in Eqs. (14.20) and (14.23) show that

$$P_z = \frac{\beta A^2}{4\omega\mu_0}\left(1 + \frac{\gamma_c^2}{\kappa_f^2}\right)\left(d + \frac{1}{\gamma_s} + \frac{1}{\gamma_c}\right) \qquad \text{TE modes} \tag{14.124}$$

$$P_z = \frac{\beta A^2}{4\omega\epsilon_0} \frac{1}{n_f^2 n_c^4 \kappa_f^2}(n_c^4 \kappa_f^2 + n_f^4 \gamma_c^2)$$

$$\times \left[d + \frac{n_f^2 n_s^2}{\gamma_s} \frac{\kappa_f^2 + \gamma_s^2}{n_s^4 \kappa_f^2 + n_f^4 \gamma_s^2} + \frac{n_f^2 n_c^2}{\gamma_c} \frac{\kappa_f^2 + \gamma_c^2}{n_c^4 \kappa_f^2 + n_f^4 \gamma_c^2} \right]$$

TM modes (14.125)

Problem 14.10: Consider a planar waveguide with the following dielectic constant distribution

$$n^2(x) = \begin{cases} n_s^2 - 2n_s \Delta n_s (x/d + b(x^2/d^2)); & x > 0 \\ n_1^2; & x < 0 \end{cases} \tag{14.126}$$

Obtain the TE modes of this waveguide. Use the WKB method and obtain the eigenvalue equation (Stewart, *et al.*, 1977; Ghatak, Khular and Thyagarajan 1978).

Problem 14.11: Consider an inhomogeneous medium having a refractive index variation along the x-direction. In Fig. 14.8, the point C represents the turning point of the ray. Show that there is an abrupt phase change of $-\tfrac{1}{2}\pi$ at the turning point.

Solution: When a wave undergoes total internal reflection at a boundary separating two media of indices n_1 and n_2 ($< n_1$), then the phase shift on reflection is given by

$$\chi = -2\Phi = -2\tan^{-1}\left[\left(\frac{\beta^2 - k_0^2 n_2^2}{k_0^2 n_1^2 - \beta^2}\right)^{\frac{1}{2}}\right] \tag{14.127}$$

where $\beta = k_0 n_1 \cos \theta$, θ being the angle made by the wave normal with the interface (see footnote on page 44), The phase shift at the turning point can be determined by a limiting procedure where we let the angle θ go to zero and let $(n_1 - n_2) \to 0$, i.e., the ray comes in at grazing incidence. In order to obtain the limiting value let us assume that

$$n_1 = n_0 + \delta$$
$$n_2 = n_0 - \delta \tag{14.128}$$

where δ is an infinitesimal quantity. We choose a ray incident at an angle θ with the interface given by

$$\cos \theta = 1 - (\delta/n_0) \tag{14.129}$$

such that

$$\beta = k_0 (n_0 + \delta) \cos \theta \simeq k_0 n_0 + O(\delta^2) \tag{14.130}$$

In the limit of $\delta \to 0$, the angle of incidence automatically tends to grazing incidence, i.e., $\theta \to 0$ as $\delta \to 0$.

Now

$$k_0^2 n_1^2 - \beta^2 = k_0^2 (n_0 + \delta)^2 - \beta^2 = 2k_0^2 n_0 \delta + O(\delta^2) \tag{14.131}$$

and

$$\beta^2 - k_0^2 n_2^2 = \beta^2 - k_0^2 (n_0 - \delta)^2 = 2k_0^2 n_0 \delta + O(\delta^2) \tag{14.132}$$

Thus in the limit of $\delta \to 0$, the refractive index difference between the two media $(n_1 - n_2 = 2\delta)$ tends to zero and the ray is incident at grazing incidence. The corresponding phase shift is

$$\chi = -2 \tan^{-1} (2k_0^2 n_0 \delta / 2k_0^2 n_0 \delta) = -2 \tan^{-1}(1)$$
$$= -\tfrac{1}{2}\pi \tag{14.133}$$

Problem 14.12: Using first order perturbation theory and considering the modes of the structure shown in Fig. 14.10(*a*) as the basis modes, show that the propagation constant β of the waveguide structure shown in Fig. 14.10(*b*) is given by

$$\beta^2 = \beta_0^2 + \beta_1^2 \tag{14.134}$$

where β_0 is the propagation constant given by Eq. (14.71)

$$\beta_1^2 = k_0^2 (n_1^2 - n_2^2) \left[1 + \left(\frac{V_1^2}{\mu_1^2} - 1 \right)^{\frac{1}{2}} \left(\frac{2\mu_1 + p \sin 2\mu_1}{1 + p \cos 2\mu_1} \right) \right]^{-1}$$
$$\times \left[1 + \left(\frac{V_2^2}{\mu_2^2} - 1 \right)^{\frac{1}{2}} \left(\frac{2\mu_2 + q \sin 2\mu_2}{1 + q \cos 2\mu_2} \right) \right]^{-1} \tag{14.135}$$

where $p = 1(-1)$ for modes symmetric (antisymmetric) in x and $q = 1(-1)$ for modes symmetric (antisymmetric) in y.

Problem 14.13: Consider an embedded strip waveguide as shown in Fig. 14.20. Construct a strip waveguide model which can be written in the form of Eq. (14.54) and which closely approximates the waveguide of Fig. 14.20.

Solution:

$$n'^2(x) = \begin{cases} n_0^2 - n_1^2/2; & x > 0 \\ n_1^2/2; & 0 > x > -a \\ n_2^2 - n_1^2/2; & x < -a \end{cases} \tag{14.136}$$

$$n''^2(y) = \begin{cases} n_1^2/2; & |y| < b/2 \\ n_2^2 - n_1^2/2; & |y| > b/2 \end{cases} \tag{14.136a}$$

Problem 14.14: Consider a directional coupler consisting of two identical wave-guides. Assuming unity power to be incident on waveguide 1, show that in the cross state if the power emerging from waveguide 1 is to be less than $-25\,\text{dB}$, the factor κL should be controlled within $\pm 3.5\%$.

Problem 14.15: A directional coupler can also be used as a 3 dB coupler in which light incident into one waveguide divides equally in both waveguides with half the power emerging from each waveguide. Consider a directional coupler with identical waveguides; (a) What should be the length of the coupler so that it behaves as a 3 dB coupler? (b) If the coupling length of the coupler is 2 mm and the 3 dB coupler is fabricated to have a length of 1.5 mm, what $\Delta\beta$ should be introduced so that it behaves as a 3 dB coupler? (c) What is the minimum value of κL required (where L is the length of the coupler) so that a 3 dB state can be obtained through introduction of a $\Delta\beta$?
(Answers: (a) $L = \pi/4\kappa$; (b) $\Delta\beta = 1.56\,\text{mm}^{-1}$; (c) $\kappa L \geqslant \frac{1}{4}\pi$.)

Problem 14.16: Consider a directional coupler with unit power incident on waveguide 1 and suppose that one requires a power P_2 in waveguide 2 after a section of length l. Show that if P_2 is to have the smallest sensitivity to changes in l, κ or $\Delta\beta$, the ratio of $\Delta\beta$ to κ should be

$$\frac{\Delta\beta}{\kappa} = 2\left(\frac{1}{P_2} - 1\right)^{\frac{1}{2}} \tag{14.137}$$

and the length l of the coupler should be

$$l = \frac{1}{2}\pi\left(\frac{1}{4}\Delta\beta^2 + \kappa^2\right)^{-\frac{1}{2}} \tag{14.138}$$

Problem 14.17: Consider a planar inhomogeneous waveguide described by

$$n(x) = \begin{cases} 1.48 + 0.02e^{-x/10}; & x > 0 \\ 1.0; & x < 0 \end{cases}$$

Fig. 14.20 An embedded strip waveguide

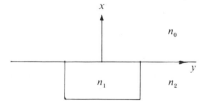

where x is measured in micrometres. A ray enters the waveguide horizontally at $x = 5\,\mu m$. Assume $\lambda_0 = 1\,\mu m$, (a) calculate the propagation constant corresponding to the ray and (b) the angle at which the ray will strike the surface at $x = 0$.
(Answer: (a) $9.375\,\mu m^{-1}$ (b) Angle with the surface $= 5.89°$)

Problem 14.18: Consider an asymmetric planar waveguide with $n_f = 1.5$, $n_c = 1.0$ and $n_s = 1.495$. If the waveguide is operated at $\lambda_0 = 1\,\mu m$, (a) in what range should the thickness lie so that only the TE_0 and TM_0 modes can propagate? (b) What would be the penetration depth of the TE_0 mode in the cover when the mode is just at its cutoff? (c) Using a prism coupling arrangement (see Sec. 2.3) for coupling light into the waveguide one measures a coupling angle of $\theta = 56.3°$ (see Fig. 2.8) for the TE_0 mode at $\lambda_0 = 1\,\mu m$. If the refractive index of the prism is 1.8 what is the thickness of the film?
(Answer: (a) $1.98\,\mu m < d < 6.07\,\mu m$, (b) $0.15\,\mu m$ (c) $4.21\,\mu m$.)

Problem 14.19: Consider a symmetric planar waveguide with $n_f = 1.5$, $n_s = 1.49$ operating in the fundamental mode with $\beta/k_0 = 1.498$ at $\lambda_0 = 1\,\mu m$. (a) Calculate the order of the penetration depth of the modal field in the substrate, (b) Calculate the angle at which rays representing the mode are travelling in the wave guide.
(Answer: (a) $1.03\,\mu m$, (b) Angle with the surface $\approx 2.96°$.)

15

The electrooptic effect

15.1 Introduction

In Chapter 3 we studied light propagation through anisotropic media and found that, in general, the state of polarization of the light beam may change as it propagates through the medium. In the present chapter we shall discuss light propagation through crystals in the presence of an externally applied electric field. This field can in general, alter the refractive indices of the crystal and thus could induce birefringence in otherwise isotropic crystals, or could alter the birefringence property of the crystal. This effect is known as the electrooptic effect. If the changes in the refractive indices are proportional to the applied electric field, such an effect is known as the Pockels effect or the linear electrooptic effect. If the changes in indices are proportional to the square of the applied electric field, the effect is referred to as the quadratic electrooptic effect or the Kerr effect.

In this chapter we shall study in detail the Pockels effect and obtain expressions for the phase shift suffered by a beam propagating through a crystal which is being acted upon by an external electric field. We will show that under certain geometrical configurations, the applied electric field acts differently on two linearly polarized light waves passing through the crystal and thus one can introduce an electric field dependent retardation between the two polarizations. We shall show in Sections 15.2 and 15.3 that such an effect can be used to control the amplitude of the light beam in accordance with the applied field. Thus this gives us a method for imposing information on a light beam through the electrooptic effect. Such a modulator is of great use in optical communications (see Chapters 13 and 14), where modulation of optical beams at very high frequencies (ranging from megahertz to gigahertz) is required; at such high frequencies mechanical shutters or moving mirrors have too much inertia to enable them to be used for modulation.

The electrooptic effect finds wide application in various devices such as directional couplers, optical switches, bistable devices, etc., in integrated optics (see Chapter 14) and also in phase, amplitude or frequency modulation of light, in Q-switching and mode locking in lasers, etc.

In Sec. 15.2 we will discuss the electrooptic effect in potassium dihydrogen phosphate (KDP) in the longitudinal mode, and will show in Sections 15.2 and 15.3 how the electrooptic effect can be used to obtain phase and amplitude modulation. In Sec. 15.3 we will discuss the transverse mode of operation of a KDP modulator. Section 15.4 discusses the electrooptic effect in lithium niobate and lithium tantalate crystals. In Sec. 15.5 we will discuss the general considerations when designing a modulator. Finally in Sec. 15.6 we will derive the equation of the index ellipsoid in the presence of the external electric field. From this equation one can determine the eigen polarization states and their corresponding velocities for any direction of propagation.

Further details, on electrooptic modulators and devices can be found in Kaminow (1974), Hartfield and Thompson (1978), Denton (1972) Yariv and Yeh (1984).

15.2   The electrooptic effect in KDP crystals: longitudinal mode

KDP, potassium dideuterium phosphate (KD*P) and ammonium dihydrogen phosphate (ADP) are some of the most widely used electrooptic crystals. In the absence of an externally applied field these crystals are uniaxial. The crystals possess a fourfold axis of symmetry (i.e., rotation of the crystal structure about this axis by an angle $2\pi/4$ leaves it invariant) and this axis is chosen as the z-axis or the c-axis (optic axis) of the crystal. In addition they also possess two mutually orthogonal axes of symmetry designated as the x and y-axes about which the crystal structure exhibits invariance for rotations of π, i.e., these axes are twofold axes of symmetry. Table 15.1 lists values of some of the important parameters for KDP, KD*P and ADP.

In this section we consider light propagating along the c-axis (chosen as the z-axis) of a KDP crystal which is acted upon by an external electric field which is also along the z-direction (see Fig. 15.1). Such a configuration is referred to as the longitudinal configuration. In the absence of the external field, since the crystal is uniaxial and its optic axis is along the z-direction, a light wave polarized normal to the z-direction (i.e., in the $x–y$ plane) travels as a principal wave with the ordinary refractive index n_o. On applying an electric field E_z along the z-direction, the crystal no longer remains uniaxial but becomes biaxial. The two eigen polarizations, i.e., the two linear polarization states which propagate through the crystal without

Table 15.1. *Electrooptic coefficients of some typical crystals*

Crystal	Optical symmetry	Refractive index		Wavelength $\lambda_0(\mu m)$	Optical transmission (μm)	Nonzero electrooptic coefficients $(10^{-12}\,\text{m/V})$
		n_o	n_e			
KDP (KH$_2$PO$_4$)	Uniaxial	1.512	1.470	0.546	0.25–1.7	$r_{41} = r_{52} = 8.77, r_{63} = 10.5$
ADP (NH$_4$H$_2$PO$_4$)	Uniaxial	1.526	1.481	0.546	0.125–1.7	$r_{41} = r_{52} = 24.5, r_{63} = 8.5$
Quartz (SiO$_2$)	Uniaxial	1.544	1.553	0.589	0.12–4.5	$r_{41} = 0.2, r_{63} = 0.93$
KD*P (KD$_2$PO$_4$)	Uniaxial	1.508	1.468	0.546	0.19–2.15	$r_{41} = r_{52} = 8.8, r_{63} = 26.4$
Lithium niobate (LiNbO$_3$)	Uniaxial	2.297	2.208	0.633	0.4–5	$r_{33} = 30.8, r_{13} = 8.6, r_{51} = 28, r_{22} = 3.4$
Lithium tantalate (LiTaO$_3$)	Uniaxial	2.183	2.188	0.60	0.45–5	$r_{33} = 33, r_{13} = 8, r_{51} = 20, r_{22} = 1$
Gallium arsenide (GaAs)	Isotropic	3.42	—	1.0	1.0–15	$r_{41} = -1.5$
Zinc sulphide (ZnS)	Isotropic	2.364	—	0.6	—	$r_{33} = 1.8, r_{13} = 0.9$

changing the state of polarization lie along $45°$ to the x and y-axes i.e., along the x' and y'-axes – see Fig. 15.1. As we shall show in Section 15.6 the refractive index $n_{x'}$ for a wave polarized along the x'-direction is given by:

$$n_{x'} = n_o - \tfrac{1}{2}n_o^3 r_{63} E_z \tag{15.1}$$

where r_{63} is a coefficient describing the electrooptic effect in this configuration. Similarly the refractive index $n_{y'}$ for a wave propagating along the z-direction and polarized along the y'-direction is given by

$$n_{y'} = n_o + \tfrac{1}{2}n_o^3 r_{63} E_z \tag{15.2}$$

If we assume that a voltage of $10\,\mathrm{kV}$ is applied across a crystal of thickness 1 cm, then the electric field produced will be:

$$E = V/d = (10^4/10^{-2})\,\mathrm{V}\,/\mathrm{m} = 10^6\,\mathrm{V/m}$$

For KDP, $n_o = 1.512$, $r_{63} = 10.5 \times 10^{-12}\,\mathrm{m/V}$ (see Table 15.1); substituting in Eqs. (15.1) and (15.2), we see that even such a large electric field produces a change in refractive index of only

$$n_o - n_{x'} = \tfrac{1}{2}n_o^3 r_{63} E_z \approx 2 \times 10^{-5} \tag{15.3}$$

Even though the change in refractive index is very small, the corresponding phase change in a light beam propagating along this direction could be significant. Thus, if light of wavelength $0.5\,\mu\mathrm{m}$ is propagating along the

Fig. 15.1 (a) Light propagating along the z-axis (c-axis) of a KDP crystal in which an external electric field is also applied along the same direction. (b) In the absence of the external field light polarized normal to the z-axis propagates as a principal wave with refractive index n_0. In the presence of an electric field E_z along the z-direction, the crystal becomes biaxial and the principal polarization directions are x' and y' which make an angle of $45°$ with the old axes.

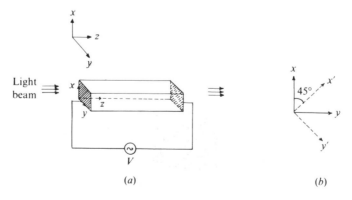

z-axis and is polarized along the x'-axis, then on applying an electric field of 10^6 V/m, the change in phase of the beam after it traverses a path of 1 cm, due to the applied external field will be

$$\Delta\phi = (2\pi/\lambda)\Delta nl$$

$$\approx 0.8\pi \qquad\qquad (15.4)$$

which is indeed a large phase shift.

15.2.1 Phase modulation

In the last section, we saw that we can alter the phase of a light beam by applying an external electric field. Let us now consider a linearly polarized plane wave propagating along the z-direction in a KDP crystal and let us assume that it is polarized along the x'-axis. An external electric field is applied along the z-direction (see Fig. 15.2). Since the x'-direction corresponds to a principal axis (whether in the absence or in the presence of the external field) the linearly polarized light wave will propagate without

Fig. 15.2 (a) Light beam linearly polarized at 45° to the x-axis (i.e., along the x'-axis) is passed along the z-direction through a KDP crystal across which an external electric field is applied along the z-direction. The resulting output beam is phase modulated through the linear electrooptic effect in the KDP crystal. (b) Photograph of the Fabry–Perot pattern observed when the phase modulated light (at $\lambda = 6328$ Å) (modulated by a KDP crystal) at 9.01 GHz is passed through an etalon of spacing 5.65 mm (corresponding to a free spectral range of 26.55 GHz). The phase modulation index ζ is 0.34 and the relative intensities of the 0 and ± 1 components are respectively $J_0^2(0.34) \approx 0.94$, $J_1^2(0.34) \approx 0.03$. Since the modulating frequency (namely 9.01 GHz) is approximately one third of the spacing between the rings (26.55 GHz), the lower side band from one ring and the upper side band from a lower ring divide the spacing into roughly equal parts (After Kaminow (1963).)

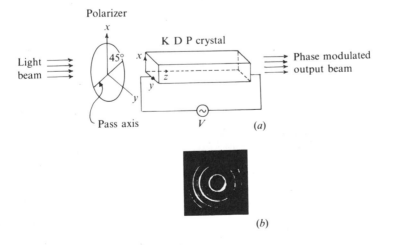

(a)

(b)

any change in state of polarization. If the crystal is of length l along the z-direction, then the wave emerging from the crystal will be given by

$$\mathscr{E}_{x'}(z=l) = \mathscr{E}_{x'}(0)\cos(\omega t - n_{x'}k_0 l)$$
$$= \mathscr{E}_{x'}(0)\cos[\omega t - n_o(\omega l/c) + (\omega/2c)n_o^3 r_{63} E_z l] \qquad (15.5)$$

where $k_0 = \omega/c$ is the free space propagation constant, $z = 0$ is assumed to be the input face of the crystal and we have used Eq. (15.1). If V is the voltage applied across the crystal then

$$V = E_z l \qquad (15.6)$$

Also, if the applied voltage is oscillatory with a frequency ω_m, then

$$V = V_0 \sin \omega_m t \qquad (15.7)$$

and we obtain the output light beam as

$$\mathscr{E}_{x'}(z) = \mathscr{E}_{x'}(0)\cos[\omega t - n_o k_0 l + (\omega/2c)n_o^3 r_{63} V_0 \sin \omega_m t]$$
$$= \mathscr{E}_{x'}(0)\cos(\omega t - n_o k_0 l + \zeta \sin \omega_m t) \qquad (15.8)$$

where

$$\zeta = (\omega/2c)n_o^3 r_{63} V_0 \qquad (15.9)$$

is referred to as the phase modulation index. Thus, a sinusoidal applied electric field leads to a sinusoidal phase variation of the output light wave with a peak value ζ. This example shows how the electrooptic effect leads to phase modulation. If we now use the following identities (see, e.g., Arfken (1970))

$$\cos(\zeta \sin \omega_m t) = J_0(\zeta) + 2J_2(\zeta)\cos 2\omega_m t + 2J_4(\zeta)\cos 4\omega_m t + \cdots \qquad (15.10)$$

$$\sin(\zeta \sin \omega_m t) = 2J_1(\zeta)\sin \omega_m t + 2J_3(\zeta)\sin 3\omega_m t + \cdots \qquad (15.11)$$

then we have

$$\begin{aligned}\mathscr{E}_{x'}(z) = \mathscr{E}_{x'}(0)\{&J_0(\zeta)\cos(\omega t - n_o k_0 l) + J_1(\zeta)\cos[(\omega + \omega_m)t - n_o k_0 l] \\ &- J_1(\zeta)\cos[(\omega - \omega_m)t - n_o k_0 l] \\ &+ J_2(\zeta)\cos[(\omega + 2\omega_m)t - n_o k_0 l] \\ &+ J_2(\zeta)\cos[(\omega - 2\omega_m)t - n_o k_0 l] + \cdots\}\end{aligned} \qquad (15.12)$$

Thus the output beam contains in addition to the wave at the fundamental frequency ω with an amplitude $J_0(\zeta)$, various sidebands at frequencies $\omega \pm \omega_m$, $\omega \pm 2\omega_m$ etc., with amplitudes $J_1(\zeta)$, $J_2(\zeta)$.... It is interesting to note that for ζ having a value such that $J_0(\zeta) = 0$, which occurs when $\zeta \approx 2.4048$, all the power in the fundamental is transferred to the harmonics.

The sideband frequencies at $\omega \pm \omega_m$ can be observed photographically by passing the phase modulated radiation through a Fabry–Perot interferometer as shown in Fig. 15.2(*b*) (Kaminow, 1963).

15.2.2 *Amplitude modulation*

In the previous section, we have shown that if we have a linearly polarized wave polarized along the x'-direction and travelling along the z-direction in a KDP crystal in which we have applied an external electric field E_z along the z-direction, then the output wave at $z = l$ would be given by

$$\mathscr{E}_{x'}(l) = \mathscr{E}_{x'}(0) \exp\left\{i[\omega t - n_o(\omega/c)l + (\omega/2c)n_o^3 r_{63} E_z l]\right\} \qquad (15.13)$$

In a similar manner, if instead of a beam polarized along the x'-direction, we have a beam polarized along the y'-direction, then the output wave at $z = l$ will be given by

$$\begin{aligned}\mathscr{E}_{y'}(l) &= \mathscr{E}_{y'}(0) \exp\left\{i[\omega t - n_{y'}(\omega/c)l]\right\} \\ &= \mathscr{E}_{y'}(0) \exp\left\{i[\omega t - n_o(\omega/c)l - (\omega/2c)n_o^3 r_{63} E_z l]\right\}\end{aligned} \qquad (15.14)$$

If we now consider an incident wave polarized along the y-direction, then it can be decomposed into two linearly polarized waves along the x' and y' directions; these two components will have equal amplitudes and will be in phase at $z = 0$, i.e., at the input of the crystal. The x' and y'-components at any value $z = l$ will be given by Eqs. (15.13) and (15.14). Thus, the two components which were in phase at $z = 0$ now develop a phase difference which is a function of the applied electric field. The retardation at $z = l$ between the two components will be

$$\gamma = (\omega/c)n_o^3 r_{63} E_z l = \omega n_o^3 r_{63} V/c \qquad (15.15)$$

where $V = E_z l$ is the voltage applied across the crystal. Observe that in this configuration, the electrooptic retardation is independent of the length of the crystal and depends only on the externally applied voltage. Also when there is no applied voltage, the phase retardation is zero since the z-axis is the optic axis of the crystal.

Recalling our discussion in Sec. 1.3, if we superpose two linearly polarized waves, polarized along two perpendicular directions, then one, in general, will get an elliptically polarized wave. The superposition will lead to a linearly polarized wave if the phase difference is an integral multiple of π and to a circularly polarized wave for a phase difference of odd-integer multiples of $\frac{1}{2}\pi$. Since in the present case, the electrooptic retardation is given by Eq. (15.15), the polarization state of the wave emerging at $z = l$ is, in general, elliptical.

We define the 'half wave' voltage V_π as the voltage required to introduce a phase shift of π between the two polarization components:

$$\gamma = \pi = (\omega/c)n_o^3 r_{63} V_\pi$$

or

$$V_\pi = \lambda_0/2n_o^3 r_{63}. \tag{15.16}$$

It can be seen that at a given wavelength, the half wave voltage is independent of the dimensions of the crystal but increases with the wavelength.

Using the values of r_{63} and n_o for KDP (Table 15.1), for $\lambda = 0.6\,\mu m$ we obtain,

$$V_\pi = \frac{0.6 \times 10^{-6}}{2 \times (1.51)^3 \times 10.5 \times 10^{-12}} \approx 8.3\,\text{kV} \tag{15.17}$$

On the other hand for KD*P we obtain

$$V_\pi = \frac{0.6 \times 10^{-6}}{2 \times (1.5)^3 \times 26.4 \times 10^{-12}} \approx 3.4\,\text{kV}$$

The half wave voltage is one of the important parameters used to compare various modulators.

As we have seen we can introduce a retardation between the components polarized along the x' and y'-directions by the application of an external field and the magnitude of the retardation is directly proportional to the magnitude of the electric field. The superposition of two retarded linearly polarized waves leads, in general, to an elliptically polarized wave. If we now pass this elliptically polarized beam through an analyser, oriented, say, perpendicular to the input polarization state (see Fig. 15.3), then the amplitude of the beam emerging from the analyser will be modulated. To

Fig. 15.3 An electrooptic amplitude modulator using KDP. The input unpolarized laser beam is passed through a polarizer oriented with its pass axis along the y-direction. The output beam is passed through an analyser with its pass axis along the x-direction.

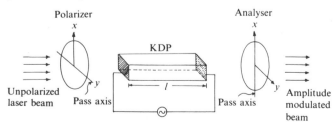

see this, let us assume that the input polarization is along the y-direction, which as before, is at $45°$ to the x'-direction (see Fig. 15.3). The analyser is now placed along the x-direction. The light wave emerging from the analyser will be given by

$$\mathscr{E} = \mathscr{E}_{x'} \cos \tfrac{1}{4}\pi - \mathscr{E}_{y'} \sin \tfrac{1}{4}\pi = (1/\sqrt{2})(\mathscr{E}_{x'} - \mathscr{E}_{y'}) \tag{15.18}$$

Since the input beam is linearly polarized along the y-axis, the amplitudes of the waves polarized along the x' and y'-directions are equal:

$$\mathscr{E}_{x'}(0) = \mathscr{E}_{y'}(0) = A/\sqrt{2} \tag{15.19}$$

Substituting the values of $\mathscr{E}_{x'}$ and $\mathscr{E}_{y'}$ given by Eqs. (15.13) and (15.14), we obtain

$$\mathscr{E} = \tfrac{1}{2}A \exp\{i[\omega t - (n_o/c)\omega l + (\omega/2c)n_o^3 r_{63} E_z l]\}[1 - \exp(-i\gamma)] \tag{15.20}$$

where

$$\gamma = (\omega/c)n_o^3 r_{63} E_z l = \pi(V/V_\pi) \tag{15.21}$$

Thus the intensity of the output beam is given by

$$I_o = \tfrac{1}{2}\mathrm{Re}[\mathscr{E}\mathscr{E}^*] = \tfrac{1}{2}A^2 \sin^2 \tfrac{1}{2}\gamma$$
$$= \tfrac{1}{2}A^2 \sin^2[(\omega/2c)n_o^3 r_{63} V]$$

where $V = E_z l$ is the applied voltage. The intensity of the input beam is given by

$$I_i = \tfrac{1}{2}A^2 \tag{15.22}$$

Thus

$$I_o/I_i = \sin^2(\tfrac{1}{2}\pi V/V_\pi) \tag{15.23}$$

where we have used Eq. (15.16). Hence the output intensity depends sinusoidally on the applied voltage. Fig. 15.4 shows the variation of I_o/I_i (also

Fig. 15.4 Variation of the transmitted intensity of the amplitude modulator versus the externally applied voltage (or equivalently the electric field).

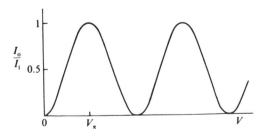

referred to as the transmission factor) as a function of the applied voltage.

Fig. 15.5 shows the output intensity variation versus time for an input sinusoidally oscillating voltage. If we write

$$V = V_o \cos \Omega t \tag{15.24}$$

then

$$I_o/I_i = \sin^2\left(\frac{\pi}{2}\frac{V_0}{V_\pi}\cos\Omega t\right) \tag{15.25}$$

If we assume $V_0 \ll V_\pi$, then we have approximately

$$\frac{I_o}{I_i} \approx \frac{\pi^2}{4}\frac{V_0^2}{V_\pi^2}\cos^2\Omega t = \frac{\pi^2}{8}\frac{V_0^2}{V_\pi^2}(1 + \cos 2\Omega t) \tag{15.26}$$

Thus if we operate the modulator around the point $V = 0$, then the output intensity modulated beam is no longer linearly related to the input signal. Indeed, a weak input signal ($V_0 \ll V_\pi$) at frequency Ω leads to an output modulated beam at twice the signal frequency, namely at 2Ω. Also if $V_0 \ll V_\pi$, the depth of modulation $\pi^2 V_0^2/8V_\pi^2$ will be very small.

In order to overcome the above problems and to have a linear amplitude modulation where the signal and the output amplitude modulated beams are linearly related, one introduces an external bias so that with no signal,

Fig. 15.5 For the amplitude modulator shown in Fig. 15.3 the output intensity variation is not linearly related to the applied voltage. Thus if the applied voltage has a frequency Ω, the output light beam is (for applied voltage $V_0 \ll V_\pi$) at a frequency 2Ω. Also since the point $V = 0$ corresponds to a minimum in the I_o/I_i versus V curve, the depth of modulation will also be small.

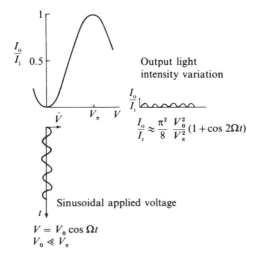

the transmission factor of the modulator is $\frac{1}{2}$. Around this point the curve is almost linear i.e., small changes in the applied voltage around the 50% transmission point lead to proportional changes in the transmissivity of the modulator. This external bias can be achieved by applying an external voltage bias of $\frac{1}{2}V_\pi$ on which the signal can be superposed. Since V_π is usually large, (for KDP $V_\pi \sim 8.3\,\text{kV}$ – see Eq. (15.17)) it is more convenient to bias the modulator optically to the 50% transmission point by placing a quarter wave plate with its fast and slow axes parallel to the x' and y'-axes of the modulator crystal, so that it would introduce an additional phase shift of $\frac{1}{2}\pi$ between $\mathscr{E}_{x'}$ and $\mathscr{E}_{y'}$ (see Fig. 15.6). This can easily be seen by noting that in the presence of the quarter wave plate, the output of the modulator for an applied voltage V will be

$$\mathscr{E} = \tfrac{1}{2}A\exp\left[i\left(\omega t - n_o\frac{\omega}{c}l - \frac{\omega}{2c}n_o^3 r_{63}E_z l\right)\right]$$
$$\cdot [1 - \exp(-i\gamma)\exp(-\tfrac{1}{2}i\pi)]$$
$$= \tfrac{1}{2}A\exp\left[i\left(\omega t - n_o\frac{\omega}{c}l - \frac{\omega}{2c}n_o^3 r_{63}E_z l\right)\right]$$
$$\cdot [1 + i\exp(-i\gamma)] \tag{15.27}$$

instead of Eq. (15.20); the factor i takes account of the additional $\frac{1}{2}\pi$ phase shift between the two components of polarization introduced by the quarter wave plate. Thus the output intensity will be

$$I_o = \tfrac{1}{2}\text{Re}(\mathscr{E}\mathscr{E}^*)$$
$$= \tfrac{1}{4}A^2(1 + \sin\gamma) \tag{15.28}$$

Since $I_i = \frac{1}{2}A^2$ (see Eq. (15.22))

$$\frac{I_o}{I_i} = \frac{1}{2}\left[1 + \sin\left(\pi\frac{V}{V_\pi}\right)\right] = \cos^2\left(\frac{\pi}{2}\frac{V}{V_\pi} - \frac{\pi}{4}\right) \tag{15.29}$$

Fig. 15.6 An electrooptic amplitude modulator using KDP. The quarter wave plate ($\frac{1}{4}\lambda$ plate) which has its axes at 45° to the x and y-axes of the KDP crystal is used for optically biassing the modulator at the linear region of the transmittance versus applied voltage curve (see Fig. 15.7).

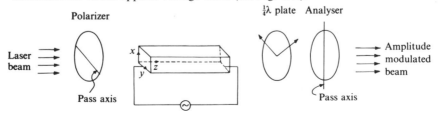

If we assume the applied signal to be given by Eq. (15.24), and again assume $V_0 \ll V_\pi$, i.e., the applied signal voltage is much less than the half wave voltage, then

$$\frac{I_o}{I_i} \approx \frac{1}{2}\left(1 + \pi \frac{V}{V_\pi}\right) = \frac{1}{2}\left(1 + \pi \frac{V_0}{V_\pi}\cos\Omega t\right) \tag{15.30}$$

which shows that the transmitted intensity is linearly related to the applied voltage (see Fig. 15.7). Also the amplitude of modulation in the output beam is $\pi V_0/V_\pi \gg \pi^2 V_0^2/V_\pi^2$ (since $V_0 \ll V_\pi$) and is greater than in the zero bias case.

A typical commercial longitudinal Pockels modulator from Quantum Technology Inc., consists of about 96% KD*P in which the crystal is in the form of a cylinder with a diameter of 12–16 mm and a length from 25 to 30 mm. The electric field is applied between two silver ring electrodes at the ends of the crystal. A laser beam with a divergence of less than 2 mrad is modulated with an extinction ratio of greater than 1000:1. The spectral range of transmission is from 400 nm to 1400 nm. The half wave voltage is typically about 3.2 kV at $\lambda_0 = 633$ nm and increases linearly with wavelength. The modulators typically operate in the 0–25 MHz frequency range. Large aperture electrooptic modulators with aperture diameters of 15–25 mm having an acceptance angle of 2° especially for modulating in-

Fig. 15.7 Using a quarter wave plate as shown in Fig. 15.6 the transmittance varies nearly linearly for $|V| \ll V_\pi$. Thus a sinusoidally applied voltage at frequency Ω leads to an intensity modulated wave at the output at the same frequency Ω as long as $V_0 \ll V_\pi$. Also since the slope of the I_o/I_i versus V curve is maximum at the value $I_o/I_i = \frac{1}{2}$, the variation of I_o for a given variation in V will be maximum.

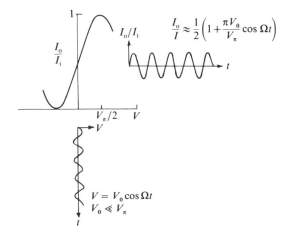

coherent narrow line light sources which are not highly collimated are also commercially available.

Problem 15.1: Using the expansion given in Eqs. (15.10) and (15.11) and assuming a signal of the form given by Eq. (15.24), show that (*a*) only odd harmonics of the signal frequency will be present in the output and (*b*) calculate the ratio of the amplitude of the third harmonic to the fundamental (see also Denton, Chen and Ballman (1967)).

15.3 The electrooptic effect in KDP crystals: transverse mode

In the above configuration, the light beam propagates along the same direction as the external applied electric field. This is referred to as the longitudinal mode of operation. In this configuration, as we have seen, the retardation is independent of the length of the crystal and depends only on the applied voltage. Further, in this configuration, one has to apply an external field along the direction of propagation of the beam; this can be achieved either by using transparent electrodes or by leaving a small aperture at the centre of the electrodes on both ends through which the beam can pass.

We will now discuss the transverse mode of modulation where the optical path is perpendicular to the direction of the applied electric field. The advantages of this configuration are that the electrodes no longer obstruct the optical beam as in the longitudinal case and, as we shall show, the retardation is proportional to the applied voltage and also to the length of the crystal and thus the half wave voltage is proportional to the ratio of the width of the crystal to its length. Thus by decreasing this ratio, one can have lower half wave voltage.

Fig. 15.8 Electrooptic KDP modulator in the transverse mode of operation. Incident light wave is polarized at 45° to the x'-direction (in the $x'-z$ plane) and propagates along the y'-direction; the electric field is applied along the z-direction. The analyser is placed in a direction normal to the polarizer. A Babinet–Soleil compensator is introduced before the analyser so as to bias the modulator in the linear region of the transmittance versus applied voltage curve.

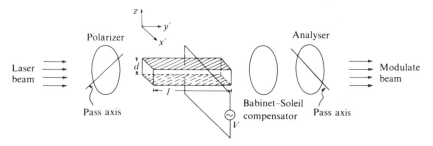

In Fig. 15.8 we have shown a transverse configuration in which the light beam propagates along the y'-direction and the field is again applied along the z-direction. In the presence of a z-directed electric field, the refractive indices for a wave propagating along the y'-direction and polarized along the x' and z-directions are respectively given by (see Eqs. (15.131) and (15.133))

$$n_{x'} = n_{\mathrm{o}} - \tfrac{1}{2} n_{\mathrm{o}}^3 r_{63} E_z \tag{15.31}$$

$$n_z = n_{\mathrm{e}} \tag{15.32}$$

Thus if the light beam incident on the crystal is linearly polarized along $45°$ to the x'-direction (in the $x'–z$ plane) then the emergent field components along the x' and z-directions after traversing a length l of the crystal will be:

$$\mathscr{E}_{x'}(y' = l) = \frac{A}{\sqrt{2}} \exp\left[i\left(\omega t - n_{\mathrm{o}} \frac{\omega}{c} l + \tfrac{1}{2} n_{\mathrm{o}}^3 \frac{\omega}{c} r_{63} E_z l \right) \right] \tag{15.33}$$

$$\mathscr{E}_z(y' = l) = \frac{A}{\sqrt{2}} \exp\left[i\left(\omega t - n_{\mathrm{e}} \frac{\omega}{c} l \right) \right] \tag{15.34}$$

where the field components at $y' = 0$ along the x' and z-directions are assumed to be

$$\mathscr{E}_{x'}(y' = 0) = \frac{A}{\sqrt{2}} e^{i\omega t}, \quad \mathscr{E}_z(y' = 0) = \frac{A}{\sqrt{2}} e^{i\omega t} \tag{15.35}$$

Hence, the retardation between the two linearly polarized components, when the beam emerges from the crystal would be

$$\begin{aligned} \gamma &= [(n_{\mathrm{o}} - n_{\mathrm{e}}) - \tfrac{1}{2} n_{\mathrm{o}}^3 r_{63} E_z](\omega/c)l \\ &= [(n_{\mathrm{o}} - n_{\mathrm{e}}) - \tfrac{1}{2} n_{\mathrm{o}}^3 r_{63}(V/d)](\omega/c)l \end{aligned} \tag{15.36}$$

where V is the voltage applied across a width d of the crystal. Notice that even when the applied voltage is zero, there is a finite retardation. This is due to the intrinsic birefringence of the material; even in the absence of a field, a light wave which is propagating along the y'-direction will be an o-wave if its polarization is along the x'-direction and an e-wave if its polarization is along the z-direction. Hence the phase shift induced by the external modulation voltage is

$$\gamma = \frac{1}{2c} n_{\mathrm{o}}^3 r_{63} \omega \left(\frac{l}{d} \right) V \tag{15.37}$$

We can once again, for this configuration, define a half wave voltage, as the voltage required to introduce an additional phase shift of π. Thus since the phase shift introduced in the absence of an external field is $(n_o - n_e)(\omega/c)l$, the half wave voltage is defined by:

$$(n_o - n_e)\frac{\omega}{c}l + \pi = \left[(n_o - n_e) - \tfrac{1}{2}n_o^3 r_{63}\frac{V}{d} \right]\frac{\omega}{c}l$$

or

$$V_\pi = \frac{\lambda_0}{n_o^3 r_{63}}\left(\frac{d}{l}\right) \tag{15.38}$$

where $\omega/c = 2\pi/\lambda_0$, λ_0 being the free space wavelength of the light wave. We have disregarded the negative sign above since it just signifies the fact that a positive voltage along the z-direction decreases the phase difference between the x' and z-polarized waves. Eq. (15.38) shows us that in the transverse configuration, the half wave voltage is not independent of the modulator length, but depends on the ratio d/l. Thus by choosing a small geometrical factor (d/l), the half wave voltage can be reduced. This is further discussed in Sec. 15.5.1. The discussions regarding the linear region of operation by external biasing etc. are the same for this configuration as those used for the longitudinal case.

Transverse electrooptic modulators based on highly deuterated KDP crystals and ADP crystals are also available commercially. Since they operate on the transverse mode, they require only low driving voltages. A typical deuterated KDP crystal transverse modulator operating over a useful electrical bandwidth of 0–100 MHz consists of a 2.5 mm diameter clear aperture requiring typical half wave voltages of 275 V at 633 nm.

15.4 Electrooptic effect in lithium niobate and lithium tantalate crystals

Another set of very important electrooptic materials are lithium niobate and lithium tantalate. Indeed, a very attractive combination of electrooptic, piezoelectric and optical properties has made lithium niobate one of the most extensively studied materials in recent years.

Both lithium niobate and lithium tantalate are trigonal crystals of point group 3m. Lithium niobate is a negative uniaxial crystal (with an ordinary refractive index greater than the extraordinary index) and lithium tantalate is a positive uniaxial crystal. The electrooptic effect in these crystals is several times that in KDP (see Table 15.1). Lithium niobate is transparent from 0.35 μm to about 5 μm. Table 15.1 gives the values of various important parameters of lithium niobate and lithium tantalate and it can be seen that the electrooptic effect in these crystals is stronger than in KDP or ADP.

We will show in Sec. 15.6.2 that when an external electric field E_z is applied along the optic axis (chosen as the z-axis) of the lithium niobate or lithium tantalate crystal then the refractive indices for a light wave polarized along the crystallographic x, y and z-directions are given by:

$$n_x = n_o - \tfrac{1}{2}n_o^3 r_{13}E_z \tag{15.39}$$

$$n_y = n_o - \tfrac{1}{2}n_o^3 r_{13}E_z \tag{15.40}$$

$$n_z = n_e - \tfrac{1}{2}n_e^3 r_{33}E_z \tag{15.41}$$

where r_{13} and r_{33} are the electrooptic coefficients (the elements of the electrooptic matrix are introduced in Sec. 15.6). Values of r_{13}, r_{33} and other coefficients for lithium niobate and lithium tantalate are given in Table 15.1. If we assume that the light beam is propagating along the y-direction, and that the incident light is linearly polarized at 45° to the z-direction in the x–z plane (see Fig. 15.8) then the retardation at a distance l from the input plane will be

$$\Delta\phi = \frac{2\pi}{\lambda_0}(n_z - n_x)l$$

$$= \frac{2\pi}{\lambda_0}(n_e - n_o)l - \frac{2\pi}{\lambda_0}\frac{(n_e^3 r_{33} - n_o^3 r_{13})}{2}E_z l \tag{15.42}$$

The first term in the above expression gives the effect of the intrinsic birefringence of the crystal. Since the zero-voltage phase shift $(2\pi/\lambda_0)(n_e - n_o)l$ depends on l, a Babinet–Soleil compensator is introduced to bias the modulator in the linear region of the transmittance versus voltage curve (see Fig. 15.8).

From Eq. (15.42), we can easily calculate the half wave voltage as

$$V_\pi = \frac{\lambda_0}{(n_e^3 r_{33} - n_o^3 r_{13})}\frac{d}{l} \tag{15.43}$$

where d represents the thickness of the crystal (see Fig. 15.8) and $E_z = V/d$.

To obtain the order of magnitude of the half wave voltage required, let us consider a lithium tantalate modulator having the following specifications

$$d = 0.25\,\text{mm}, \qquad l = 9.5\,\text{mm}$$

For light of $\lambda_0 = 0.6328\,\mu\text{m}$, using the values of n_e, n_o, r_{33} and r_{13} given in Table 15.1 in Eq. (15.43) one can calculate the half wave voltage to be

$$V_\pi \approx 64\,\text{V} \tag{15.44}$$

The above parameters correspond to the experiment of Biazzo (1971) who has obtained a measured half wave voltage of about 80 V peak to peak.

On the other hand if the incident light was polarized along the z-direction then the application of an electric field along z will lead to phase modulation of the beam and the output light will still be polarized along the z-direction. The phase of the emerging beam will be given by

$$\phi = \frac{2\pi}{\lambda_0} n_e l - \frac{2\pi}{\lambda_0} \frac{n_e^3 r_{33}}{2} E_z l$$

Thus the voltage required to change the phase of the output beam by π will be, assuming as before, $E_z = V/d$,

$$\frac{2\pi}{\lambda_0} \frac{n_e^3 r_{33}}{2} V_\pi \frac{l}{d} = \pi$$

or

$$V_\pi = \frac{\lambda_0}{n_e^3 r_{33}} \frac{d}{l} \tag{15.45}$$

Comparing Eqs. (15.43) and (15.45) we find that since in lithium niobate and lithium tantalate r_{33} and r_{13} have the same sign, it is much easier to obtain phase modulation than intensity modulation. Indeed for the same parameters as before the half wave voltage for a π phase shift in the phase modulator is

$$V_\pi \approx 48 \text{ V}.$$

Problem 15.2: Show that for a lithium niobate crystal used as a phase modulator with light propagating along the x-direction and polarized along z and the electric field applied along the c-axis, the voltage for half wave retardation for a crystal with $d = l$ is about 1910 V and the peak voltage for converting all power from the fundamental to the sidebands is about 1461 V. (Use the parameters listed in Table 15.1 for lithium niobate).

15.5 General considerations on modulator design

In this section we shall briefly discuss the various factors encountered in connection with an actual modulator design.

15.5.1 Geometrical considerations

In Sec. 15.3 we have seen that when a modulator crystal is operated in the transverse configuration, the half wave voltage is proportional to the thickness to length ratio (d/l) of the crystal. In the analysis we assumed d and l to be independent of one another. In actual practice, for a given length l of the crystal, the minimum value of d permissible will be determined by diffraction, and will correspond to the case in which the beam can just pass through the sample.

We consider a Gaussian beam passing through the crystal as shown in Fig. 15.9. If w_0 represents the waist size of the Gaussian beam, then the size of the beam at any plane at a distance s from the waist will be (see Eq. (5.11))

$$w(s) = w_0(1 + (\lambda'^2 s^2/\pi^2 w_0^4))^{\frac{1}{2}} \tag{15.46}$$

where $\lambda' = \lambda_0/n$ is the wavelength of the beam in the medium. In order to obtain a small value of d/l, one can arrange to have the beam traverse the modulator crystal as shown in Fig. 15.9, i.e., to have the waist of the Gaussian beam at the centre of the crystal. In such a case the diameter of the beam at the ends of the crystal of total length l will be

$$d = 2w(s = \tfrac{1}{2}l) = 2w_0(1 + (\lambda'^2 l^2/4\pi^2 w_0^4))^{\frac{1}{2}} \tag{15.47}$$

Let us now find the optimum value of w_0 which will give us the minimum d. Thus we put

$$\frac{\mathrm{d}}{\mathrm{d}w_0}(d) = 0 = 2\left(1 + \frac{\lambda'^2 l^2}{4\pi^2 w_0^4}\right)^{\frac{1}{2}} + \frac{2w_0}{2(1 + (\lambda'^2 l^2/4\pi^2 w_0^4))^{\frac{1}{2}}}\left(\frac{\lambda'^2 l^2}{4\pi^2}\right)\left(-\frac{4}{w_0^5}\right)$$

Solving the above equation for w_0, we obtain the optimum w_0 for minimum value of d as

$$w_0 = (\lambda' l/2\pi)^{\frac{1}{2}} \tag{15.48}$$

Substituting this value of w_0 in Eq. (15.47), we obtain

$$d = 2(\lambda' l/\pi)^{\frac{1}{2}} \tag{15.49}$$

Thus for a given l, the minimum value of d is given by Eq. (15.49). For a practical modulator, the length l of the modulator crystal is limited to a few centimetres due to the problems associated with growing large crystals. In practical modulator design, one usually introduces a safety factor S

Fig. 15.9 A Gaussian beam having waist radius w_0 passes through a crystal of length l. The figure shows the optimum passage of the Gaussian beam through a crystal of length l and width d.

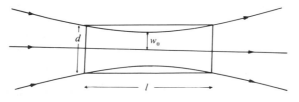

($\geqslant 1$) such that the minimum value of d is

$$d = 2S(\lambda'l/\pi)^{\frac{1}{2}} \tag{15.50}$$

S is typically in the range 3–6.

As an example we consider a lithium tantalate modulator crystal of length $l = 9.5$ mm. Using the value of $n_o(\approx n_e)$ as given in Table 15.1 we obtain for $\lambda_0 = 0.6328\,\mu$m, the minimum d as

$$d = 2\left(\frac{0.6328 \times 10^{-4} \times 0.95}{\pi \times 2.188}\right)^{\frac{1}{2}}$$

$$\approx 0.06 \text{ mm}$$

The modulator crystal considered in Sec. 15.4 had a $d = 0.25$ mm corresponding to a safety factor $S \approx 4$.

As another example, we choose a KDP crystal of length 2 cm. Substituting in Eq. (15.50) and using a safety factor $S = 3$ we obtain the minimum value of d as

$$d \approx 3.8 \times 10^{-2} \text{ cm}$$

The corresponding half wave voltage (at $\lambda_0 = 0.6\,\mu$m) would be given by

$$V_\pi = \frac{\lambda_0}{n_o^3 r_{63}}\left(\frac{d}{l}\right) \approx 314 \text{ V} \tag{15.51}$$

which is about $1/25^{\text{th}}$ of the longitudinal configuration for which the half wave voltage is ~ 8.3 kV (see Eq. (15.17)).

15.5.2 *Transit time limitations*

In this section we will show that because of the finite transit time of the optical beam (as it passes through the length of the modulator) there are limitations on the frequencies of the modulating signal. Indeed we will show that the transit time plays no part for $\omega_m \ll 2/\tau_d$, where τ_d is the time taken by the optical beam to pass through the crystal and is given by

$$\tau_d = nl/c \tag{15.52}$$

where l represents the length of the modulator crystal and n is either n_e or n_o. In earlier sections, we have derived expressions for the phase shift suffered by the optical beam as it propagates through the modulator crystal, and we have assumed that the applied electric field indeed remains constant for the propagation time τ_d. For a crystal of length 1 cm, taking $n \approx 1.5$, we get

$$\tau_d \approx 0.5 \times 10^{-10}\,\text{s} \tag{15.53}$$

If the modulating voltage that is applied to the electrodes is oscillating with a frequency of (say) 2×10^{10} Hz then it can be seen that as the light beam propagates through the modulator crystal, the applied electric field no longer remains constant within the time taken by the light to travel through the modulator crystal and the earlier analysis will not remain valid. In order to take into account the finite transit time of the light beam, we assume that the applied electric field is sinusoidal with an angular frequency ω_m, i.e. the applied electric field is given by

$$E = E_0 \cos \omega_m t \tag{15.54}$$

Let l represent the length of the crystal and let us choose the input plane of the crystal as $\zeta = 0$ and the output plane as $\zeta = l$; the ζ-axis is chosen along the length of the crystal (see Fig. 15.10). In order to calculate the time variation of the retardation at the output of the crystal, we consider the beam of light which is emerging from the crystal at a time t. This beam will have been at the plane ζ at a time $t - (l - \zeta)n/c$, and at such a time, the applied electric field will have had a value

$$E = E_0 \cos \left\{ \omega_m [t - (l - \zeta)n/c] \right\} \tag{15.55}$$

where n is the refractive index corresponding to the particular state of polarization.

For specificity we consider the longitudinal configuration of a KDP crystal modulator (a similar analysis can also be performed for any other configuration). We first consider the phase shift suffered by the x'-polarized wave as it travels from the plane ζ to the plane $\zeta + d\zeta$, which is given by (see Eq. (15.5))

$$d\phi_o = \frac{\omega}{c} n_o \, d\zeta - \frac{\omega}{c} \frac{n_o^3 r_{63}}{2} E_0 \cos \left\{ \omega_m \left[t - \frac{(l - \zeta)}{c} n_{x'} \right] \right\} d\zeta \tag{15.56}$$

Thus the total phase shift suffered by the (x'-polarized) wave as it propagates

Fig. 15.10 Light beam propagating along the ζ-direction in a modulator crystal.

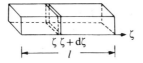

from the input plane $\zeta = 0$ to the output plane $\zeta = l$ will be

$$\phi_o = \int_{\zeta=0}^{\zeta=l} d\phi_o = \frac{\omega}{c} n_o l - \frac{\omega}{2c} n_o^3 r_{63} E_0 l \cos\left(\omega_m t - \frac{\omega_m l}{2c} n_{x'}\right)$$

$$\times \sin\left(\frac{\omega_m l}{2c} n_{x'}\right) \bigg/ \left(\frac{\omega_m l}{2c} n_{x'}\right) \tag{15.57}$$

Similarly the phase shift suffered by the y'-polarized wave will be

$$\phi_e = \frac{\omega}{c} n_o l + \frac{\omega}{2c} n_o^3 r_{63} E_0 l \cos\left(\omega_m t - \frac{\omega_m l}{2c} n_{y'}\right)$$

$$\times \sin\left(\frac{\omega_m l}{2c} n_{y'}\right) \bigg/ \left(\frac{\omega_m l}{2c} n_{y'}\right) \tag{15.58}$$

As the phase difference between the x' and y'-polarized waves at the output of the crystal is of the order of π, we have

$$(\omega/c)(n_{x'} - n_{y'})l \sim \pi \tag{15.59}$$

Thus

$$\frac{\omega_m l}{c}(n_{x'} - n_{y'}) = \frac{\omega_m}{\omega} \pi \approx \frac{10^{10}}{10^{15}} \pi \ll 1 \tag{15.60}$$

Consequently the retardation is given by

$$\phi = \phi_e - \phi_o \approx \phi^0 \cos\left(\omega_m t - \tfrac{1}{2}\omega_m \tau_d\right)\left(\sin \tfrac{1}{2}\omega_m \tau_d / \tfrac{1}{2}\omega_m \tau_d\right) \tag{15.61}$$

where $\tau_d = n_o l/c$ and

$$\phi^0 = (\omega/c)n_o^3 r_{63} E_0 l \tag{15.62}$$

represents the peak retardation corresponding to $\omega_m \tau_d \ll 1$. Indeed for $\omega_m \tau_d \ll 1$, the variation of retardation with time will be $\phi = \phi^0 \cos \omega_m t$, i.e., the retardation will be modulated at the modulating frequency ω_m. The condition $\omega_m \tau_d \ll 1$ implies that the time taken by the light beam to propagate through the crystal is much less than the time period of the modulating voltage and in such a case our earlier analysis of Sec. 15.2 is valid. The factor

$$\frac{\sin \tfrac{1}{2}\omega_m \tau_d}{\tfrac{1}{2}\omega_m \tau_d} \tag{15.63}$$

gives the reduction in the peak retardation due to the finite transit time. Indeed if $\omega_m \tau_d = 2\pi$ or $\tau_d = 2\pi/\omega_m$, i.e., the time taken by the light beam to traverse the crystal is exactly equal to the time period of the applied electric field, then there is no retardation. Physically, this happens because the retardation introduced in the beam in the first half of the crystal is exactly cancelled by the retardation introduced in the second half of the crystal.

Thus for a given length of the crystal i.e., for a given τ_d, the transit time plays no part for modulation frequencies $\omega_m \ll 2/\tau_d$. If we define the highest useful modulation frequency as the frequency when $\frac{1}{2}\omega_m\tau_d = \frac{1}{4}\pi$, at such a frequency

$$\frac{\sin\frac{1}{2}\omega_m\tau_d}{\frac{1}{2}\omega_m\tau_d} \approx 0.9$$

and

$$(\omega_m)_{max} = \pi/2\tau_d = \pi c/2nl$$

or

$$(\nu_m)_{max} = c/4nl \tag{15.64}$$

For a **KDP** crystal of length $l = 1$ cm, and $n \approx 1.5$, we get the highest modulation frequency as

$$(\nu_m)_{max} = \frac{3 \times 10^{10}}{4 \times 1.5 \times 1} = 5\,\text{GHz} \tag{15.65}$$

15.5.3 *Travelling wave modulators*

The above limitation on the maximum modulation frequency due to transit time effects can be overcome by having the applied electric field also propagate along the length of the crystal. Such modulators are referred to as travelling wave modulators. As we shall now show, the transit time limitation can be completely overcome by making use of this geometry and by adjusting the phase velocity of the modulation signal to be equal to the phase velocity of the light wave propagating through the crystal.

Fig. 15.11 A typical travelling wave electrooptic modulator. The electrode pair forms a transmission line and the applied electric field also travels as a wave in the crystal.

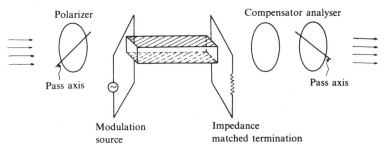

Under such a condition a particular portion of the light wave will experience the same electric field as it propagates along the modulator.

Fig. 15.11 shows a typical configuration of a travelling wave modulator in which the modulator crystal is placed between a pair of electrodes which forms a transmission line. The modulating field also propagates along the crystal and the two velocities are matched by choosing particular geometries for the transmission line (see e.g., Peters (1963, 1965); Kaminow and Liu (1963)). We shall now calculate the variation of the output light amplitude with time when the applied electric field on the electrodes is also a travelling wave.

For a travelling wave, the applied electric field is no longer given by Eq. (15.54) but by the following equation

$$E = E_0 \cos(\omega_m t - k_m \zeta) \tag{15.66}$$

where ω_m is the modulation frequency, k_m represents the wave vector of the applied electric field, and the wave is also assumed to travel along the ζ-axis, along which the light beam is also travelling. The velocity of propagation of the applied field will be

$$v_m = \frac{\omega_m}{k_m} \tag{15.67}$$

Proceeding as in the case of a nonpropagating applied electric field, we see that if we are observing the output at a time t, then the wave will have been at a plane at a distance ζ from the input plane (see Fig. 15.10) at a time

$$\tau = t - \frac{(l - \zeta)}{c} n \tag{15.68}$$

The applied electric field at this value of ζ at a such a time τ will have been

$$\begin{aligned} E &= E_0 \cos\{\omega_m[t - (l - \zeta)n/c] - k_m \zeta\} \\ &= E_0 \cos[\omega_m(t - (nl/c)) + \zeta(\omega_m(n/c) - k_m)] \end{aligned} \tag{15.69}$$

Thus the phase shift in travelling from the plane ζ to $\zeta + d\zeta$ will be[†]

$$\Delta\phi = (\omega/c)\Delta n \, d\zeta \tag{15.70}$$

where Δn is the difference in the refractive indices of the two polarization states induced by the applied electric field. Since Δn is proportional to the

---

[†] As discussed in the previous section, n appearing in Eq. (15.68) should either be $n_{x'}$ or $n_{y'}$; however, because of Eq. (15.60) we do not differentiate between $n_{x'}$ and $n_{y'}$ in the analysis.

applied electric field, we have

$$d\phi = \alpha \cos\left[\omega_m(t - ln/c) + \zeta((\omega_m n/c) - k_m)\right] d\zeta \qquad (15.71)$$

where α is the retardation per unit length due to a constant applied electric field of strength E_0 and in the longitudinal configuration is given by $\alpha = (\omega/2c)n_o^3 r_{63} E_0$. Thus, the total retardation in the wave as it propagates through the length l of the crystal will be

$$\phi = \int_{\zeta=0}^{\zeta=l} d\phi = \alpha \int_0^l \cos\left[\omega_m(t - ln/c) + \zeta((\omega_m n/c) - k_m)\right] d\zeta$$

$$= \alpha \frac{\{\sin(\omega_m t - k_m l) - \sin\left[\omega_m(t - ln/c)\right]\}}{(n/c)\omega_m - k_m}$$

$$= \phi_0 \cos\left[\omega_m t - \tfrac{1}{2}l(k_m + \omega_m(n/c)\right] \frac{\sin\left[\tfrac{1}{2}l(\omega_m(n/c) - k_m)\right]}{\tfrac{1}{2}l(\omega_m(n/c) - k_m)} \qquad (15.72)$$

where

$$\phi_0 = \alpha l \qquad (15.73)$$

is the peak induced retardation over the length l of the crystal. If $v(=c/n)$ and $v_m(=\omega_m/k_m)$ represent the phase velocities of the light beam and the modulation field respectively in the crystal we obtain from Eq. (15.72)

$$\phi = \phi_0 \cos\left[\omega_m t - \frac{n\omega_m l}{2c}\left(1 + \frac{v}{v_m}\right)\right] \frac{\sin\left[\tfrac{1}{2}\omega_m \tau_d(1 - (v/v_m))\right]}{\tfrac{1}{2}\omega_m \tau_d(1 - (v/v_m))}$$

$$(15.74)$$

where τ_d is again given by Eq. (15.52) and gives the time taken by the light wave to traverse the length of the crystal. In the present case, the reduction factor is given by

$$\frac{\sin\left[\tfrac{1}{2}\omega_m \tau_d(1 - (v/v_m))\right]}{\tfrac{1}{2}\omega_m \tau_d(1 - (v/v_m))} \qquad (15.75)$$

which is similar to Eq. (15.63) except for the factor $(1 - (v/v_m))$. We see now that if the phase velocities of the two waves are matched, i.e., if $v = v_m$, then the finite τ_d has no restriction on the modulation and maximum retardation is achieved irrespective of the length of the crystal.

15.5.4 *Circuit aspect of modulators*

In deriving the expressions for retardation, modulation etc. in the earlier sections we have tacitly assumed that the modulating field in the electrooptic crystal was uniform throughout its length L and also that the field was constant during the time it takes for the light beam to propagate through the crystal. The first condition is satisfied if the wavelength λ_m of the modulating field inside the crystal is much larger than the length L of the crystal, i.e.,

$$\lambda_m = \frac{2\pi v_m}{\omega_m} = \frac{2\pi c}{\omega_m \kappa^{\frac{1}{2}}} \gg L \quad \text{or} \quad L \ll \frac{2\pi c}{\omega_m \kappa^{\frac{1}{2}}} \tag{15.76}$$

where $v_m = c/\kappa^{\frac{1}{2}}$ is the velocity of the modulating field inside the crystal and ω_m represents the maximum modulation frequency, κ is the dielectric constant of the modulating crystal at ω_m. The second condition regarding transit time is given by (see Sec. 15.5.2)

$$\tau_d = \frac{nL}{c} \ll \frac{2\pi}{\omega_m} \quad \text{or} \quad L \ll \frac{2\pi c}{\omega_m n} \tag{15.77}$$

Since $\kappa^{\frac{1}{2}} \gg n$, the condition given by Eq. (15.76) on the length of the modulating crystal is dominant. If the length of the crystal satisfies Eq. (15.76) then the modulating crystal may be regarded as a lumped capacitance. To apply the modulating signal efficiently to a lumped device, one requires the addition of other passive circuit elements so as to match the impedance of the power supply.

Thus let us consider the simple configuration shown in Fig. 15.12 where C represents the capacitance due to the modulator crystal and R_s denotes the resistance in the circuit including the internal resistance of the modulating source represented by V_m. The total impedance of the circuit shown in Fig. 15.12 at a frequency ω_m is

$$Z = R_s + \frac{1}{j\omega_m C} = \frac{j\omega_m C R_s + 1}{j\omega_m C} \tag{15.78}$$

The voltage across the capacitance will be given by

$$V_C = V_m \frac{j\omega_m C}{1 + j\omega_m C R_s} \frac{1}{j\omega_m C} = \frac{V_m}{1 + j\omega_m C R_s} \tag{15.79}$$

Fig. 15.12 The modulator is represented by a lumped capacitance C and R_s denotes the internal resistance of the modulating source.

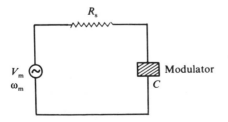

At very low frequencies $\omega_m \ll 1/CR_s$, the voltage drop across the capacitor is very nearly equal to the applied modulating signal and there is very little drop in voltage across the resistance. As the modulation frequency increases, the voltage drop across the modulating crystal decreases and at a frequency

$$\omega_0 = 1/CR_s \tag{15.80}$$

the voltage across the modulating crystal is only $V_m/\sqrt{2}$. By definition, we define this frequency as the cutoff frequency of the modulator. This method of driving the modulator corresponds to a low pass configuration for operation over a band from 0 to ω_0. We also call this frequency ω_0 as the bandwidth of this modulator under such a configuration.

The modulator crystal can also be operated under bandpass operation by resonating the crystal capacitance with a parallel inductance L as shown in Fig. 15.13. For the circuit shown in Fig. 15.13 the total impedance of the circuit is

$$Z = R_s + \left(j\omega_m C + \frac{1}{j\omega_m L} \right)^{-1} = R_s + \frac{j\omega_m L}{1 - \omega_m^2 LC} \tag{15.81}$$

and the voltage drop across the modulator crystal will be

$$
\begin{aligned}
V &= \frac{V_m}{R_s + (j\omega_m L/(1 - \omega_m^2 LC))} \frac{j\omega_m L}{1 - \omega_m^2 LC} \\
&= \frac{j\omega_m L V_m}{R_s(1 - \omega_m^2 LC) + j\omega_m L} \\
&= \frac{\omega_m L V_m}{[R_s^2(1 - (\omega_m^2/\omega_0^2))^2 + \omega_m^2 L^2]^{\frac{1}{2}}} \\
&\quad \times \exp\left\{ \tfrac{1}{2}i\pi - i\tan^{-1}\left[\frac{\omega_m L}{R_s(1 - \omega_m^2 LC)} \right] \right\}
\end{aligned}
\tag{15.82}
$$

where $\omega_0^2 = 1/LC$. The maximum voltage drop V_m across the modulator

Fig. 15.13 Modulator operated under bandpass operation by introducing an inductance L in parallel to the modulator.

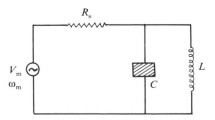

crystal appears at the frequency

$$\omega_m = \omega_0 = 1/(LC)^{\frac{1}{2}} \tag{15.83}$$

It can be shown from Eq. (15.82) that the voltage drop across the modulating crystal drops to approximately $V_m/\sqrt{2}$ at frequencies $\omega_0 \pm \Delta\omega_m/2$ where

$$\Delta\omega_m \approx 1/CR_s \tag{15.84}$$

represents the bandwidth of the modulator. Thus for applied modulating frequencies ω_m satisfying $\omega_0 - (\Delta\omega_m/2) < \omega_m < \omega_0 + (\Delta\omega_m/2)$, more than $V_m/\sqrt{2}$ of the applied modulating voltage drops across the modulating crystal. The resonant frequency ω_0 can be chosen by selecting an appropriate value for L.

Bandwidth is also an important parameter characterizing a modulator. A large bandwidth would imply the possibility of placing a larger amount of information on a light beam. In fact, to use the large information-carrying capacity of light waves (see Chapter 13) it is essential to be able to modulate the light beam over a wide bandwidth.

To obtain a prescribed degree of modulation or retardation, one requires a certain peak applied modulating voltage. Thus from Eq. (15.15) it is clear that to obtain a peak retardation of ϕ_m one requires a peak modulating voltage V_m given by

$$V_m = (c/\omega n_o^3 r_{63})\phi_m \tag{15.85}$$

Hence if we use a modulating source with this peak modulating voltage in either the low pass configuration or the bandpass configuration, then the peak retardation ϕ_m will appear at low applied frequencies in the low pass configuration, and at the resonant frequency $\omega_0 = 1/(LC)^{\frac{1}{2}}$ in the bandpass configuration.

One of the important parameters characterizing a modulator is the drive power required per unit bandwidth of operation of the modulator to obtain a specified modulation depth. This quantity $P/\Delta f$ is usually expressed in milliwatts per megahertz.

Since the modulating crystal is a pure capacitance it dissipates no power but the resistances in the circuit do dissipate power and this must be supplied by the modulating source. If V_m represents the peak modulation voltage applied, then the power dissipated in the resistance R_s will be approximately

$$P = V_m^2/2R_s \tag{15.86}$$

Substituting for R_s in terms of the bandwidth of the modulator device from Eq. (15.80) or Eq. (15.84) we get

$$P = \tfrac{1}{2}V_m^2 C\Delta\omega = \pi C V_m^2 \Delta f \tag{15.87}$$

or

$$P/\Delta f = \pi C V_m^2 \tag{15.88}$$

Thus the power per unit bandwidth (also called the specific energy) depends only on the modulator crystal parameters (through C) and the peak modulation voltage V_m which is determined for a given depth of modulation. The drive power necessary for a certain depth of modulation increases linearly with bandwidth since the resistance has to be decreased to obtain a larger bandwidth and if the resistance decreases, the power dissipated increases proportionally.

Let us first consider an intensity modulator which is optically biased (by a $\tfrac{1}{4}\lambda$ plate) to operate in the linear region (see Sec. 15.2). Then for 100% intensity modulation (i.e., output intensity varying between the input intensity and zero intensity) one requires a peak modulating voltage of $\tfrac{1}{2}V_\pi$. Thus the specific energy for 100% intensity modulation would be

$$P/\Delta f = \tfrac{1}{4}\pi C V_\pi^2 \tag{15.89}$$

We now consider a square cross section crystal of transverse dimension $d \times d$ and of length l. Then neglecting the fringing fields, the capacitance of the modulator crystal in the transverse configuration will be given by

$$C = \epsilon_0 \kappa(dl/d) = \epsilon_0 \kappa l \tag{15.90}$$

where $\epsilon_0(= 8.854 \times 10^{-12}\,\mathrm{C^2/N\,m^2})$ is the permittivity of free space and κ is the dielectric constant of the crystal.

For a transverse intensity modulator using KDP in the configuration discussed in Sec. 15.3 for which V_π is given by Eq. (15.38), we obtain the specific energy as

$$\frac{P}{\Delta f} = \pi\,\frac{\epsilon_0 \kappa l}{4}\,\frac{\lambda_0^2}{n_o^6 r_{63}^2}\,\frac{d^2}{l^2}$$

$$= \frac{\pi\epsilon_0 \kappa \lambda_0^2}{4 n_o^6 r_{63}^2}\,\frac{d^2}{l} \tag{15.91}$$

As already discussed in Sec. 15.5.1 we cannot choose d and l independently; they are related through

$$\frac{d^2}{l} = S^2\,\frac{4\lambda'}{\pi} = S^2\,\frac{4\lambda_0}{n_o \pi} \tag{15.92}$$

where S is the safety factor. Substituting from Eq. (15.92) in Eq. (15.91) we obtain

$$\frac{P}{\Delta f} = \frac{\pi \epsilon_0 \kappa \lambda_0^2}{4n_o^3 r_{63}^2} S^2 \frac{4\lambda}{n_o \pi} \tag{15.93}$$

or

$$\frac{P}{\Delta f} = \frac{\epsilon_0 \kappa \lambda_0^3}{n_o^7 r_{63}^2} S^2 \tag{15.94}$$

For a lithium niobate or lithium tantalate transverse modulator, the half wave voltage V_π is given by Eq. (15.43) and carrying out a analysis similar to that for KDP we obtain the specific energy for a transverse lithium niobate or lithium tantalate modulator as

$$\frac{P}{\Delta f} = \frac{\epsilon_0 \kappa \lambda_0^3}{n_o (n_e^3 r_{33} - n_o^3 r_{13})^2} S^2 \tag{15.95}$$

It can be seen from above that the lower bound on the specific energy is independent of the modulator dimensions chosen and depends only on the material properties and the wavelength of operation.

Substituting the values of various parameters for KDP and for lithium niobate from Table 15.1, we obtain the respective specific energies as

$$\left(\frac{P}{\Delta f} \right)_{\text{KDP}} \approx 24 S^2 \, \text{mW/MHz} \tag{15.96}$$

$$\left(\frac{P}{\Delta f} \right)_{\text{LiNbO}_3} \approx 0.54 S^2 \, \text{mW/MHz} \tag{15.97}$$

where we have used $\kappa = 21$ for KDP and $\kappa = 28$ for lithium niobate. The above shows the superiority of lithium niobate over KDP. For a safety factor of $S = 5$, a lithium niobate modulator would require about 14 mW/MHz for 100% intensity modulation. Thus a lithium niobate modulator operating at 50 MHz bandwidth would require about 0.7 W of drive power at about 100 V.

A basic limitation in the modulators that we have discussed above comes about because the diffraction of the beam results in a minimum requirement for the cross sectional area ($\propto d^2$) of the modulator for a given length. This basic limitation can be overcome by using optical waveguide modulators where cross sectional dimensions of the order of the wavelength of light can be obtained and the interaction length can be chosen to be essentially unbounded. This comes about because of the waveguiding phenomenon which counteracts the diffraction divergence of the beam. Thus using optical waveguide modulators, 1–2 orders of magnitude reduction in drive power requirements and operating voltage are feasible. Experimental waveguide modulators have been fabricated showing very low drive power and drive voltages (see Chapter 14).

Problem 15.3. In the transverse configuration discussed in Sec. 15.3 we found that even in the absence of an applied electric field, a retardation between the two polarization states was introduced due to the static birefringence of the crystal. The temperature dependence of this static birefringence can lead to fluctuations in the depth of intensity modulation and/or poor extinction. For example for a lithium tantalate crystal of length 9.5 mm, the change in phase at the output could be as much as 180° for only 1° C change in temperature (Biazzo, 1971). Indeed for lithium tantalate (Denton, *et al.*, 1967).

$$\frac{1}{l}\frac{\partial}{\partial T}[l(n_e - n_o)]|_{T = 40\,°C} \approx 4.7 \times 10^{-5}/°C \tag{15.98}$$

Show that this temperature dependence of static birefringence can be compensated by using a configuration consisting of two identical lithium tantalate crystals placed one after the other with their *c*-axes in opposite directions and with a half wave plate introduced between the crystals with its axis at 45° to the *c*-axis of the first crystal. Show that the half wave voltage of the resultant modulator is also reduced by a factor of $\frac{1}{2}$ with respect to that using a single crystal (Biazzo, 1971).

Problem 15.4: Eq. (15.29) gives the intensity passing through an analyser whose pass axis is perpendicular to the pass axis of the polarizer. If the pass axis of the analyser is parallel to that of the input polarizer, show that the output intensity would be given by

$$\frac{I_o}{I_i} = \frac{1}{2}\left[1 - \sin\left(\frac{\pi V}{V_\pi}\right)\right] \tag{15.99}$$

Example In this example we shall discuss the basic principle behind the operation of an analog to digital (A/D) converter using electrooptic modulators (Taylor, 1975, 1979; Taylor, Taylor and Bauer, 1978; Takizawa and Okada, 1979). A/D converters are widely used to translate analog sensor measurements into the digital language for computing, information processing and control systems and in communications. The function of such an A/D converter is to sample an input analog signal repetitively and to convert each analog sampled signal value into a series of digital numbers ('zeros' and 'ones') to approximate the sampled value.

The electrooptic A/D converter is based on the fact that the output of an intensity modulator is a periodic function of the applied voltage. This is similar to the periodic variation of each bit in a binary representation of an analog quantity.

Let us consider a set of four transverse electrooptic modulators with electrodes of length

$$L_n = 2^{n-1} L_0 \tag{15.100}$$

i.e., with lengths

$$L_0, 2L_0, 4L_0 \text{ and } 8L_0 \tag{15.101}$$

where L_0 is the length corresponding to the shortest modulator. Since the length of the electrodes is different in each modulator the electrooptic retardations for the same applied voltage will be different in each of the modulators. Each modulator is also assumed to be followed by a compensator which can be adjusted for different additional phase shifts between the two polarization components. Let V_π represent the half wave voltage corresponding to the modulator of length L_0. In such a case if we give an additional phase shift of $\frac{1}{2}\pi$ to the light emerging from the first modulator, then the output intensity will be given by (see Eq. 15.29)

$$\frac{I_1}{I_i} = \cos^2 \left(\frac{\pi V}{2V_\pi} - \frac{\pi}{4} \right) \tag{15.102}$$

Fig. 15.14 Variation of the output intensity of four electro-optic modulators as a function of applied voltage V. For a threshold intensity of $\frac{1}{2}I_0$, the binary output corresponding to the input voltage V is shown at the bottom.

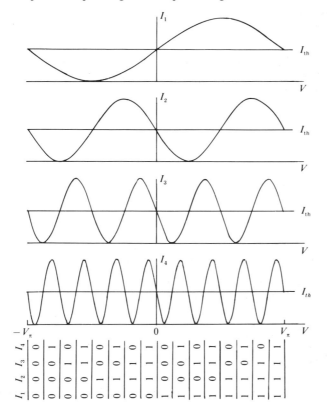

Observe that we are borrowing the analysis presented for the longitudinal configuration although we are here considering a transverse configuration. Indeed the analysis given in Sec. 15.2 is still valid; γ represents the phase shift between the two polarization components.

We now consider the second modulator whose length is $2L_0$. If instead of introducing a phase difference of $+\frac{1}{2}\pi$ between $\mathscr{E}_{x'}$ and $\mathscr{E}_{y'}$ components we introduce a phase difference of $-\frac{1}{2}\pi$ then following an analysis similar to that used to derive Eq. (15.27) one obtains for the output intensity

$$I_2 = \tfrac{1}{8}A^2 \operatorname{Re}[(1 - i\,e^{-i2\gamma})(1 + i\,e^{i2\gamma})]$$

$$= \tfrac{1}{4}A^2(1 - \sin 2\gamma)$$

or

$$\frac{I_2}{I_i} = \cos^2\left(\frac{\pi V}{V_\pi} + \frac{\pi}{4}\right)$$

where γ is again given by Eq. (15.36), and V_π corresponds to the half wave voltage of the modulator with length L_0. Observe that the phase shift introduced by the electrooptic effect is now 2γ since in the transverse configuration, the retardation introduced is proportional to the length of the modulator (see Eq. (15.37)) which is now $2L_0$. Similarly, the output light intensity from the third and fourth modulators of length $4L_0$ and $8L_0$ will be

$$\frac{I_3}{I_i} = \cos^2\left(\frac{2\pi V}{V_\pi} + \frac{\pi}{4}\right) \tag{15.104}$$

$$\frac{I_4}{I_i} = \cos^2\left(\frac{4\pi V}{V_\pi} + \frac{\pi}{4}\right) \tag{15.105}$$

The light emerging from each of the modulators is now detected and amplified and a binary representation of the analog voltage which is applied to all the modulators is obtained by electronically comparing the output intensity with a threshold I_{th} and generating a 'one' or a 'zero' if $I_n > I_{th}$ or $I_n < I_{th}$ respectively.

In order to understand this, in Fig. 15.14 we have plotted the output intensities from the various modulators as a function of the applied voltage. At the bottom of the figure we have also given the binary output for a threshold of $I_{th} = I_0/2$.

15.6   The index ellipsoid in the presence of an external electric field

In Chapter 3 we discussed in detail light propagation through anisotropic media. We found that for any given direction of propagation one has two linearly polarized modes of propagation. The index ellipsoid (see Eq. (3.78)) describes the birefringence property of the crystal. For any given direction of propagation, in order to determine the two normal modes of polarization, we found that if we draw a plane perpendicular to the direction of propagation which passes through the centre of the index ellipsoid then it will intersect the ellipsoid in an ellipse; the directions of the major and minor axes of the ellipse define the polarization direction of the two normal modes. The half lengths of these axes give the refractive indices seen by the correspondingly polarized modes. Any other polarization state propagating along the same direction does not retain its state of polarization but changes as it propagates (see Chapter 3).

If we refer to the principal axis system of the crystal, a general biaxial crystal can be described by an equation of the form

$$\frac{x^2}{n_1^2} + \frac{y^2}{n_2^2} + \frac{z^2}{n_3^2} = 1 \tag{15.106}$$

where n_1, n_2 and n_3 are the three principal indices of the crystal and (x, y, z) represents the principal axis coordinate system (see Fig. 3.12). For a uniaxial crystal with its c-axis parallel to the z-direction $n_1 = n_2 = n_o$ and $n_3 = n_e$; n_o and n_e represent the ordinary and extraordinary refractive indices of the crystal. In such a case Eq. (15.106) reduces to

$$\frac{x^2 + y^2}{n_o^2} + \frac{z^2}{n_e^2} = 1 \tag{15.107}$$

The above form of the equation for the index ellipsoid refers to the principal axis system. In any other coordinate system (x', y', z'), the equation for the index ellipsoid will be

$$\frac{x'^2}{n_1'^2} + \frac{y'^2}{n_2'^2} + \frac{z'^2}{n_3'^2} + 2\frac{y'z'}{n_4'^2} + 2\frac{z'x'}{n_5'^2} + 2\frac{x'y'}{n_6'^2} = 1 \tag{15.108}$$

where $n_1', n_2', \ldots, n_6'$ are constants. In order to write the above equation in a more compact form we represent the coordinate axes by (x_1', x_2', x_3') and in such a case Eq. (15.108) can be rewritten as

$$\sum_{i,j=1,2,3} x_i' x_j' / n_{ij}'^2 = 1 \tag{15.109}$$

where

$$\left.\begin{array}{l} n_{11}' = n_1', \quad n_{22}' = n_2', \quad n_{33}' = n_3' \\ n_{32}' = n_{23}' = n_4', \quad n_{31}' = n_{13}' = n_5', \quad n_{12}' = n_{21}' = n_6' \end{array}\right\} \tag{15.110}$$

If we denote the principal axis system by (x_1, x_2, x_3) then Eq. (15.109) becomes

$$\sum_{i,j=1,2,3} x_i x_j / n_{ij}^2 = 1 \tag{15.111}$$

with

$$n_{11} = n_1, \quad n_{22} = n_2, \quad n_{33} = n_3 \tag{15.112}$$

$$n_{23}^{-1} = n_{32}^{-1} = 0, \quad n_{31}^{-1} = n_{13}^{-1} = 0, \quad n_{12}^{-1} = n_{21}^{-1} = 0 \qquad (15.113)$$

The electrooptic effect refers to the change in the refractive indices of the crystal due to an applied external electric field. The Pockels effect refers to changes in the refractive indices proportional to the first power of the applied electric field. If the change is proportional to the square of the applied electric field then the effect is referred to as the quadratic electrooptic effect or the Kerr effect. The change in the crystal refractive indices can be described in terms of the changes in the shape, dimensions and orientation of the crystal index ellipsoid.

The Pockels effect can exist only in those crystals that do not possess an inversion symmetry. A crystal is said to possess an inversion symmetry if inversion about any of its lattice points (i.e., replacing $\mathbf{r}$ by $-\mathbf{r}$, $\mathbf{r}$ being the position vector of any lattice point measured from the point about which inversion is being performed) leaves the crystal structure invariant. In order to show that such a crystal cannot exhibit a linear electrooptic effect let $\Delta n_1 = rE$ be the change in refractive index caused by an applied electric field, r being the electrooptic coefficient. If we now reverse the direction of the applied electric field, then the change in index becomes $\Delta n_2 = -rE$. Since the crystal possesses inversion symmetry the applied fields E and $-E$ are physically identical so that $\Delta n_1 = \Delta n_2$ which gives us $r = -r$. This can be satisfied only if $r = 0$. Thus, there can be no linear electrooptic effect in crystals that possess inversion symmetry.

Instead of specifying the changes in the refractive indices, it is more convenient to consider the changes in $(1/n^2)$ due to the external electric field; these can always be related to the changes in n. The index ellipsoid referred to the principal axis system, in the absence of an external electric field is described by Eq. (15.111). The changes in $1/n^2$ due to the Pockel effect are described by the following equation:

$$\Delta(1/n^2)_{ij} = \sum_{k=1,2,3} r_{ijk} E_k \qquad (15.114)$$

where $E_k(k = 1, 2, 3)$ represents the components of the applied electric field and $r_{ijk}(i, j, k = 1, 2, 3)$ are the elements of the electrooptic tensor.

Since the indices i and j commute we can replace the indices ij by a single index taking values from 1 to 6. By convention we replace the various indices as follows:

$$
\begin{aligned}
11 \to 1, \quad 22 \to 2, \quad 33 \to 3, \quad 32, \quad 23 \to 4, \\
13, \quad 31 \to 5, \quad 12, \quad 21 \to 6
\end{aligned} \qquad (15.115)
$$

Thus Eq. (15.114) becomes

$$\Delta(1/n^2)_i = \sum_{k=1,2,3} r_{ik}E_k; \quad i = 1, 2\ldots, 6 \tag{15.116}$$

The elements $r_{ik}(i = 1, 2\ldots, 6, k = 1, 2, 3)$ together form a 6×3 matrix as follows:

$$[r] = \begin{pmatrix} r_{11} & r_{12} & r_{13} \\ r_{21} & r_{22} & r_{23} \\ r_{31} & r_{32} & r_{33} \\ r_{41} & r_{42} & r_{43} \\ r_{51} & r_{52} & r_{53} \\ r_{61} & r_{62} & r_{63} \end{pmatrix} \tag{15.117}$$

The above matrix $[r]$ is referred to as the linear electrooptic tensor. As discussed before all the elements r_{ij} vanish for a crystal possessing inversion symmetry. The coefficients r_{ij} have typical values of $10^{-12}\,\text{m/V}$ (see Table 15.1).

As is well known a crystal may be described by various symmetry operations under which it remains invariant. Thus a crystal may be classified as belonging to one of the seven crystal classes: triclinic, monoclinic, orthorhombic, tetragonal, cubic, trigonal and hexagonal. The form of the electrooptic tensor is determined by the symmetry operations of the particular crystal class to which the crystal belongs. These symmetry operations can be used to specify which of the various elements must be zero, which of them must be equal etc. The electrooptic tensors for all crystal classes are discussed in Yariv and Yeh (1984).

15.6.1 Index ellipsoid of KDP

As an example we consider a KDP crystal for which the only nonzero elements of the electrooptic tensor are $r_{41}, r_{52} = r_{41}$ and r_{63}:

$$r_{ij} = \begin{pmatrix} 0 & 0 & 0 \\ 0 & 0 & 0 \\ 0 & 0 & 0 \\ r_{41} & 0 & 0 \\ 0 & r_{41} & 0 \\ 0 & 0 & r_{63} \end{pmatrix} \tag{15.118}$$

The z-axis is chosen as the c-axis of the crystal and the x and y-axes are the two crystal axes along which the crystal exhibits twofold symmetry. Since KDP is a uniaxial crystal, if the z-axis is assumed to be along the c-axis, then the index ellipsoid in the absence of an external field can be written as

$$\frac{x^2 + y^2}{n_o^2} + \frac{z^2}{n_e^2} = 1 \tag{15.119}$$

where n_o and n_e are the ordinary and extraordinary refractive indices. If an external field $\mathbf{E}$ is applied then using Eq. (15.116) we have

$$\Delta(1/n^2)_1 = 0, \quad \Delta(1/n^2)_2 = 0, \quad \Delta(1/n^2)_3 = 0 \tag{15.120}$$
$$\Delta(1/n^2)_4 = r_{41}E_x, \quad \Delta(1/n^2)_5 = r_{41}E_y, \quad \Delta(1/n^2)_6 = r_{63}E_z$$

Thus the index ellipsoid in the presence of the external field becomes

$$\frac{x^2 + y^2}{n_o^2} + \frac{z^2}{n_e^2} + 2r_{41}E_x yz + 2r_{41}E_y zx + 2r_{63}E_z xy = 1 \tag{15.121}$$

Thus in the presence of an external electric field the index ellipsoid of the crystal is no longer described by an equation of the form of Eq. (15.119) but now contains product terms between x, y and z. Thus on application of an external field the index ellipsoid changes its orientation.

Let us first assume that the external field is applied along the c-axis, i.e., along the z-axis. In such a case, the index ellipsoid is described by

$$\frac{x^2 + y^2}{n_o^2} + \frac{z^2}{n_e^2} + 2r_{63}E_z xy = 1 \tag{15.122}$$

i.e., only one cross term xy is present. In order to find the new refractive indices and the eigen polarizations, we transform from the principal axes x, y, z to another set x', y', z' in which the index ellipsoid would be diagonal. Since there is no cross term involving z, it means that the new transformed z-axis, namely the z'-axis is identical to the old z-axis i.e., $z = z'$. In addition, since the terms in x and y are symmetric, the new x', y'-axis must be related to the old x, y-axes through a $45°$ rotation about the z-axis (see Problem 15.5). Thus we choose the new coordinates x', y' to be

$$x' = x\cos 45° + y\sin 45° = (x + y)/\sqrt{2} \tag{15.123}$$

$$y' = -x\sin 45° + y\cos 45° = (-x + y)/\sqrt{2} \tag{15.124}$$

Since we would like to write Eq. (15.122) in terms of x', y' we solve Eqs. (15.123) and (15.124) for x and y in terms of x' and y' and obtain

$$x = (x' - y')/\sqrt{2} \tag{15.125}$$

$$y = (x' + y')/\sqrt{2} \tag{15.126}$$

Substituting these values of x and y in Eq. (15.122), we obtain

$$x'^2((1/n_o^2) + r_{63}E_z) + y'^2((1/n_o^2) - r_{63}E_z) + z'^2/n_e^2 = 1 \qquad (15.127)$$

With respect to the new coordinate system, the equation of the index ellipsoid contains only square terms and thus (x', y', z') represents the principal axis system in the presence of an external field along the z-direction.

It should be noted here that the direction of the axes in the presence of an external field along the z-direction is independent of the magnitude of the applied field and is at $45°$ to the crystallographic axes.

The new crystal indices are $n_{x'}$, $n_{y'}$ and $n_{z'}$ where

$$1/n_{x'}^2 = 1/n_o^2 + r_{63}E_z \qquad (15.128)$$

$$1/n_{y'}^2 = 1/n_o^2 - r_{63}E_z \qquad (15.129)$$

$$1/n_{z'}^2 = 1/n_e^2 \qquad (15.130)$$

It may be noted that the crystal which is uniaxial in the absence of an external field, becomes biaxial when a field is applied along the z-direction.

If we assume $r_{63}E_z \ll 1/n_o^2$ then we have

$$n_{x'} = n_o(1 + n_o^2 r_{63}E_z)^{-\frac{1}{2}} \approx n_o(1 - \tfrac{1}{2}n_o^2 r_{63}E_z)$$

or

$$n_{x'} = n_o - \tfrac{1}{2}n_o^3 r_{63}E_z \qquad (15.131)$$

Similarly

$$n_{y'} = n_o + \tfrac{1}{2}n_o^3 r_{63}E_z \qquad (15.132)$$

$$n_{z'} = n_e \qquad (15.133)$$

Table 15.1 gives the values of various r_{ij} coefficients for KDP and ADP.

15.6.2 *Index ellipsoid of lithium niobate, lithium tantalate*

The linear electrooptic matrices for lithium niobate and lithium tantalate are of the form

$$[r] = \begin{pmatrix} 0 & -r_{22} & r_{13} \\ 0 & r_{22} & r_{13} \\ 0 & 0 & r_{33} \\ 0 & r_{51} & 0 \\ r_{51} & 0 & 0 \\ -r_{22} & 0 & 0 \end{pmatrix} \qquad (15.134)$$

Table 15.1 gives the values of various important characteristics of these crystals and also the values of the various electrooptic coefficients r_{ij}. As is apparent from Table 15.1 the electrooptic coefficients in lithium niobate and lithium tantalate are several times those in KDP or ADP.

As can be seen from Table 15.1 the largest electrooptic coefficient in lithium niobate is r_{33}. Hence to get the maximum refractive index change, one makes use of this coefficient. Fig. 15.8 shows a transverse configuration in which the external electric field is applied along the c-axis of the crystal and we assume that light propagates along the crystal x or y-axis. In the absence of the applied electric field the index ellipsoid will be described by

$$\frac{x^2 + y^2}{n_o^2} + \frac{z^2}{n_e^2} = 1 \tag{15.135}$$

where n_o and n_e are respectively the ordinary and extraordinary refractive indices. Since the external applied field is along the z-direction (c-axis) the index ellipsoid in the presence of the external field will be given by

$$\frac{x^2 + y^2}{n_o^2} + \frac{z^2}{n_e^2} + r_{13}E_z x^2 + r_{13}E_z y^2 + r_{33}E_z z^2 = 1 \tag{15.136}$$

where we have used the fact that for lithium niobate,

$$r_{23} = r_{13}, \quad r_{43} = r_{53} = r_{63} = 0 \tag{15.137}$$

Eq. (15.136) can be rewritten as

$$x^2((1/n_o^2) + r_{13}E_z) + y^2((1/n_o^2) + r_{13}E_z) + z^2((1/n_e^2) + r_{33}E_z) = 1 \tag{15.138}$$

Thus under the application of the electric field along z only the lengths of the axes of the ellipsoid change, the ellipsoid does not undergo any rotation such as that seen with KDP since there are no cross terms in Eq. (15.138).

Assuming as in the case of KDP, $r_{33}E_z \ll n_e$, $r_{13}E_z \ll n_o$, the refractive indices for x, y and z-polarized waves will be given by

$$n_x = n_o - \tfrac{1}{2}n_o^3 r_{13}E_z \tag{15.139}$$

$$n_y = n_o - \tfrac{1}{2}n_o^3 r_{13}E_z \tag{15.140}$$

$$n_z = n_e - \tfrac{1}{2}n_e^3 r_{33}E_z \tag{15.141}$$

Additional problems

Problem 15.5: Starting from Eq. (15.122) show that the principal axes of the index ellipsoid in the presence of an electric field along the z-direction are at $45°$ to the x and y-directions.

Solution: Let us consider a rotated coordinate system (x', y', z') which is obtained from the crystallographic axes (x, y, z) by a rotation by an angle ψ about the z-axis. In such a case we will have

$$\left.\begin{array}{l} x' = x \cos\psi + y \sin\psi \\ y' = -x \sin\psi + y \cos\psi \\ z' = z \end{array}\right\} \qquad (15.142)$$

Inverting the above equations, we obtain

$$\left.\begin{array}{l} x = x' \cos\psi - y' \sin\psi \\ y = x' \sin\psi + y' \cos\psi \\ z = z' \end{array}\right\} \qquad (15.143)$$

Substituting x, y and z in terms of x', y' and z' in Eq. (15.122) we obtain

$$\frac{x'^2 + y'^2}{n_o^2} + \frac{z^2}{n_e^2} + 2r_{63}E_z(x'^2 \cos\psi \sin\psi - y'^2 \cos\psi \sin\psi)$$

$$+ 2r_{63}E_z x'y'(\cos^2\psi - \sin^2\psi) = 1 \qquad (15.144)$$

In order to find the principal axes, the angle ψ must be so chosen as to make the cross term $x'y'$ vanish. This happens if

$$\cos^2\psi - \sin^2\psi = \cos 2\psi = 0$$

i.e.

$$\psi = \tfrac{1}{4}\pi = 45° \qquad (15.145)$$

Using this value of ψ in Eq. (15.144), we get Eq. (15.127).

Problem 15.6: In Sec. 15.3 we considered a transverse configuration of a KDP modulator where the light wave propagated along the y'-direction and the field was applied along the z-direction. Consider light propagation along the z-direction and assume that the external field is applied along the x-direction. Calculate the directions of the axes of the index ellipsoid in the presence of a field and the principal indices of refraction.

Solution: If the applied electric field is only along the x-direction, then the equation of the index ellipsoid becomes (see Eq. (15.121))

$$\frac{x^2 + y^2}{n_o^2} + \frac{z^2}{n_e^2} + 2r_{41}E_x yz = 1 \qquad (15.146)$$

We have now to transform this into its diagonal form where cross terms are absent. Since there is no cross term involving x, the new x-axis will coincide with the old axis. In order to determine the new y and z-axes, we rotate the coordinate axes by an angle α about the x-axis. In such a case we may write (see Eq. (15.143))

$$\left.\begin{array}{l} y = y' \cos\alpha - z' \sin\alpha \\ z = y' \sin\alpha + z' \cos\alpha \end{array}\right\} \qquad (15.147)$$

If we substitute the above expressions into Eq. (15.146) and make the cross product term $y'z'$ zero we get

$$\tan 2\alpha = -\frac{2r_{41}E_x}{(1/n_e^2 - 1/n_o^2)} \tag{15.148}$$

Thus in the presence of a field in the x-direction the principal y' and z'-axes are rotated through an angle α in the y–z plane where α is given by Eq. (15.148). The angle of rotation is dependent on the value of the field E_x applied.

As a typical example we take a KDP crystal and assume an applied field of 10^6 V/m. Using the values of the other coefficients given in Table 15.1 we get

$$\tan 2\alpha = 7 \times 10^{-4}$$

or

$$\alpha \approx 2 \times 10^{-2} \text{ degrees} \tag{15.149}$$

Thus even for such large fields the rotation of the axes is very small.

Substituting the expressions for y and z in Eq. (15.146) the equation for the index ellipsoid can be shown to be given by

$$\frac{x'^2}{n_o^2} + y'^2 \left(\frac{\cos^2 \alpha}{n_o^2} + \frac{\sin^2 \alpha}{n_e^2} + 2r_{41}E_x \sin \alpha \cos \alpha \right)$$

$$+ z'^2 \left(\frac{\sin^2 \alpha}{n_o^2} + \frac{\cos^2 \alpha}{n_e^2} - 2r_{41}E_x \sin \alpha \cos \alpha \right) = 1 \tag{15.150}$$

Thus the new principal indices $n_{x'}$, $n_{y'}$ and $n_{z'}$ are given by

$$n_{x'} = n_o \tag{15.151}$$

$$n_{y'} = \left(\frac{\cos^2 \alpha}{n_o^2} + \frac{\sin^2 \alpha}{n_e^2} + 2r_{41}E_x \sin \alpha \cos \alpha \right)^{-\frac{1}{2}} \tag{15.152}$$

$$n_{z'} = \left(\frac{\sin^2 \alpha}{n_o^2} + \frac{\cos^2 \alpha}{n_e^2} - 2r_{41}E_x \sin \alpha \cos \alpha \right)^{-\frac{1}{2}} \tag{15.153}$$

If we use Eq. (15.148) and retain terms to order E_x^2 we will get, after some algebra,

$$n_{y'} \approx n_o + \frac{n_o^5 n_e^2 r_{41}^2 E_x^2}{2(n_o^2 - n_e^2)} \tag{15.154}$$

$$n_{z'} \simeq n_e - \frac{n_e^5 n_o^2 r_{41}^2 E_x^2}{2(n_o^2 - n_e^2)} \tag{15.155}$$

where we have assumed $\sin^2 \alpha \approx \alpha^2$, $\cos^2 \alpha \approx 1 - \alpha^2$, $\tan 2\alpha \approx 2\alpha$.

16

The strain optic tensor

16.1 Introduction

The mechanical strain produced by a propagating acoustic wave in a medium generates a periodic refractive index grating and when an optical beam interacts with this refractive index grating we have what is known as acoustooptic interaction. This interaction is a subject of tremendous importance and is used in many devices such as modulators, deflectors, frequency shifters, etc. In this chapter we introduce the strain optic tensor and discuss how one may obtain the refractive index variation produced by the propagating acoustic wave. This will be used in the following two chapters to study acoustooptic interactions and in Chapter 19 we will discuss some of the important acoustooptic devices.

16.2 The strain optic tensor

In Chapter 15 we discussed the electrooptic effect in which an applied electric field resulted in the variation of the refractive index of the medium. Just like in the electrooptic effect, the effect of the strain on the optical properties of the medium is described in terms of changes in the index ellipsoid. The index ellipsoid of a medium in the absence of strain can, in general, be written in the form (see Chapter 3)

$$\frac{x^2}{n_1'^2} + \frac{y^2}{n_2'^2} + \frac{z^2}{n_3'^2} + \frac{2yz}{n_4'^2} + \frac{2xz}{n_5'^2} + \frac{2xy}{n_6'^2} = 1 \tag{16.1}$$

If (x, y, z) corresponds to the principal axis system, then the index ellipsoid will be represented by

$$\frac{x^2}{n_1^2} + \frac{y^2}{n_2^2} + \frac{z^2}{n_3^2} = 1 \tag{16.2}$$

where n_1, n_2, n_3 are the principal indices of refraction.

The components of the strain tensor are defined by the following equations

$$S_{xx} = \partial u/\partial x, \quad S_{yy} = \partial v/\partial y, \quad S_{zz} = \partial w/\partial z \tag{16.3}$$

$$\left.\begin{array}{l} S_{xy} = (\partial u/\partial y) + (\partial v/\partial x) = S_{yx} \\ S_{yz} = (\partial v/\partial z) + (\partial w/\partial y) = S_{zy} \\ S_{xz} = (\partial u/\partial z) + (\partial w/\partial x) = S_{zx} \end{array}\right\} \tag{16.4}$$

where u, v and w represent the displacements along the x, y and z-directions respectively. The first three components, S_{xx}, S_{yy} and S_{zz} define *normal strain*. They represent change in length per unit length in the three directions specified by x, y and z-axes. The other three components S_{xy}, S_{yz} and S_{xz} represent *shear strains*. It is convenient to introduce a one-index symbol for the strain components. Accordingly we define the following

$$\left.\begin{array}{l} S_1 = S_{xx}, \quad S_2 = S_{yy}, \quad S_3 = S_{zz} \\ S_4 = S_{yz}, \quad S_5 = S_{zx}, \quad S_6 = S_{xy} \end{array}\right\} \tag{16.5}$$

On application of mechanical strain, the changes in $(1/n^2)_i$ are given by

$$\Delta(1/n^2)_i = \sum_{j=1}^{6} p_{ij}S_j; \quad i = 1, \ldots, 6 \tag{16.6}$$

where p_{ij} represent the strain optic coefficients and in the presence of strain, the new index ellipsoid becomes

$$x^2\left((1/n_1^2) + \sum_j p_{1j}S_j\right) + y^2\left((1/n_2^2) + \sum_j p_{2j}S_j\right)$$

$$+ z^2\left((1/n_3^2) + \sum_j p_{3j}S_j\right) + 2yz\sum_j p_{4j}S_j$$

$$+ 2xz\sum_j p_{5j}S_j + 2xy\sum_j p_{6j}S_j = 1 \tag{16.7}$$

Just like the electrooptic tensor, the form (but not the magnitude) of the strain optic tensor can be derived from consideration of the symmetry of the medium.

For example for isotropic media,

$$p = \begin{bmatrix} p_{11} & p_{12} & p_{12} & 0 & 0 & 0 \\ p_{12} & p_{11} & p_{12} & 0 & 0 & 0 \\ p_{12} & p_{12} & p_{11} & 0 & 0 & 0 \\ 0 & 0 & 0 & \tfrac{1}{2}(p_{11} - p_{12}) & 0 & 0 \\ 0 & 0 & 0 & 0 & \tfrac{1}{2}(p_{11} - p_{12}) & 0 \\ 0 & 0 & 0 & 0 & 0 & \tfrac{1}{2}(p_{11} - p_{12}) \end{bmatrix} \tag{16.8}$$

Table 16.1. *Photoelastic properties of some materials*

Material	Photoelastic coefficients		Figure of merit $M_2(\times 10^{-15}\,\text{s}^3/\text{kg})$
Fused silica	$p_{11} = +0.121,$	$p_{12} = +0.270$	1.51
Water	$p_{11} = 0.31,$	$p_{12} = 0.31$	160
Dense flint SF-4	$p_{11} = +0.232,$	$p_{12} = +0.256$	4.53
Lithium niobate (LiNbO$_3$)	$p_{11} = -0.02,$ $p_{13} = +0.13,$ $p_{31} = +0.17,$ $p_{41} = -0.15,$	$p_{12} = +0.08$ $p_{14} = -0.08$ $p_{33} = +0.07$ $p_{44} = +0.12$	13.6
Lithium tantalate (LiTaO$_3$)	$p_{11} = 0.08,$ $p_{13} = 0.09,$ $p_{31} = 0.09,$ $p_{41} = 0.02,$	$p_{12} = 0.08$ $p_{14} = 0.03$ $p_{33} = 0.15$ $p_{44} = 0.02$	1.37
α-Quartz (SiO$_2$)	$p_{11} = +0.16,$ $p_{13} = +0.27,$ $p_{31} = +0.29,$ $p_{41} = -0.047,$	$p_{12} = +0.27$ $p_{14} = -0.03$ $p_{33} = +0.10$ $p_{44} = -0.079$	
Tellurium dioxide (TeO$_2$)	$p_{11} = 0.0074,$ $p_{13} = +0.340,$ $p_{33} = +0.240,$ $p_{66} = -0.046$	$p_{12} = +0.187$ $p_{31} = +0.090$ $p_{44} = -0.17$	793
KDP	$p_{11} = +0.251,$ $p_{13} = +0.246,$ $p_{33} = +0.221,$ $p_{66} = 0.058$	$p_{12} = +0.249$ $p_{31} = +0.225$ $p_{44} = ?$	3.8

Adapted from Pinnow (1972). All data correspond to room temperature and $\lambda_0 = 0.6328\,\mu\text{m}$ except for α-quartz for which $\lambda_0 = 0.59\,\mu\text{m}$.
When the sign of p_{ij} is not specified it implies only the absolute value is known.
The values of M_2 (defined by Eq. (18.61)) are the maximum known.
For data on other photoelastic materials readers may refer to Pinnow (1972).

Thus there are only two independent components p_{11} and p_{12}. The values of p_{11}, and p_{12} for some isotropic media are given in Table 16.1. We have also tabulated the values of M_2 which is known as the figure of merit and will be defined in Chapter 18.

Similarly for lithium niobate, lithium tantalate and quartz (SiO$_2$) belonging to the trigonal system, the strain optic tensor is given by

$$p = \begin{bmatrix} p_{11} & p_{12} & p_{13} & p_{14} & 0 & 0 \\ p_{12} & p_{11} & p_{13} & -p_{14} & 0 & 0 \\ p_{31} & p_{31} & p_{33} & 0 & 0 & 0 \\ p_{41} & -p_{41} & 0 & p_{44} & 0 & 0 \\ 0 & 0 & 0 & 0 & p_{44} & p_{41} \\ 0 & 0 & 0 & 0 & p_{14} & \frac{1}{2}(p_{11}-p_{12}) \end{bmatrix} \tag{16.9}$$

Thus there are eight independent coefficients, the values of which are given in Table 16.1.

Crystals like tellurium dioxide (TeO_2) and KDP have seven independent coefficients and the strain optic tensor is given by

$$p = \begin{bmatrix} p_{11} & p_{12} & p_{13} & 0 & 0 & 0 \\ p_{12} & p_{11} & p_{13} & 0 & 0 & 0 \\ p_{31} & p_{31} & p_{33} & 0 & 0 & 0 \\ 0 & 0 & 0 & p_{44} & 0 & 0 \\ 0 & 0 & 0 & 0 & p_{44} & 0 \\ 0 & 0 & 0 & 0 & 0 & p_{66} \end{bmatrix} \tag{16.10}$$

The corresponding values are also tabulated in Table 16.1.

Now, for a given propagating acoustic wave $S_1, \ldots, S_6$ would be known. Thus knowing the strain optic tensor the six components of $\Delta(1/n^2)$ can be calculated by using Eq. (16.6) which should be written in the form of a 3×3 matrix. Finally, the change in dielectric permittivity can be calculated by using the following formula which will be derived in Problem 16.3:

$$\overline{\overline{\Delta\epsilon}} = -\frac{1}{\epsilon_0} \bar{\epsilon} \Delta\left(\frac{1}{n^2}\right) \bar{\epsilon} \tag{16.11}$$

We will illustrate the procedure by considering specific examples. The derived expressions for $\overline{\overline{\Delta\epsilon}}$ will be used in Sec. 18.8.

16.3   Calculation of $\overline{\overline{\Delta\epsilon}}$ for a longitudinal acoustic wave propagating in an isotropic medium

We consider a longitudinal acoustic wave propagating along the z-direction in an isotropic medium such as water. Such a wave will create a normal strain along the z-direction and therefore the only nonvanishing strain component will be given by

$$S_{zz} = S_3 = S_0 \sin(\Omega t - Kz) \tag{16.12}$$

where S_0 is the peak strain and Ω and K are the angular frequency and propagation constant of the acoustic wave. Using Eqs. (16.6) and (16.8) we may write

$$\Delta(1/n^2)_2 = \Delta(1/n^2)_1 = p_{12}S_0 \sin(\Omega t - Kz) \qquad (16.13)$$

$$\Delta(1/n^2)_3 = p_{11}S_0 \sin(\Omega t - Kz) \qquad (16.14)$$

$$\Delta(1/n^2)_4 = \Delta(1/n^2)_5 = \Delta(1/n^2)_6 = 0 \qquad (16.15)$$

The above is a contracted representation of

$$\Delta(1/n^2) = \begin{pmatrix} p_{12}S_3 & 0 & 0 \\ 0 & p_{12}S_3 & 0 \\ 0 & 0 & p_{11}S_3 \end{pmatrix} \qquad (16.16)$$

Now for an isotropic medium,

$$\bar{\bar{\epsilon}} = \begin{pmatrix} \epsilon & 0 & 0 \\ 0 & \epsilon & 0 \\ 0 & 0 & \epsilon \end{pmatrix} = \epsilon_0 \begin{pmatrix} n^2 & 0 & 0 \\ 0 & n^2 & 0 \\ 0 & 0 & n^2 \end{pmatrix} \qquad (16.17)$$

where n is the refractive index of the medium. Hence using Eqs. (16.16) and (16.17) in Eq. (16.11) we obtain

$$\overline{\overline{\Delta\epsilon}} = -\epsilon_0 n^4 \begin{pmatrix} p_{12}S_3 & 0 & 0 \\ 0 & p_{12}S_3 & 0 \\ 0 & 0 & p_{11}S_3 \end{pmatrix} \qquad (16.18)$$

We will use the above equation in Sec. 18.8.1 to study acoustooptic coupling. We may, however, note that using Eq. (16.16), the index ellipsoid is given by

$$(x^2 + y^2)\left(\frac{1}{n^2} + p_{12}S_3\right) + z^2\left(\frac{1}{n^2} + p_{11}S_3\right) = 1 \qquad (16.19)$$

The above equation implies that in the presence of the acoustic wave the medium becomes uniaxial with the optic axis along the direction of propagation of the acoustic wave and because of Eq. (16.12), the refractive index varies periodically along the z-direction with a period $\Lambda(=2\pi/K)$.

16.4 Calculation of $\overline{\overline{\Delta\epsilon}}$ for a shear wave propagating along the z-direction in lithium niobate

We next consider lithium niobate (or lithium tantalate) for which

$$\bar{\bar{\epsilon}} = \epsilon_0 \begin{pmatrix} n_0^2 & 0 & 0 \\ 0 & n_0^2 & 0 \\ 0 & 0 & n_e^2 \end{pmatrix} \qquad (16.20)$$

where n_0 and n_e represent the ordinary and extraordinary refractive indices. We consider the propagation of a shear wave polarized along y and propagating along the z-direction. For such a case, the only nonvanishing strain components will be

$$S_{zy} = S_{yz} = S_4 \tag{16.21}$$

Thus

$$
\left.
\begin{aligned}
\Delta(1/n^2)_1 &= \sum_{j=1}^{6} p_{1j}S_j = p_{14}S_4 \\
\Delta(1/n^2)_2 &= p_{24}S_4 = -p_{14}S_4 \\
\Delta(1/n^2)_4 &= p_{44}S_4
\end{aligned}
\right\} \tag{16.22}
$$

and

$$\Delta(1/n^2)_3 = \Delta(1/n^2)_5 = \Delta(1/n^2)_6 = 0$$

The above is a contracted representation of

$$
\Delta(1/n^2) =
\begin{pmatrix}
p_{14}S_4 & 0 & 0 \\
0 & -p_{14}S_4 & p_{44}S_4 \\
0 & p_{44}S_4 & 0
\end{pmatrix}
\tag{16.23}
$$

Thus in the presence of the acoustic wave the index ellipsoid will become

$$x^2\left(\frac{1}{n_o^2} + p_{14}S_4\right) + y^2\left(\frac{1}{n_o^2} - p_{14}S_4\right) + \frac{z^2}{n_e^2} + 2yzp_{44}S_4 = 1 \tag{16.24}$$

and hence the new principal axis will be rotated about the x-axis. Using Eqs. (16.20) and (16.23) in Eq. (16.11) we obtain

$$
\overline{\overline{\Delta\epsilon}} = -\epsilon_0
\begin{pmatrix}
p_{14}S_4n_o^4 & 0 & 0 \\
0 & -p_{14}S_4n_o^4 & p_{44}S_4n_o^2n_e^2 \\
0 & p_{44}S_4n_o^2n_e^2 & 0
\end{pmatrix}
\tag{16.25}
$$

Problems

Problem 16.1: Consider an acoustic shear wave polarized along x and propagating along y in lithium niobate. Obtain the corresponding $\overline{\overline{\Delta\epsilon}}$.

Solution: Since the acoustic wave is a shear wave polarized along x and propagating along y, the only nonvanishing strain components are $S_{xy} = S_{yx} = S_6$. Thus using Eqs. (16.6) and (16.9) we obtain

$$
\Delta(1/n^2) =
\begin{pmatrix}
0 & \frac{1}{2}(p_{11}-p_{12})S_6 & p_{41}S_6 \\
\frac{1}{2}(p_{11}-p_{12})S_6 & 0 & 0 \\
p_{41}S_6 & 0 & 0
\end{pmatrix}
\tag{16.26}
$$

If we now use Eq. (16.11) we get

$$
\overline{\overline{\Delta\epsilon}} = -\epsilon_0
\begin{pmatrix}
0 & \frac{1}{2}n_o^4(p_{11}-p_{12})S_6 & n_o^2n_e^2p_{41}S_6 \\
\frac{1}{2}n_o^4(p_{11}-p_{12})S_6 & 0 & 0 \\
n_o^2n_e^2p_{41}S_6 & 0 & 0
\end{pmatrix}
\tag{16.27}
$$

Problem 16.2: Show that for a longitudinal acoustic wave propagating along the z-direction in an isotropic medium, the principal refractive indices are approximately given by

$$n_x = n_y \approx n - \tfrac{1}{2} n^3 p_{12} S_0 \sin(\Omega t - Kz) \tag{16.28}$$

$$n_z \approx n - \tfrac{1}{2} n^3 p_{11} S_0 \sin(\Omega t - Kz) \tag{16.29}$$

Problem 16.3: The impermeability tensor

$$\eta_{ij} = (1/n^2)_{ij} \tag{16.30}$$

is defined by the following equation

$$\epsilon_{ik}\eta_{kj} = \epsilon_0 \delta_{ij} \tag{16.31}$$

where $[\epsilon_{ik}]$ is the dielectric permittivity tensor and

$$\delta_{ij} = \begin{cases} 1; & i = j \\ 0; & i \neq j \end{cases} \tag{16.32}$$

Since

$$\Delta\eta_{ij} = \Delta(1/n^2)_{ij} = \sum_{k,l} p_{ijkl} S_{kl}; \quad i, j, k, l = x, y, z \tag{16.33}$$

show that the change in the dielectric permittivity tensor is given by

$$\Delta\epsilon_{ij} = -\frac{\epsilon_{im} p_{mk\alpha\beta} S_{\alpha\beta} \epsilon_{kj}}{\epsilon_0} \tag{16.34}$$

Here we are assuming summation over repeated indices.

Solution: From Eq. (16.31) it follows that

$$\Delta\epsilon_{ik}\eta_{kj} + \epsilon_{ik}\Delta\eta_{kj} = 0 \tag{16.35}$$

Post multiplying by ϵ_{jl} and using Eq. (16.31), we have

$$\Delta\epsilon_{ik}\epsilon_0\delta_{kl} = -\epsilon_{ik}\Delta\eta_{kj}\epsilon_{jl}$$

or

$$\Delta\epsilon_{il} = -\frac{1}{\epsilon_0}\epsilon_{ik}\Delta\eta_{kj}\epsilon_{jl} = -\frac{1}{\epsilon_0}\epsilon_{ik}\Delta\left(\frac{1}{n^2}\right)_{kj}\epsilon_{jl} \tag{16.36}$$

which on using Eq. (16.33) will give us Eq. (16.34). Eq. (16.36) can also be written in the form

$$\overline{\overline{\Delta\epsilon}} = -\frac{1}{\epsilon_0}\bar{\epsilon}\Delta\left(\frac{1}{n^2}\right)\bar{\epsilon} \tag{16.37}$$

17

Acoustooptic effect: Raman–Nath diffraction

17.1 Introduction

The acoustooptic effect is the change in the refractive index of a medium caused by the mechanical strain produced by an acoustic wave. Since the strain varies periodically in the acoustic wave, the refractive index of the medium also varies periodically leading to a refractive index grating. When a light beam is incident on such a refractive index grating, diffraction takes place and this produces either multiple order diffraction or only single order diffraction. The former is referred to as Raman–Nath diffraction and is usually observed at low acoustic frequencies. The latter is analogous to Bragg diffraction of X-rays in crystals and is referred to here also as Bragg diffraction; this is usually observed at high acoustic frequencies.

The interaction between acoustic waves and light waves is used in a number of applications such as in acoustooptic modulators, deflectors, frequency shifters for heterodyning, spectrum analysers, Q-switching and mode locking in lasers. In this chapter we will discuss the basic principle of Raman–Nath diffraction and in the next chapter we will discuss Bragg diffraction.

17.2 Raman–Nath and Bragg regimes of diffraction

As discussed in the previous section when an acoustic wave propagates in a medium, the periodic strain associated with the acoustic wave generates a periodic refractive index variation in the medium. This periodic refractive index grating has the same period as the acoustic wave and is also propagating at the same velocity as the acoustic wave. Typically the refractive index variations are about 10^{-4} around the mean refractive index value (see Sec. 18.16). Even though this is a very small change, the effect of the acoustooptic interaction can be quite large due to the interaction length between the optical and acoustic waves being very large compared to the wavelength of the light wave. This is very similar to what was observed in electrooptic interactions (see Chapter 15).

When a light wave is allowed to fall on such a refractive index grating, it undergoes diffraction and depending on the wavelength of the optical and acoustic waves and the length of interaction, one may have either multiple order (Raman–Nath) or single order (Bragg) diffraction.

We shall show in Sec. 17.3 that if the length of interaction L between the optical and acoustic waves satisfies (see Fig. 17.1)

$$L \ll k/K^2 = \Lambda^2 n_0/2\pi\lambda_0$$
$$K = 2\pi/\Lambda, \quad k = 2\pi n_0/\lambda_0 \tag{17.1}$$

(where Λ is the acoustic wavelength, n_0 is the refractive index of the medium and λ_0 is the free space optical wavelength) then the incident light wave diffracts into multiple orders. This is referred to as Raman–Nath diffraction.

On the other hand, if

$$L \gg k/K^2 \tag{17.2}$$

then only one diffraction order is produced and that too only when the so-called Bragg condition is satisfied; the corresponding angle of incidence θ is close to

$$\theta_B = \sin^{-1}(\lambda_0/2n_0\Lambda) \tag{17.3}$$

(see Fig. 17.2). This is referred to as Bragg diffraction.

In the Raman–Nath regime the acoustically perturbed medium acts as a thin phase grating (see Sec. 17.4) while in the Bragg regime, the medium acts as a volume grating much like atomic planes in a crystal. In both the

Fig. 17.1 Raman–Nath diffraction corresponds to the condition given by Eq. (17.1) and leads to multiple order diffraction.

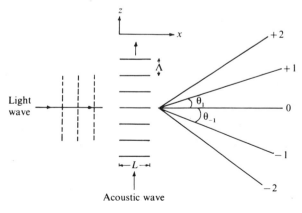

cases, since the acoustic wave generates a moving refractive index grating, the frequencies of the diffracted waves are different from those of the incident wave. This may be interpreted as due to a Doppler shift in frequency from a moving grating.

17.3   A simple experimental set up to observe Raman–Nath diffraction

A typical experimental arrangement for the observation of Raman–Nath diffraction is shown in Fig. 17.3. On one end of a cell (containing a liquid such as water) an acoustic transducer is bonded. The acoustic transducer is typically a piezoelectric crystal in which an applied electric field generates strain. Thus on applying an oscillating electric field, the crystal expands and contracts at the same frequency. This leads to the generation of waves of compression and rarefaction in water which in turn

Fig. 17.2 Bragg diffraction corresponds to the condition given by Eq. (17.2) and leads to only a single diffraction order.

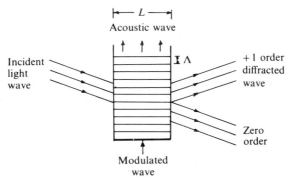

Fig. 17.3 A typical experimental arrangement to observe Raman–Nath diffraction. Acoustic waves are propagating along the z-direction and the light beam propagates along the x-direction consistent with Fig. 17.1.

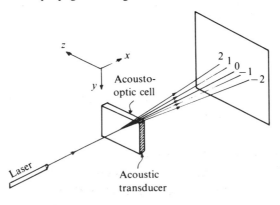

produces a periodic refractive index variation. At the other end of the cell an acoustic absorber in placed so that there are no reflected waves. Light from a laser is allowed to pass through the cell perpendicular to the direction of propagation of the acoustic waves. The diffraction effect is observed on a screen placed at a distance of about a metre from the acoustooptic cell.

As the electrical signal on the transducer is increased, the power in the acoustic waves will initially be very low and one will just observe the direct beam. As the acoustic power increases, one would start to observe two diffracted spots on either side of the direct beam. These correspond to the $+1$ and -1 order diffracted beams. If the power is further increased one can observe even higher orders of diffraction namely the $+2$ and -2 orders etc. (see Fig. 17.3) just as in a normal diffraction grating (see Sec. 4.10.2) one will observe diffracted spots symmetrically located about the zero order direct beam.

As an example if we take a 6 MHz acoustic wave in water, taking the velocity of acoustic waves in water to be 1500 m/s, we obtain for the acoustic wavelength

$$\Lambda = \frac{1500 \, \text{m/s}}{6 \times 10^6 \, \text{s}^{-1}} = 250 \, \mu\text{m}$$

If one uses a He–Ne laser, then $\lambda_0 = 6328 \, \text{Å}$ and for Raman–Nath diffraction the width L of the cell over which the interaction takes place must satisfy the following condition:

$$L \ll \frac{\Lambda^2 n_0}{2\pi\lambda_0} = \frac{(0.025)^2 \times 1.33}{2\pi \times 6.328 \times 10^{-5}} \, \text{cm}$$

$$\approx 2 \, \text{cm}$$

Thus a cell width of $\lesssim 1 \, \text{cm}$ would enable one to observe Raman–Nath diffraction.

We shall show in Sec. 17.4 that the angle between the first order and the direct beam is given by (see Eq. (17.17))

$$\theta_1 = \sin^{-1}(\lambda_0/n_0\Lambda)$$

which for our example corresponds to $\sim 0.11°$. Thus on a screen at a distance of a metre, the zero and first orders will be separated by a distance of $\sim 2 \, \text{mm}$ which can be resolved by the eye.

17.4 Theory of Raman–Nath diffraction

We first consider the interaction of a plane light wave with a propagating acoustic wave which produces a moving refractive index grating. If Ω and K represent the angular frequency and propagation constant of the acoustic wave (which is assumed to propagate along the z-direction) then we may write for the moving periodic refractive index grating (see Problem 16.2).

$$n(z, t) = n_0 + \Delta n \sin (\Omega t - Kz) \qquad (17.4)$$

where n_0 is the refractive index of the medium in the absence of the acoustic wave and Δn is the peak change in refractive index due to the acoustic wave. Let L be the width of the acoustic beam (see Fig. 17.1) and let a plane light wave be incident normally as shown in Fig. 17.1. We consider that L is small enough so that the medium behaves as a thin phase grating and the phase change introduced by propagation through a length L of the medium at any value of z is

$$\Delta\phi = (2\pi/\lambda_0)n(z, t)L = \phi_0 + \phi_1 \sin (\Omega t - Kz) \qquad (17.5)$$

where λ_0 is the free space wavelength and

$$\phi_0 = (2\pi/\lambda_0)n_0 L, \quad \phi_1 = (2\pi/\lambda_0)\Delta n L \qquad (17.6)$$

Let us now consider the incidence of a plane wave of constant amplitude propagating along the x-direction (see Fig. 17.1). On passing through the acoustically perturbed medium, the phase across the wavefront is modulated due to different phase changes across the phase front. Thus the variation of the transmitted field on the plane $x = L$ can be written as

$$E_t = E_0 e^{i[\omega t - \phi_0 - \phi_1 \sin(\Omega t - Kz)]} \qquad (17.7)$$

We now use the following identity (see e.g., Watson (1958))

$$e^{-i\zeta \sin \theta} = J_0(\zeta) + 2 \sum_{n=1}^{\infty} J_{2n}(\zeta) \cos 2n\theta$$

$$- 2i \sum_{n=1}^{\infty} J_{2n-1}(\zeta) \sin [(2n-1)\theta] \qquad (17.8)$$

Thus the field variation on the plane $x = L$ can be written as

$$E_t = E_0 e^{i(\omega t - \phi_0)}[J_0(\phi_1) - J_1(\phi_1)\{e^{i\theta} - e^{-i\theta}\}$$
$$+ J_2(\phi_1)\{e^{2i\theta} + e^{-2i\theta}\} + \cdots]$$

where $\theta = (\Omega t - Kz)$. Thus

$$E_t = E_0 J_0(\phi_1)e^{i(\omega t - \phi_0)}$$
$$- E_0 J_1(\phi_1)[e^{i[(\omega + \Omega)t - Kz - \phi_0]} - e^{i[(\omega - \Omega)t + Kz - \phi_0]}]$$
$$+ E_0 J_2(\phi_1)[e^{i[(\omega + 2\Omega)t - 2Kz - \phi_0]} + e^{i[(\omega - 2\Omega)t + 2Kz - \phi_0]}] - \cdots \quad (17.9)$$

From above we note the following:

(a) The first term represents the zero order diffracted wave which is characterized by a constant phase along the z-direction. Hence this term would correspond to a plane wave propagating along the x-direction with a frequency which is the same as that of the incident wave. Its amplitude, however, is now reduced by a factor $J_0(\phi_1)$. The electric field (in the region $x > L$) associated with this zero order wave would be given by

$$E_t^0 = E_0 J_0(\phi_1) e^{i[\omega t - k(x-L) - \phi_0]} \tag{17.10}$$

(b) The second and third terms on the RHS of Eq. (17.9) represent the $+1$ and -1 diffracted orders respectively. The $+1$ order is upshifted in frequency by the acoustic frequency Ω, i.e.,

$$\omega_{+1} = \omega + \Omega \tag{17.11}$$

and the -1 order is downshifted in frequency by Ω, i.e.,

$$\omega_{-1} = \omega - \Omega \tag{17.12}$$

Now, the second and third terms represent plane waves of frequencies $\omega + \Omega$ and $\omega - \Omega$ propagating along directions such that the corresponding z-components of their propagation vectors are $+K$ and $-K$ respectively. Therefore the electric fields associated with the $+1$ and -1 diffracted orders would be given by

$$- E_0 J_1(\phi_1) e^{i[(\omega + \Omega)t - k_2^+(x-L) - Kz - \phi_0]} \tag{17.13}$$

and

$$+ E_0 J_1(\phi_1) e^{i[(\omega - \Omega)t - k_2^-(x-L) + Kz - \phi_0]} \tag{17.14}$$

respectively. Here[†]

$$k_2^+ = [(\omega + \Omega)^2/c^2 - K^2]^{\frac{1}{2}}; \quad k_2^- = [(\omega - \Omega)^2/c^2 - K^2]^{\frac{1}{2}} \tag{17.15}$$

represent the corresponding x-components of the propagation vector. Since $\Omega/\omega \sim 10^{-5} \ll 1$

$$(\omega \pm \Omega)/c \approx \omega/c = k_0 \tag{17.16}$$

i.e., the magnitude of the propagation vector of the diffracted wave is almost equal to that of the undiffracted wave (this is true in isotropic media). Hence if the direction of propagation of the $+1$ order wave makes an angle θ_1 with the x-direction (see Fig. 17.1) then

$$\sin \theta_1 = K/k = \lambda_0/n_0 \Lambda \tag{17.17}$$

[†] This follows from the fact that for a plane wave $k_x^2 + k_y^2 + k_z^2$ must necessarily be equal to $(\omega \pm \Omega)^2/c^2$ and had there been a y-component of **k** it would have appeared as a y dependent phase term in Eq. (17.9).

Similarly the -1 order propagates along a direction making an angle θ_{-1} with the x-axis such that

$$\sin \theta_{-1} = -K/k = -\lambda_0/n_0\Lambda \tag{17.18}$$

The amplitude of the $+1$ and -1 order diffracted waves are $\pm E_0 J_1(\phi_1)$.

In general, the mth order diffracted wave has an amplitude $E_0 J_m(\phi_1)$, has a frequency $\omega + m\Omega$ and propagates along a direction making an angle θ_m with the x-axis given by

$$\sin \theta_m = m(\lambda_0/n_0\Lambda) \tag{17.19}$$

Fig. 17.4 shows the variation of the relative intensities of various orders with ϕ_1. The above diffraction is referred to as Raman–Nath diffraction.

(c) Since $J_0(x) = 0$ when $x \approx 2.405,\ 5.520,\ 8.654, \ldots$, etc. the zero order is absent when

$$\phi_1 (= (2\pi/\lambda_0)\Delta nL) \approx 2.405, 5.520, 8.654, \ldots \tag{17.20}$$

i.e., for such values of ϕ_1, all the incident light is converted into various diffraction orders.

(d) Intensity of the first order will be maximum when $J_1(\phi_1)$ is maximum which occurs when

$$\phi_1 \approx 1.85 \tag{17.21}$$

The maximum value of $J_1(\phi_1) \approx 0.582$ and hence the maximum diffraction efficiency in the first order is $\approx 33.9\%$.

Fig. 17.4 Variation of the relative intensities of various orders corresponding to the Raman–Nath diffraction.

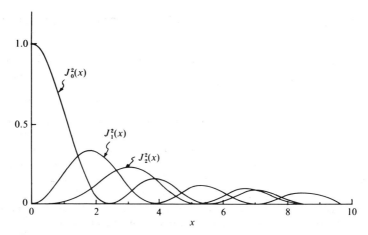

(*e*) For small acoustic powers $\phi_1 \ll \frac{1}{2}\pi$ and since

$$J_n(\phi_1) \approx \frac{1}{n!}\left(\frac{\phi_1}{2}\right)^n \quad \text{for } \phi_1 \ll 1, \tag{17.22}$$

we can neglect the second and higher diffraction orders and write for the transmitted field

$$E_t \sim E_0 e^{i(\omega t - \phi_0)} - \tfrac{1}{2}E_0\phi_1 e^{i[(\omega + \Omega)t - Kz - \phi_0]}$$
$$+ \tfrac{1}{2}E_0\phi_1 e^{i[(\omega - \Omega)t + Kz - \phi_0)]} \tag{17.23}$$

Hence for small acoustic powers, the relative intensity in the first order is

$$\eta \approx \tfrac{1}{4}\phi_1^2 = \pi^2(\Delta n)^2 L^2/\lambda_0^2 \tag{17.24}$$

Now $(\Delta n)^2$ is proportional to the acoustic power (see Eqs. (18.56) and (18.59)). Thus if the incident acoustic wave is amplitude modulated the first order diffracted light will be intensity modulated. This is the basic principle behind the use of Raman–Nath diffraction for acoustooptic modulation (see Sec. 19.2).

(*f*) In the analysis given above we have assumed L to be small enough so that the phase delay of a wave propagating through a distance L is $(2\pi/\lambda_0)n(z)L$. This would be valid if[†]

$$L \ll \Lambda^2/\lambda$$

Usually the condition for Raman–Nath diffraction is written as

$$Q = K^2L/k = 2\pi\lambda L/\Lambda^2 \ll 1 \tag{17.25}$$

In the other limit

$$Q \gg 1 \tag{17.26}$$

We have Bragg diffraction where as we will discuss in Chapter 18 only one order is produced.

As an example if we consider the propagation of acoustic waves in water (acoustic velocity $= 1500\,\text{m/s}$) at a frequency of $5\,\text{MHz}$, for Raman–Nath

---

[†] This can be physically understood if we refer to the Fresnel diffraction by a circular aperture (see Sec. 5.6) where for a point very close to the aperture (i.e., when $z \ll a^2/\lambda$) the aperture will contain a very large number of half period zones and the field will be determined by the first half period zone only. Thus the field will not see any transverse variation of refractive index if there is no variation of refractive index within the first half period zone. Thus referring to Fig. 17.1, if the distance Λ contains a very large number of half period zones the amplitude at any value of z at $x = L$ will be determined only by the field at the same value of z at $x = 0$.

diffraction to occur at an optical wavelength of $0.6\,\mu m$,

$$L \ll \Lambda^2/2\pi\lambda = 2.4\,\text{cm}$$

Thus an acoustic wave of width $\sim 1\,\text{cm}$ would lead to Raman–Nath diffraction. The restriction on length becomes more severe at higher frequencies since the maximum allowable length L reduces as $1/\Omega^2$. Thus at $50\,\text{MHz}$, the length of the acoustic column has to be much less than $0.03\,\text{cm}$!

One of the disadvantages of using the Raman–Nath diffraction configuration is the limitation on the interaction length L due to Eq. (17.25). Because this limitation reduces the interaction length considerably at high acoustic frequencies, for a large diffracted light intensity the sound power must be very large. Hence for higher frequencies, Bragg diffraction modulators are preferred.

Problem 17.1: Consider a light wave incident at an angle θ with the acoustic wave front as shown in Fig. 17.4. Obtain the intensity of the nth order diffraction assuming the Raman–Nath diffraction configuration.

Solution: Consider the portion of the light wave incident at P and propagating along PQ as shown in Fig. 17.5. The refractive index variation along the z-direction is again described by Eq. (17.4).

Since in the Raman–Nath regime we can assume the medium to be a phase grating, we first calculate the optical path length of PQ assuming the ray to propagate straight from P to Q. The refractive index at a point $R(x, z')$ on the path PQ will be

$$n(z') = n(z + x\tan\theta) = n_0 + n_1\sin\left[\Omega t - Kz - Kx\tan\theta\right] \tag{17.27}$$

Fig. 17.5 Light wave incident at an angle θ with the acoustic wavefront.

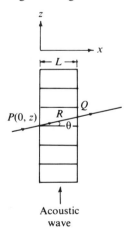

Hence the optical path length of *PQ* will be

$$\Delta = \int_{x=0}^{x=L} n(z')\,\mathrm{d}s$$

$$= (1/\cos\theta) \int_0^L [n_0 + n_1 \sin(\Omega t - Kz - Kx\tan\theta)]\,\mathrm{d}x$$

where we have used $\mathrm{d}s = \mathrm{d}x/\cos\theta$. Thus

$$\Delta = \frac{1}{\cos\theta}\left[n_0 L + \frac{2n_1}{K\tan\theta} \sin(\tfrac{1}{2}KL\tan\theta)\sin(\Omega t - Kz - \tfrac{1}{2}KL\tan\theta)\right]$$

$$(17.28)$$

The electric field variation at $x = 0$ is given by

$$E_\mathrm{i} = E_0 \exp[\mathrm{i}(\omega t - kz\sin\theta)] \tag{17.29}$$

then the electric field variation on the plane $x = L$ will be

$$E_\mathrm{t} = E_0 \exp[\mathrm{i}(\omega t - kz\sin\theta - k_0\Delta)]$$

$$= E_0 \exp\left[\mathrm{i}(\omega t - kz\sin\theta) - \frac{\mathrm{i}n_0 L}{\cos\theta}k_0\right.$$

$$\left. - \delta_0 \sin(\Omega t - Kz - \tfrac{1}{2}KL\tan\theta)\right] \tag{17.30}$$

where

$$\delta_0 = \frac{2n_1 k_0}{K\sin\theta} \sin(\tfrac{1}{2}KL\tan\theta) \tag{17.31}$$

Using Eq. (17.8) we can again show the existence of various orders of diffraction. The intensity of the nth order diffraction will be

$$I_n \propto J_n^2\left[\frac{2n_1 k_0}{K\sin\theta} \sin(\tfrac{1}{2}KL\tan\theta)\right]$$

$$= J_n^2\left[\phi_1 \frac{\sin(\tfrac{1}{2}KL\tan\theta)}{\tfrac{1}{2}KL\tan\theta}\right] \tag{17.32}$$

where

$$\phi_1 = n_1 k_0 L/\cos\theta \tag{17.33}$$

It may be noted from Eq. (17.32) that the diffraction pattern is symmetric since $J_{-n}^2 = J_n^2$. Also maximum diffraction occurs when $\theta = 0$ i.e., for normal incidence. For oblique incidence the effective ϕ_1 reduces by a factor $\sin(\tfrac{1}{2}KL\tan\theta)/(\tfrac{1}{2}KL\tan\theta)$. Thus when

$$KL\tan\theta = 2m\pi, \quad m = 1, 2, 3\ldots \tag{17.34}$$

all diffraction effects disappear. In order to understand what this represents, if we replace K by $2\pi/\Lambda$ in Eq. (17.34) we obtain

$$L\tan\theta = m\Lambda \tag{17.35}$$

Now, $L \tan \theta$ represents the transverse displacement of the light wave when it propagates over the length L of the medium (see Fig. 17.5). For θ satisfying Eq. (17.35), a portion of the light wave propagates through both positive and negative changes in refractive index and the integrated phase shift becomes zero.

Problem 17.2: Consider a 5 MHz acoustic wave travelling in water. Calculate the acoustic wavelength. For light incident normally at a wavelength of $0.6328 \, \mu m$, calculate the angular separation between different orders. For $\Delta n = 10^{-4}$, calculate the fractional intensity in the zero and first order diffracted light if $L = 5 \, cm$.

We must mention here that the above analysis can also be used in the study of thin phase gratings used in holography. Since the holographic gratings represent stationary index perturbations, no frequency shifts will be observed. Otherwise, the multiple order diffraction, their angles of diffraction, their intensities etc. are also applicable to thin transmission phase holograms.

18

Acoustooptic effect:
Bragg–diffraction

18.1 Introduction

In the previous chapter we discussed the interaction of a light wave with a periodic refractive index variation produced by a propagating acoustic wave. We assumed that

$$\text{The interaction length } L \ll \frac{k}{K^2} = \frac{\Lambda^2 n_0}{2\pi\lambda_0} \tag{18.1}$$

which leads to what is known as Raman–Nath diffraction. If on the other hand

$$L \gg k/K^2 = \Lambda^2 n_0/2\pi\lambda_0 \tag{18.2}$$

then we have the so-called Bragg regime of diffraction where we can no longer consider the refractive index perturbation to act as a thin phase grating. In this chapter we will develop the coupled wave analysis which describes the variation in amplitude of the incident and diffracted light waves. We will consider two extreme cases: one in which the light wave is incident almost normally on the propagating acoustic wave; this leads to what is known as small Bragg angle diffraction and is usually used at low acoustic frequencies. The other case corresponds to the light wave being incident almost along the (or opposite to the) direction of propagation of the acoustic wave; this leads to what is known as large Bragg angle diffraction and is usually used at large acoustic frequencies. In both cases we will find that coupling takes place mainly between an incident and a diffracted wave.

The interaction of light waves with periodic refractive index variations is a subject of considerable importance and finds applications in many fields such as holography, optical information storage, integrated optic devices etc. We will show that the theory developed in this chapter can be directly applied to, for example, reflection and transmission volume holograms which is a topic of great importance.

18.2   Small Bragg angle diffraction

We consider an acoustic wave propagating along the z-direction in a medium having a dielectric permittivity ϵ_u. The propagating acoustic wave generates a refractive index grating and the dielectric permittivity in the presence of the acoustic wave can be assumed to be given by

$$\epsilon = \epsilon_u + \Delta\epsilon \sin(\Omega t - Kz) \tag{18.3}$$

where the second term on the RHS is due to acoustooptic interaction and Ω and K represent the angular frequency and propagation constant of the propagating acoustic wave. For small Bragg angle diffraction, the light wave will be propagating almost along the x-direction as shown in Fig. 18.1(a). The electric field associated with the incident wave is written in the form

$$\begin{aligned}\mathscr{E}_0 &= \hat{\mathbf{e}} A_0 e^{i(\omega t - \mathbf{k}\cdot\mathbf{r})} \\ &= \hat{\mathbf{e}} A_0 e^{i(\omega t - \alpha x - \beta z)}\end{aligned} \tag{18.4}$$

where A_0 represents the amplitude of the wave; α and β represent the x and z-components of the propagation vector $\mathbf{k}$ of the incident wave and $\hat{\mathbf{e}}$ represents the corresponding direction of polarization. We will show in Sec. 18.4 that such an incident plane wave (propagating almost along the x-direction) will be coupled to the following two diffracted plane waves.

Fig. 18.1 Generation of a $+1$ order diffracted wave in small Bragg angle diffraction. If $\theta = \theta_\mathrm{B}$, $\mathbf{k}, \mathbf{k}_+$ and $\mathbf{K}$ form an isosceles triangle as shown in (b) and the diffraction efficiency is maximum. If $\theta \neq \theta_\mathrm{B}$, β_+ is still equal to $\beta + K$ but $\alpha_+ \neq \alpha$ and the triangle is no longer complete as shown in (c).

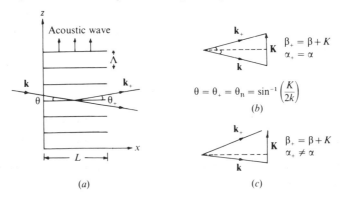

+1 *Order.* As will be shown in Sec. 18.4, for efficient generation of +1 order diffracted waves, the incident light wave must propagate in a direction as shown in Fig. 18.1(*a*), making an angle

$$\theta = \theta_B = \sin^{-1}(\lambda_0/2n_0\Lambda)$$

with the *x*-axis. Here $n_0 = (\epsilon_u/\epsilon_0)^{\frac{1}{2}}$ represents the refractive index of the unperturbed medium. The field associated with the diffracted wave is given by

$$\begin{aligned} \mathscr{E}_+ &= \hat{\mathbf{e}}_+ A_+ e^{i(\omega_+ t - \mathbf{k}_+ \cdot \mathbf{r})} \\ &= \hat{\mathbf{e}}_+ A_+ e^{i(\omega_+ t - \alpha_+ x - \beta_+ z)} \end{aligned} \tag{18.5}$$

where

$$\omega_+ = \omega + \Omega \tag{18.6}$$

$$\alpha_+ = k_+ \cos\theta_+ \tag{18.7}$$

$$\beta_+ = k_+ \sin\theta_+ = \beta + K \tag{18.8}$$

$$k_+^2 = \alpha_+^2 + \beta_+^2 = (\omega + \Omega)^2 \epsilon_u \mu_0$$

In general, the transfer of power to the diffracted wave is very small unless α_+ is very close to α. Maximum power transfer takes place when (see Sec. 18.4)

$$\Delta\alpha = \alpha - \alpha_+ = 0 \tag{18.9}$$

Since $\Omega/\omega \ll 1$, in an isotropic medium

$$k_+ \approx k$$

and Eq. (18.9) implies $\theta_+ = \theta$. If we now use the fact that

$$\beta = -k \sin\theta$$

then Eq. (18.8) will give us

$$\theta = \theta_+ = \theta_B = \sin^{-1}(K/2k) = \sin^{-1}(\lambda_0/2n_0\Lambda) \tag{18.10}$$

which is known as the Bragg condition and the angle θ_B is known as the Bragg angle. When Eqs. (18.8) and (18.9) are satisfied $\mathbf{k}$, $\mathbf{k}_+$ and $\mathbf{K}$ form a triangle as shown in[†] Fig. 18.1(*b*). If the angle of incidence deviates from the Bragg angle then β_+ and α_+ are still given by Eqs. (18.7) and (18.8) and the triangle is no longer complete as shown in Fig. 18.1(*c*).

−1 *Order.* For efficient generation of −1 order diffracted waves, the incident light wave must propagate in a direction as shown in Fig. 18.2

[†] Since $\Omega \ll \omega$, in an isotropic medium $|\mathbf{k}_+| \approx |\mathbf{k}|$ and the triangle shown in Fig. 18.1(*b*) is almost isosceles.

and making the same angle as given by Eq. (18.10) with the x-axis. The field associated with the diffracted wave is given by

$$\mathscr{E}_- = \hat{\mathbf{e}}_- A_- e^{i(\omega_- t - \mathbf{k}_- \cdot \mathbf{r})}$$
$$= \hat{\mathbf{e}}_- A_- e^{i(\omega_- t - \alpha_- x - \beta_- z)} \tag{18.11}$$

where

$$\omega_- = \omega - \Omega \tag{18.12}$$

$$\alpha_- = k_- \cos\theta_- \tag{18.13}$$

$$\beta_- = k_- \sin\theta_- = \beta - K \tag{18.14}$$

and

$$k_-^2 = \alpha_-^2 + \beta_-^2 = (\omega - \Omega)^2 \epsilon_u \mu_0$$

Once again, the transfer of power is maximum when

$$\Delta\alpha = \alpha_- - \alpha = 0 \tag{18.15}$$

The angle of incidence for maximum diffraction is again given by Eq. (18.10).

We should mention here that for a given value of K one cannot satisfy both Eqs. (18.9) and (18.15) simultaneously and therefore when the Bragg condition (Eq. (18.10)) is satisfied one will observe only one of the diffracted orders as shown in Figs. 18.1(a) and 18.2(a); the intensity of the other diffracted order will be negligibly small because the corresponding $\Delta\alpha$ will be large.

In order to understand the transition to the Raman–Nath regime we consider a plane light wave (propagating along the $+x$-direction) incident on an acoustic wave as shown in Fig. 18.3. In order for the light wave to

Fig. 18.2 Generation of the -1 order diffracted wave in small Bragg angle diffraction. If $\theta = \theta_B$, $\mathbf{k}$, $\mathbf{k}_-$ and $\mathbf{K}$ form an isosceles triangle as shown in (b) and the diffraction efficiency is maximum. If $\theta \neq \theta_B$, β_- is still equal to $\beta - K$ but $\alpha_- \neq \alpha$.

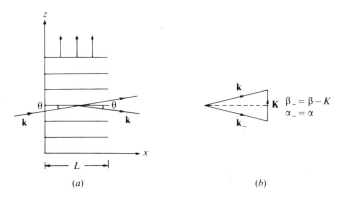

(a)   (b)

undergo Bragg diffraction, the acoustic wave must propagate such that its wavefront makes an angle (see Eq. (18.10))

$$\theta = \pm \sin^{-1}(\lambda_0/2n_0\Lambda) = \pm \sin^{-1}(K/2k) \qquad (18.16)$$

with the propagation direction of the light wave which is the x-axis. Such an acoustic wave has a propagation vector whose x-component is given by

$$K_x = K \sin\theta = \pm K^2/2k \qquad (18.17)$$

We now go back to Fig. 17.1 and consider the effect of reducing the width L of the acoustic wavefront from a large value. Obviously, the effect of reducing L is to introduce diffraction in the acoustic wave. In other words the propagating acoustic wave will consist of a spectrum of plane acoustic waves propagating along different directions. Indeed for an acoustic wavefront of width L, the spread in the value of K_x will be given by

$$\Delta K_x \approx 1/L \qquad (18.18)$$

Thus if the spectral component of the propagating acoustic wave has a substantial amplitude along a direction which has K_x given by Eq. (18.17) then the acoustic wave will generate both $+1$ and -1 order Bragg diffracted waves. This would happen when

$$\Delta K_x \approx 1/L \gg K^2/2k$$

i.e., when

$$L \ll 2k/K^2 = \Lambda^2 n_0/\pi\lambda_0 \qquad (18.19)$$

which is roughly the same condition as given by Eq. (18.1). The $+1$ and

Fig. 18.3 A light wave (propagating along the x-direction) is incident on an acoustic wave propagating along a direction which makes an angle θ with the z-axis. For maximum diffraction θ must be equal to the Bragg angle θ_B.

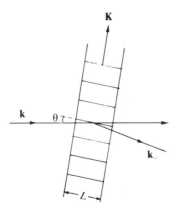

−1 order diffracted waves may undergo further Bragg diffraction to give the multiple order spectrum as discussed in the previous chapter.

18.3   Basic equations governing Bragg diffraction

We assume a scalar theory in which the electric field $\mathscr{E}$ satisfies the scalar wave equation:

$$\nabla^2 \mathscr{E} = \mu_0 \frac{\partial^2 \mathscr{D}}{\partial t^2} = \mu_0 \frac{\partial^2}{\partial t^2} [\epsilon_u + \Delta\epsilon \sin(\Omega t - Kz)]\mathscr{E}$$

or

$$\left(\nabla^2 - \epsilon_u \mu_0 \frac{\partial^2}{\partial t^2} \right) \mathscr{E} \approx \mu_0 \Delta\epsilon \sin(\Omega t - Kz) \frac{\partial^2 \mathscr{E}}{\partial t^2} \tag{18.20}$$

where we have used Eq. (18.3) and in the last step we have assumed $\Omega \ll \omega$; the latter condition physically implies that the light wave sees an almost stationary refractive index distribution. The validity of the scalar theory will be discussed in Sec. 18.6. We will, without any loss of generality, assume propagation in the x–z plane so that the field will have no y dependence. Thus the above equation will become

$$\frac{\partial^2 \mathscr{E}}{\partial x^2} + \frac{\partial^2 \mathscr{E}}{\partial z^2} - \epsilon_u \mu_0 \frac{\partial^2 \mathscr{E}}{\partial t^2}$$

$$= \frac{1}{2i} \mu_0 \Delta\epsilon (e^{i(\Omega t - Kz)} - e^{-i(\Omega t - Kz)}) \frac{\partial^2 \mathscr{E}}{\partial t^2} \tag{18.21}$$

Now, if we assume the electric field to be of the form $e^{i\omega t}$ Then on the RHS we will have terms proportional to $e^{i(\omega + \Omega)t}$ and $e^{i(\omega - \Omega)t}$. Therefore, because of the periodic refractive index grating, diffracted waves of frequencies $\omega + \Omega$ and $\omega - \Omega$ will be generated[†]. We therefore assume the total electric field to be given by

$$\mathscr{E} = \mathscr{E}_0 + \mathscr{E}_+ + \mathscr{E}_- \tag{18.22}$$

where

$$\mathscr{E}_0 = A_0(x, z)e^{i(\omega t - \mathbf{k} \cdot \mathbf{r})} = A_0(x, z)e^{i(\omega t - \alpha x - \beta z)} \tag{18.23}$$

$$\mathscr{E}_\pm = A_\pm(x, z)e^{i(\omega_\pm t - \mathbf{k}_\pm \cdot \mathbf{r})}$$

$$= A_\pm(x, z)e^{i[(\omega \pm \Omega)t - \alpha_\pm x - \beta_\pm z]} \tag{18.24}$$

Now, the plane wave solutions correspond to the unperturbed medium

---

[†] The two frequencies in turn may generate frequencies $\omega + 2\Omega$ and $\omega - 2\Omega$ etc. as happens in the Raman–Nath regime; however, in the Bragg regime this can be neglected.

and therefore

$$\omega^2 \epsilon_u \mu_0 = k^2 = \alpha^2 + \beta^2 \tag{18.25}$$

and

$$(\omega \pm \Omega)^2 \epsilon_u \mu_0 = k_\pm^2 = \alpha_\pm^2 + \beta_\pm^2 \tag{18.26}$$

Substituting Eq. (18.22) in Eq. (18.21) and neglecting terms proportional[†] to $\partial^2 A / \partial x^2$ and $\partial^2 A / \partial z^2$ we obtain

$$-2i\left(\alpha \frac{\partial A_0}{\partial x} + \beta \frac{\partial A_0}{\partial z}\right) e^{i(\omega t - \alpha x - \beta z)}$$

$$-2i\left(\alpha_+ \frac{\partial A_+}{\partial x} + \beta_+ \frac{\partial A_+}{\partial z}\right) e^{i[(\omega + \Omega)t - \alpha_+ x - \beta_+ z]}$$

$$-2i\left(\alpha_- \frac{\partial A_-}{\partial x} + \beta_- \frac{\partial A_-}{\partial z}\right) e^{i[(\omega - \Omega)t - \alpha_- x - \beta_- z]}$$

$$= -\frac{1}{2i}\mu_0 \Delta \epsilon \left[e^{i(\Omega t - Kz)} - e^{-i(\Omega t - Kz)} \right] \left[\omega^2 A_0 e^{i(\omega t - \alpha x - \beta z)} \right.$$

$$+ (\omega + \Omega)^2 A_+ e^{i[(\omega + \Omega)t - \alpha_+ x - \beta_+ z]}$$

$$\left. + (\omega - \Omega)^2 A_- e^{i[(\omega - \Omega)t - \alpha_- x - \beta_- z]} \right] \tag{18.27}$$

Since the above equation is valid for all times, the coefficients of $e^{i\omega t}$, $e^{i(\omega \pm \Omega)t}$ may be equated to obtain[‡]

$$-2i\left(\alpha \frac{\partial A_0}{\partial x} + \beta \frac{\partial A_0}{\partial z}\right) e^{-i(\alpha x + \beta z)}$$

$$= -\frac{1}{2i}\omega^2 \mu_0 \Delta \epsilon \left[- A_+ e^{-i[\alpha_+ x + (\beta_+ - K)z]} + A_- e^{-i[\alpha_- x + (\beta_- + K)z]} \right] \tag{18.28}$$

$$-2i\left(\alpha_\pm \frac{\partial A_\pm}{\partial x} + \beta_\pm \frac{\partial A_\pm}{\partial z}\right) e^{-i(\alpha_\pm x + \beta_\pm z)}$$

$$= \mp \frac{1}{2i}\omega^2 \mu_0 \Delta \epsilon A_0 e^{-i[\alpha x + (\beta \pm K)z]} \tag{18.29}$$

where in Eq. (18.27) we have assumed $(\omega \pm \Omega)^2 \approx \omega^2$.

In general, the above differential equations which contain both the x

---

[†] This is valid when $|\partial^2 A / \partial x^2| \ll |k \partial A / \partial x|$ i.e., when A does not vary appreciably over a wavelength of the light wave. Similarly for $\partial^2 A / \partial z^2$.

[‡] It may be noted that there are terms like $e^{i(\omega \pm 2\Omega)t}$ etc. which appear only on the RHS of Eq. (18.27). These can be considered only if additional terms like $\mathscr{E}_{2\pm} \sim e^{i(\omega \pm 2\Omega)t}$ etc. are included in Eq. (18.22).

and z dependences of the field amplitudes are difficult to solve. Hence we consider two extreme cases: one in which the light wave is incident almost normally on the propagating acoustic wave (see Figs. 18.1 and 18.2); this is known as small Bragg angle diffraction and we may neglect the z dependences of A_0 and A_+ and the resulting equations are solved in the next section. The other case corresponds to what is known as large Bragg angle diffraction where the light wave propagates almost along the direction of propagation of the acoustic wave. In this case we may neglect the x dependence of A and the resulting equations are solved in Sec. 18.5.

We should mention here that that in general $\mathscr{E}_0$ and $\mathscr{E}_+$ will not have the same state of polarization. For example, $\mathscr{E}_0$ may be z-polarized and $\mathscr{E}_+$ may be y-polarized. Nevertheless the theory developed in this section will remain valid if ε_u and $\Delta\varepsilon$ are defined appropriately (see Sec. 18.6).

18.4   Coupled wave analysis for small Bragg angle diffraction

As discussed in the previous section, in this case we may assume $\partial A/\partial z \approx 0$ so that A_0, A_+ and A_- are functions of x only. Thus Eqs. (18.28) and (18.29) become

$$-2i\alpha \frac{dA_0}{dx} e^{-i(\alpha x + \beta z)} = -\frac{1}{2i}\mu_0\omega^2\Delta\epsilon(-A_+ e^{-i[\alpha_+ x + (\beta_+ - K)z]}$$

$$+ A_- e^{-i[\alpha_- x + (\beta_- + K)z]}) \tag{18.30}$$

$$-2i\alpha_\pm \frac{dA_\pm}{dx} e^{-i[\alpha_\pm x + \beta_\pm z)} = \mp\frac{1}{2i}\omega^2\mu_0\Delta\epsilon A_0 e^{-i[\alpha x + (\beta \pm K)z]} \tag{18.31}$$

The z dependent factors should cancel out and therefore we must have

$$\beta_+ = \beta + K \tag{18.32}$$

and

$$\beta_- = \beta - K \tag{18.33}$$

which are the Bragg conditions. We will show that the amplitude A_+ will be negligibly small unless $\alpha_+ \approx \alpha$ and similarly for A_-. Since one cannot satisfy both $\alpha_+ \approx \alpha$ and $\alpha_- \approx \alpha$ simultaneously, we first assume $\alpha_+ \approx \alpha$ and neglect A_- in the above equations. Thus we obtain the following coupled wave equations

$$d\tilde{A}_0/dx = \kappa\tilde{A}_+ e^{ix\Delta\alpha} \tag{18.34}$$

$$d\tilde{A}_+/dx = -\kappa\tilde{A}_0 e^{-ix\Delta\alpha} \tag{18.35}$$

where

$$\kappa = \omega^2\mu_0\Delta\epsilon/4(\alpha\alpha_+)^{\frac{1}{2}} \tag{18.36}$$

$$\Delta\alpha = \alpha - \alpha_+ \tag{18.37}$$

$$\tilde{A}_0 = (\alpha/2\omega\mu_0)^{\frac{1}{2}} A_0 \tag{18.38}$$

$$\tilde{A}_+ = (\alpha_+/2\omega\mu_0)^{\frac{1}{2}} A_+ \tag{18.39}$$

The factors multiplying A_0 and A_+ are such that the powers carried by the incident and diffracted waves are $|\tilde{A}_0|^2$ and $|\tilde{A}_+|^2$ respectively. The above equations correspond to the case in which the incident wave couples to the $+1$ order diffracted wave (see Fig. 18.1). In order to solve Eqs. (18.34) and (18.35), we first differentiate Eq. (18.34) with respect to x and use Eqs. (18.34) and (18.35) to obtain

$$\frac{d^2\tilde{A}_0}{dx^2} - i(\Delta\alpha)\frac{d\tilde{A}_0}{dx} + \kappa^2\tilde{A}_0 = 0 \tag{18.40}$$

The solution of the above equation is

$$\tilde{A}_0(x) = C_0 e^{ix(\frac{1}{2}\Delta\alpha + \delta)} + D_0 e^{ix(\frac{1}{2}\Delta\alpha - \delta)} \tag{18.41}$$

where
$$\delta = [\kappa^2 + \tfrac{1}{4}(\Delta\alpha)^2]^{\frac{1}{2}} \tag{18.42}$$

From Eqs. (18.41) and (18.34) we get

$$A_+(x) = (C_+ e^{ix(\frac{1}{2}\Delta\alpha + \delta)} + D_+ e^{ix(\frac{1}{2}\Delta\alpha - \delta)})e^{-ix\Delta\alpha} \tag{18.43}$$

where

$$C_+ = (i/\kappa)(\tfrac{1}{2}\Delta\alpha + \delta)C_0 \tag{18.44}$$

$$D_+ = (i/\kappa)(\tfrac{1}{2}\Delta\alpha - \delta)D_0 \tag{18.45}$$

The quantities C_0 and D_0 are determined by the boundary conditions.

We now consider a specific example when a wave of unit power is incident at $x = 0$ on the grating produced by the acoustic wave (see Fig. 18.1). In such a case

$$\tilde{A}_0(x = 0) = 1, \quad \tilde{A}_+(x = 0) = 0 \tag{18.46}$$

Using these in Eqs. (18.41) and (18.43), we get

$$C_0 = \tfrac{1}{2} - (1/4\delta)\Delta\alpha, \quad D_0 = \tfrac{1}{2} + (1/4\delta)\Delta\alpha \tag{18.47}$$

Substituting for C_0 and D_0 in Eqs. (18.41) and (18.43) we get for the power in the incident and in the $+1$ order diffracted wave as

$$P_0(x) = |\tilde{A}_0(x)|^2 = \cos^2(\delta x) + (\Delta\alpha/2\delta)^2 \sin^2(\delta x) \tag{18.48}$$

$$P_+(x) = |\tilde{A}_+(x)|^2 = (\kappa/\delta)^2 \sin^2(\delta x) \tag{18.49}$$

These equations are similar to the ones obtained while considering coupling between two waveguides (see Sec. 14.6.3). Fig. 18.4 shows a typical variation of $P_+(x)$ with κx for $\Delta\alpha = 0$ and for $\Delta\alpha = 2\kappa$. From Eqs. (18.48) and (18.49) we may conclude the following:

(*a*) It can readily be seen that

$$P_0(x) + P_+(x) = 1 \tag{18.50}$$

which is just the conservation of power.

(*b*) Complete power transfer from the incident wave to the diffracted wave is possible only if

$$\Delta\alpha = 0 \Rightarrow \alpha = \alpha_+ \tag{18.51}$$

For such a case, we have

$$P_0(x) = \cos^2(\kappa x) \tag{18.52}$$

$$P_+(x) = \sin^2(\kappa x) \tag{18.53}$$

Thus the fractional power coupled from the incident wave into the $+1$ order diffracted wave in a length $x = L$ is given by

$$\eta = P_+(L) = \sin^2(\kappa L) \tag{18.54}$$

and complete power transfer requires an interaction length of

$$L = \pi/2\kappa, \quad 3\pi/2\kappa, \dots \text{ etc.} \tag{18.55}$$

In order to apply the above results to a specific problem, we first note that the value of $\Delta\epsilon$ depends on the polarization state of the incident and acoustic waves and also on the polarization state of the diffracted wave; we will explicitly show this in Sec. 18.6. In general, we may write

$$\Delta\epsilon = \epsilon_0 n^4 \bar{p}\bar{S} \tag{18.56}$$

Fig. 18.4 Variation of the fractional power coupled from the incident wave into the $+1$ order diffracted wave as a function of κx for $\Delta\alpha = 0$ and $\Delta\alpha = 2\kappa$. Notice that when $\Delta\alpha = 0$, complete power transfer occurs.

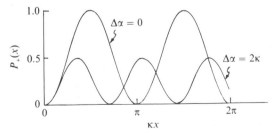

where ϵ_0 is the free space permittivity, $\bar{p}$ and $\bar{S}$ are the effective acoustooptic coefficient and the effective strain respectively and n represents the effective refractive index. Thus

$$\kappa = \frac{\omega}{4c}\frac{n^3\bar{p}\bar{S}}{\cos\theta_B} \tag{18.57}$$

where we have assumed incidence at the Bragg angle so that

$$\alpha = \alpha_+ = (\omega/c)n\cos\theta_B \tag{18.58}$$

Now we can express the acoustic intensity I_a as

$$I_a = \tfrac{1}{2}\rho v_a^3 \bar{S}^2 \tag{18.59}$$

where ρ is the density of the medium and v_a is the acoustic velocity in the medium. Thus in terms of I_a, Eq. (18.57) becomes

$$\kappa = \frac{\pi}{\sqrt{2\lambda_0}\cos\theta_B}(M_2 I_a)^{\frac{1}{2}} \tag{18.60}$$

where

$$M_2 = n^6\bar{p}^2/\rho v_a^3 \tag{18.61}$$

is known as the figure of merit of the acoustooptic device. The corresponding diffraction efficiency is given by

$$\eta = \sin^2\left[\frac{\pi}{\sqrt{2\lambda_0}\cos\theta_B}(M_2 I_a)^{\frac{1}{2}}L\right] \tag{18.62}$$

For unity diffraction efficiency we must have $\kappa L = \tfrac{1}{2}\pi$, i.e.,

$$I_a = \frac{\lambda_0^2\cos^2\theta_B}{2M_2 L^2} \tag{18.63}$$

If the width of the acoustic transducer is H (along the y-direction) then the acoustic power required for complete diffraction is

$$P_a = I_a LH = \frac{\lambda_0^2\cos^2\theta_B}{2M_2}\left(\frac{H}{L}\right) \tag{18.64}$$

From Eq. (18.64) it can be seen that the acoustic power required for complete power transfer depends inversely on the acoustooptic figure of merit M_2. Thus the larger the value of M_2, the smaller will be the required acoustic power. From Eq. (18.61) it follows that for a good figure of merit, the material must have a high refractive index, low acoustic velocity, a low density and a large photoelastic constant. The figure of merit of some important acoustooptic materials is given in Table 18.1.

Table 18.1. *Figure of merit M_2 of typical materials used in acoustooptic devices.*

Material	Optical transmission range (μm)	Figure of merit, $M_2(10^{-15}\,s^3/kg)$
Tellurium dioxide	0.35–5	34.5
Lead Molybdate	0.42–2.5	36.3
SF–59 (glass)	0.46–2.5	19.0
Lithium niobate	0.4–4.5	7.0
Fused Quartz	0.2–4.5	1.56
Germanium	2–20	8.40

Adapted from *Laser Focus* (1983)

It can also be seen from Eq. (18.62) that the diffraction efficiency depends inversely on wavelength. Thus the acoustic power requirements at infrared wavelengths may be much more than those at visible ones. In addition for a large diffraction efficiency a large aspect ratio L/H is desirable. As an example we consider a typical acoustic transducer with $H = 2\,\text{mm}$, $L = 50\,\text{mm}$, $\Omega = 2\pi \times 40\,\text{MHz}$ in extra dense flint glass for which $n = 1.92$, $\bar{p} \approx 0.25$, $v_a = 3.1 \times 10^3\,\text{m/s}$, $\rho = 6.3 \times 10^3\,\text{kg/m}^3$. Thus $M_2 \approx 1.7 \times 10^{-14}$ s^3/kg and for $\lambda_0 = 6328\,\text{Å}$, the Bragg angle is

$$\theta_B = \sin^{-1}(\lambda_0/2n\Lambda) \approx 0.12°$$

where we have used

$$\Lambda = 2\pi v_a/\Omega \approx 0.78 \times 10^{-4}\,\text{m}$$

Thus from Eq. (18.63) we have

$$P_a \approx 0.47\,\text{W}$$

We may note that for the above acoustic power

$$\Delta\epsilon/\epsilon_0 \approx n^4\bar{p}\bar{S} = n^4\bar{p}(2P_a/\rho v_a^3 LH)^{\frac{1}{2}} \approx 2.4 \times 10^{-5}$$

which indeed represents a small perturbation. We may also note that for the present example

$$k/LK^2 = \Lambda^2 n_0/2\pi\lambda_0 L \approx 0.06$$

and therefore Eq. (18.2) is satisfied.

Problem 18.1: Show that for very low acoustic powers

$$\eta \approx \pi^2 M_2 L^2 I_a/2\lambda_0^2 \cos^2\theta_B \tag{18.65}$$

which for $\theta_B = 0$ is the same as obtained for Raman–Nath diffraction (see Sec. 19.2). (Hint: from Eq. (18.56) we get $\Delta n = n^3 \bar{p} \bar{S}/2$.)

Problem 18.2: Show that if the angle of incidence deviates by $\Delta\theta$ from the Bragg angle then the diffraction efficiency is given by

$$\eta = \frac{1}{[1 + (K\Delta\theta/2\kappa)^2]} \sin^2 \{\kappa L[1 + (K\Delta\theta/2\kappa)^2]^{\frac{1}{2}}\} \qquad (18.66)$$

Solution: Referring to Fig. 18.1(*a*) we note that $\beta = -k\sin\theta$, $\beta_+ = k\sin\theta_+$ so that Eq. (18.8) becomes

$$\sin\theta_+ = -\sin\theta + K/k \qquad (18.67)$$

Thus if $\theta < \theta_B$ then $\theta_+ > \theta_B$ and thus the diffraction is no longer along the direction of specular reflection! From Eq. (18.67), we obtain

$$d\theta_+/d\theta = -1 \quad \text{at} \quad \theta = \theta_B = \theta_+$$

Thus if $\theta = \theta_B + \Delta\theta$ then $\theta_+ = \theta_B - \Delta\theta$ and

$$\Delta\alpha = \alpha - \alpha_+ = k[\cos(\theta_B + \Delta\theta) - \cos(\theta_B - \Delta\theta)]$$
$$\approx -K\Delta\theta \qquad (18.68)$$

Substituting in Eq. (18.49) we get Eq. (18.65). The deviation $\Delta\theta$ from the Bragg angle could be caused by the misalignment of the incident light wave with respect to the acoustic wave. We should also mention here that because of the finite value of L, the acoustic wave will undergo diffraction spreading ($\Delta K \sim 1/L$) and therefore the diffracted light wave will have an angular spread $\Delta\theta_+$ which will be a measure of ΔK.

In the numerical example discussed in p. 530 we had

$$\lambda_0 = 0.6328 \,\mu\text{m}, \quad \Lambda = 78 \,\mu\text{m}, \quad n = 1.92$$
$$L = 50 \,\text{mm} \quad \text{and} \quad \kappa L = \tfrac{1}{2}\pi$$

so that $\eta = 1$ and the corresponding acoustic power was 0.47 W. The Bragg angle was 0.12°. Now if the angle of incidence changes from the Bragg angle to say 0.15° then the corresponding diffraction efficiency reduces from 1.0 to 0.62. This can easily be seen by noting that

$$\Delta\theta = 0.03° \approx 5.24 \times 10^{-4} \,\text{rad}$$

Thus

$$\frac{K\Delta\theta}{2\kappa} = \frac{2\pi}{\Lambda} \frac{\Delta\theta}{2} \frac{2L}{\pi} \approx 0.67$$

giving $\eta = 0.62$ which shows that if the angle of incidence changes very slightly from the Bragg angle, the diffraction efficiency reduces considerably. We can use Eq. (18.66) to study the wavelength selectivity of small

Bragg angle diffraction. We consider two wavelengths differing by 10% ($\lambda_0 = 6328$ Å and $\lambda_1 = 6961$ Å) incident at the Bragg angle corresponding to λ_0, i.e.,

$$\theta = \sin^{-1}(\lambda_0/2n\Lambda) \approx 0.1211°$$

Now the Bragg angle corresponding to λ_1 is

$$\theta = \sin^{-1}(0.6961/2 \times 1.92 \times 78) \approx 0.1332°$$

Thus for a wavelength of 6961Å, $\Delta\theta = 0.0121°$ and using $\kappa(\lambda_0)L = \frac{1}{2}\pi$, we find

$$\frac{K\Delta\theta}{2\kappa(\lambda)} = \frac{2\pi}{\Lambda}\frac{\Delta\theta}{2}\frac{\lambda}{\kappa(\lambda_0)\lambda_0} \approx 0.298$$

where we have used the fact that κ depends inversely on wavelength. Substituting in Eq. (18.65) we obtain $\eta = 0.92$ which shows the very small wavelength selectivity of small Bragg angle scattering.

18.5   Large Bragg angle diffraction

We next consider the case of large Bragg angle diffraction where the light wave makes a small angle with the direction of propagation of the acoustic wave (see Fig. 18.5). In such a case we can neglect the x dependence of the field amplitudes so that Eqs. (18.28) and (18.29) reduce to

$$\beta\frac{dA_0}{dz}e^{-i(\alpha x + \beta z)} = \frac{1}{4}\omega^2\mu_0\Delta\epsilon(A_+ e^{-i[\alpha_+ x + (\beta_+ - K)z]}$$

$$- A_- e^{-i[\alpha_- x + (\beta_- + K)z]}) \tag{18.69}$$

$$\beta_\pm\frac{dA_\pm}{dz}e^{-i(\alpha_\pm x + \beta_\pm z)} = \mp\frac{1}{4}\omega^2\mu_0\Delta\epsilon A_0 e^{-i[\alpha x + (\beta \pm K)z]} \tag{18.70}$$

The x dependent factors should cancel out and therefore we will have

$$\alpha_\pm = \alpha \tag{18.71}$$

We will show that the amplitude A_+ is negligibly small unless $\beta_+ = \beta + K$ and similarly for A_-. Since one cannot satisfy both $\beta_+ = \beta + K$ and $\beta_- = \beta - K$ simultaneously, we first assume $\beta_+ \approx \beta + K$ and neglect A_- in the above equations to obtain the following coupled wave equations:

$$\frac{d\tilde{A}_0}{dz} = \frac{\beta}{|\beta|}\sigma\tilde{A}_+ e^{i(\Delta\beta)z} \tag{18.72}$$

$$\frac{d\tilde{A}_+}{dz} = -\frac{\beta_+}{|\beta_+|}\sigma\tilde{A}_0 e^{-i(\Delta\beta)z} \tag{18.73}$$

where

$$\sigma = \frac{\omega^2 \mu_0 \Delta\epsilon}{4(|\beta||\beta_+|)^{\frac{1}{2}}} = \frac{\omega^2}{4c^2} \frac{\Delta\epsilon}{\epsilon_0} \frac{1}{(|\beta||\beta_+|)^{\frac{1}{2}}} \qquad (18.74)$$

$$\tilde{A}_0 = (|\beta|/2\omega\mu_0)^{\frac{1}{2}} A_0 \qquad (18.75)$$

$$\tilde{A}_+ = (|\beta_+|/2\omega\mu_0)^{\frac{1}{2}} A_+ \qquad (18.76)$$

and

$$\Delta\beta = \beta - \beta_+ + K \qquad (18.77)$$

The factors multiplying A_0 and A_+ are such that the powers carried by the incident and diffracted waves are $|\tilde{A}_0|^2$ and $|\tilde{A}_+|^2$ respectively. We can have two different configurations: (a) the incident and diffracted waves propagate almost along the same direction as the incident wave; this is known as codirectional coupling (see Fig. 18.5(a)); (b) the diffracted wave propagates in a direction which is almost opposite to that of the incident wave; this is known as contradirectional coupling (see Fig. 18.5(b)). We will consider the two cases separately.

Fig. 18.5 When the light wave is incident almost along (or opposite to) the direction of propagation of the acoustic wave, then we may have either (a) codirectional or (b) contradirectional coupling.

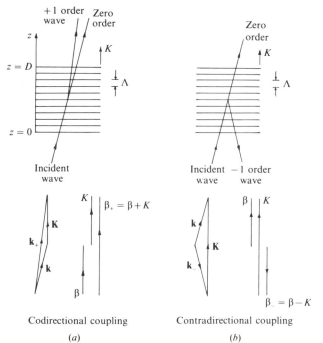

Codirectional coupling   Contradirectional coupling

(a)   (b)

18.5.1 Codirectional coupling

For the configuration shown in Fig. 18.5(*a*), we have

$$\beta/|\beta| = 1 = \beta_+/|\beta_+| \tag{18.78}$$

and Eqs. (18.72) and (18.73) result in an equation which is similar to Eq. (18.40) with κ, $\Delta\alpha$ and x replaced by σ, $\Delta\beta$ and z respectively. Thus we may carry out an analysis similar to that in Sec. 18.4 and if we use the boundary conditions

$$\tilde{A}_0(z = 0) = 1, \quad \tilde{A}_+(z = 0) = 0 \tag{18.79}$$

we will obtain expressions for the power in the incident and $+1$ order diffracted waves which are of the same form as Eqs. (18.48) and (18.49) and the corresponding diffraction efficiency will be given by

$$\eta = \frac{1}{[1 + (\Delta\beta)^2/4\sigma^2]} \sin^2 \left\{ \sigma L \left[1 + \frac{(\Delta\beta)^2}{4\sigma^2} \right]^{\frac{1}{2}} \right\} \tag{18.80}$$

where $\Delta\beta = \beta - \beta_+ + K$. Complete power transfer to the $+1$ diffracted wave can occur only if $\Delta\beta = 0$ i.e.,

$$\beta_+ = \beta + K \tag{18.81}$$

and if the length of interaction is $\pi/2\sigma$.

As is evident from Fig. 18.5(*a*), in order to satisfy the conditions $\alpha_+ = \alpha$ and $\beta_+ = \beta + K$, β_+ must be greater than β and therefore the diffracted wave must correspond to a different value of ϵ_u; for example, the incident and diffracted waves may correspond to the o and e-waves of an anisotropic medium. Obviously the roles can be reversed, i.e., we may have an incident e-wave along $\mathbf{k}_+$ and the diffracted wave is an o-wave propagating along $\mathbf{k}$, this would then correspond to the -1 order of codirectional coupling. As will be evident from the following two examples, the codirectional interaction is highly wavelength selective and thus can be used for making tunable acoustooptic filters.

Problem 18.3: (*a*) Show that for colinear codirectional coupling the free space wavelength λ_0 for maximum diffraction efficiency is given by

$$\lambda_0 = \lambda_{0c} = (n_2 - n_1)\Lambda \tag{18.82}$$

where n_1 and n_2 are the refractive indices seen by the incident and diffracted waves respectively. (*b*) Show that when $\sigma L = \frac{1}{2}\pi$, for wavelengths

$$\lambda_0 = \lambda_{0c} \pm \Delta\lambda = \lambda_{0c} \pm \frac{\sqrt{3}}{2} \frac{\Lambda}{L} \lambda_0 \tag{18.83}$$

the diffraction efficiency becomes zero. The quantity $\Delta\lambda$ is known as the bandwidth of the filter. (c) For lithium niobate at $\lambda_0 = 0.6\,\mu m$, $n_1 = n_e = 2.2082$, $n_2 = n_o = 2.2967$ where the subscripts o and e refer to the o and e-waves respectively. Calculate Λ required for a λ_{oc} of $0.6\,\mu m$ and obtain the corresponding bandwidth for $L = 3.6\,cm$. (We should note in the above problem the highly wavelength selective nature of the filter and furthermore from Eq. (18.82) it is obvious that by varying the acoustic frequency the centre wavelength can be tuned.)

Solution: (a) Since the light wave is incident normally on the acoustic grating,

$$\beta = k = (2\pi/\lambda_0)n_1$$
$$\beta_+ = k_+ = (2\pi/\lambda_0)n_2$$

implying

$$\Delta\beta = (2\pi/\lambda_0)(n_1 - n_2) + 2\pi/\Lambda \qquad (18.84)$$

For maximum diffraction efficiency $\Delta\beta = 0$ and Eq. (18.82) follows. (b) $\eta = 0$ when $\Delta\beta = \pm\sqrt{3}\pi/L$ from which Eq. (18.83) follows.
(Answer: (c) $\Lambda = 6.78\,\mu m$, $\Delta\lambda \sim 1\,\text{Å}$)

Problem 18.4: In Fig. 18.5(a) assume the incident wave to be propagating along the $-z$-direction. Consider codirectional coupling and show that the magnitude of the wave vector of the diffracted wave will be $k - K$ and $k + K$ for $+1$ order and -1 order waves respectively. Thus the incident and diffracted waves should correspond to different values of ϵ_u.

18.5.2 Contradirectional coupling

For the configuration shown in Fig. 18.5(b), $\beta(= \beta_-)$ of the diffracted wave equals $\beta - K$ and therefore it corresponds to -1 order diffraction with

$$\beta/|\beta| = +1 \quad\text{and}\quad \beta_-/|\beta_-| = -1 \qquad (18.85)$$

The corresponding coupled wave equations are

$$d\tilde{A}_0/dz = -\sigma\tilde{A}_- e^{i(\Delta\beta)z} \qquad (18.86)$$

$$d\tilde{A}_-/dz = -\sigma\tilde{A}_0 e^{-i(\Delta\beta)z} \qquad (18.87)$$

where now

$$\Delta\beta = \beta - \beta_- - K \qquad (18.88)$$

If we multiply Eq. (18.86) by $\tilde{A}_0$ and Eq. (18.87) by $\tilde{A}_-$ and subtract we will obtain

$$\frac{d}{dz}(|\tilde{A}_0|^2 - |\tilde{A}_-|^2) = 0 \qquad (18.89)$$

which implies conservation of energy flowing along the z-direction.

If we solve Eqs. (18.86) and (18.87) using a procedure similar to that used in Sec. 18.4 we will obtain

$$\tilde{A}_0(z) = e^{i(\Delta\beta)z/2}(P_1 e^{gz} + Q_1 e^{-gz}) \tag{18.90}$$

$$\tilde{A}_-(z) = -(1/\sigma)e^{-i(\Delta\beta)z/2}[P_1(g + \tfrac{1}{2}i\Delta\beta)e^{gz} - Q_1(g - \tfrac{1}{2}i\Delta\beta)e^{-gz}] \tag{18.91}$$

where

$$g = (\sigma^2 - \tfrac{1}{4}(\Delta\beta)^2)^{\frac{1}{2}} \tag{18.92}$$

If D represents the length of the medium (along the z-direction) then for a wave incident at $z = 0$, the boundary conditions will be[†]

$$\tilde{A}_0(z = 0) = 1, \quad \tilde{A}_-(z = D) = 0 \tag{18.93}$$

Using the above boundary conditions and straightforward algebra, we obtain the following expressions for the power of the incident wave at $z = D$ and the power of the diffracted wave at $z = 0$:

$$P_0(D) = |\tilde{A}_0|^2 = \frac{g^2}{g^2 \cosh^2 gD + (\tfrac{1}{2}\Delta\beta)^2 \sinh^2 gD} \tag{18.94}$$

$$P_-(0) = |\tilde{A}_-|^2 = \frac{[g^2 + (\Delta\beta/2)^2]\sinh^2 gD}{g^2 \cosh^2 gD + (\tfrac{1}{2}\Delta\beta)^2 \sinh^2 gD} \tag{18.95}$$

The power in the diffracted wave is maximum only when $\Delta\beta = 0$ and for such a case

$$P_-(0) = \tanh^2 \sigma D \tag{18.96}$$

Fig. 18.6 The variation of the incident and the contradirectionally coupled diffracted power with z.

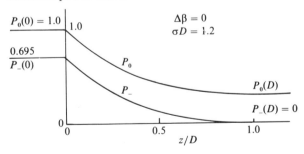

[†] Since we are assuming that the diffracted wave is propagating along the $-z$-direction and there is no periodic perturbation beyond $z = D$, the amplitude of the diffracted wave must be zero at $z = D$. This is quite similar to what we did in Sec. 2.4 where we put $E_n^- = 0$ at the last interface.

Fig. 18.6 shows the variation of P_0 and P_- with z for $\Delta\beta = 0$ and $\sigma D = 1.2$ and as can be seen for this case only about 69% of the incident power is coupled to the diffracted wave.

It follows from above that a periodic index perturbation can be used to reflect an incident wave and Eq. (18.96) gives the maximum reflectivity. Such reflectors are called Bragg reflectors and are used in distributed Bragg reflector (DBR) lasers and in distributed feedback (DFB) lasers. We may note that as $D \rightarrow \infty$, $P_-(0) \rightarrow 1$ and $P_0(D) \rightarrow 0$.

Problem 18.5: Referring to Fig. 18.5(b), assume $\lambda_0 = 6328\,\text{Å}$, $\Delta\epsilon/\epsilon_0 = 3 \times 10^{-4}$, $\Delta\beta = 0$ and colinear contradirectional coupling (i.e., $\beta = k = \frac{1}{2}K$) in a medium of refractive index 1.5. Calculate the wavelength of the acoustic wave and obtain the length of interaction required for a reflection efficiency of 90%.
(Answer: $\Lambda \approx 0.21\,\mu\text{m}$ $D = 3.7\,\text{mm}$)

Problem 18.6: In spite of the low figure of merit M_2 of fused quartz, the high optical quality, low optical absorption and high damage threshold make it a superior acoustooptic modulator for Q-switching in lasers (see Sec. 9.6). Assuming an optical wavelength of $1.064\,\mu\text{m}$ and an acoustic frequency of 50 MHz, calculate the Bragg angle θ_B and also the angle θ' between the undiffracted and diffracted beams as they emerge from the quartz. Also determine if a cell of 50 mm width would act in the Bragg regime or in the Raman–Nath regime. The refractive index of quartz is 1.45 and the acoustic velocity is 3.76×10^5 cm/s.
(Answer: Acoustic wavelength $= 75.2\,\mu\text{m}$, $\theta_B \approx 16.8'$, $\theta' \simeq 48.7'$).

Problem 18.7: The acoustic velocities in lithium niobate of a longitudinal wave propagating along z and a shear wave propagating along z with particle motion (polarization) along x are $v_l = 7.33 \times 10^5$ cm/s and $v_s = 3.59 \times 10^5$ cm/s respectively. Calculate the acoustic wavelength at a frequency of 1 GHz. What are the corresponding Bragg angles for $\lambda_0 = 6328\,\text{Å}$? (Spencer, Lenzo and Ballman, 1967).
(Answer: $\Lambda_l = 7.43\,\mu\text{m}$, $\Lambda_s = 3.71\,\mu\text{m}$, $\theta_{Bl} = 2.44°$, $\theta_{Bs} = 4.88°$)

18.6 Application to periodic media

In Sec. 2.7 we considered the reflection of a light wave from a periodic layer using the matrix method. Here we shall apply the coupled mode theory to obtain the reflection coefficient of the same periodic medium which is assumed to consist of alternating layers of refractive index $n_0 + \Delta n$ and $n_0 - \Delta n$ and whose periodicity is d (see Fig. 2.19(a)). Unlike our analysis up to now which assumes a sinusoidally periodic medium (as described by Eq. (18.3)), the present medium is not sinusoidally periodic. The dielectric constant variation in such a medium may be described by

$$\epsilon(z) = \epsilon_0[n_0 + \Delta n f(z)]^2 \approx \epsilon_u + \Delta\epsilon' f(z) \qquad (18.97)$$

where $\epsilon_u = \epsilon_0 n_0^2$, $\Delta\epsilon' = 2\epsilon_0 n_0 \Delta n$

$$f(z) = \begin{cases} 1; & 0 < |z| < d/4 \\ -1; & d/4 < |z| < d/2 \end{cases} \tag{18.98}$$

and $f(z + d) = f(z)$. If we make a Fourier expansion of $f(z)$ we can show that

$$f(z) = \sum_{m=1,2,\dots} \frac{4}{m\pi} \sin\frac{m\pi}{2} \cos\frac{2m\pi z}{d} \tag{18.99}$$

and

$$\epsilon(z) = \epsilon_u + \sum_{m=1,2,\dots} \Delta\epsilon_m \cos\frac{2m\pi z}{d}$$

where

$$\Delta\epsilon_m = \frac{4}{m\pi} \Delta\epsilon' \sin\frac{m\pi}{2} \tag{18.100}$$

represents the peak dielectric permittivity change corresponding to the mth Fourier coefficient. Thus the periodic medium can be considered to be a superposition of various sinusoidally varying dielectric constant media of periodicities d, $d/3$, $d/5, \dots$, etc. If we consider the lightwave to be incident normally along the z-direction then in order to study reflection we must consider contradirectional coupling. Since in the present case the grating is stationary, we may consider it to be a limiting form of Eq. (18.3) with $\Omega \to 0$. Thus the Bragg condition can be satisfied only for the -1 order and as such we must have $\alpha = \alpha_- = 0$,

$$\beta_- = -\beta = -(\omega/c)n_0 \tag{18.101}$$

and[†]

$$\Delta\beta = \beta - \beta_- - K = \left(\frac{4\pi}{\lambda_0}\right)n_0 - 2\pi m/d \tag{18.102}$$

As discussed earlier maximum coupling occurs when $\Delta\beta = 0$, i.e., at wavelengths given by

$$\lambda_0 = \lambda_{0c} = (2n_0 d/m); \quad m = 1, 3, 5, \dots \tag{18.103}$$

Further the coupling coefficient σ (see Eq. (18.74)) due to the sinusoidal grating component of period d/m is given by

$$\sigma_m = \frac{\omega^2 \mu_0 \Delta\epsilon_m}{4(|\beta||\beta_-|)^{\frac{1}{2}}} = \frac{4\Delta n}{m\lambda_0} \sin\frac{m\pi}{2}$$

[†] We could as well have taken the acoustic wave to propagate in the $-z$-direction, then we should have considered contradirectional coupling to the $+1$ order wave with $\beta_+ = \beta + K$ where $K = -2m\pi/d$.

Thus at the wavelengths given by Eq. (18.103)

$$\sigma = \frac{2\Delta n}{n_0 d} \sin \frac{m\pi}{2} \tag{18.104}$$

and the reflectivity is given by (see Eq. (18.96))

$$R = \tanh^2 \left(\frac{2\Delta n}{n_0 d} D \sin \frac{m\pi}{2} \right) \tag{18.105}$$

where D is the thickness of the periodic medium. It may be noted that the peak reflectivity has the value

$$R = \tanh^2 \left(\frac{2\Delta n}{n_0 d} D \right) \tag{18.106}$$

for all the following wavelengths

$$\lambda_0 = \lambda_{0c} = 2n_0 d/m; \quad m = 1, 3, 5, \ldots \tag{18.107}$$

We next determine the variation of the reflectivity around λ_{0c} and obtain the bandwidth. For $\Delta\beta < 2\sigma$, the reflectivity variation is given by Eq. (18.95). When $\Delta\beta > 2\sigma$, g becomes imaginary and the reflectivity becomes

$$R = \frac{\sigma^2 \sin^2 bD}{b^2 \cos^2 bD + \frac{1}{4}(\Delta\beta)^2 \sin^2 bd} \tag{18.108}$$

where

$$b = ig = [\tfrac{1}{4}(\Delta\beta)^2 - \sigma^2]^{\frac{1}{2}} \tag{18.109}$$

Thus $R = 0$ when $bD = \pi, 2\pi, 3\pi\ldots$. The first zero on either side of the maximum occurs when

$$bD = \pm \pi \Rightarrow \Delta\beta = \pm 2\sigma(1 + (\pi^2/\sigma^2 D^2))^{\frac{1}{2}} \tag{18.110}$$

Now using Eqs. (18.102) and (18.103), we obtain

$$\Delta\beta = -4\pi n_0 \Delta\lambda/\lambda_{0c}^2 \tag{18.111}$$

From Eqs. (18.110) and (18.111) we obtain

$$\Delta\lambda = \frac{\lambda_{0c}}{m\pi} \frac{d}{D} (\sigma D) \left(1 + \frac{\pi^2}{\sigma^2 D^2} \right)^{\frac{1}{2}} \tag{18.112}$$

As an example, we consider a periodic medium with

$$n_0 = 1.5, \quad \Delta n = 0.01, \quad d = 2000 \,\text{Å}, \quad D = 0.2 \,\text{mm}.$$

For such a case we readily obtain $\lambda_{0c} = 6000 \,\text{Å}, 2000 \,\text{Å}, \ldots$, etc. $R \simeq 1$ and the bandwidth corresponding to the peak at $6000 \,\text{Å}$ is $\approx 25 \,\text{Å}$.

18.6.1 Application to reflection holograms

We now consider the application of contradirectional coupling to reflection holograms. Such holograms are formed by interference between two waves travelling in almost opposite directions in the recording medium. In the simplest case of two propagating plane waves, the hologram consists of a sinusoidal refractive index variation. An incident reconstruction wave is coupled to a diffracted wave propagating in the opposite direction.

Since in the present case, the grating has pure sinusoidal variation, we may use the analysis of Sec. 18.6 directly with $m = 1$. Thus the peak reflectivity will occur at $\lambda_{0c} = 2n_0 d$ and the coupling coefficient is given by $\sigma = 2\Delta n/n_0 d$.

As a typical example we consider a reflection hologram formed in an emulsion of refractive index $n_0 = 1.50$ and thickness $D = 20\ \mu m$. If the fringe spacing is $d = 0.163\ \mu m$ and $\Delta n = 0.012$, we readily obtain the centre wavelength $\lambda_{0c} \approx 0.488\ \mu m$ and the diffraction efficiency at λ_{0c} is $\approx 92\%$. Using Eq. (18.112) we obtain for the bandwidth $\Delta \lambda \approx 50\ \mathring{A}$. This shows the high degree of wavelength selectivity of reflection holograms which permits the use of white light for their reconstruction.

18.7   Transition to the Raman–Nath regime

In the Bragg regime discussed above we have shown that appreciable coupling occurs between only two waves. In the previous chapter we discussed the Raman–Nath diffraction where we obtained multiple diffraction orders for interaction lengths satisfying Eq. (18.1). We shall now show that under the condition specified by Eq. (18.1), we have multiple order diffraction with the amplitude of the nth order proportional to $J_n(\phi_1)$ (see Eq. (17.9)).

Let us consider normal incidence of a plane wave as shown in Fig. 17.1. Referring to the analysis given in Sec. 18.4, for the present case we have

$$\beta = 0, \quad \alpha = k = \omega(\epsilon_u \mu_0)^{\frac{1}{2}} \tag{18.113}$$

and hence from Eqs. (18.32) and (18.33), we obtain

$$\beta_+ = K, \quad \beta_- = -K \tag{18.114}$$

and

$$\alpha_\pm^2 = \omega^2 \epsilon_u \mu_0 - K^2 \tag{18.115}$$

where we have assumed $\Omega \ll \omega$. Thus Eq. (18.30) becomes

$$d\tilde{A}_0/dx \approx \kappa e^{i(\Delta\alpha)x}(\tilde{A}_+ - \tilde{A}_-) \tag{18.116}$$

We now consider the field variation at $x = L$ where L is assumed to be small so that

$$L\Delta\alpha \ll 1 \tag{18.117}$$

and $e^{i(\Delta\alpha)x} \approx 1$. Thus

$$d\tilde{A}_0/dx \approx \kappa(\tilde{A}_+ - \tilde{A}_-) \tag{18.118}$$

In an exactly similar manner, the first order waves will generate the corresponding second order waves etc. Thus if $\tilde{A}_n$ represents the amplitude of the nth order diffracted wave then we may write

$$d\tilde{A}_n/d\zeta = \tfrac{1}{2}(\tilde{A}_{n+1} - \tilde{A}_{n-1}) \tag{18.119}$$

where

$$\zeta = 2\kappa x = k_0\Delta nx \tag{18.120}$$

and we have used $\Delta\epsilon/\epsilon_0 = 2n\Delta n$. Now the Bessel functions $J_n(x)$ satisfy the following recurrence relation

$$dJ_n/dx = \tfrac{1}{2}(J_{n-1} - J_{n+1}) \tag{18.121}$$

Thus the amplitude of the nth order diffracted wave is proportional to $J_n(k_0\Delta nL)$ which is consistent with the findings of Sec. 17.4. Using Eqs. (18.113)–(18.115), Eq. (18.117) can easily be shown to be consistent with Eq. (18.1)

18.8 Vector approach to coupled wave equations

In this section we give the rigorous vector approach and justify the use of the scalar model used in Sec. 18.2. We start with the vector wave equation which is given by

$$(\nabla^2 - \mu_0\bar{\bar{\epsilon}}_u(\partial^2/\partial t^2))\mathscr{E} \simeq \mu_0\bar{\bar{\Delta\epsilon}}\sin(\Omega t - Kz)\partial^2\mathscr{E}/\partial t^2 \tag{18.122}$$

where

$$\bar{\bar{\epsilon}} = \bar{\bar{\epsilon}}_u + \overline{\overline{\Delta\epsilon}}\sin(\Omega t - Kz) \tag{18.123}$$

represents the permittivity tensor of the medium in the presence of the acoustic wave (cf. Eq. (18.3))

Following an argument given in Sec. 18.3, we assume the electric field to be given by

$$\mathscr{E} = \hat{e}_0\mathscr{E}_0 e^{i\omega t} + \hat{e}_+\mathscr{E}_+ e^{i(\omega+\Omega)t} + \hat{e}_-\mathscr{E}_- e^{i(\omega-\Omega)t} \tag{18.124}$$

We substitute this in Eq. (18.122) and separate out terms corresponding to $e^{i\omega t}$ and $e^{i(\omega\pm\Omega)t}$ to obtain

$$(\nabla^2 + \omega^2\mu_0\bar{\bar{\epsilon}}_u)\hat{e}_0\mathscr{E}_0$$
$$= (1/2i)\mu_0\overline{\overline{\Delta\epsilon}}[-\hat{e}_-(\omega-\Omega)^2\mathscr{E}_- e^{-iKz} + \hat{e}_+(\omega+\Omega)^2\mathscr{E}_+ e^{iKz}] \tag{18.125}$$

and

$$(\nabla^2 + (\omega \pm \Omega)^2 \mu_0 \epsilon_u)\hat{\mathbf{e}}_\pm \mathscr{E}_\pm = \mp (\omega^2/2i)\mu_0 \Delta \epsilon \hat{\mathbf{e}}_0 \mathscr{E}_0 e^{\mp iKz} \qquad (18.126)$$

We assume $\hat{\mathbf{e}}_0$ and $\hat{\mathbf{e}}_\pm$ to be Cartesian components so that

$$\nabla^2 \hat{\mathbf{e}}_0 \mathscr{E}_0 = \hat{\mathbf{e}}_0 \nabla^2 \mathscr{E}_0 \qquad (18.127)$$

$$\nabla^2 \hat{\mathbf{e}}_\pm \mathscr{E}_\pm = \hat{\mathbf{e}}_\pm \nabla^2 \mathscr{E}_\pm \qquad (18.128)$$

It may be noted that for the above equations to be consistent $\hat{\mathbf{e}}_0$ and $\hat{\mathbf{e}}_\pm$ should be eigen polarizations of $\bar{\bar{\epsilon}}_u$ and $\hat{\mathbf{e}}_\pm$ should be along the direction of $\bar{\bar{\Delta\epsilon}}\hat{\mathbf{e}}_0$. If we now premultiply Eq. (18.125) by $\hat{\mathbf{e}}_0^\dagger$ and Eq. (18.126) by $\hat{\mathbf{e}}_\pm^\dagger$ we will obtain

$$(\nabla^2 + \omega^2 \mu_0 \epsilon_u^0)\mathscr{E}_0 = -\frac{1}{2i}\mu_0 \omega^2 (\Delta\epsilon_- \mathscr{E}_- e^{-iKz} - \Delta\epsilon_+ \mathscr{E}_+ e^{iKz}) \quad (18.129)$$

and

$$(\nabla^2 + (\omega \pm \Omega)^2 \mu_0 \epsilon_u^\pm)\mathscr{E}_\pm = \pm \frac{1}{2i}\mu_0 \Delta\epsilon_\pm \omega^2 \mathscr{E}_0 e^{\mp iKz} \qquad (18.130)$$

where

$$\epsilon_u^0 = \hat{\mathbf{e}}_0^\dagger \bar{\bar{\epsilon}}_u \hat{\mathbf{e}}_0; \quad \epsilon_u^\pm = \hat{\mathbf{e}}_\pm^\dagger \bar{\bar{\epsilon}}_u \hat{\mathbf{e}}_\pm \qquad (18.131)$$

and

$$\Delta\epsilon_\pm = \hat{\mathbf{e}}_\pm^\dagger \bar{\bar{\Delta\epsilon}} \hat{\mathbf{e}}_0 = \hat{\mathbf{e}}_0^\dagger \Delta\epsilon \hat{\mathbf{e}}_\pm \qquad (18.132)$$

are all scalar quantities. We may now proceed in a manner very similar to that used in Sec. 18.3 and obtain the Bragg conditions. The coupling coefficient κ is given by

$$\kappa = \frac{\omega^2 \mu_0}{4(\alpha\alpha_\pm)^{\frac{1}{2}}} (\hat{\mathbf{e}}_\pm^\dagger \bar{\bar{\Delta\epsilon}} \hat{\mathbf{e}}_0) \qquad (18.133)$$

Notice that, in general, the incident and diffracted waves see different unperturbed refractive indices. Thus for small Bragg angle scattering, $\beta_\pm$ will be rigorously determined from Eqs. (18.32) and (18.33), while α and $\alpha_\pm$ will be determined from the following equations:

$$\alpha^2 = \omega^2 \epsilon_u^0 \mu_0 - \beta^2 \qquad (18.134)$$

$$\alpha_\pm^2 = \omega^2 \epsilon_u^\pm \mu_0 - \beta_\pm^2 \qquad (18.135)$$

Eqs (18.32) and (18.33) and the coupled wave analysis of Section 18.4 will remain valid and maximum diffraction efficiency would occur when $\alpha_\pm = \alpha$.

18.9   Evaluation of $\hat{\mathbf{e}}^\dagger \bar{\bar{\Delta\epsilon}} \hat{\mathbf{e}}_0$

As a simple example, we consider the propagation of a longitudinal acoustic wave along the z-direction in an isotropic medium for which $\bar{\bar{\Delta\epsilon}}$

is given by (see Sec. 16.3)

$$\bar{\bar{\Delta\epsilon}} = -\epsilon_0 n^4 \begin{pmatrix} p_{12}S_3 & 0 & 0 \\ 0 & p_{12}S_3 & 0 \\ 0 & 0 & p_{11}S_3 \end{pmatrix} \tag{18.136}$$

Let us consider an incident wave propagating almost along the x-direction and polarized along the y-direction. Thus

$$\hat{e}_0 = \begin{pmatrix} 0 \\ 1 \\ 0 \end{pmatrix} \tag{18.137}$$

and

$$\bar{\bar{\Delta\epsilon}}\hat{e}_0 = -\epsilon_0 n^4 \begin{pmatrix} 0 \\ p_{12}S_3 \\ 0 \end{pmatrix} \tag{18.138}$$

Thus the diffracted wave will be y-polarized for which

$$\hat{e}_+^{\dagger} = (0\ \ 1\ \ 0) \tag{18.139}$$

Thus

$$\hat{e}_+^{\dagger}\bar{\bar{\Delta\epsilon}}\hat{e}_0 = -\epsilon_0 n^4 p_{12}S_3 \tag{18.140}$$

and

$$\kappa_+ = -(\omega^2 \epsilon_0 \mu_0 / 4\alpha_0)n^4 p_{12}S_3 \tag{18.141}$$

Problem 18.8: Consider an acoustic shear wave polarized along the y-direction and propagating along the z-direction in an isotropic medium: (a) show that

$$\bar{\bar{\Delta\epsilon}} = -\epsilon_0 n^2 \begin{pmatrix} 0 & 0 & 0 \\ 0 & 0 & \frac{1}{2}(p_{11}-p_{12})S_4 \\ 0 & \frac{1}{2}(p_{11}-p_{12})S_4 & 0 \end{pmatrix} \tag{18.142}$$

where the symbols have the same meaning as in Sec. 16.3; (b) show that if the incident wave is z-polarized and propagating along the x-direction, the diffracted wave will be y-polarized and conversely.

We next consider large Bragg angle diffraction in lithium niobate. To be consistent with the notation in the literature we assume the optic axis to be along the z-direction. Thus in Fig. 18.7, x, y and z must be replaced by z, x and y respectively. We now consider an acoustic shear wave polarized along x and propagating along the y-axis in a lithium niobate crystal. If we consider an incident e-wave polarized along the z-direction then using Eq. (16.27) we may write

$$\bar{\bar{\Delta\epsilon}}\hat{e}_0 = -\epsilon_0 \begin{pmatrix} 0 & \Gamma_1 & \Gamma_2 \\ \Gamma_1 & 0 & 0 \\ \Gamma_2 & 0 & 0 \end{pmatrix} \begin{pmatrix} 0 \\ 0 \\ 1 \end{pmatrix} = -\epsilon_0 \begin{pmatrix} \Gamma_2 \\ 0 \\ 0 \end{pmatrix}$$

Table 18.2. *Refractive indices and the corresponding acoustic wavelengths for codirectional coupling in lithium niobate*

$\lambda(\mu m)$	n_o	n_e	$\Lambda = \dfrac{\lambda}{(n_o - n_e)}$ (μm)	f (GHz)
0.45	2.3780	2.2772	4.464	0.90
0.50	2.3410	2.2457	5.247	0.76
0.60	2.2967	2.2082	6.780	0.59
0.70	2.2716	2.1874	8.314	0.48
0.80	2.2571	2.1745	9.685	0.41
1.00	2.2370	2.1567	12.453	0.32

Table partially adapted from Hartfield and Thompson (1978).

where

$$\Gamma_1 = \tfrac{1}{2} n_o^4 (p_{11} - p_{12}) S_6, \quad \Gamma_2 = n_o^2 n_e^2 p_{41} S_6$$

Thus the diffracted wave is polarized along the x-direction and hence is an o-wave. Thus the acoustic wave couples the incident e-wave to the o-wave.

As an example we consider codirectional colinear coupling at $\lambda_0 = 6000 \, \text{Å}$ for which (see Table 18.2)

$$n_o = 2.2967, \quad n_e = 2.2082$$

If the incident wave is an e-wave, then for codirectional coupling to the

Fig. 18.7 Codirectional colinear coupling in Lithium niobate.

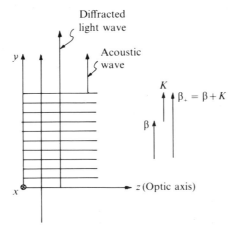

+1 order mode (see Fig. 18.5(a)) we must have

$$K = \beta_+ - \beta$$
$$= (2\pi/\lambda_0)(n_o - n_e) \tag{18.143}$$

Thus the required acoustic wavelength is

$$\Lambda = \lambda_0/(n_o - n_e) \approx 6.8\,\mu\text{m} \tag{18.144}$$

This coupling between the two polarizations can be used in the design of acoustooptic tunable filters.

In Table 18.2 we have given the optical wavelength dependence of n_o and n_e in lithium niobate. The corresponding acoustic frequencies required for codirectional coupling are also given in the table. We have assumed $v_a \approx 4 \times 10^3\,\text{m/s}$ for the shear wave.

Problem 18.9: At $\lambda_0 = 0.6\,\mu\text{m}$ for lithium tantalate, $n_o = 2.1834$, $n_e = 2.1878$. (a) What should be the incident polarization for codirectional colinear coupling to $+1$ and -1 order waves? (b) Obtain the corresponding acoustic wavelength which will now be found to be much larger.

Problem 18.10: Using the results of Problem 16.1, show that for 100% power transfer in the colinear codirectional coupling in lithium niobate, the strain required is given by

$$S_6 = -\frac{\lambda_0}{p_{41}L(n_o n_e)^{\frac{3}{2}}} \tag{18.145}$$

Using Eq. (18.59) and

$$p_{41} = -0.151, \quad L = 5\,\text{cm}, \quad \rho = 4.64\,\text{g/cm}^3$$

Calculate the acoustic intensity required for complete coupling.

In general, the value of $\hat{e}^\dagger_+ \bar{\bar{\Delta\epsilon}}\hat{e}_0$ depends on the polarization state of the incident light wave and that of the acoustic wave and the coefficients p_{ij}. We may symbolically write

$$\kappa_+ = -(\omega^2\epsilon_0\mu_0/4\alpha_0)\bar{n}^4 \bar{p}\bar{S} \tag{18.146}$$

where $\bar{p}, \bar{S}$ and $\bar{n}$ can be assumed to represent the effective acoustooptic coefficient, the effective strain and the effective refractive index of the medium.

19

Acoustooptic devices

19.1 Introduction

In the previous two chapters we have discussed the basic principle behind acoustooptic interaction. In this chapter we will discuss some important acoustooptic devices. We have seen in the previous two chapters that the intensity of the diffracted light depends on the acoustic power. Thus by changing the acoustic power one can correspondingly modulate the intensity of the diffracted light beam. This is the basic principle behind an acoustooptic modulator. We have also seen that the angle of diffraction depends on the acoustic frequency. Thus if we have an acoustic beam whose frequency is changed, the corresponding diffracted light beam will appear along different directions. This is the basic principle behind the acoustooptic deflector. If the acoustic transducer is given an input signal whose frequency increases with time, then the corresponding diffracted light beam will scan along different directions leading to an acoustooptic scanner. If on the other hand the acoustic transducer is fed simultaneously with a signal containing different frequencies then corresponding to each frequency, the diffracted light appears along different directions and this principle is used in the acoustooptic spectrum analyser. In this chapter we shall discuss the operation of a Raman–Nath modulator, a Bragg modulator, a deflector and a spectrum analyser.

19.2 Raman–Nath acoustooptic modulator

Fig. 19.1 shows an acoustooptic modulator based on Raman–Nath diffraction. The signal carrying the information modulates the amplitude of the acoustic wave propagating through the medium. The light beam incident on the cell is diffracted and at the output, the zero order beam is blocked using a stop as shown in Fig. 19.1. We have shown in Sec. 17.4 that the diffraction efficiency in the first order is given by (see

Eq. (17.24))

$$\eta = J_1^2(k_0 \Delta n L) \approx \pi^2 (\Delta n)^2 L^2 / \lambda_0^2 \tag{19.1}$$

where Δn is the peak change in refractive index caused by the propagating acoustic wave and L is the length of interaction and we have assumed $k_0 \Delta n L \ll 1$. Now using Eq. (18.56) we have

$$2n\Delta n = \Delta\epsilon/\epsilon_0 = n^4 \bar{p} \bar{S} \tag{19.2}$$

Thus

$$\eta = (\pi^2/4\lambda_0^2) n^6 \bar{p}^2 \bar{S}^2 L^2 \tag{19.3}$$

If we now use (see Eqs. (18.59) and (18.64))

$$P_a = I_a L H$$
$$\quad = \tfrac{1}{2}\rho v_a^3 \bar{S}^2 L H \tag{19.4}$$

we will obtain

$$\eta = \frac{\pi^2 M_2}{2\lambda_0^2}\left(\frac{L}{H}\right)P_a \tag{19.5}$$

where M_2 is the figure of merit given by Eq. (18.61). Thus from Eq. (19.5) we see that for small acoustic powers η is proportional to P_a and hence as the acoustic beam is modulated, the diffracted power will also be correspondingly modulated. To obtain low acoustic powers M_2 should be large and the aspect ratio L/H should also be large.

One of the main drawbacks of using Raman–Nath interaction is the restriction on the length of interaction. We recall that for Raman–Nath diffraction, we must have

$$L \ll n\Lambda^2/2\pi\lambda_0 = nv_a^2/2\pi\lambda_0 f^2 \tag{19.6}$$

Fig. 19.1 An acoustooptic modulator based on Raman–Nath diffraction. The stop blocks the light due to the zero order.

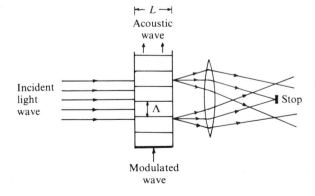

where $\Lambda(=v_a/f)$ is the acoustic wavelength and v_a and f are the acoustic velocity and acoustic frequency respectively. For low acoustic frequencies ($\lesssim 10\,\text{MHz}$), the restriction imposed by Eq. (19.6) is not too severe but for higher frequencies the maximum allowable L becomes too small. Thus if we consider fused quartz as the medium of propagation, then

$$n = 1.46, \quad v_a = 5.95 \times 10^3\,\text{m/s} \tag{19.7}$$

and for $\lambda_0 = 0.6328\,\mu\text{m}$,

$$\begin{aligned} &\text{at} \quad f = 10\,\text{MHz}, \quad L \ll 13\,\text{cm}\\ &\text{at} \quad f = 50\,\text{MHz}, \quad L \ll 5\,\text{mm} \end{aligned}$$

Thus for operation at higher frequencies L become very small and from Eq. (19.5) we see that the acoustic powers required for an effective operation will become very large. In addition as we will show in Sec. 19.3, the maximum modulation bandwidth also increases with acoustic frequency. Thus Raman–Nath modulators can be used at relatively low acoustic frequencies and hence with limited modulation bandwidths. For higher carrier frequencies and higher bandwidths, Bragg diffraction modulators are preferred. These will be discussed in the following section.

19.3   Bragg modulator

In Sec. 18.2 we have shown that when the interaction length satisfies the condition

$$L \gg nv_a^2/2\pi\lambda_0 f^2 \tag{19.8}$$

the acoustooptic interaction is in the Bragg regime. In this regime the light wave must be incident at the Bragg angle θ_B given by Eq. (18.10) in order that it may be coupled to the diffracted wave. The diffraction efficiency for

Fig. 19.2 An acoustooptic modulator based on Bragg diffraction.

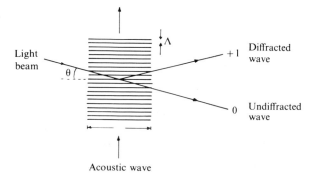

incidence at the Bragg angle is given by (see Eq. (18.62))

$$\eta = \sin^2 \left[\frac{\pi}{\lambda_0 \cos \theta_\mathrm{B}} \left(\frac{M_2}{2} \frac{L}{H} P_\mathrm{a} \right)^{\frac{1}{2}} \right] \tag{19.9}$$

For small acoustic powers the above equation becomes

$$\eta \approx \frac{\pi^2 M_2}{2 \lambda_0^2 \cos^2 \theta_\mathrm{B}} \left(\frac{L}{H} \right) P_\mathrm{a} \tag{19.10}$$

which is almost the same as Eq. (19.5). Thus, the intensity of the diffracted light is proportional to the acoustic power and hence variations in the acoustic power will lead to corresponding variations in the diffracted beam intensity. Fig. 19.2 shows a typical configuration of a Bragg diffraction modulator. As before, for a low acoustic power requirement the material must have a high figure of merit and the aspect ratio (L/H) should be large.

From Eq. (19.9) one can also calculate the acoustic power required for complete coupling from the incident to the diffracted wave (see Eq. (18.64)).

19.3.1 *Bandwidth of the modulator*

Bandwidth is one of the most important characteristics of a modulator. We shall now calculate the bandwidth of an acoustooptic modulator operating in the Bragg regime.

We first consider the interaction of an optical wave with an acoustic wave (at a frequency f) both of which are of infinite spatial extent. Thus both the waves are very well directed and are hence characterized by well defined propagation vectors. Thus if the Bragg condition is to be satisfied then the angle of incidence of the wave must be the Bragg angle defined by Eq. (18.10) (see Fig. 19.3). Now let us consider the incidence of a light wave having an angular divergence $\delta\theta_0$ on a well directed acoustic wave (see Fig. 19.4(a)). It is obvious that there will be a range of incidence angles

Fig. 19.3 A well directed light beam undergoing Bragg diffraction by a well directed acoustic wave. For Bragg diffraction to occur, the angle of incidence must be the Bragg angle θ_B.

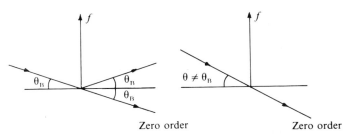

for which only one of the incident plane wave components of the light wave is incident at the Bragg angle. Thus this incident plane wave component of light will undergo Bragg diffraction and the other plane wave components will not undergo Bragg diffraction. Similarly if we have a light wave with a very small angular divergence incident on an acoustic wave having a finite divergence $\delta\theta_a$ (see Fig. 19.4(b)), then there will be a range of incident angles for which the incident light wave will be Bragg diffracted by one acoustic plane wave component making the Bragg angle corresponding to the incident light wave. If we now consider the case when both the incident light wave and the acoustic wave have finite angular divergences of $\delta\theta_0$ and $\delta\theta_a$ respectively (see Fig. 19.5), then it follows from Fig. 19.5(b) that in order that all acoustic wave components be used in Bragg diffraction and also that for every incident light wave component, there be a corresponding acoustic wave component at Bragg angle, we must choose $\delta\theta_a \gtrsim \delta\theta_0$.

In the above discussion the acoustic wave was assumed to have only a single frequency. It is well known that a modulated acoustic wave is a superposition of acoustic wave components at different frequencies. Thus

Fig. 19.4 (a) Interaction of a light beam having an angular convergence/divergence $\delta\theta_0$ with a well directed acoustic wave. The plane wave component of light incident at the Bragg angle is diffracted. Notice that not all incident light wave components undergo Bragg diffraction. (b) Interaction of a well directed light beam with an acoustic wave having a divergence of $\delta\theta_a$. The incident light wave picks up an acoustic wave component satisfying the Bragg angle and undergoes Bragg diffraction.

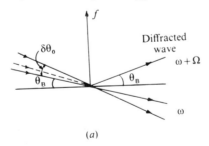

(a)

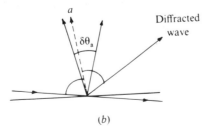

(b)

when the incident light wave interacts with an amplitude modulated acoustic wave, then the incident light wave interacts with each of the component acoustic frequencies and Bragg diffraction will occur if the Bragg condition corresponding to the various frequency components is satisfied. Since the diffracted light waves are shifted in frequency by the corresponding acoustic wave frequencies, when the diffracted light waves

Fig. 19.5 Interaction of a light beam which has a divergence of $\delta\theta_0$ with an acoustic beam which has a divergence $\delta\theta_a$; (a) with $\delta\theta_0 \ll \delta\theta_a$ and (b) $\delta\theta_0 \approx \delta\theta_a$. In (b) almost all acoustic wave components are being used in Bragg diffraction and for every incident light wave component there is a corresponding acoustic wave component at the Bragg angle.

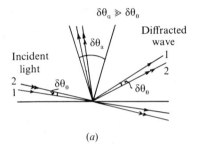

(a)

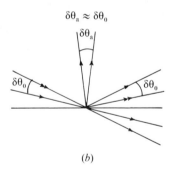

(b)

Fig. 19.6 A well collimated light beam interacting with acoustic waves at frequencies f_0 and f' each of which is diverging over an angular region $\delta\theta_a$. Notice that the diffracted waves of frequencies $v + f_0$ and $v + f'$ are nonoverlapping and thus there is no intensity modulation in the diffracted light.

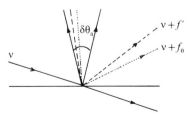

interfere, there will be a beating between the waves which ultimately results in the intensity modulation of the diffracted light. In order to simplify the discussion, we first consider two acoustic waves at frequencies f_0 and f' which are propagating simultaneously. This is equivalent to an amplitude modulated acoustic wave with a modulation frequency which is equal to the beat frequency $(f' - f_0)$. Now if a light wave with a very small angular divergence interacts with such an acoustic wave with an angular divergence $\delta\theta_a$ (see Fig. 19.6), then the incident light wave selects those components out of the various plane wave components of the acoustic wave which satisfy the Bragg conditions corresponding to f_0 and f' and undergoes diffraction. Now from Fig. 19.6 it follows that the two diffracted light wave components at frequencies $v + f_0$ and $v + f'$ (where v is the incident light frequency) are propagating along different directions and there is no overlap. Thus the diffracted light will not exhibit any intensity modulation. Observe that since the amplitudes of the acoustic wave components at f_0 and f' are constant, the amplitudes of the diffracted light wave components at $v + f_0$ and $v + f'$ are constant. It is the beating among the two waves that will lead to an amplitude modulation; this is accomplished by having an incident light wave with a finite angular divergence.

In Fig. 19.7 we have shown an incident light wave which has an angular divergence $\delta\theta_0$ interacting with acoustic waves at frequencies f_0 and f'

Fig. 19.7 Light waves incident over an angular region $\delta\theta_0$ interacting with acoustic waves at frequencies f_0 and f' and having a divergence $\delta\theta_0$: light waves incident along the directions PO and QO are diffracted along OP' and OQ' respectively by acoustic waves of frequency f_0.

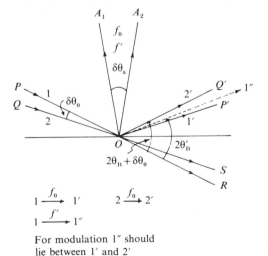

For modulation $1''$ should lie between $1'$ and $2'$

(each having an angular divergence $\delta\theta_a$). Incident light waves which have their propagation vectors along 1 and 2 interact with the acoustic waves at frequency f_0 and are diffracted along the 1′ and 2′ directions. If we consider $f' > f_0$ then since the Bragg angle increases with acoustic frequency (see Eq. (18.10)), the light wave incident along direction 1 may pick an acoustic wave at f' which satisfies the Bragg condition and be diffracted along a different direction (say 1″ – see Fig. 19.7). Some of the incident light waves incident between directions 1 and 2 will pick corresponding acoustic waves at f' and give the diffracted light wave. Now, it can be seen from Fig. 19.7 that the largest f' such that the light waves diffracted by acoustic waves at f_0 and f' overlap angularly is such that the diffracted light wave corresponding to $v + f'$ lies along the direction 2′.

In order to calculate this largest f' we first notice that the angle between the undiffracted and diffracted wave is twice the Bragg angle (see Fig. 19.3). Hence according to the above argument overlap will occur if $2\theta'_B < 2\theta_B + \delta\theta_0$, Thus the largest f' such that the diffracted wave is modulated corresponds to

$$\theta'_B = \sin^{-1}(\lambda_0 f'/2n_0 v_a) = \theta_B + \tfrac{1}{2}\delta\theta_0$$

or

$$f' = (2n_0 v_a/\lambda_0)\sin(\theta_B + \tfrac{1}{2}\delta\theta_0)$$
$$\approx f_0 + (n_0 v_a/\lambda_0)\cos\theta_B\,\delta\theta_0 \tag{19.11}$$

where we have assumed $\delta\theta_0 \ll 1$ and have used the relation

$$f_0 = (2n_0 v_a/\lambda_0)\sin\theta_B \tag{19.12}$$

Since an acoustic wave containing both frequencies f_0 and f' simultaneously corresponds to amplitude modulation at a frequency $f' - f_0$, from Eq.

Fig. 19.8 A typical configuration of a Bragg diffraction modulator.

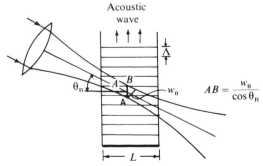

(19.11) we obtain for the maximum modulation frequency

$$\Delta f_{max} = f' - f_0 = (n_0 v_a / \lambda_0) \cos \theta_B \delta \theta_0 \qquad (19.13)$$

Thus the larger $\delta\theta_0$ is, the larger Δf_{max} will be. In Fig. 19.8 we have shown a typical configuration of a Bragg diffraction modulator in which the incident light beam is focussed, with the help of a lens, into the acoustooptic cell. If w_0 represents the minimum width of the light beam inside the medium, then

$$\delta\theta_0 \approx \lambda_0 / n_0 w_0 \qquad (19.14)$$

Thus

$$\Delta f_{max} \approx \frac{v_a}{w_0 / \cos \theta_B} = \frac{1}{\tau} \qquad (19.15)$$

where τ is the transit time for the acoustic wave to cross the light beam. Thus the modulation bandwidth of the acoustooptic modulator is the inverse of the acoustic propagation time through the light beam. Now, $\delta\theta_0$ cannot be arbitrarily increased to increase the bandwidth since if $\delta\theta_0$ becomes too large, the diffracted and undiffracted waves may start overlapping. Hence the maximum modulation bandwidth will be such that there is no such overlap. In Fig. 19.9 we have shown the diffraction of a light beam with an angular divergence $\delta\theta_0$ by an acoustic wave with an angular

Fig. 19.9 Diffraction of a light beam by acoustic waves at frequencies f_0, $f_0 + \Delta f_{max}$ and $f_0 - \Delta f_{max}$ where Δf_{max} is the maximum modulation bandwidth given by Eq. (19.15). In order that there be no overlap of the diffracted and undiffracted beams $\angle DOR > 0$, giving the maximum possible modulation bandwidth (Eq. (19.16)).

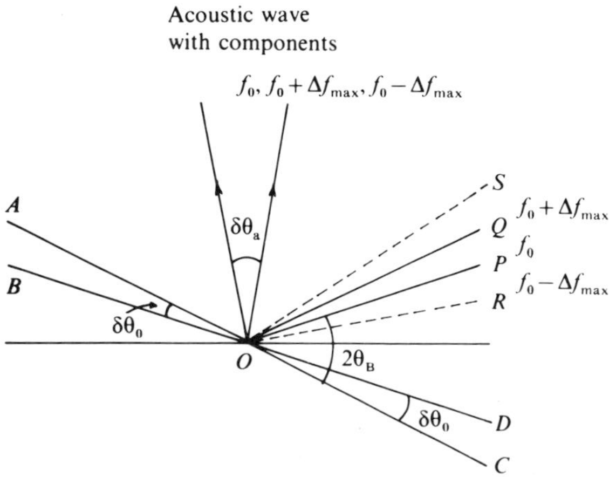

divergence $\delta\theta_a > \delta\theta_0$ and which contains frequencies f_0, $f_0 + \Delta f_{max}$ and $f_0 - \Delta f_{max}$ where Δf_{max} is the maximum modulation bandwidth given by Eq. (19.9) for the given $\delta\theta_0$. It follows from the angles shown in Fig. 19.9 that in order for the undiffracted beam lying in the angular region between OC and OD not to overlap with the diffracted beam lying between OR and OS, we must have $\angle DOR > 0$ or

$$2\theta_B - 2\delta\theta_0 > 0 \Rightarrow \delta\theta_0 < \theta_B$$

which using Eqs. (19.12) and (19.13) becomes

$$\Delta f_{max} < \tfrac{1}{2} f_0 \tag{19.16}$$

where we have assumed $\sin\theta_B \approx \theta_B$ and $\cos\theta_B \approx 1$. Thus it follows from Eq. (19.16) that the largest modulation bandwidth is half the centre

Fig. 19.10 An amplitude modulated acoustic wave propagating across a light beam of width w_0; (a) and (b) correspond to $t = 0$ and $t = \frac{1}{2}\Delta f$ for the case when $\Delta f \ll v_a/w_0$; Δf is the modulation frequency of the acoustic wave. Since the amplitude of the acoustic wave varies with time across the light wave, the diffracted light wave is also modulated. (c) When $\Delta f \sim v_a/w_0 \approx 1/\tau$, there will be no intensity modulation of the diffracted wave.

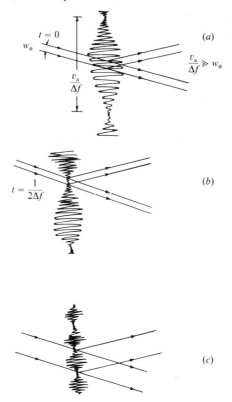

frequency. Hence for large bandwidths, f_0 must be large implying Bragg regime of operation.

The limit on the maximum modulation bandwidth given by Eq. (19.15) can also be understood from the following: for low acoustic modulation frequencies the sound intensity is almost uniform throughout the cross section of the optical beam (see Figs. 19.10(a) and (b)). As the modulation frequency increases, the sound intensity is no longer uniform across the light beam cross section and the depth of modulation will reduce. If the modulation frequency is so high that one full modulation cycle is present in the light beam cross section (see Fig. 19.10(c)) then there will no longer be any change in the intensity of the diffracted beam. Thus the maximum modulation bandwidth will be the inverse of the transit time which is the same as Eq. (19.15).

As a typical example we consider an acoustooptic modulator with As_2S_3 as the acoustooptic medium for which

$$n_0 = 2.61, \quad v_a = 2.6 \times 10^3 \, \text{m/s}$$

Let us assume that the modulator is to operate with a bandwidth of 120 MHz at $\lambda_0 = 0.63 \, \mu\text{m}$. Using Eq. (19.15), we find that the corresponding transit time of the acoustic wave through the optical beam is

$$\tau \sim 1/\Delta f \approx 8.3 \, \text{ns}$$

Since $\tau \approx w_0/v_a$, we obtain the corresponding optical beam width

$$w_0 \approx \tau v_a \approx 22 \, \mu\text{m}$$

Thus

$$\delta\theta_0 \approx \lambda_0/n_0 w_0 \approx 1.1 \times 10^{-2} \, \text{rad}$$

As discussed before the angular divergence of the acoustic wave must be chosen such that $\delta\theta_a \gtrsim \delta\theta_0$. We choose

$$\delta\theta_a \approx \Lambda/L \approx 1.5\delta\theta_0 \approx 1.65 \times 10^{-2} \, \text{rad}$$

Since for Bragg diffraction $Q \gg 1$, we arbitrarily choose

$$Q = 2\pi\lambda_0 L/n_0\Lambda^2 = 12 \tag{19.17}$$

Using the values of λ_0, n_0 and Λ/L in Eq. (19.17), we obtain

$$\Lambda \approx 7.7 \, \mu\text{m}$$

which corresponds to a centre frequency of

$$f_0 = v_a/\Lambda \approx 340 \, \text{MHz}$$

and the corresponding interaction length is

$$L \approx 0.5 \, \text{mm}$$

An actual device reported by Maydan (1970) operates at $f_0 = 350 \, \text{MHz}$ with a focussed beam size of $19 \, \mu\text{m}$ and rise time of $\sim 6 \, \text{ns}$.

For more detailed analysis taking into account the actual light beam profile and the modulation frequency response, readers are referred to Young and Yao (1981).

Problem 19.1: Consider an acoustooptic modulator in tellurium dioxide for which $n_0 \approx 2.26$, $v_a \approx 4.2 \times 10^3 \, \text{m/s}$. If the modulator is required to have a bandwidth of $100 \, \text{MHz}$ for operation at $\lambda_0 = 5145 \, \text{Å}$, calculate w_0, the centre frequency and the length of interaction.
(Answer: $w_0 \approx 42 \, \mu\text{m}$, $f_0 \approx 280 \, \text{MHz}$, $L \approx 1.9 \, \text{mm}$.)

Problem 19.2: Consider a Bragg diffraction modulator operating with acoustic waves of frequency 50 MHz. (*a*) Is the diffracted light amplitude modulated? (*b*) If the diffracted light is optically mixed with the undiffracted light on a photodetector through a pair of beam splitters as shown in Fig. 19.11, what will be the time dependence of the current in the photodetector? Such mixing leads to what is known as optical heterodyning.

From the above discussion we have seen that for a given acoustooptic modulator, the bandwidth may be increased by focussing the light beam inside the modulator. However, too much focussing will result in a reduction in diffraction efficiency due to the nonavailability of acoustic waves at the Bragg angle for all the incident light wave components, Fig. 19.12 shows a typical variation of the diffraction efficiency versus rise time for commercial acoustooptic modulator from Newport Research Corporation (USA). In addition greater focussing of the beam results in beam distortion and may even lead to optical damage in the material. Table 19.1 shows the

Fig. 19.11 Mixing of the diffracted and the undiffracted light beams on a photodetector.

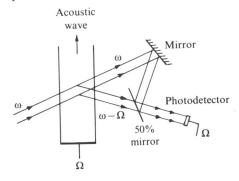

Table 19.1. *Typical acoustooptic modulator specifications*

Spectral range (μm)	Material	Bandwidth (MHz)	Diffraction efficiency (% at power in W)
0.3–1.5	Fused quartz	50	>40 at 5 W
0.4–1.1	Tellurium dioxide	125	65 at 1.5 W
0.4–1.2	Glass	10	85 at 1.2 W
0.4–2.0	Lithium niobate	500	2 at 1 W
10.6	Germanium	5	>80 at 15 W

characteristics of some typical commercially available acoustooptic modulators.

There are some applications in which one requires light modulation at much higher frequencies, for example, around the acoustic wave frequency itself. Such a modulation can indeed be achieved using different configurations. For example, Fig. 19.13 corresponds to a light beam undergoing Bragg diffraction by standing acoustic waves in the medium. The standing wave can be considered to be a superposition of acoustic waves propagating

Fig. 19.12 A typical variation of the diffraction efficiency versus rise time of an acoustooptic modulator (Newport Research Corporation, USA).

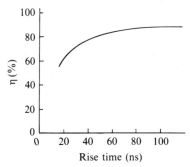

Fig. 19.13 Diffraction of a light beam from a standing acoustic wave.

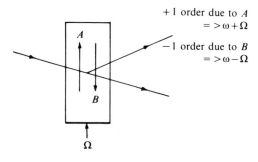

in opposite directions as shown in Fig. 19.13. The acoustic wave propagating along direction A will diffract the incident light wave into the $+1$ order which will have a frequency $\omega + \Omega$ where ω and Ω are the frequencies of the incident light wave and the acoustic wave. Similarly, the acoustic wave propagating along B generates the -1 order which has a frequency $\omega - \Omega$ and which is propagating along the same direction as the $+1$ order. The interference between these two diffracted waves will lead to an intensity modulation at the frequency 2Ω. Such a standing wave acoustooptic modulator is used in the mode locking of lasers (see Sec. 9.7).

19.4 Acoustooptic deflectors

We have seen that if a light wave is incident on an acoustic wave at the Bragg angle θ_B then the angle of deflection θ_d (which is the angle between the diffracted and the incident wave) is $2\theta_B$ (see Fig. 19.3). Thus assuming $\theta_B \ll 1$, we have

$$\theta_d = 2\theta_B \approx (\lambda_0/n_0 v_a)f \qquad (19.18)$$

where f is the acoustic frequency. Thus angle of defection θ_d is directly proportional to the acoustic frequency f (see Fig. 19.14). Hence if the Bragg condition can be satisfied over a frequency range Δf, then the corresponding range over which the light beam may be deflected will be

$$\Delta\theta_d = (\lambda_0/n_0 v_a)\Delta f \qquad (19.19)$$

Since the diffracted light wave will have a finite angular spread, the quantity of interest in a deflector is the number of resolvable spots N, i.e., the number of independently addressable directions. If $\delta\theta_0$ is the angular divergence of the optical beam then we have

$$N \approx \frac{\Delta\theta_d}{\delta\theta_0} \approx \frac{\lambda_0}{n_0 v_a}\Delta f \frac{n_0 w_0}{\lambda_0}$$

$$= \tau\Delta f \qquad (19.20)$$

Fig. 19.14 The deflection of a well collimated light beam by an acoustic wave with an angular divergence $\delta\theta_a$.

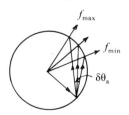

where we have used Eq. (19.10) for $\delta\theta_0$ and as before $\tau(\approx w_0/v_a)$ represents the acoustic propagation time through the light beam which is also the access time of the beam deflector, i.e., it represents the minimum time required to change the position of the deflected spot randomly. Since for a good resolution $\delta\theta_0$ must be small, the light wave must have a large width w_0. Thus in order to be able to satisfy the Bragg condition over a range of frequencies Δf, the acoustic beam is chosen to have an angular divergence

$$\delta\theta_a \approx \Lambda/L \tag{19.21}$$

As can be seen from Fig. 19.4(*b*) for such an angular spread of the acoustic waves, the Bragg conditions can be satisfied over a frequency range of

$$\Delta f \approx (2n_0 v_a/\lambda_0)\delta\theta_a$$

$$= \left(\frac{2n_0 f}{\lambda_0 L}\right)\Lambda^2 \tag{19.22}$$

Thus a large Δf requires a small acoustic transducer length. At the same time a reduction of L also leads to a reduction in diffraction efficiency (see Eq. (19.10)). A small L also leads to wastage of acoustic power since for any given deflection only a portion of the acoustic energy which satisfies the Bragg condition is being used. In order to overcome this, various beam steering techniques have been devised in which the sound beam changes the angle of propagation as the frequency changes so as to satisfy the corresponding Bragg angle (Korpel, *et al*, 1966; Gottlieb, Ireland and Ley, 1983).

As an example of an acoustooptic beam deflector we consider extra dense flint glass for which

$$n_0 = 1.92, \quad v_a = 3.1 \times 10^3 \, \text{m/s}$$

For an acoustic transducer of length $L = 5\,\text{cm}$ and operating around a centre frequency of 40 MHz, the corresponding acoustic wavelength is

$$\Lambda = v_a/f \approx 77.5 \, \mu\text{m}$$

Thus

$$\delta\theta_a \approx \Lambda/L \approx 1.55 \times 10^{-3} \, \text{rad}$$

Using Eq. (19.22) we obtain for the bandwidth for $\lambda_0 = 0.63 \, \mu\text{m}$

$$\Delta f \approx 29 \, \text{MHz}$$

We consider an incident optical beam of width $w_0 = 2.5$ cm so that the access time is

$$\tau \approx w_0/v_a \approx 8 \,\mu s$$

Thus the number of resolvable spots is

$$N = \tau \Delta f \approx 230$$

For a given deflected beam angle, the angular width of the diffracted light beam is

$$\delta\theta_0 \approx \lambda_0/w_0 \approx 2.5 \times 10^{-5} \,\text{rad}$$

Problem 19.3: If the device discussed above is to be used as a modulator, what optical beam size would one choose? Obtain the corresponding bandwidth of the modulator. Assume $\lambda_0 = 0.63 \,\mu m$.

Solution: As discussed in Sec. 19.3 we choose w_0 such that $\delta\theta_0 \approx \delta\theta_a/1.5$. Thus

$$\lambda_0/n_0 w_0 = \Lambda/1.5L \approx 10^{-3} \,\text{rad}$$

or

$$w_0 \approx 0.33 \,\text{mm}$$

The corresponding bandwidth is

$$\Delta f \sim 1/\tau = v_a/w_0 \approx 9 \,\text{MHz}$$

Problem 19.4: Consider an acoustooptic deflector with lead molybdate as the interaction medium. If the interaction length is 10 mm and the centre frequency is 150 MHz, calculate the number of resolvable spots and the access time of the deflector if the optical beam width is 2.5 cm and $\lambda_0 = 0.63 \,\mu m$. For lead molybdate you may use $n_0 = 2.4$, $v_a = 3.98 \times 10^3$ m/s.

Solution: A centre frequency of 150 MHz corresponds to

$$\Lambda = v_a/f \approx 26.5 \,\mu m$$

Thus from Eq. (19.19) we have

$$\Delta f \approx 80 \,\text{MHz}$$

and

$$\tau \approx w_0/v_a \approx 6.3 \,\mu s$$

Thus

$$N = \Delta f \tau = 500$$

19.5   Acoustooptic spectrum analyser

As discussed in the previous section, in an acoustooptic deflector, the angle at which the diffracted beam appears is proportional to the acoustic frequency (see Eq. (19.18)). If the acoustic transducer is fed simultaneously with different frequencies $f_1, f_2, f_3, \ldots$, etc., then the Bragg diffracted light waves will propagate along different angles. Thus if we have the configuration shown in Fig. 19.15, the different diffracted light beams will focus at different points on the focal plane of the lens. In addition, since the diffraction efficiency is approximately proportional to the power in the corresponding acoustic wave component, the diffracted light intensity pattern on the output plane will give the Fourier spectrum of the signal feeding the acoustic transducer. Hence the configuration shown in Fig. 19.15 leads to a real time spectral analysis of wide band signals and is referred to as a spectrum analyser.

Since the spectrum analyser is essentially a deflector we can use the results obtained in the previous section. We have seen that the change in the angle of deflection $\Delta\theta_d$ for a change in acoustic frequency of Δf is given by Eq. (19.19). If $\delta\theta_0$ as given by Eq. (19.14) represents the angular

Fig. 19.15 An acoustooptic spectrum analyser: the intensity distribution on the plane P is proportional to the power spectrum of the signal fed to the acoustic transducer.

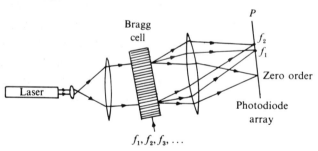

Fig. 19.16 An integrated optic spectrum analyser.

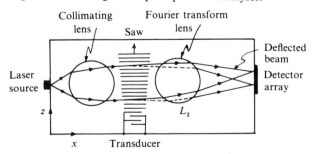

divergence of the optical beam then the minimum resolvable frequency difference will be the value of Δf when $\Delta \theta_d = \delta \theta_0$. Calling this frequency resolution δf, we obtain from Eqs. (19.14) and (19.19),

$$(\lambda_0/n_0 v_a)\delta f = \lambda_0/n_0 w_0$$

or

$$\delta f = v_a/w_0 = 1/\tau \tag{19.23}$$

i.e., the frequency resolution is essentially the inverse of the acoustic transit time through the optical beam. The bandwidth of the device will essentially be the acoustic frequency range Δf over which the Bragg condition could be satisfied.

Fig. 19.16 shows what is known as an integrated optic spectrum analyser (Anderson, 1978). In Chapter 14 we discussed how a light beam may be guided near the surface of a substrate by forming an optical waveguide. One can also have an acoustic wave which is essentially propagating along the surface of the substrate; such an acoustic wave is called a surface acoustic wave. Thus one can have interaction between the guided optical wave and a surface acoustic wave very similar to that which we discussed earlier. In Fig. 19.16 light from a laser diode is coupled into a planar optical waveguide formed on a surface of a suitable substrate such as lithium niobate. The light beam is collimated by a lens on the waveguide, and the collimated light beam interacts with the surface acoustic wave which is produced by a transducer on the surface of the substrate. The deflected beam is further focussed by another waveguide lens on the plane of a detector array whose output gives the complete spectrum of the signal feeding the transducer. Typical reported performances are a bandwidth of 1 GHz with frequency resolutions of 4–5 MHz (Suhara and Nishihara, 1986).

20

Nonlinear optics

20.1   Introduction

In all that has been discussed in earlier chapters we have assumed
that when a light beam propagates through a material, the properties of
the material are not affected by the light beam itself. However, if the intensity
of the light beam is large enough, the properties of the medium (such as
refractive index etc.) are affected and the study of the propagation of a
light beam becomes quite involved. For one thing the principle of
superposition does not remain valid.[†] This is the domain of nonlinear
optics where many new effects are observed. Basically, the nonlinear effects
are due to the dependence of properties such as the refractive index on the
electric and magnetic fields associated with light beam. Before the advent
of lasers, the electric fields associated with light beams were so weak that
nonlinear effects could not easily be observed. With the advent of laser
beams, it is now possible to have electric fields which are strong enough
for many interesting non-linear effects to be observed. It is of interest to
mention that the fact that intense electric and magnetic fields change the
properties of a medium has been known for a very long time. In 1845
Faraday discovered that the plane of polarization of a light beam
propagating through glass is rotated if a magnetic field is applied along
the direction of propagation of the light beam. This is known as the Faraday
effect and is perhaps the first experiment which demonstrated that light is
closely related to electromagnetism. In another experiment Kerr (around
1875) showed by applying a strong electric field in glass that light undergoes

---

[†] The superposition principle follows from the linearity of Maxwell's equations which
implies that if E_1 and E_2 are solutions of Maxwell's equations then $E_1 + E_2$ is also a
solution. However, when parameters such as the dielectric constant depend on the
electric field itself then Maxwell's equations become nonlinear and the superposition principle breaks down.

double refraction. This is known as the Kerr effect. Both the Faraday and Kerr effects demonstrate that electric and magnetic fields change the optical properties of a medium. Thus if the electric and magnetic fields associated with the light beam are strong enough, the properties of the medium will be affected which, in turn, will affect the propagation of the light beam.

In this chapter we will present brief discussions on two important nonlinear phenomena: the phenomenon of self focussing and second harmonic generation. Both of these phenomena are extremely interesting and have considerable applications.

20.2   The self focussing phenomenon

The self focussing (or defocussing) of light beam is due to the dependence of the refractive index on the intensity of the beam. In order to understand the self focussing phenomenon physically we assume the nonlinear dependence of the refractive index on the intensity to be of the form

$$n = n_1 + \tfrac{1}{2} n' E_0^2 \tag{20.1}$$

where n_1 is the refractive index of the medium in the absence of the electromagnetic field, n' is a constant representing the nonlinear effect[†] and

---

[†] This dependence may arise from a variety of mechanisms such as the Kerr effect, electrostriction, the thermal effect etc. The simplest to understand is the thermal effect which is due to the fact that when an intense optical beam which has a transverse distribution of intensity propagates through an absorbing medium, a temperature gradient is set up. For example, if the beam has a Gaussian transverse intensity variation, (i.e., of the form e^{-r^2/a^2}; the direction of propagation being along the z-axis), then the temperature will be maximum on the axis (i.e., $r = 0$) and will decrease with an increase in the value of r. If $dn/dT > 0$, the refractive index will be maximum on the axis and the beam will undergo focussing; on the other hand if $dn/dT < 0$, the beam will undergo defocussing (see e.g., Sodha (1973); Ghatak and Thyagarajan (1978)).

The Kerr effect is due to the anisotropic polarizability of liquid molecules (such as carbon disulphide). An intense light wave will tend to orient the anisotropically polarized molecules such that the direction of maximum polarizability is along the direction of the electric vector; this changes the dielectric constant of the medium. On the other hand, electrostriction (which is important in solids) is the force which a nonuniform electric field exerts on a material medium; this force affects the density of the material, which in turn affects the refractive index. Thus, a beam with a nonuniform intensity distribution along its wavefront will give rise to a refractive index variation leading to the focussing (or defocussing) of the beam. For a detailed discussion on electrostriction and the Kerr effect the reader is referred to Panofsky and Phillips (1962), Wagner, Haus and Marburger (1968), Sodha, Ghatak and Tripathi (1974).

E_0 represents the amplitude of the electric field. As an example, we consider the incidence of a laser beam (propagating in the z-direction) which has a Gaussian intensity distribution in the transverse direction i.e., we assume

$$E(x, y, z = 0, t) \approx E_0 \cos \omega t \qquad (20.2)$$

with

$$E_0 = E_{00} e^{-r^2/a^2} \qquad (20.3)$$

where a represents the width of the Gaussian beam and r represents the cylindrical coordinate. In the absence of any nonlinear effects the beam will undergo diffraction divergence (see Sec. 5.4). However, if the beam is incident on a medium characterized by a positive value of n', the intensity distribution will create a refractive index distribution which will have a maximum value on the axis (i.e., at $r = 0$) and will gradually decrease with r. Indeed, using Eqs. (20.1) and (20.3) we will have

$$\begin{aligned} n &= n_1 + \tfrac{1}{2} n' E_{00}^2 e^{-2r^2/a^2} \\ &\approx (n_1 + \tfrac{1}{2} n' E_{00}^2) - \tfrac{1}{2} (\alpha/n_1) r^2 \end{aligned} \qquad (20.4)$$

where

$$\alpha = (2n_1 n'/a^2) E_{00}^2 \qquad (20.5)$$

and in writing Eq. (20.4) we have expanded the exponential term and have retained only the first two terms, in other words, we are restricting ourselves to small values of r which is just the paraxial approximation. The term $\tfrac{1}{2} n' E_{00}^2$ is usually very small compared to n_1 so we may write (after squaring)

$$n^2 \approx n_1^2 - \alpha r^2 \qquad (20.6)$$

We may recall that in Sec. 11.8 we considered propagation in a medium whose refractive index decreased parabolically from the axis and showed that the beam would undergo periodic focussing. Indeed the medium behaves like a converging lens of focal length $\pi n_1/2\alpha^{\frac{1}{2}}$ (see also Fig. 13.5). In the present case also because of nonlinear effects (with $n' > 0$), the medium will act as a converging lens of focal length given approximately by

$$f_{NL} \approx \frac{\pi n_1}{2\alpha^{\frac{1}{2}}} \approx \frac{\pi}{2} \left(\frac{n_1}{2n' E_{00}^2} \right)^{\frac{1}{2}} a \qquad (20.7)$$

the subscript NL signifying that the effect is due to nonlinear phenomenon. Thus because of nonlinear effects the beam is said to undergo *self focusing*, the word *self* signifies the fact that the beam creates its own refractive index gradient resulting in the focussing of the beam[†].

In the above analysis we have neglected diffraction effects. Now, in the absence of any nonlinear effects, the beam will spread due to diffraction and the angle of divergence will be approximately given by (see Fig. 4.11)

$$\theta_d \approx \frac{\lambda}{\pi a} = \frac{\lambda_0/n_1}{\pi a} \tag{20.8}$$

where λ_0 is the free space wavelength. Thus the phenomenon of diffraction can be approximated by a diverging lens of focal length (see Fig. 20.11)

$$f_d \approx a/\theta_d \approx \tfrac{1}{2}ka^2 \tag{20.9}$$

where

$$k = 2\pi/\lambda = 2\pi n_1/\lambda_0 \tag{20.10}$$

Clearly if $f_{NL} > f_d$, the diffraction divergence will dominate and the beam will diverge. On the other hand, if $f_{NL} < f_d$, the nonlinear focussing effects will dominate and the beam will undergo self focussing. For $f_{NL} \approx f_d$, the two effects will cancel each other and the beam will propagate without any focussing or defocussing. This is the condition of *uniform waveguidelike propagation*. In order to determine the critical power of the beam we note that the condition $f_{NL} \approx f_d$ implies

$$\tfrac{1}{2}\pi(n_1/2n'E_{00}^2)^{\frac{1}{2}} a \approx \tfrac{1}{2}ka^2$$

or

$$E_{00}^2 \approx \frac{1}{n_1 n'}\frac{\lambda_0^2}{8a^2} \tag{20.11}$$

Fig. 20.1 The divergence of a parallel beam incident on a diverging lens.

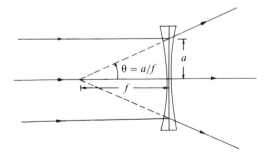

Now the total power of the beam is given by

$$P = \int_0^\infty \text{velocity} \times (\text{energy/unit volume}) \times 2\pi r \, dr$$

$$= \int_0^\infty (c/n_1)(\tfrac{1}{2}\epsilon E_0^2) 2\pi r \, dr$$

$$= (c/n_1)(\tfrac{1}{2}n_1^2\epsilon_0 E_{00}^2) \int_0^\infty e^{-2r^2/a^2} 2\pi r \, dr \tag{20.12}$$

$$\simeq \tfrac{1}{4}\pi n_1 c \epsilon_0 E_{00}^2 a^2$$

where $\epsilon (= n_1^2\epsilon_0)$ is the dielectric permittivity of the medium and $\epsilon_0 (= 8.85 \times 10^{-12}\,\text{C}^2/\text{N m}^2)$ is the dielectric permittivity of free space. Substituting the expression for E_{00}^2 from Eq. (20.11) in Eq. (20.12) we obtain the following expression for the critical power:

$$P_{cr} \approx \frac{\pi}{32}(c\epsilon_0)\frac{\lambda_0^2}{n'} \tag{20.13}$$

Garmire, Chiao and Townes (1966) carried out experiments on the self focussing of a ruby laser beam ($\lambda_0 = 0.6943\,\mu\text{m}$) in carbon disulphide and found that the critical power was $25 \pm 5\,\text{kW}$. Eq. (20.13) gives us

$$P_{cr} \approx \frac{3.14}{32} \times 3 \times 10^8 \times 8.85 \times 10^{-12} \times \frac{(0.6943 \times 10^{-6})^2}{2 \times 10^{-20}}$$

$$\approx 6.3\,\text{kW} \tag{20.14}$$

where we have used the following parameters for carbon disulphide: $n_1 \approx 1.6276$, $n' \approx 1.8 \times 10^{-11}$ cgs units $\approx 2 \times 10^{-20}$ mks units. (The mks unit for n' is $(\text{meter/volt})^2$.) Although Eq. (20.13) gives a result which is wrong by a factor of about 4, one does obtain the correct order. A more rigorous theory based on the solution of the wave equation gives us (see, e.g., Akhmanov, Sukhorukov and Khoklov (1968), Ghatak and Thyagarajan (1978), Sodha, *et al.* (1974)):

$$P_{cr} \approx 0.039\,(c\epsilon_0)\,\lambda_0^2/n' \tag{20.15}$$

which gives a critical power of about $2.5\,\text{kW}$. The discrepency is probably due to the saturation of the nonlinearity and uncertainty in the value of n'. Notice that except for the difference in the numerical factor, the expressions given by Eqs. (20.13) and (20.15) are identical.

Obviously, for $P < P_{cr}$, the diffraction divergence will dominate and the beam will diverge. For $P > P_{cr}$, the nonlinear effects will dominate and the beam will self focus producing intense electric fields which may cause breakdown of the medium. The self focussing phenomenon is of importance in many areas; for a detailed account we refer to Akhmanov, *et al.* (1968), Sodha, Ghatak and Tripathi (1976), Svelto (1974) and references therein.

Problem 20.1: In Sec. 13.7 we showed that for a parabolic index medium described by an equation of the form given by Eq. (20.6), the fundamental mode has a transverse field variation of the form e^{-r^2/w^2} with $w = 2/k_0\alpha^{\frac{1}{2}}$. Thus for uniform waveguide like propagation $a = w_0$. Show that this leads to

$$P_{\text{cr}} = 0.039(c\epsilon_0)\lambda_0^2/n'$$

consistent with Eq. (20.15).

20.3 Second harmonic generation

Another very interesting nonlinear optical effect is the phenomenon of second harmonic generation (usually abbreviated as SHG) in which one generates an optical beam of frequency 2ω from the interaction of a high power laser beam (of frequency ω) with a suitable crystal. The first demonstration of SHG was made by Franken and his coworkers in 1961 by focussing a 3 kW ruby laser pulse ($\lambda_0 = 6943$ Å) on a quartz crystal. In Fig 20.2 the incident beam from the left is a ruby laser beam (red light) passing through a crystal of **KDP** and a portion of the energy being converted into blue light which is the second harmonic. The remaining ruby light is suppressed by using a glass filter after the crystal.

Physically, SHG is due to the nonlinear dependence of the polarization on the electric field; by polarization we imply the dipole moment induced

Fig. 20.2 SHG of a ruby laser beam passing through a KDP crystal (photograph courtesy Dr R.W. Terhune).

by the electric field and this induced dipole moment is due to the relative displacement of the centre of the negative charge from that of the nucleus.

In order to understand the generation of the second harmonic we start with a charge free ($\rho = 0$), nonconducting ($\mathbf{J} = 0$) and nonmagnetic ($\mathscr{B} = \mu_0 \mathscr{H}$) homogeneous medium so that Maxwell's equations (see Sec. 1.2) lead to the following equation

$$\nabla^2 \mathscr{E} = \mu_0 \frac{\partial}{\partial t} (\nabla \times \mathscr{H}) = \mu_0 \frac{\partial^2 \mathscr{D}}{\partial t^2} \tag{20.16}$$

Now the displacement vector $\mathscr{D}$ is given by

$$\mathscr{D} = \epsilon_0 \mathscr{E} + \mathscr{P} \tag{20.17}$$

where $\mathscr{P}$ represents the polarization (or dipole moment per unit volume). For an isotropic medium and for weak fields, $\mathscr{P}$ is proportional to $\mathscr{E}$:

$$\mathscr{P} = \epsilon_0 \chi \mathscr{E} \tag{20.18}$$

and we have

$$\mathscr{D} = \epsilon \mathscr{E} \tag{20.19a}$$

where

$$\epsilon = \epsilon_0 (1 + \chi) \tag{20.19b}$$

represents the dielectric permittivity and χ the electric susceptibility of the material. For very high fields the polarization has a nonlinear dependence on the electric field and we may write

$$\mathscr{D} = \epsilon \mathscr{E} + \mathscr{P}_{\mathrm{NL}} \tag{20.20}$$

where $\mathscr{P}_{\mathrm{NL}}$ represents the nonlinear part of the polarization which becomes negligible for weak fields. Substituting in Eq. (20.16) we get

$$\nabla^2 \mathscr{E} - \epsilon \mu_0 \frac{\partial^2 \mathscr{E}}{\partial t^2} = \mu_0 \frac{\partial^2}{\partial t^2} \mathscr{P}_{\mathrm{NL}} \tag{20.21}$$

We should mention here that in the derivation of Eq. (20.21), we have used the condition that $\nabla \cdot \mathscr{E} = 0$ which is strictly valid only in isotropic media. In anisotropic media, this condition is not valid, however, for practical crystals $\mathscr{E}$ is almost perpendicular to $\mathbf{k}$ and therefore we may assume $\nabla \cdot \mathscr{E} \approx 0$. For example, in Problem 3.16 we have shown that for KDP the angle between $\mathscr{E}$ and $\mathscr{D}$ (the latter being perpendicular to $\mathbf{k}$) is about $2°$.

At this point we must mention that in nonlinear optics we must always be very careful in using the complex notation[†]. In Eq. (20.21) $\mathscr{E}$ represents the actual electric field and therefore should either be assumed to be of the form

$$\mathscr{E} = E \cos(\omega t - kz + \phi) \tag{20.22}$$

where E and ϕ are real quantities or should be written in the form

$$\mathscr{E} = \text{Re}(E e^{i(\omega t - kz)})$$
$$= \tfrac{1}{2}(E e^{i(\omega t - kz)} + \text{cc}) \tag{20.23}$$

where cc stands for the complex conjugate of the quantity preceding it; the amplitude E may be complex.

We consider a wave of frequency ω propagating through a medium and consider the generation of the second harmonic frequency 2ω as the beam propagates through the medium. Now, the field at ω generates a polarization at 2ω which acts as a source for the generation of an electromagnetic wave at 2ω. Corresponding to the frequencies ω and 2ω the electric field is assumed to be given by

$$\mathscr{E}^{(\omega)} = \tfrac{1}{2}(E_1(z)e^{i(\omega t - k_1 z)} + \text{cc}) \tag{20.24}$$

and

$$\mathscr{E}^{(2\omega)} = \tfrac{1}{2}(E_2(z)e^{i(2\omega t - k_2 z)} + \text{cc}) \tag{20.25}$$

respectively. The quantities

$$k_1 = \omega(\epsilon_1 \mu_0)^{\frac{1}{2}} = (\omega/c)n_1 \tag{20.26}$$

and

$$k_2 = 2\omega(\epsilon_2 \mu_0)^{\frac{1}{2}} = (2\omega/c)n_2 \tag{20.27}$$

represent the propagation vectors at ω and 2ω respectively; ϵ_1 and ϵ_2 represent the dielectric permittivities at ω and 2ω, and n_1 and n_2 represent the corresponding wave refractive indices. To be consistent with the notation used in Chapter 3, we should use another subscript w and write them

---

[†] In linear optics we can always superpose complex fields such as $e^{i(\omega t - kz)}$ and then take the real part. This is because of the fact that

$$\text{Re}(\mathscr{E}_1 + \mathscr{E}_2) = \text{Re}\mathscr{E}_1 + \text{Re}\,\mathscr{E}_2$$

However, in any nonlinear term (such as the RHS of Eq. (20.30)) we must take the square of the *real* electric field because

$$(\text{Re}\,\mathscr{E})^2 \neq \text{Re}\,\mathscr{E}^2$$

(see also the discussion in Sec. 1.5 in the calculation of the Poynting vector).

as n_{w_1} and n_{w2}; however, for brevity, we will drop the subscript w but should remember that the refractive indices used in this chapter refer to wave refractive indices rather than to ray refractive indices.

It should be noted that the amplitudes E_1 and E_2 are assumed to be z dependent – this is because at $z = 0$ (where the beam is incident on the medium) the amplitude E_2 is zero and this would develop as the beam propagates through the medium. We will now develop an approximate theory for the generation of the second harmonic.

In order to consider SHG, we write the wave equation corresponding to 2ω

$$\nabla^2 \mathscr{E}^{(2\omega)} - \epsilon_2 \mu_0 \frac{\partial^2 \mathscr{E}^{(2\omega)}}{\partial t^2} = \mu_0 \frac{\partial^2}{\partial t^2} \mathscr{P}_{NL}^{(2\omega)} \tag{20.28}$$

where, as will be shown in Sec. 20.3.1, the nonlinear polarization at 2ω can be written as

$$\mathscr{P}_{NL}^{(2\omega)} = \tfrac{1}{2}(P_{NL}^{(2\omega)} e^{2i(\omega t - k_1 z)} + cc) \tag{20.29}$$

with

$$P_{NL}^{(2\omega)} = \alpha E_1 E_1 \tag{20.30}$$

and the parameter α depends on the nonlinear material. Substituting for $\mathscr{E}^{(2\omega)}$ and $\mathscr{P}_{NL}^{(2\omega)}$ from Eqs. (20.25) and (20.29) in Eq. (20.28) and equating terms proportional to $e^{2i\omega t}$ we obtain

$$[\nabla^2 + \epsilon_2 \mu_0 (2\omega)^2] E_2(z) e^{i(2\omega t - k_2 z)}$$
$$= -\mu_0 \alpha (4\omega^2) E_1(z) E_1(z) e^{2i(\omega t - k_1 z)} \tag{20.31}$$

Now

$$(\nabla^2 + 4\epsilon_2 \mu_0 \omega^2) E_2(z) e^{i(2\omega t - k_2 z)}$$
$$\approx [-k_2^2 E_2 - 2ik_2(dE_2/dz) + (2\omega)^2 \epsilon_2 \mu_0 E_2] e^{i(2\omega t - k_2 z)}$$
$$\approx -2ik_2(dE_2/dz) e^{i(2\omega t - k_2 z)} \tag{20.32}$$

where we have used the relation $k_2^2 = (2\omega)^2 \epsilon_2 \mu_0$ and have neglected the term proportional to $d^2 E_2/dz^2$, i.e., we have assumed

$$d^2 E_2/dz^2 \ll k_2(dE_2/dz)$$

or

$$\frac{1}{k_2(dE_2/dz)} \frac{d}{dz}\left(\frac{dE_2}{dz}\right) \ll 1 \tag{20.33}$$

The above equation implies that the distance over which dE_2/dz changes appreciably is large compared with the wavelength which is indeed true for all practical cases. Substituting from Eq. (20.32) in Eq. (20.31) we obtain

$$\frac{dE_2}{dz} = -\frac{i\mu_0\alpha c\omega}{n_2} E_1^2(z)e^{i(\Delta k)z} \tag{20.34}$$

where

$$\Delta k = k_2 - 2k_1 = (2\omega/c)(n_2 - n_1) \tag{20.35}$$

In order to solve Eq. (20.34) we assume that the field $E_1(z)$ depletes very slightly (with z) so that the quantity E_1^2 on the RHS can be assumed to be independent of z. If we now integrate Eq. (20.34), we obtain

$$E_2(z) = -\frac{i\mu_0\alpha c\omega}{n_2} z E_1^2 e^{i\beta} \frac{\sin\beta}{\beta} \tag{20.36}$$

where

$$\beta = \tfrac{1}{2}(\Delta k)z = (\omega/c)(n_2 - n_1)z \tag{20.37}$$

Now, the intensities associated with the beams corresponding to ω and 2ω are given by (see Sec. 1.4):

$$I_1 = (k_1/2\omega\mu_0)|E_1|^2 = (n_1/2c\mu_0)|E_1|^2 \tag{20.38}$$

and

$$I_2 = (k_2/4\omega\mu_0)|E_2|^2 = (n_2/2c\mu_0)|E_2|^2 \tag{20.39}$$

If A represents the area of cross section of the beam then the powers associated with the waves ω and 2ω will be given by

$$P_1 = \tfrac{1}{2}A(n_1/c\mu_0)|E_1|^2 \tag{20.40}$$

$$P_2 = \tfrac{1}{2}A(n_2/c\mu_0)|E_2|^2 \tag{20.41}$$

Substituting for $|E_2|^2$ from Eq. (20.36) and then using Eq. (20.40) we get after some elementary simplifications

$$\eta = \frac{P_2}{P_1} = \frac{2c^3\mu_0^3\alpha^2\omega^2}{n_1^2 n_2} z^2 \frac{P_1}{A}\left(\frac{\sin\beta}{\beta}\right)^2 \tag{20.42}$$

where η represents the SHG efficiency. It may be noted that η increases if the nonlinear coefficient α increases (obviously) and if the frequency increases. However, for a given power P_1 the conversion efficiency increases if the area of the beam decreases – thus a focussed beam will have a greater SHG efficiency. The most important factor is $(\sin\beta/\beta)^2$ which is a sharply peaked function around $\beta = 0$, attaining a maximum value of 1 at $\beta = 0$. Thus for maximum SHG efficiency

$$\beta = 0 \Rightarrow n_2 = n_1 \tag{20.43}$$

i.e., the refractive index at 2ω must be equal to the refractive index at ω – this is known as the *phase matching condition*.

We see from Eq. (20.42) that the z dependence of η is of the form

$$F = z^2 \frac{\sin^2 \beta}{\beta^2} = \left[\frac{\sin (\Delta kz/2)}{(\Delta k/2)} \right]^2 \tag{20.44}$$

For a given Δk, the above function is periodic in z; the smallest z for which the function is maximum is

$$z = L_c = \frac{\pi}{\Delta k} = \frac{\pi c}{2\omega(n_2 - n_1)} \tag{20.45}$$

where we have used Eq. (20.35) for Δk. The length L_c is called the phase coherence length and represents the maximum crystal length up to which the second harmonic power increases. Thus, if the length of the crystal is less than L_c, the second harmonic power increases almost quadratically with z. For $z > L_c$, the second harmonic power begins to reduce again.

To show how important phase matching is, we consider the reduction in SHG efficiency due to nonphase matched operation. The only term in η which depends on z and Δk is given by F (Eq. (20.44)). For phase matched operation, $F = z^2$ and as discussed above for nonphase matched operation the maximum crystal length for useful SHG is L_c and for this value of z,

$$F = 4/\Delta k^2 = 4L_c^2/\pi^2 \tag{2.46}$$

Thus the reduction in efficiency for nonphase matched operation is

$$R = 4L_c^2/\pi^2 z^2 \tag{20.47}$$

If we take the difference in refractive indices for 2ω and ω to be 0.01, i.e., $n_2 - n_1 = 0.01$, then

$$L_c \approx \frac{10^{-4}}{4 \times 0.01} = 25 \times 10^{-4} \, \text{cm}$$

where we have taken the fundamental wavelength to be 1 μm. For $z = 1$ cm, we have

$$R = \frac{4 \times 625 \times 10^{-8}}{\pi^2 \times 1} \approx 2.5 \times 10^{-6}$$

which gives an enormous reduction in efficiency. Thus for efficient SHG it is very important to have the phase matching condition very closely satisfied.

In general, because of dispersion, it is very difficult to satisfy the phase matching condition. However, in a birefringent medium with $n_o > n_e$ it may

Table 20.1 *Sellemier coefficients corresponding to the equation*

$$n^2 = A - \frac{B_1}{(B_2 - \lambda_0^2)} - \frac{C_1 \lambda_0^2}{C_2 - \lambda_0^2}$$

		KDP	ADP
A	o	2.259276	2.302842
	e	2.132668	2.163510
B_1	o	0.01008956	0.011125165
	e	0.008637494	0.009616676
B_2	o	0.012942625	0.013253659
	e	0.012281043	0.012989120
C_1	o	13.00522	15.102464
	e	3.227992	5.919896
C_2	o	400	400
	e	400	400

Table adapted from Kurtz (1972); λ_0 is measured in micrometres.

be possible to find a direction along which the refractive index of the o-wave for ω equals the refractive index of the e-wave for 2ω. (For media with $n_e > n_o$, the direction would correspond to that along which the refractive index of the e-wave for ω equals the refractive index for the o-wave for 2ω.) This can readily be understood by considering a specific example. We consider the SHG in KDP corresponding to the incident ruby laser wavelength ($\lambda_0 = 0.6943\ \mu m$, $\omega = 2.7150 \times 10^{15}$ Hz). Using Table 20.1 we obtain

$$\left.\begin{array}{ll} n_o^\omega = 1.50502, & n_e^\omega = 1.46532 \\ n_o^{2\omega} = 1.53269, & n_e^{2\omega} = 1.48711 \end{array}\right\} \tag{20.48}$$

The refractive index variation for the e-wave is given by (see Eq. (3.53))[†]

$$n_e(\theta) = \left(\frac{\sin^2 \theta}{n_e^2} + \frac{\cos^2 \theta}{n_o^2}\right)^{-\frac{1}{2}} \tag{20.49}$$

where θ is the angle that the wave makes with the optic axis. Now as can be seen from the above equations,

$$n_e < n_e(\theta) < n_o \tag{20.50}$$

[†] Notice that $n_e(\theta)$ and n_e are different quantities. Therefore, one should always explicitly write the θ dependence when writing $n_e(\theta)$.

Since n_o at $0.6943 \, \mu m$ lies between n_e and n_o at $0.34715 \, \mu m$, there will always exist an angle θ_m along which

$$n_e^{2\omega}(\theta_m) = n_o^\omega = 1.50502 \tag{20.51}$$

Thus θ_m will satisfy

$$\left[\frac{\sin^2 \theta_m}{(n_e^{2\omega})^2} + \frac{\cos^2 \theta_m}{(n_o^{2\omega})^2} \right]^{-\frac{1}{2}} = n_o^\omega$$

or

$$\frac{1 - \cos^2 \theta_m}{(n_e^{2\omega})^2} + \frac{\cos^2 \theta_m}{(n_o^{2\omega})^2} = \frac{1}{(n_o^\omega)^2}$$

or

$$\cos^2 \theta_m = \left[\frac{(n_o^\omega)^2 - (n_e^{2\omega})^2}{(n_o^{2\omega})^2 - (n_e^{2\omega})^2} \right] \frac{(n_o^{2\omega})^2}{(n_o^\omega)^2} \tag{20.52}$$

If we substitute the values given by Eq. (20.48) we will get

$$\theta_m \approx 50.5° \tag{20.53}$$

which represents the angle that the incident wave should make with the optic axis for a maximum value of η (see Fig. 20.3). The corresponding value of θ_m measured experimentally is $50.4 \pm 1°$.

The above technique in which phase matching is achieved by choosing a direction of propagation which is not at $90°$ to the crystal optic axis is called critical phase matching. We have seen in Chapter 3 that in a uniaxial

Fig. 20.3 The phase matching condition corresponds to the incident o-wave (fundamental) propagating at an angle θ_m with the optic axis. The X, Y and Z-axes are the principal axes of the crystal. The fundamental wave is an o-wave with its electric field perpendicular to the optic axis (Z-axis). The second harmonic is an e-wave and hence lies in the z–Z plane nearly perpendicular to the z-axis.

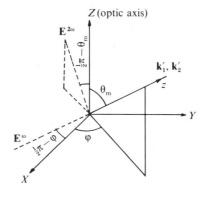

medium, the power flow of the e-wave is not along the direction of **k**. Thus as the fundamental (o-wave) and the second harmonic propagate along the crystal, the power generated in the second harmonic will separate from the fundamental leading to what is referred to as 'walk off'. The angle between the Poynting vector **S** and **k** of the e-wave (at frequency 2ω) is given by (see Eq. (3.114))

$$\tan\phi = \frac{[n_e^{2\omega}(\theta_m)]^2}{2}\left[\frac{1}{(n_e^{2\omega})^2} - \frac{1}{(n_o^{2\omega})^2}\right]\sin 2\theta_m \qquad (20.54)$$

If the focussed fundamental beam has a transverse dimension $\sim w$ then the fundamental and second harmonic powers will separate after a distance of approximately

$$L_w \approx w/\tan\phi \qquad (20.55)$$

As an example, if we again consider KDP, $\phi \approx 2°$ (see Problem 3.16) and for $w = 100\,\mu m$,

$$L_w \approx 0.25\,\text{cm}$$

which may be less than the actual crystal length.

Notice that for phase matching at $\theta_m = 90°$, $\phi = 0$ and there is no walk off. Hence when possible, $90°$ phase matching is desirable. This is referred to as noncritical phase matching. As we will show later, $90°$ phase matching also has additional advantages in terms of the acceptance angle.

20.3.1 Calculation of nonlinear polarization

The ith component of the nonlinear polarization induced in a medium is usually described by the following equation

$$\mathcal{P}_i = 2\sum_{j,k} d_{ijk}\mathcal{E}_j\mathcal{E}_k; \quad i,j,k = 1,2,3 \qquad (20.56)$$

where $\mathcal{E}_j$ is the jth component of the instantaneous electric field and 1, 2, 3 refer to the principal axes direction of the crystal; these directions will be denoted by X, Y and Z. For SHG, since the field contains components at ω and 2ω we get

$$\mathcal{E}_j = \tfrac{1}{2}(E_j^{(\omega)}e^{i(\omega t - k_1 z)} + E_j^{(2\omega)}e^{i(2\omega t - k_2 z)} + \text{cc}) \qquad (20.57)$$

where propagation is assumed to be along the z-direction. Thus

$$\mathcal{P}_i = \tfrac{1}{2}d_{ijk}(E_j^{(\omega)}e^{i(\omega t - k_1 z)} + E_j^{(2\omega)}e^{i(2\omega t - k_2 z)} + \text{cc})$$
$$\times (E_k^{(\omega)}e^{i(\omega t - k_1 z)} + E_k^{(2\omega)}e^{i(2\omega t - k_2 z)} + \text{cc}) \qquad (20.58)$$

and hence the ith component of the nonlinear polarization at frequency 2ω is

$$\mathscr{P}_i^{(2\omega)} = \tfrac{1}{2}d_{ijk}(E_j^{(\omega)}E_k^{(\omega)}e^{2i(\omega t - k_1 z)} + \text{cc})$$
$$= \tfrac{1}{2}(P_i^{(2\omega)}e^{2i(\omega t - k_1 z)} + \text{cc}) \tag{20.59}$$

where

$$P_i^{(2\omega)} = d_{ijk}E_j^{(\omega)}E_k^{(\omega)} \tag{20.60}$$

We again consider SHG in KDP. As mentioned earlier, for phase matching we consider the fundamental wave to be an o-wave and the second harmonic to be an e-wave. The d_{ijk} coefficients are specified with respect to the principal axes fixed in the crystal which we denote by X, Y, Z. For KDP, the only nonzero components[†] are

$$\left.\begin{array}{l} d_{123} = d_{132} = d_{14} \approx 5 \times 10^{-24}\,\text{mks units} \\ d_{231} = d_{213} = d_{25} = d_{14} \\ d_{312} = d_{321} = d_{36} \approx 5 \times 10^{-24}\,\text{mks units} \end{array}\right\} \tag{20.61}$$

Thus for KDP, we have

$$\left.\begin{array}{l} P_X^{(2\omega)} = d_{XYZ}E_Y^{(\omega)}E_Z^{(\omega)} + d_{XZY}E_Z^{(\omega)}E_Y^{(\omega)} \\ \qquad = 2d_{14}E_Z^{(\omega)}E_Y^{(\omega)} \\ P_Y^{(2\omega)} = 2d_{14}E_Z^{(\omega)}E_X^{(\omega)} \\ P_Z^{(2\omega)} = 2d_{36}E_X^{(\omega)}E_Y^{(\omega)} \end{array}\right\} \tag{20.62}$$

Since KDP is a uniaxial crystal, the Z-axis is the optic axis and since the fundamental wave at ω is an o-wave, $E_Z^{(\omega)} = 0$. Thus the induced polarization has only the component

$$P_Z^{(2\omega)} = 2d_{36}E_X^{(\omega)}E_Y^{(\omega)} \tag{20.63}$$

Since the ordinary polarization lies in the $X–Y$ plane, if the projection of **k** in the $X–Y$ plane makes an angle ϕ with the X-axis (see Fig. 20.3),

$$E_X^{(\omega)} = E_1 \sin\phi, \quad E_Y^{(\omega)} = -E_1 \cos\phi \tag{20.64}$$

Thus

$$P_Z^{(2\omega)} = -d_{36}\sin 2\phi\, E_1^2 \tag{20.65}$$

Now the second harmonic is an e-wave which is also propagating along the z-direction; the electric field associated with this e-wave will lie in the $z–Z$ plane and will be almost perpendicular to **k**. Thus if θ_m is the angle between **k** and the optic axis (Z-axis), then the component of $P_{\text{NL}}^{(2\omega)}$ which

[†] We are using the correspondence $X = 1$, $Y = 2$, $Z = 3$.

excites the second harmonic is given by

$$P_{NL}^{(2\omega)} = P_Z^{(2\omega)} \cos\left(\tfrac{1}{2}\pi - \theta_m\right)$$
$$= -d_{36} \sin\theta_m \sin 2\phi E_1^2 \tag{20.66}$$

Thus

$$\mathscr{P}_{NL}^{(2\omega)} = \tfrac{1}{2}\left(-d_{36} \sin\theta_m \sin 2\phi\, E_1 E_1 e^{2i(\omega t - k_1 z)} + \text{cc}\right) \tag{20.67}$$

Comparing with Eq. (20.29), we obtain

$$\alpha = -d_{36} \sin\theta_m \sin 2\phi \tag{20.68}$$

To maximize α, we choose $\phi = \tfrac{1}{4}\pi$; the value of θ_m is determined by the phase matching condition; thus

$$\alpha = -d_{36} \sin\theta_m \tag{20.69}$$

and we obtain from Eq. (20.42)

$$\eta = \frac{P_2(z)}{P_1(0)} = \frac{2c^3\mu_0^3(d_{36}\sin\theta_m)^2\,\omega^2}{n_1^2 n_2} z^2 \frac{P_1}{A}\left(\frac{\sin\beta}{\beta}\right)^2 \tag{20.70}$$

In order to appreciate this numerically we assume the phase matching condition to be satisfied at $\lambda_0 = 0.6943\,\mu m$ for which (see Eq. (20.53))

$$\theta_m \approx 50.5°$$

and

$$n_e(\theta_m)|_{\lambda_0 = 0.34715\,\mu m} = n_o|_{\lambda_0 = 0.6943\,\mu m} \approx 1.5 \tag{20.71}$$

Corresponding to the phase matching angle $\beta = 0$ and assuming

$$P_1/A \approx 10^{11}\,\text{W/m}^2; \quad z \approx 1\,\text{cm}$$

we get

$$\eta \approx 0.035 \tag{20.72}$$

implying a 3.5% conversion efficiency for a 1 cm length of the crystal. Obviously the present theory will be valid only when η is small compared to unity.

Problem 20.2: Consider SHG in ADP at $\lambda_0 = 1.0\,\mu m\,(\omega = 1.885 \times 10^{15}\,\text{s}^{-1})$ for which

$$\left.\begin{array}{ll} n_o^\omega = 1.495628, & n_e^\omega = 1.460590 \\ n_o^{2\omega} = 1.514498, & n_e^{2\omega} = 1.472068 \end{array}\right\} \tag{20.73}$$

Show that $\theta_m \approx 39.6°$ which compares well with the experimental value of $41.9 \pm 1°$ at $\lambda_0 = 1.0582 \, \mu m$.

20.3.2 *Effect of deviation from the phase matching angle*

We will next show that if the propagation direction differs slightly from the phase matching direction $\theta = \theta_m$, the SHG efficiency η decreases considerably. Now, referring to Eq. (20.37)

$$\beta = (\omega/c)[n_e^{2\omega}(\theta) - n_o^\omega]z \tag{20.74}$$

We make a Taylor series expansion of $n_e^{2\omega}(\theta)$ around $\theta = \theta_m$:

$$n_e^{2\omega}(\theta) \approx n_e^{2\omega}(\theta_m) + \left.\frac{dn_e^{2\omega}(\theta)}{d\theta}\right|_{\theta=\theta_m} (\theta - \theta_m)$$

$$\approx n_o^\omega + \left.\frac{dn_e^{2\omega}(\theta)}{d\theta}\right|_{\theta=\theta_m} (\theta - \theta_m) \tag{20.75}$$

where we have used the phase matching condition (see Eq. (20.51)):

$$n_e^{2\omega}(\theta_m) = n_o^\omega \tag{20.76}$$

Now

$$\frac{1}{n_e^2(\theta)} = \frac{\cos^2 \theta}{n_o^2} + \frac{\sin^2 \theta}{n_e^2}$$

Thus

$$-\frac{2}{n_e^3(\theta)} \frac{dn_e(\theta)}{d\theta} = \sin 2\theta \left(\frac{1}{n_e^2} - \frac{1}{n_o^2}\right)$$

or

$$\left.\frac{dn_e^{2\omega}(\theta)}{d\theta}\right|_{\theta=\theta_m} = -\frac{(n_o^\omega)^3}{2} \sin 2\theta_m \left[\frac{1}{(n_e^{2\omega})^2} - \frac{1}{(n_o^{2\omega})^2}\right] \tag{20.77}$$

Substituting in Eq. (20.75) we obtain

$$\beta \approx \Gamma(\theta - \theta_m) \tag{20.78}$$

where

$$\Gamma = \frac{\omega}{2c} z(n_o^\omega)^3 \left[\frac{1}{(n_o^{2\omega})^2} - \frac{1}{(n_e^{2\omega})^2}\right] \sin 2\theta_m \tag{20.79}$$

Since the second harmonic power goes to zero at $\beta = \pi$ (see Eq. (20.59)) the corresponding angular deviation is given by

$$\Delta\theta = \theta - \theta_m = \pi/\Gamma \tag{20.80}$$

If we substitute the values corresponding to $6943\,\text{Å}$ and $3471\,\text{Å}$ (see Eq. (20.73)), we obtain

$$\Gamma \approx 0.414 \times 10^4$$

for $z = 1$ cm. Thus for $\beta = \pi$,

$$\theta - \theta_m = \pi/\Gamma = 7.6 \times 10^{-4}\,\text{rad}$$
$$\approx 0.04°$$

Thus if the direction of propagation deviates from the phase matching angle by $0.04°$ then the conversion efficiency becomes zero. Fig. 20.4 shows an experimental plot of the variation of the SHG power as a function of the direction of propagation and, as can be seen, it closely fits with the $(\sin \beta/\beta)^2$ curve.

We have mentioned earlier that one usually focusses the beam to increase the conversion efficiency. We must also note that the focussing increases the divergence of the beam and hence this tends to reduce the efficiency since a portion of the incident beam now deviates from the phase matched direction.

Problem 20.3: From Eqs. (20.79) and (20.80) we see that if the phase matching angle $\theta_m = \tfrac{1}{2}\pi$, then $\Delta\theta \to \infty$; this happens due to the neglect of higher powers of $\Delta\theta$ in Eq. (20.75). Retaining the next highest order term in Eq. (20.75), show that for $\theta_m = \tfrac{1}{2}\pi$, the corresponding value is

$$\Delta\theta = \left\{ \frac{2\pi c}{\omega z}\, \frac{(n_o^{2\omega})^2}{n_e^{2\omega}[(n_o^{2\omega})^2 - (n_e^{2\omega})^2]} \right\}^{\frac{1}{2}} \approx \left[\frac{\pi c}{\omega z}\, \frac{1}{(n_o^{2\omega} - n_e^{2\omega})} \right]^{\frac{1}{2}} \tag{20.81}$$

The last expression is true for $n_o^{2\omega} \approx n_e^{2\omega}$.

Taking the example of **KDP** and $\lambda_0 = 1.06\,\mu\text{m}$ obtain the corresponding value of the acceptance angle $\Delta\theta$ and compare with the case in which $\theta_m = 50.5°$. This

Fig. 20.4 Variation of the SHG power with phase mismatch (after Ashkin, Boyd and Dziedic (1963)).

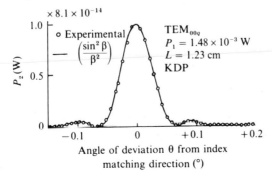

$\times 8.1 \times 10^{-14}$

o Experimental

$-\ \left(\dfrac{\sin^2 \beta}{\beta^2}\right)$

TEM_{00q}
$P_1 = 1.48 \times 10^{-3}\,\text{W}$
$L = 1.23$ cm
KDP

P_2 (W)

Angle of deviation θ from index matching direction (°)

shows the greater tolerance available with noncritical phase matching (i.e., with $\theta_m = 90°$).

20.3.3 Calculation of nonlinear polarization at ω

Once the field corresponding to 2ω is generated, it can interact with the incident wave at frequency ω to generate a nonlinear polarization at ω. This can easily be seen by referring to Eq. (20.58) and considering the terms leading to frequency ω. From Eq. (20.58) we obtain

$$\mathscr{P}_i^{(\omega)} = \tfrac{1}{2}(P_i^{(\omega)}e^{i[\omega t - (k_2 - k_1)z]} + \text{cc}) \tag{20.82}$$

where

$$\begin{aligned} P_i^{(\omega)} &= [d_{ijk}E_j^{(2\omega)}E_k^{(\omega)*} + d_{ijk}E_k^{(2\omega)}E_j^{(\omega)*}] \\ &= (d_{ijk} + d_{ikj})E_j^{(2\omega)}E_k^{(\omega)*} \\ &= 2d_{ijk}E_j^{(2\omega)}E_k^{(\omega)*} \end{aligned} \tag{20.83}$$

In the above equations it is assumed that the summation is performed over repeated indices and we have used $d_{ijk} = d_{ikj}$ for a lossless system. Notice a factor of 2 difference between Eqs. (20.60) and (20.83). If we again consider the incident wave at ω to be an o-wave and the second harmonic to be an e-wave and assume the direction of propagation to be as shown in Fig. 20.3, then we have[†]

$$E_X^{(\omega)} = E_1 \sin \phi, \quad E_Y^{(\omega)} = -E_1 \cos \phi, \quad E_Z^{(\omega)} = 0, \tag{20.84}$$

$$\left. \begin{aligned} E_X^{(2\omega)} &= -E_2 \cos \theta_m \cos \phi \\ E_Y^{(2\omega)} &= -E_2 \cos \theta_m \sin \phi \\ E_Z^{(2\omega)} &= E_2 \sin \theta_m \end{aligned} \right\} \tag{20.85}$$

Thus

$$\begin{aligned} P_X^{(\omega)} &= 2d_{123}E_Y^{(2\omega)}E_Z^{(\omega)*} + 2d_{132}E_Z^{(2\omega)}E_Y^{(\omega)*} \\ &= -2d_{14}E_1^*E_2 \cos \phi \sin \theta_m \end{aligned} \tag{20.86}$$

$$\begin{aligned} P_Y^{(\omega)} &= 2d_{213}E_X^{(2\omega)}E_Z^{(\omega)*} + 2d_{231}E_Z^{(2\omega)}E_X^{(\omega)*} \\ &= 2d_{14}E_1^*E_2 \sin \phi \sin \theta_m \end{aligned} \tag{20.87}$$

where we have used the fact that for KDP the only nonzero d_{ijk} elements are d_{14}, $d_{25} = d_{14}$ and d_{36}. Hence the component of the nonlinear

[†] Since the second harmonic is an e-wave, it is the $\mathbf{D}^{(2\omega)}$ (and not $\mathbf{E}^{(2\omega)}$) which is perpendicular to $\mathbf{k}$. We are assuming that the birefringence is small so that $\mathbf{D}$ and $\mathbf{E}$ are nearly parallel.

polarization parallel to the o-wave polarization is

$$P_{NL}^{(\omega)} = P_X^{(\omega)} \sin\phi - P_Y^{(\omega)} \cos\phi$$
$$= -4d_{14}E_1^*E_2 \sin\theta_m \cos\phi \sin\phi$$
$$= -2d_{14} \sin\theta_m \sin 2\phi E_1^*E_2 \qquad (20.88)$$

Thus we may write

$$\mathscr{P}_{NL}^{(\omega)} = \tfrac{1}{2}[2\alpha'E_1^*E_2 e^{i[\omega t - (k_2 - k_1)z]} + \text{cc}] \qquad (20.89)$$

where α' is given by

$$\alpha' = -d_{14}\sin\theta_m \sin 2\phi = \alpha \qquad (20.90)$$

and we have used the symmetry condition $d_{14} = d_{36}$.

20.3.4 Coupled equations and their solution

In the last section we showed that the nonlinear polarization generated at ω is given by

$$\mathscr{P}_{NL}^{(\omega)} = \tfrac{1}{2}(2\alpha E_2 E_1^* e^{i[\omega t - (k_2 - k_1)z]} + \text{cc}) \qquad (20.91)$$

Thus the wave equation at frequency ω will be

$$\tfrac{1}{2}(\nabla^2 + \epsilon_1\mu_0\omega^2)E_1(z)e^{i(\omega t - k_1 z)} = \mu_0\alpha E_2 E_1^*(-\omega^2)e^{i[\omega t - (k_2 - k_1)z]} \qquad (20.92)$$

Carrying out a procedure similar to the one used earlier we obtain

$$dE_1/dz = -i\kappa E_2(z)E_1^*(z)e^{-i(\Delta k)z} \qquad (20.93)$$

where

$$\kappa = \alpha c\omega\mu_0/n_1 \qquad (20.94)$$

and as before $\Delta k = k_2 - 2k_1$. We recall Eq. (20.34)

$$dE_2/dz = -i\kappa(n_1/n_2)E_1^2 e^{i\Delta k z} \qquad (20.95)$$

Eqs. (20.93) and (20.95) represent coupled equations and together determine the variation of power in the fundamental and the second harmonic.

For propagation along the phase matched direction, $n_1 = n_2$, $\Delta k = 0$ and Eqs. (20.93) and (20.95) become

$$dE_2/dz = -i\kappa E_1^2 \qquad (20.96)$$

$$dE_1/dz = -i\kappa E_2 E_1^* \qquad (20.97)$$

From Eqs. (20.96) and (20.97) and their complex conjugates, it follows that

$$\frac{d}{dz}(|E_1|^2 + |E_2|^2) = 0 \qquad (20.98)$$

which is just the conservation of power. Substituting for $|E_1|^2$ and $|E_2|^2$ from Eqs. (20.40) and (20.41) in Eq. (20.98), we also have

$$\frac{d}{dz}\left(\frac{P_1}{\hbar\omega} + 2\frac{P_2}{2\hbar\omega}\right) = 0 \qquad (20.99)$$

Since P_1 and P_2 represent the powers of the fundamental at ω and the second harmonic at 2ω, Eq. (20.99) implies that the number of photons generated at 2ω is half the number of photons annihilated at ω.

If we assume that at $z = 0$, there is no second harmonic wave, i.e.,

$$E_2(z = 0) = 0, \qquad (20.100)$$

then we may solve the coupled wave equations given by Eqs. (20.96) and (20.97) (see Problem 20.3) and obtain for the SHG efficiency

$$\eta = P_2(z)/P_1(0) = \tanh^2[\Gamma(P_1(0)/A)^{\frac{1}{2}}z] \qquad (20.101)$$

where

$$\Gamma = (2c^3\omega^2\mu_0^3\alpha^2/n_1^3)^{\frac{1}{4}} \qquad (20.102)$$

For small values of η, Eq. (20.101) reduces to Eq. (20.70) with $\Delta k = 0$.

Problem 20.3: Assuming $E_2(z) = 0$ at $z = 0$, solve the coupled equations (Eqs. (20.96) and (20.97)) for E_1 and E_2 and obtain Eq. (20.101).

Solution: To solve Eqs. (20.96) and (20.97), we define

$$E_1(z) = u_1(z)e^{i\phi_1(z)} \qquad (20.103)$$

$$E_2(z) = u_2(z)e^{i\phi_2(z)} \qquad (20.104)$$

where u_1, u_2, ϕ_1 and ϕ_2 are real functions of z. Substituting in Eqs. (20.96) and (20.97) and equating real and imaginary parts of the two equations, we obtain

$$du_1/dz = -\kappa u_1 u_2 \sin\theta \qquad (20.105)$$

$$du_2/dz = \kappa u_1^2 \sin\theta \qquad (20.106)$$

$$d\theta/dz = \cot\theta\,(d/dz)[\ln(u_1^2 u_2)] \qquad (20.107)$$

where

$$\theta(z) = 2\phi_1(z) - \phi_2(z) \qquad (20.108)$$

Integrating Eq. (20.107) we obtain

$$u_1^2 u_2 \cos\theta = K \qquad (20.109)$$

where K is a constant independent of z. Since $E_2(z = 0) = 0$, $u_2(z = 0) = 0$ and thus $K = 0$. For $z \neq 0$ since $u_1 \neq 0$, $u_2 \neq 0$, θ must be $\frac{1}{2}\pi$. Also if $E_1(0)$ is the field of the fundamental wave at $z = 0$, we have from Eqs. (20.103), (20.104), (20.105) and (20.106)

$$u_1^2(z) + u_2^2(z) = \text{const.} = u_1^2(0) \tag{20.110}$$

where we have arbitrarily set $\phi_1 = 0$ and $u_1(0)$ represents the amplitude of the fundamental wave at $z = 0$. In writing Eq. (20.110) we have also assumed $u_2(0) = 0$. Thus Eq. (20.106) becomes

$$du_2/dz = \kappa[u_1^2(0) - u_2^2(z)] \tag{20.111}$$

which can be integrated to give

$$u_2(z) = u_1(0) \tanh[u_1(0)\kappa z] \tag{20.112}$$

This can be written in terms of power as

$$P_2(z) = P_1(0) \tanh^2[\Gamma(P_1(0)/A)^{\frac{1}{2}}z] \tag{20.113}$$

with Γ given by Eq. (20.102).

20.3.5 Generation of sum and difference frequencies

The nonlinear interaction which leads to SHG can also lead to the generation of sum and difference frequencies when two beams of different frequencies interact through the nonlinear medium. Consider interaction between three frequencies ω_1, ω_2 and ω_3 satisfying the relation

$$\omega_3 - \omega_2 - \omega_1 = 0 \tag{20.114}$$

We assume the nonlinear polarization to be given by

$$\mathscr{P}_{\text{NL}} = \alpha \mathscr{E} \mathscr{E} \tag{20.115}$$

where

$$\mathscr{E} = \tfrac{1}{2}(E_1 e^{i(\omega_1 t - k_1 z)} + E_2 e^{i(\omega_2 t - k_2 z)} + E_3 e^{i(\omega_3 t - k_3 z)} + \text{cc}) \tag{20.116}$$

represents the total electric field. Proceeding in a manner similar to that used in SHG, we can show that the complex fields E_1, E_2 and E_3 (which represent the fields at ω_1, ω_2 and ω_3) satisfy the following coupled equations:

$$\left.\begin{array}{l} dE_3/dz = -i\kappa_3 E_2 E_1 e^{i\Delta kz} \\ dE_2/dz = -i\kappa_2 E_3 E_1^* e^{-i\Delta kz} \\ dE_1/dz = -i\kappa_1 E_3 E_2^* e^{-i\Delta kz} \end{array}\right\} \tag{20.117}$$

where

$$\kappa_i = \frac{\mu_0 c\alpha}{2}\frac{\omega_i}{n_i}; \qquad i = 1, 2, 3 \tag{20.118}$$

and

$$\Delta k = k_3 - k_2 - k_1 \tag{20.119}$$

If the input consists of ω_1 and ω_2 then the nonlinear interaction will generate the sum frequency $\omega_1 + \omega_2 = \omega_3$ and for efficient generation, the following phase matching condition should be satisfied:

$$k_3 = k_1 + k_2 \tag{20.120}$$

or

$$\omega_3 n_3 = \omega_1 n_1 + \omega_2 n_2 \tag{20.121}$$

where n_1, n_2 and n_3 are the refractive indices at ω_1, ω_2 and ω_3 respectively. Similarly, if the input consists of ω_3 and ω_1, then the difference frequency $\omega_3 - \omega_1 = \omega_2$ will be generated efficiently if the same phase matching condition (see Eq. (20.120)) is satisfied.

We should mention here that one should exercise caution in using the above analysis for studying SHG by setting $\omega_1 = \omega_2$ (implying $\omega_3 = 2\omega_1$). This can be seen by putting $E_1 = E_2$, $\omega_1 = \omega_2$, $k_1 = k_2$ in Eq. (20.116) to obtain

$$\mathscr{E} = \tfrac{1}{2}[2E_1 e^{i(\omega_1 t - k_1 z)} + E_3 e^{i(\omega_3 t - k_3 z)} + cc] \tag{20.122}$$

Notice the additional factor of 2 in the above equation in front of E_1 which does not appear in Eq. (20.57).

A simple application of sum frequency generation is in the upconversion of the 10.6 μm CO_2 laser beam ($\omega_2 = 1.7783 \times 10^{14}\,s^{-1}$) by letting it interact with a Nd:YAG laser beam at 1.06 μm ($\omega_1 = 17.783 \times 10^{14}\,s^{-1}$). The interaction can occur if both the beams are allowed to propagate simultaneously through a suitable nonlinear crystal (such as Ag_3AsS_3) and thereby generate the sum frequency $\omega_3 = \omega_2 + \omega_1 = 19.561 \times 10^{14}\,s^{-1}$ which corresponds to the wavelength of 0.964 μm. One of the advantages of this upconversion lies in the fact that radiations in the wavelength region of 10 μm are difficult to detect and one may, therefore, upconvert it in the 1 μm region and detect it with many of the available detectors.

Appendices

A

Wave equation and its solutions

We first consider the one-dimensional wave equation

$$\frac{\partial^2 \psi}{\partial x^2} = \frac{1}{v^2}\frac{\partial^2 \psi}{\partial t^2} \tag{A1}$$

The most general solution of the above equation is (see e.g., Ghatak (1977) Sec. 8.9)

$$\psi = f(x - vt) + g(x + vt) \tag{A2}$$

where f and g are arbitrary functions of their argument. The first term on the RHS of the above equation represents an undistorted pulse propagating in the $+x$-direction with speed v; similarly the second term represents a pulse propagating in the $-x$-direction with the same speed v. Thus if we are able to obtain an equation of the type

$$\frac{\partial^2 \psi}{\partial x^2} = \epsilon\mu\frac{\partial^2 \psi}{\partial t^2} \tag{A3}$$

then we are sure to have waves whose velocity is given by

$$v = 1/(\epsilon\mu)^{\frac{1}{2}} \tag{A4}$$

The plane wave solutions of Eq. (A1) are given by

$$\psi(x, t) = e^{i(\omega t \pm kx)} \tag{A5}$$

where

$$\omega/k = v \tag{A6}$$

We next consider the three-dimensional wave equation

$$\nabla^2 \Psi = \frac{1}{v^2}\frac{\partial^2 \Psi}{\partial t^2} \tag{A7}$$

We solve it by the method of separation of variables

$$\Psi(x, y, z, t) = X(x)Y(y)Z(z)T(t) \tag{A8}$$

Substituting in Eq. (A7) and dividing by Ψ we obtain

$$\left(\frac{1}{X}\frac{d^2X}{dx^2}\right) + \left(\frac{1}{Y}\frac{d^2Y}{dy^2}\right) + \left(\frac{1}{Z}\frac{d^2Z}{dz^2}\right) = \frac{1}{v^2}\left(\frac{1}{T}\frac{d^2T}{dt^2}\right) \tag{A9}$$

The variables have indeed separated out and we may write

$$\frac{1}{X}\frac{d^2X}{dx^2} = -k_x^2$$

$$\frac{1}{Y}\frac{d^2Y}{dy^2} = -k_y^2$$

$$\frac{1}{Z}\frac{d^2Z}{dz^2} = -k_z^2 \tag{A10}$$

$$\frac{1}{T}\frac{d^2T}{dt^2} = -\omega^2$$

where

$$\omega^2/v^2 = k_x^2 + k_y^2 + k_z^2 = k^2 \tag{A11}$$

The solutions of the above equations can be written in terms of exponential functions and we may write

$$\Psi(x, y, z, t) = A e^{i(\omega t \mp \mathbf{k} \cdot \mathbf{r})}$$
$$= A e^{i[\omega t \mp (k_x x + k_y y + k_z z)]} \tag{A12}$$

the upper and lower signs correspond to waves propagating along $\mathbf{k}$ and opposite to $\mathbf{k}$ respectively. In Fig. 1.1 the plane shown is perpendicular to $\mathbf{k}$ and for an arbitrary point P on this plane

$$\mathbf{k} \cdot \mathbf{r} = k(OP) \cos \theta = k(OL) \tag{A13}$$

where L is the foot of the perpendicular on the plane perpendicular to $\mathbf{k}$. Obviously for *all* points on this plane $\mathbf{k} \cdot \mathbf{r}$ and therefore the phase is constant. Thus the constant phase front is perpendicular to $\mathbf{k}$.

As an example, we assume

$$k_x = k \cos 30° = \frac{\sqrt{3}}{2}k, \quad k_y = k \sin 30° = \tfrac{1}{2}k, \quad k_z = 0 \tag{A14}$$

so that the angles that $\mathbf{k}$ makes with x, y and z-axes are 30°, 60° and 90° respectively.

We finally note that if we assume spherical symmetry of the solution then Eq. (A7) becomes

$$\frac{1}{r^2}\frac{\partial}{\partial r}\left(r^2\frac{\partial\Psi}{\partial r}\right)=\frac{1}{v^2}\frac{\partial^2\Psi}{\partial t^2} \tag{A15}$$

Substituting $\Psi=(1/r)u(r)$ we get

$$\frac{\partial^2 u}{\partial r^2}=\frac{1}{v^2}\frac{\partial^2 u}{\partial t^2} \tag{A16}$$

which is of the same form as Eq. (A1). Thus the general solution of Eq. (A15) is

$$\Psi=\frac{f(r-vt)}{r}+\frac{g(r+vt)}{r} \tag{A17}$$

which represents outgoing and incoming spherical waves.

B

The index ellipsoid

In this appendix we shall introduce a geometrical construction which will permit us to determine the velocities of propagation and the polarization states of the two waves that can propagate unchanged along a given direction. The electrical energy density is given by (see Eq. (1.61))

$$w_e = \tfrac{1}{2}(\mathbf{E}\cdot\mathbf{D}) = \tfrac{1}{2}(E_x D_x + E_y D_y + E_z D_z) \tag{B1}$$

In the principal axis system

$$E_x = D_x/\epsilon_0 K_x, \quad E_y = D_y/\epsilon_0 K_y, \quad E_z = D_z/\epsilon_0 K_z \tag{B2}$$

and thus

$$(D_x^2/K_x) + (D_y^2/K_y) + (D_z^2/K_z) = 2\epsilon_0 w_e = G \tag{B3}$$

Since the electric energy density w_e is positive, G is a positive quantity. Thus Eq. (B3) represents an ellipsoid in D_x, D_y, D_z space. Writing

$$x = D_x/G^{\frac{1}{2}}, \quad y = D_y/G^{\frac{1}{2}}, \quad z = D_z/G^{\frac{1}{2}} \tag{B4}$$

we can write Eq. (B3) as

$$(x^2/K_x) + (y^2/K_y) + (z^2/K_z) = 1 \tag{B5}$$

The above equation representing an ellipsoid with semiaxes $K_x^{\frac{1}{2}}$, $K_y^{\frac{1}{2}}$ and $K_z^{\frac{1}{2}}$ is referred to as the index ellipsoid (or optical indicatrix or Fletcher's ellipsoid) – see Fig. 3.12. We will now show that one can find the directions of $\mathbf{D}$ and the corresponding refractive indices for a given direction of propagation $\hat{\boldsymbol{\kappa}}$ from the index ellipsoid. For this we draw a plane normal to $\hat{\boldsymbol{\kappa}}$ i.e., normal to the propagation direction and passing through the centre of the ellipsoid. The plane will intersect the ellipsoid in an ellipse. In order to find the length of the semiminor and semimajor axes of the ellipse we see that the curve of intersection will satisfy Eq. (B5) for the index ellipsoid as well as the equation

$$x\kappa_x + y\kappa_y + z\kappa_z = 0 \tag{B6}$$

which is the equation of the plane, perpendicular to $\hat{\kappa}$ which passes through the origin. The major and minor axes of the ellipse will correspond to the maximum and minimum values of

$$r^2 = x^2 + y^2 + z^2 \tag{B7}$$

subject to (x, y, z) satisfying Eqs. (B5) and (B6). In order to find the extremum value of r subject to Eqs. (B5) and (B6) we use the Lagrange method of undetermined multipliers. We introduce two multipliers α_1 and α_2 and consider the function

$$F(x, y, z) = x^2 + y^2 + z^2 + \alpha_1(x\kappa_x + y\kappa_y + z\kappa_z)$$
$$+ \alpha_2((x^2/K_x) + (y^2/K_y) + (z^2/K_z)) \tag{B8}$$

The extremum values will now correspond to

$$\partial F/\partial x = 0, \quad \partial F/\partial y = 0, \quad \partial F/\partial z = 0 \tag{B9}$$

with no other subsidiary condition. Thus we obtain

$$\partial F/\partial x = 2x + \alpha_1\kappa_x + (2\alpha_2 x/K_x) = 0 \tag{B10}$$

$$\partial F/\partial y = 2y + \alpha_1\kappa_y + (2\alpha_2 y/K_y) = 0 \tag{B11}$$

$$\partial F/\partial z = 2z + \alpha_1\kappa_z + (2\alpha_2 z/K_z) = 0 \tag{B12}$$

In order to determine α_2, we multiply Eq. (B10) by x Eq. (B11) by y, Eq. (B12) by z and add and using Eqs. (B6) and (B7) get

$$2r^2 + 2\alpha_2 = 0$$

or

$$\alpha_2 = -r^2 \tag{B13}$$

To determine α_1 we multiply Eqs. (B10), (B11) and (B12) by κ_x, κ_y and κ_z respectively and add to get

$$\alpha_1 + 2\alpha_2\left(\frac{x\kappa_x}{K_x} + \frac{y\kappa_y}{K_y} + \frac{z\kappa_z}{K_z}\right) = 0 \tag{B14}$$

Substituting the values of α_1 and α_2 from Eqs. (B13) and (B14) in Eqs. (B10)–(B12) we get

$$x\left(1 - \frac{r^2}{K_x}\right) + \kappa_x r^2\left(\frac{x\kappa_x}{K_x} + \frac{y\kappa_y}{K_y} + \frac{z\kappa_z}{K_z}\right) = 0 \tag{B15}$$

with similar equations for y and z. In Eq. (B15), r corresponds to the semimajor or the semiminor axis. For a given direction of propagation, i.e., for given κ_x, κ_y and κ_z, the above set of equations form a homogeneous set in x, y and z. For nontrivial solutions, the determinant of the their coefficients must vanish; this determinant would give us an equation for r^2, the roots of which will give us the semimajor and the semiminor axes. Dividing Eq. (B15) throughout by r^2, it may be recast in the form

$$\left(\frac{K_x}{r^2} - \kappa_y^2 - \kappa_z^2 \right) \frac{x}{K_x} + \kappa_x \kappa_y \frac{y}{K_y} + \kappa_x \kappa_z \frac{z}{K_z} = 0 \tag{B16}$$

and two similar equations. Comparing Eq. (B16) with Eq. (3.39) it is obvious that except for E_x, E_y and E_z being replaced by x/K_x, y/K_y and z/K_z, they are identical. Since the solution of the determinant Eq. (3.43) gives us the two allowed values of n, the semimajor and semiminor axes obtained from the determinantal equation in r^2 are proportional to the two refractive indices. In addition, the directions of these axes correspond to the directions of the **D** of the two possible solutions. Also since the major and minor axes of an ellipse are normal to each other, the **D**s corresponding to the two polarization states are orthogonal to each other.

Thus in order to find the velocities of propagation and the **D**-directions corresponding to a plane wave propagating in an anisotropic medium, one constructs a plane perpendicular to the propagation direction which passes through the centre of the index ellipsoid. This plane intersects the ellipsoid in an ellipse; the orientation of the semiminor and semimajor axes gives the directions of the **D**s corresponding to the two independent eigen plane waves and the refractive indices seen by these waves would just be the lengths of the semiminor and semimajor axes. We should mention that the state and direction of polarization of the wave is described by **D** rather than by **E** since it is the **D** which is perpendicular to the propagation direction **k**, and **E** is not, in general, perpendicular to **k**.

C

Density of modes

In Sec. 9.2 we showed that the allowed values of k_x, k_y and k_z (inside a closed rectangular cavity of dimensions $2a \times 2b \times d$) are given by

$$k_x = m\pi/2a, \quad k_y = n\pi/2b, \quad k_z = q\pi/d; \quad m, n, q = 0, 1, 2, \ldots \qquad \text{(C1)}$$

Now, the number of modes for which k_x lies between k_x and $k_x + dk_x$ is equal to the number of integers lying between $2ak_x/\pi$ and $2a(k_x + dk_x)/\pi$; this number is approximately equal to

$$(2a/\pi)dk_x$$

and similarly for k_y and k_z. Thus the number of modes in the volume element $dk_x dk_y dk_z$ (of the $\mathbf{k}$ space) is given by

$$\left(\frac{2a}{\pi}dk_x\right)\left(\frac{2b}{\pi}dk_y\right)\left(\frac{d}{\pi}dk_z\right) = \frac{V}{\pi^3}dk_x dk_y dk_z \qquad \text{(C2)}$$

where $V = 4abd$ represents the volume of the cavity. Thus the number of modes per unit volume in the $\mathbf{k}$ space is

$$2V/\pi^3$$

where the factor of 2 is due to the two independent states of polarization. If $P(k)dk$ represents the modes for which the magnitude of $\mathbf{k}$ lies between k and $k + dk$, then

$$P(k)dk = (2V/\pi^3)4\pi k^2 dk \qquad \text{(C3)}$$

Since

$$k = (2\pi v/c)n_0 \qquad \text{(C4)}$$

we obtain

$$P(v)dv = \frac{8\pi n_0^3 v^2}{c^3} V\, dv \qquad \text{(C5)}$$

where n_0 is the refractive index of the medium filling the cavity.

D

Solution of the scalar wave equation for an infinite square law medium

For an infinitely extended square law medium (see Eq. (11.112))

$$n^2(x) = n_1^2[1 - 2\Delta(x/a)^2] \tag{D1}$$

the scalar wave equation

$$d^2\psi/dx^2 + [k_0^2 n^2(x) - \beta^2]\psi(x) = 0 \tag{D2}$$

can be written in the form

$$d^2\psi/d\xi^2 + [\Lambda - \xi^2]\psi(\xi) = 0 \tag{D3}$$

where

$$\xi = \alpha x, \quad \alpha = [k_0^2 n_1^2(2\Delta)/a^2]^{\frac{1}{4}} \tag{D4}$$

and

$$\Lambda = \frac{k_0^2 n_1^2 - \beta^2}{\alpha^2} = \frac{k_0^2 n_1^2 - \beta^2}{(k_0 n_1/a)(2\Delta)^{\frac{1}{2}}} \tag{D5}$$

Eq. (D3) is the same as one obtains in the linear harmonic oscillator problem in quantum mechanics (see, e.g., Ghatak and Lokanathan (1984), Chapter 3). In order to solve Eq. (D3) we write

$$\psi(\xi) = e^{-\xi^2/2}u(\xi) \tag{D6}$$

Elementary manipulations give us

$$\frac{d^2u}{d\xi^2} - 2\xi\frac{du}{d\xi} + (\Lambda - 1)u(\xi) = 0 \tag{D7}$$

We solve the above equation by the power series method:

$$u(\xi) = \sum_r a_r \xi^{r+s} = \xi^s[a_0 + a_1\xi + a_2\xi^2 + \cdots] \tag{D8}$$

Substituting in Eq. (D7) we get

$$s(s-1)a_0 = 0 \tag{D9}$$

$$s(s + 1)a_1 = 0 \tag{D10}$$

and

$$\frac{a_{r+2}}{a_r} = \frac{2r + 2s + 1 - \Lambda}{(r + s + 2)(r + s + 1)} \tag{D11}$$

According to one of the theorems in the theory of differential equations since the root $s = 0$ (of Eq. (D10)) makes a_1 indeterminate, it should determine both the solutions. Further in the limit of $r \to \infty$, a_{r+2}/a_r tends to $2/r$ and therefore it behaves (for large r) as e^{ξ^2}. Thus for $u \to 0$ as $x \to \pm \infty$ (the condition for a mode to be guided) the infinite series in Eq. (D8) should become a polynomial and therefore we must have

$$\Lambda = 2m + 1; \quad m = 0, 1, 2, \ldots \tag{D12}$$

For $m = 0, 2, 4, \ldots$ the series involving even powers of ξ becomes a polynomial and we must set $a_1 = 0$. Similarly, for $m = 1, 3, 5, \ldots$ the series involving odd powers of ξ becomes a polynomial and we must set $a_0 = 0$. These polynomials are the Hermite polynomials and because of Eq. (D6) the field patterns are the Hermite–Gauss functions. If we normalise these functions we obtain Eqs. (11.113)–(11.117). Substituting Eq. (D12) in Eq. (D5) we obtain

$$\beta_m^2 = k_0^2 n_1^2 - \frac{k_0 n_1}{a} (2\Delta)^{\frac{1}{2}} (2m + 1); \quad m = 0, 1, 2 \ldots \tag{D13}$$

which represents the allowed values of the propagation constant.

E

Leakage calculations of a packet of radiation modes

The fractional power that remains inside the core at z is approximately given by the overlap integral

$$W(z) \approx \left| \int_0^\infty \psi^*(x, 0)\psi(x, z)dx \right|^2 \tag{E1}$$

Now

$$\psi(x, z) = \int d\beta\, \phi(\beta)\psi_\beta(x)e^{i\beta z} \tag{E2}$$

(see Eq. (12.17)). Thus

$$W(z) \approx \left| \int dx \left[\int d\beta\, \phi^*(\beta)\psi_\beta^*(x) \right] \left[\int d\beta'\, \phi(\beta')\psi_{\beta'}(x)e^{i\beta' z} \right] \right|^2$$

$$\approx \left| \int d\beta\, \phi^*(\beta) \int d\beta'\, \phi(\beta')e^{i\beta' z} \int dx\, \psi_{\beta'}^*(x)\psi_\beta(x) \right|^2 \tag{E3}$$

The last integral is $\delta(\beta - \beta')$ (see Eq. (E31)). Since

$$\int d\beta'\, \phi(\beta')e^{i\beta' z}\delta(\beta - \beta') = \phi(\beta)e^{i\beta z} \tag{E4}$$

we obtain

$$W(z) = \left| \int |\phi(\beta)|^2 e^{i\beta z} d\beta \right|^2 \tag{E5}$$

In order to evaluate the above integral we must evaluate $|\phi(\beta)|^2$ which is given by (see Eq. (12.21)):

$$|\phi(\beta)|^2 \approx |A/A_g|^2 \tag{E6}$$

We know A_g (see Eq. (12.14)). In order to evaluate A, we express it in terms of D and then use Eq. (12.7). Since the wave packet is a superposition of radiation modes around the 'quasi-mode', $|\phi(\beta)|^2$ is very sharply peaked around $\beta \approx \beta_g$ and therefore all calculations will be carried out near $\beta = \beta_g$.

We begin with the calculation of C around $\beta = \beta_g$:

$$C = \tfrac{1}{2} A [\sin \alpha x_1 - (\alpha/\gamma) \cos \alpha x_1]_{\substack{\alpha \approx \alpha_g \\ \beta \approx \beta_g}} \quad \text{(see Eq. (12.4b))}$$

$$\approx \tfrac{1}{2} A [\sin \alpha_g x_1 - (\alpha_g/\gamma_g) \cos \alpha_g x_1]$$

$$\approx \tfrac{1}{2} A \left[\frac{\alpha_g}{\delta} + \frac{\alpha_g}{\gamma_g} \frac{\gamma_g}{\delta} \right]$$

where the subscript g refers to the 'guided mode' and use has been made of the relations

$$\sin \alpha_g x_1 = \alpha_g/\delta, \quad \cos \alpha_g x_1 = -\gamma_g/\delta \tag{E7}$$

$$\delta^2 = \alpha^2 + \gamma^2 = k_0^2(n_1^2 - n_2^2) \tag{E8}$$

Thus

$$C \approx (A/\delta)\alpha_g \tag{E9}$$

Since

$$B|_{\beta = \beta_g} = 0 \tag{E10}$$

we must make a Taylor series expansion of B around $\beta = \beta_g$:

$$B \approx \frac{dB}{d\beta}\bigg|_{\beta = \beta_g} (\beta - \beta_g) = \left[\frac{dB}{d\alpha} \frac{d\alpha}{d\beta} \right]_{\substack{\alpha = \alpha_g \\ \beta = \beta_g}} (\beta - \beta_g)$$

$$\approx \tfrac{1}{2} A \bigg[\left(x_1 \cos \alpha x_1 - \frac{\alpha}{\gamma} x_1 \sin \alpha x_1 \right.$$

$$+ \frac{1}{\gamma} \cos \alpha x_1 - \frac{\alpha}{\gamma^2} \frac{d\gamma}{d\alpha} \cos \alpha x_1 \bigg) \frac{d\alpha}{d\beta} \bigg]_{\substack{\alpha = \alpha_g \\ \beta = \beta_g}} (\beta - \beta_g)$$

where we have used Eq. (12.4a). If we use Eq. (E8) and the relation

$$\alpha^2 = k_0^2 n_1^2 - \beta^2 \tag{E11}$$

we obtain

$$d\gamma/d\alpha = -\alpha/\gamma \quad \text{and} \quad d\alpha/d\beta = -\beta/\alpha \tag{E12}$$

On substitution in the expression for B and simplifying we obtain

$$B \approx \tfrac{1}{2} A \frac{\beta_g \delta}{\gamma_g^2 \alpha_g} (1 + \gamma_g x_1)(\beta - \beta_g) \tag{E13}$$

Substituting for B and C in Eq. (12.4c) we get

$$|D_\pm| \approx \tfrac{1}{4}|A| \frac{\delta\beta_g}{\gamma_g^2 \alpha_g}(1 + \gamma_g x_1) \left|1 + \frac{\gamma_g}{i\alpha_g}\right| e^{\gamma_g(x_2 - x_1)}$$

$$\times \left|(\beta - \beta_g) + \frac{2\gamma_g^2 \alpha_g^2}{\beta_g \delta^2 (1 + \gamma_g x_1)} \frac{1 - (\gamma g/i\alpha_g)}{1 + (\gamma_g/i\alpha_g)} e^{-2\gamma_g(x_2 - x_1)}\right|$$

$$\approx \tfrac{1}{4}|A|^2 \frac{\beta_g \delta^2}{\gamma_g^2 \alpha_g^2}(1 + \gamma_g x_1) e^{\gamma_g(x_2 - x_1)}[(\beta - \beta_g')^2 + \Gamma^2]^{1/2} \qquad (E14)$$

where Γ and β_g' are given by Eqs. (12.22)–(12.24). Equating the RHS of the above equation to the RHS of Eq. (12.7) we get

$$|A| \approx \frac{4\gamma_g^2 \alpha_g^2 e^{-\gamma_g(x_2 - x_1)}}{\delta^2 (2\pi\alpha_g \beta_g)^{\frac{1}{2}}(1 + \gamma_g x_1)[(\beta - \beta_g')^2 + \Gamma^2]^{\frac{1}{2}}} \qquad (E15)$$

Using Eq. (12.14) and simplifying we get

$$|\phi(\beta)|^2 = \left|\frac{A}{A_g}\right|^2 \approx \frac{\Gamma}{\pi} \frac{1}{(\beta - \beta_g')^2 + \Gamma^2} \qquad (E16)$$

Thus Eq. (E5) becomes

$$W(z) \approx \left|\frac{\Gamma}{\pi} \int \frac{e^{i\beta z}\, d\beta}{(\beta - \beta_g')^2 + \Gamma^2}\right|^2 \qquad (E17)$$

We may evaluate the integral from $-\infty$ to $+\infty$ since most of the contribution will come from the region near resonance ($\beta \approx \beta_g$). We introduce the variable

$$\xi = \beta - \beta_g' \qquad (E18)$$

to write

$$W(z) \approx \left|\frac{\Gamma}{\pi} \int_{-\infty}^{+\infty} \frac{e^{i\xi z}}{(\xi + i\Gamma)(\xi - i\Gamma)}\right|^2 \qquad (E19)$$

For $z > 0$, the integral may be evaluated by using complex variable techniques and Jordan's lemma. In the complex ξ-plane we choose a contour which consists of the real axis and a semi-circle in the upper half plane where the integral vanishes. There is a simple pole within the contour at

$$\xi = i\Gamma$$

so that

$$W(z) \approx \left|\frac{\Gamma}{2\pi} 2\pi i \frac{e^{-\Gamma z}}{2i\Gamma}\right|^2 = e^{-2\Gamma z} \qquad (E20)$$

which shows how the power inside the 'core' decays exponentially. Further use of Eqs. (E14) and (12.22) gives us Eq. (12.27).

Orthonormalization of radiation modes

We first note that

$$\int_0^\infty \psi_{\beta'}^*(x)\psi_\beta(x)dx = 0 \quad \text{if} \quad \beta' \neq \beta \tag{E21}$$

One can understand the above condition by putting a fictitious boundary at $x = x_3$ in Fig. 12.3(a), i.e.,

$$n(x) = \begin{cases} n_1; & 0 < x < x_1 \\ n_2; & x_1 < x < x_2 \\ n_1; & x_2 < x < x_3 \\ n_2; & x > x_3 \end{cases} \tag{E22}$$

For $n_2 < \beta/k_0 < n_1$, the modes are now discrete and therefore should be orthogonal (see Problem 11.7). We now let $x_3 \to \infty$ in which case the modes become very closely spaced but retain their orthogonality. In the limit of $x_3 \to \infty$, we get Eq. (E21).

In order to consider the case $\beta = \beta'$, we start with the following integral:

$$\int_0^\infty \psi_{\beta'}^*(x)\psi_\beta(x)dx = \int_0^{x_2} \psi_{\beta'}^*(x)\psi_\beta(x)dx + \int_{x_2}^\infty \psi_{\beta'}^*(x)\psi_\beta(x)dx \tag{E23}$$

Now

$$I = \int_{x_2}^\infty \psi_{\beta'}^*(x)\psi_\beta(x)dx$$

$$= \int_0^\infty [D_-e^{-i\alpha'\xi} + D_+e^{i\alpha'\xi}][D_+e^{i\alpha\xi} + D_-e^{-i\alpha\xi}]d\xi \tag{E24}$$

where $\xi = x - x_2$ and we have used Eq. (12.2c) with the fact that $D_+ = D_-^*$. Thus

$$I = D_-D_+\left[\int_0^\infty e^{i(\alpha - \alpha')\xi}d\xi + \int_0^\infty e^{-i(\alpha - \alpha')\xi}d\xi\right]$$

$$+ D_+^2\int_0^\infty e^{i(\alpha + \alpha')\xi}d\xi + D_-^2\int_0^\infty e^{-i(\alpha + \alpha')\xi}d\xi \tag{E25}$$

It is well known that

$$\delta(x) = \frac{1}{\pi}\,\underset{g \to \infty}{\mathscr{L}t}\,\frac{\sin gx}{x} \tag{E26}$$

Thus

$$\int_0^\infty e^{i(\alpha+\alpha')\xi}\,d\xi = \underset{g\to\infty}{\mathscr{L}t}\int_0^g e^{i(\alpha+\alpha')\xi}\,d\xi$$

$$= \underset{g\to\infty}{\mathscr{L}t}\, 2e^{i(\alpha+\alpha')g/2}\frac{\sin\left[(\alpha+\alpha')g/2\right]}{(\alpha+\alpha')}$$

$$= 0 \tag{E27}$$

since $\alpha \neq -\alpha'$. Similarly the last integral in Eq. (E25) is also zero. Furthermore,

$$\int_0^\infty e^{i(\alpha-\alpha')\xi}\,d\xi + \int_0^\infty e^{-i(\alpha-\alpha')\xi}\,d\xi = \int_{-\infty}^{+\infty} e^{i(\alpha-\alpha')\xi}\,d\xi = 2\pi\delta(\alpha-\alpha') \tag{E28}$$

Now

$$\delta(\beta'^2 - \beta^2) = \delta(\alpha^2 - \alpha'^2) = \delta[(\alpha-\alpha')(\alpha+\alpha')] = (1/2\alpha)\cdot\delta(\alpha-\alpha') \tag{E29}$$

and

$$\delta(\beta'^2 - \beta^2) = \delta[(\beta'-\beta)(\beta'+\beta)] = (1/2\beta)\delta(\beta'-\beta) \tag{E30}$$

Thus

$$\int_{x_2}^\infty \psi_{\beta'}^*(x)\psi_\beta(x)\,dx = |D_\pm|^2\, 2\pi(\alpha/\beta)\delta(\beta'-\beta) \tag{E31}$$

When $\beta = \beta'$, the contribution of the integral from 0 to x_2 can be neglected and we may therefore write

$$\int_0^\infty \psi_{\beta'}^*(x)\psi_\beta(x) = (2\pi\alpha/\beta)|D_\pm|^2\,\delta(\beta'-\beta) \tag{E32}$$

Thus if $|D_\pm| = (\beta/2\pi\alpha)^{\frac{1}{2}}$, the modes satisfy the orthonormality condition.

F

WKB analysis of multimode fibres

The WKB method is applicable to multimode fibres which have profiles such that the wave equation is separable to one-dimensional equations and when the variation of the refractive index is small in distances $\sim \lambda$. In this appendix we will use the WKB method to study the propagation characteristics of multimode fibres and determine the time taken by the various modes to propagate through a certain distance of the fibre. We will follow the analysis of Gloge and Marcatili (1973).

The propagation constants

We may recall that in Sec. 14.4 we used the WKB method to study the propagation characteristics of slab waveguides for which the scalar wave equation is given by

$$\mathrm{d}^2\psi/\mathrm{d}x^2 + [k_0^2 n^2(x) - \beta^2]\psi(x) = 0 \tag{F1}$$

The propagation constants β_m are determined from the relation

$$\int_{x_1}^{x_2} [k_0^2 n^2(x) - \beta_m^2]^{\frac{1}{2}} \, \mathrm{d}x = (m + \tfrac{1}{2})\pi; \quad m = 0, 1, 2, \ldots \tag{F2}$$

Now, for a cylindrically symmetric profile (i.e., for n depending on the cylindrical coordinate r only) the propagation constants are determined by solving the radial part of the scalar wave equation (see Eq. (13.21)) which, on making the transformation

$$r = \mathrm{e}^x \tag{F3}$$

takes the form

$$\mathrm{d}^2 R/\mathrm{d}x^2 + (\{[n^2(x)k_0^2 - \beta^2]\}\mathrm{e}^{2x} - l^2) \quad R = 0 \tag{F4}$$

Observe that even though r goes from 0 to ∞, x goes from $-\infty$ to $+\infty$.

Eq. (F4) resembles the one-dimensional wave equation and the quantization condition is

$$\int_{x_1}^{x_2} \{[n^2(x)k_0^2 - \beta^2]e^{2x} - l^2\}^{\frac{1}{2}}\,dx = (m + \tfrac{1}{2})\pi \tag{F5}$$

where x_1 and x_2 are the turning points where the quantity in brace-style brackets vanishes. Using the transformation in Eq. (F3), we can write Eq. (F5) as

$$\int_{r_1}^{r_2} [n^2(r)k_0^2 - \beta^2 - l^2/r^2]^{\frac{1}{2}}\,dr \approx m\pi \tag{F6}$$

where r_1 and r_2 are the turning points and we have assumed $m \gg 1$ so that on the RHS we replace $m + \tfrac{1}{2}$ by m; only then is it possible to obtain an analytical expression for the propagation constant for a power law profile. Further, the assumption of $l \gg 1$ and $m \gg 1$ implies that we are dealing with highly multimoded fibres.

For a smooth profile of the type shown in Fig. 13.14 the discrete values of β will lie between $k_0 n_1$ and $k_0 n_2$ i.e.,

$$k_0 n_1 > \beta > k_0 n_2 \tag{F7}$$

In Fig. F1 we have given qualitative plots of $k_0^2 n^2(r) - \beta^2$ and l^2/r^2; the points of intersection are the turning points r_1 and r_2. Now, for a given value of l, the number of modes will be equal to the maximum value of m (which we denote by m') and the value m' will correspond to the minimum

Fig. F1 Variation of $k_0^2 n^2(r) - \beta^2$ and l^2/r^2 versus r for a smooth profile. Since for guided modes $\beta^2 > k_0^2 n_2^2$, the two curves can intersect at most at two points r_1 and r_2 which are the turning points. [Adapted from Gloge and Marcatili (1973).]

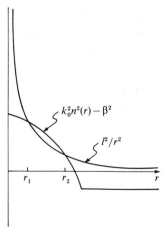

value of β^2:

$$m'(l) = (1/\pi) \int_{r_1}^{r_2} [k_0^2 n^2(r) - \beta_{\min}^2 - l^2/r^2]^{\frac{1}{2}} \, dr \tag{F8}$$

Obviously,

$$\beta_{\min} \approx k_0 n_2 \tag{F9}$$

The number of modes whose propagation constants are greater than a certain value of β (say equal to β') will be the value of m *corresponding to* β'. Thus if the number of modes is designated as $m'(l, \beta')$. Then

$$m'(l, \beta') \approx (1/\pi) \int_{r_1}^{r_2} [k_0^2 n^2(r) - \beta'^2 - l^2/r^2]^{\frac{1}{2}} \, dr \tag{F10}$$

The *total* number of modes (whose propagation constants are greater than β') will be given by

$$v(\beta') = 2m'(l=0, \beta') + 4m'(l=1, \beta') + 4m'(l=2, \beta') + \cdots + 4m'(l=l_{\max}, \beta') \tag{F11}$$

where $l_{\max}$ denotes the maximum value of l corresponding to a given value of β'. In writing the above equation, we have used the fact that the $l = 0$ mode is two fold degenerate and $l \geqslant 1$ modes are four fold degenerate. Replacing the sum in Eq. (F11) by an integral we get

$$v(\beta') \approx \frac{4}{\pi} \int_0^{l_{\max}} m'(l, \beta') \, dl$$

$$\approx \frac{4}{\pi} \int_0^{l_{\max}} \int_{r_1}^{r_2} [k_0^2 n^2(r) - \beta'^2 - l^2/r^2]^{\frac{1}{2}} \, dr \, dl \tag{F12}$$

In order to evaluate the above integral we interchange the order of integration. The domain of integration is shown in Fig. F2. Obviously, for a given value of r, l goes from 0 to $r(k_0^2 n^2 - \beta^2)^{\frac{1}{2}}$; further the value of r

Fig. F2 The domain of integration of the integral appearing in Eq. (F12). For a given l value the points of intersection with the curve give the turning points. [Adapted from Gloge and Marcatili (1973).]

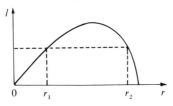

goes from 0 to r_{max} where

$$k_0^2 n^2(r_{max}) - \beta^2 = 0 \tag{F13}$$

Thus

$$v(\beta) \approx \frac{4}{\pi} \int_0^{r_{max}} \int_0^{r[k_0^2 n^2(r) - \beta^2]^{\frac{1}{2}}} [k_0^2 n^2(r) - \beta^2 - l^2/r^2]^{\frac{1}{2}} \, dl \, dr$$

The integration over l is very easy to carry out and gives

$$v(\beta) = \int_0^{r_{max}} [k_0^2 n^2(r) - \beta^2] r \, dr \tag{F14}$$

We now consider the power law profile (see Eq. (13.117)) so that

$$v(\beta) \approx \int_0^{r_{max}} [(k_0^2 n_1^2 - \beta^2) - k_0^2 n_1^2 2\Delta(r/a)^q] r \, dr$$

with

$$r_{max} = a \left[\frac{k_0^2 n_1^2 - \beta^2}{2\Delta k_0^2 n_1^2} \right]^{1/q} \tag{F15}$$

On evaluating the integral we obtain

$$v(\beta) = a^2 k_0^2 n_1^2 \Delta \frac{q}{q+2} \left(\frac{k_0^2 n_1^2 - \beta^2}{2\Delta k_0^2 n_1^2} \right)^{(q+2)/q} \tag{F16}$$

Since the minimum value of β is $k_0 n_2$, the *total* number of guided modes will approximately be

$$N = a^2 k_0^2 n_1^2 \Delta \frac{q}{q+2} = \frac{1}{2} \frac{q}{q+2} V^2 \tag{F17}$$

where we have used Eq. (13.120) and the fact that

$$\frac{k_0^2 n_1^2 - \beta_{min}^2}{2 k_0^2 n_1^2 \Delta} = \frac{k_0^2 (n_1^2 - n_2^2)}{2 k_0^2 n_1^2 \Delta} = 1$$

For a typical multimode graded index fibre we have $q \approx 2$, $V \approx 50$ and the total number of guided modes will be approximately 600. Eq. (F17) also tells us that for a given value of the waveguide parameter V, the total number of guided modes in a step index ($q = \infty$) fibre is twice as many as in a parabolic index ($q = 2$) fibre.

Now, if we label the propagation constants as $\beta_1, \beta_2, \ldots$ (β_1 corresponding to the maximum value of β) then Eq. (F16) gives us

$$\beta_v = k\left[1 - 2\Delta\left(\frac{q+2}{q}\frac{v}{a^2 k^2 \Delta}\right)^{q/(q+2)}\right]^{\frac{1}{2}} = k\left[1 - 2\Delta\left(\frac{v}{N}\right)^{q/(q+2)}\right]^{\frac{1}{2}}$$

(F18)

where $k \equiv k_0 n_1$. We should mention here that the label v stands for the composite pair (l, m). The above equation can be rewritten in the form

$$\beta_v^2 = k^2 - \Gamma\Delta^{2/(q+2)}k^{4/(q+2)} = k^2 - 2k^2\delta$$

(F19)

where

$$\Gamma = 2\left[\frac{q+2}{q}\frac{v}{a^2}\right]^{q/(q+2)}, \quad \delta = \Delta\left(\frac{v}{N}\right)^{q/(q+2)} = \frac{1}{2k^2}(\Gamma\Delta^{2/(q+2)}k^{4/(q+2)})$$

(F20)

Since $v < N$, the value of δ lies between 0 and Δ

$$0 < \delta < \Delta$$

(F21)

Group velocity and group delay per unit length
In order to evaluate the group velocity, we evaluate $d\beta/dk$:

$$2\beta_v\frac{d\beta_v}{dk} = 2k - \frac{4}{(q+2)k}(\Gamma\Delta^{2/(q+2)}k^{4/(q+2)})$$

$$- \frac{2}{q+2}\frac{1}{\Delta}\frac{d\Delta}{dk}(\Gamma\Delta^{2/(q+2)}k^{4/(q+2)})$$

or

$$\frac{d\beta_v}{dk} = (1 - 2\delta)^{-\frac{1}{2}}\left(1 - \frac{4}{q+2}\delta - \frac{\epsilon}{q+2}\delta\right)$$

(F22)

where

$$\epsilon = \frac{2k}{\Delta}\frac{d\Delta}{dk} = -\frac{2n_1}{N_1}\left(\frac{\lambda_0\Delta'}{\Delta}\right)$$

(F23)

$$N_1 = n_1 - \lambda_0 dn_1/d\lambda_0 = n_1 - \lambda_0 n_1'$$

(F24)

and primes denote differentiation with respect to the free space wavelength λ_0. In writing the last step of Eq. (F23), we have made use of the relation

$$\frac{dk}{d\lambda_0} = \frac{d}{d\lambda_0}\left(n_1\frac{2\pi}{\lambda_0}\right) = -\frac{2\pi}{\lambda_0^2}[n_1 - \lambda_0 n_1'] = -\frac{2\pi N_1}{\lambda_0^2}$$

(F25)

Thus

$$\frac{k}{\Delta}\frac{d\Delta}{dk} = n_1 \frac{2\pi}{\lambda_0}\frac{1}{\Delta}\left(\frac{d\Delta}{d\lambda_0}\right)\left(\frac{dk}{d\lambda_0}\right)^{-1} = -\frac{n_1}{N_1}\left(\frac{\lambda_0\Delta'}{\Delta}\right)$$

Making the binomial expansion in Eq. (F22) and retaining terms up to $O(\Delta^2)$ we get

$$\frac{d\beta_v}{dk} \approx \left[1 - \frac{q-2-\epsilon}{q+2}\delta + \frac{3q-2-2\epsilon}{2(q+2)}\delta^2\right] + O(\delta^3) \qquad \text{(F26)}$$

Now the group delay per unit length (which is the inverse of the group velocity) is given by

$$\tau_v = \frac{1}{v_v} = \frac{d\beta_v}{d\omega} = -\frac{\lambda_0^2}{2\pi c}\frac{d\beta_v}{d\lambda_0}$$

$$= -\frac{\lambda_0^2}{2\pi c}\frac{d\beta_v}{dk}\frac{dk}{d\lambda_0} = \frac{N_1}{c}\frac{d\beta_v}{dk} \qquad \text{(F27)}$$

Thus the time taken for the vth mode to propagate through a distance z of the fibre will be given by

$$t_v = \frac{z}{v_v} = \frac{N_1 z}{c}\left[1 + \frac{q-2-\epsilon}{q+2}\delta + \frac{3q-2-2\epsilon}{2(q+2)}\delta^2\right] + O(\delta^3)$$

$$\text{(F28)}$$

The above equation is in a form identical to the one given by Olshansky and Keck (1976).

G

Coupled mode equations

In this appendix we will derive the coupled mode equations which describe the variation in the amplitude of the waves propagating in each individual waveguide of a directional coupler. Let $n_1(x, y)$ and $n_2(x, y)$ represent the refractive index variation in the transverse plane of waveguide 1 in the absence of waveguide 2 and that of waveguide 2 in the absence of waveguide 1. Let $n(x, y)$ represent the refractive index variation of the directional coupler consisting of the waveguides 1 and 2. For example, for a directional coupler consisting of two step-index planer waveguides, $n_1(x), n_2(x)$ and $n(x)$ are shown in Fig. G1.

If β_1 and β_2 represent the propagation constants of the modes of

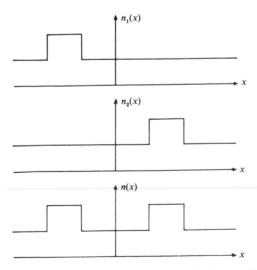

Fig. G1. (a) and (b) The refractive index profiles corresponding to two isolated step index planar waveguides and (c) The refractive index profile corresponding to a directional coupler formed by the two waveguides.

waveguides 1 and 2 in the absence of the other then we may write

$$\nabla_t^2 \psi_1 + [k_0^2 n_1^2(x, y) - \beta_1^2]\psi_1 = 0 \tag{G1}$$

$$\nabla_t^2 \psi_2 + [k_0^2 n_2^2(x, y) - \beta_2^2]\psi_2 = 0 \tag{G2}$$

where

$$\nabla_t^2 = \nabla^2 - \frac{\partial^2}{\partial z^2} = \frac{\partial^2}{\partial x^2} + \frac{\partial^2}{\partial y^2} \tag{G3}$$

and $\psi_1(x, y)$ and $\psi_2(x, y)$ represent the transverse mode field patterns of waveguides 1 and 2 respectively in the absence of the other.

If $\Psi(x, y, z)$ represents the total field of the directional coupler structure then we have

$$\nabla_t^2 \Psi + \frac{\partial^2 \Psi}{\partial z^2} + k_0^2 n^2(x, y)\Psi = 0 \tag{G4}$$

We now approximate Ψ as follows

$$\Psi(x, y, z) = A(z)\psi_1(x, y)e^{-i\beta_1 z} + B(z)\psi_2(x, y)e^{-i\beta_2 z} \tag{G5}$$

which is valid when the two waveguides are not very strongly interacting. In Eq. (G5) we have written the total field as a superposition of the fields in the first and second waveguides with amplitudes $A(z)$ and $B(z)$ which are functions of z. For infinite separation between the two waveguides, obviously the waveguides are noninteracting and A and B would then be independent of z. The coupling between the two waveguides leads to z dependent amplitudes. Substituting for Ψ in Eq. (G4) we obtain

$$Ae^{-i\beta_1 z}(\nabla_t^2 \psi_1 - \beta_1^2 \psi_1 + k_0^2 n^2 \psi_1) + Be^{-i\beta_2 z}(\nabla_t^2 \psi_2 - \beta_2^2 \psi_2 + k_0^2 n^2 \psi_2)$$
$$- 2i\beta_1(dA/dz)\psi_1 e^{-i\beta_1 z} - 2i\beta_2(dB/dz)\psi_2 e^{-i\beta_2 z} = 0 \tag{G6}$$

where we have neglected terms proportional to $d^2 A/dz^2$ and $d^2 B/dz^2$ which is justified when $A(z)$ and $B(z)$ are slowly varying functions of z. Using Eqs. (G1) and (G2), Eq. (G6) becomes

$$k_0^2 \Delta n_1^2 A\psi_1 + k_0^2 \Delta n_2^2 B\psi_2 e^{i\Delta\beta z} - 2i\beta_1(dA/dz)\psi_1$$
$$- 2i\beta_2(dB/dz)\psi_2 e^{i\Delta\beta z} = 0 \tag{G7}$$

where

$$\Delta n_1^2 = n^2(x, y) - n_1^2(x, y) \tag{G8}$$

$$\Delta n_2^2 = n^2(x, y) - n_2^2(x, y) \tag{G9}$$

$$\Delta\beta = \beta_1 - \beta_2 \tag{G10}$$

Multiplying Eq. (G7) by ψ_1^* and integrating over the whole cross section, we obtain

$$dA/dz = -i\kappa_{11}A(z) - i\kappa_{12}Be^{i\Delta\beta z} \tag{G11}$$

where

$$\kappa_{11} = \frac{k_0^2}{2\beta_1} \frac{\displaystyle\iint_{-\infty}^{\infty} \psi_1^* \Delta n_1^2 \psi_1 \, dxdy}{\displaystyle\iint_{-\infty}^{\infty} \psi_1^* \psi_1 \, dxdy} \tag{G12}$$

$$\kappa_{12} = \frac{k_0^2}{2\beta_1} \frac{\displaystyle\iint_{-\infty}^{\infty} \psi_1^* \Delta n_2^2 \psi_2 \, dxdy}{\displaystyle\iint_{-\infty}^{\infty} \psi_1^* \psi_1 \, dxdy} \tag{G13}$$

In writing Eq. (G11) we have neglected the overlap integral of the modes i.e. we assume

$$\iint_{-\infty}^{\infty} \psi_1^* \psi_2 \, dxdy \ll \iint_{-\infty}^{\infty} \psi_1^* \psi_1 \, dxdy$$

which is valid for weak coupling between the waveguides.

    Similarly if we multiply Eq. (G7) by ψ_2^* and integrate we would obtain

$$dB/dz = -i\kappa_{22}B - i\kappa_{21}Ae^{-i\Delta\beta z} \tag{G14}$$

where

$$\kappa_{22} = \frac{k_0^2}{2\beta_2} \frac{\displaystyle\iint_{-\infty}^{\infty} \psi_2^* \Delta n_2^2 \psi_2 \, dxdy}{\displaystyle\iint_{-\infty}^{\infty} \psi_2^* \psi_2 \, dxdy} \tag{G15}$$

$$\kappa_{21} = \frac{k_0^2}{2\beta_2} \frac{\displaystyle\iint_{-\infty}^{\infty} \psi_2^* \Delta n_1^2 \psi_1 \, dxdy}{\displaystyle\iint_{-\infty}^{\infty} \psi_2^* \psi_2 \, dxdy} \tag{G16}$$

We can write Eqs. (G11) and (G14) in a different form if we define

$$a(z) = A(z)e^{-i\beta_1 z} \tag{G17}$$

$$b(z) = B(z)e^{-i\beta_2 z} \tag{G18}$$

Substituting from Eqs. (G17) and (G18) in Eqs. (G11) and (G14), we have

$$da/dz = -i(\beta_1 + \kappa_{11})a - i\kappa_{12}b \tag{G19}$$

$$db/dz = -i(\beta_2 + \kappa_{22})b - i\kappa_{21}a \tag{G20}$$

The above two equations represent the coupled mode equations. It follows from Eqs. (G19) and (G20) that κ_{11} and κ_{22} represent the corrections to the propagation constants of each individual waveguide mode due to the presence of the other waveguide. These correction factors are normally neglected in the analysis although one can very easily incorporate them. Thus the coupled equations may be written as

$$da/dz = -i\beta_1 a - i\kappa_{12}b \tag{G21}$$

$$db/dz = -i\beta_2 b - i\kappa_{21}a \tag{G22}$$

These are the coupled mode equations which have been used in Chapter 14.

References and suggested reading

Chapters 1–5

Born, M. and Wolf, E. (1975) *Principles of Optics*, Pergamon Press, Oxford.

Casey, H.C. and Panish, M.B. (1978) *Heterostructure Lasers*, Academic Press, New York.

Ghatak, A. (1977) *Optics*, Tata McGraw Hill, New Delhi.

Ghatak, A. and Thyagarajan, K. (1978) *Contemporary Optics*, Plenum Pub. Corp., New York.

Hartfield, E. and Thompson, B.J. (1978) Optical Modulators, in *Handbook of Optics* (Eds. W.J. Driscoll and W. Vaughan), McGraw Hill, New York.

Heavens, O.S. (1937) *Optical Properties of Thin Solid Films*, Butterworth, London.

Irving, J. and Mullineux, I. (1959) *Mathematics in Physics and Engineering*, Academic Press, New York.

Jenkins, F.A. and White, H.E. (1957) *Fundamentals of Optics*, McGraw Hill, New York.

Johnk, C.T.A. (1975) *Engineering electromagnetic fields and waves*, John Wiley, New York.

Kressel, H. and Butler, J.K. (1977) *Semiconductor Lasers and Heterojunction LEDs*, Academic Press, New York.

Lynch, D.W. and Hunter, W.R. (1985) Comments on the Optical constants of metals and an introduction to the data for several metals, in *Handbook of Optical Constants of Solids* (Ed. E. D. Palik) Academic Press, Orlando.

Nussbaum, A. and Philips, R.A. (1976) *Contemporary Optics for Scientists and Engineers*, Prentice Hall, Englewood Cliffs, New Jersey.

Panofsky, W.K.H. and Phillips, M. (1962) *Classical Electricity and Magnetism*, Addison-Wesley, Reading, Massachusetts.

Sommerfeld, A. (1964) *Optics*, Academic Press, New York.

Thyagarajan, K. and Ghatak, A.K. (1981) *Lasers: Theory and Applications*, Plenum Press, New York.

Wolfe, W. (1978) Properties of optical materials, in *Handbook of Optics* (Eds. W.J. Driscoll and W. Vaughan), McGraw Hill, New York.

Chapters 6, 7

Casasent, D. (Ed) (1978) *Optical Data Processing*, Springer Verlag, Berlin.

Cathey, W.T. (1974) *Optical Information Processing and Holography*, Wiley, New York.

Collier, R.J., Burckhardt, C.B., and Lin, L.H. (1971) *Optical Holography*, Academic Press, New York.

Ghatak, A. (1977) *Optics*, Tata McGraw Hill, New Delhi.

Ghatak, A. and Thyagarajan, K. (1978) *Contemporary Optics*, Plenum Pub. Corp., New York.

Goodman, J.W. (1968) *Introduction to Fourier Optics*, McGraw Hill, New York.

Lee, S.H. (Ed.) (1981) *Optical Information Processing*, Springer Verlag, Berlin.

Powell, R.L. and Stetson, K.A. (1965) Interferometric vibrational analysis by wavefront reconstruction, *J. Opt. Soc. Am.*. **55**, 1593

Stroke, G.W. (1969) *An Introduction to Coherent Optics and Holography*, Academic Press, New York.

Stroke, G.W., Halioua, M. and Srinivasan, V. (1975) 'Holographic image restoration using Fourier spectrum analysis of blurred photographs in computer aided synthesis of Wiener filters, *Phys. Letts* **51A**, 383.

Tsujiuchi, J., Matsuda, K. and Takeya, N. (1971) Correlation techniques by holography and its application to fingerprint identification, in *Applications of Holography* (Eds. E.S. Barrekette, W.E. Kock, T. Ose, J. Tsujiuchi and G.W. Stroke) p. 247, Plenum Press, New York.

Wuerker, R.F. (1971) Experimental aspects of holographic interferometry, in *Applications of Holography* (Eds. E.S. Barrekette, W.E. Kock, T. Ose, J. Tsujiuchi and G.W. Stroke), p. 127, Plenum Press, New York.

Chapters 8–10

Agarwal, G.P. and Dutta N.K. (1986) Distributed feedback InGaAs P lasers, *J. Inst. Electronics Telecom. Engs. (India)* (special issue on opto electronics and optical communication) **32**, 187.

Davis, L.W. (1968) Effects of transverse-mode selection on the wavefront of a ruby laser, *J. Appl. Phys.* **39**, 5331.

Demtroder, W. (1981) *Laser Spectroscopy*, Springer Series in Chemical Physics, Vol. 5, Springer Verlag, Berlin.

Fox, A.G. and Li, T. (1961) Resonant modes in a maser interferometer, *Bell Syst. Tech. J.* **40**, 453.

Fox, A.G., Schwarz, S.E., and Smith, P.W. (1968) Use of neon as a non-linear absorber for mode locking a He–Ne laser, *Appl. Phys. Letts* **12**, 371.

Ghatak, A and Thyagarajan, K. (1978) *Contemporary Optics*, Plenum Press, New York.

Gopal, E.S.R. (1974) *Statistical Mechanics and Properties of Matter*, Wiley, New York.

Gordon, J.P., Zeiger, H.J. and Townes, C.H. (1955) The maser – new type of microwave amplifier, frequency standard and spectrometer, *Phys. Rev.* **99**, 1264.

Hecht, J. (1986) *The Laser Guidebook*, McGraw Hill, New York.

Jacobs, S.F. (1979) How monochromatic is laser light, *Am. J. Phys.* **47**, 597.

Koechner, W. (1976) *Solid State Laser Engineering*, Springer Verlag, New York.

Kogelnik, H. and Li, T. (1966) Laser beam and resonators, *Appl. Opt.* **5**, 1550.

Kressel, H. (Ed). (1982) *Semiconductor Devices for Optical Communications*, Vol. 39, Topics in Applied Physics, Springer Verlag, Berlin.

Kressel, H. and Bulter, J.K. (1977) *Semiconductor Lasers and Heterojunction LEDs*, Academic Press, New York.

Maitland, A. and Dunn, M.H. (1969) *Laser Physics*, North Holland, Amsterdam.

Mooradian, A. (1985) Laser linewidth, *Physcis Today* **38**, 43.

Schawlow, A.L., and Townes, C.H. (1958) Infrared and optical masers, *Phys. Rev.* **112**, 1940.

Seigman, A.E. (1986) *Lasers*, Oxford University Press, Oxford.

Shapiro, S.L. (1977) *Ultrashort Light Pulses: Picosecond Techniques and Applications*, Springer Verlag, Berlin.

Slepian, D. and Pollack, H.O. (1961) Prolate spheroidal wave functions – Fourier analysis and uncertainty -I, *Bell Syst. Tech. J.* **40**, 43.

Smith, P.W. (1972) Mode selection in lasers, *Proc. IEEE* **60**, 422.

Snavely, B.B. (1969) Flashlamp excited organic dye lasers, *Proc. IEEE* **57**, 1374.

Suematsu, Y., Kishino, K., Arai, S. and Koyama, F. (1985) Dynamic single mode semiconductor laser with a distributed reflector, in *Semiconductors and Semimetals*, Vol 22, Part B (Ed. W.T. Tsang) Academic Press.

Thyagarajan, K and Ghatak, A.K. (1981) *Lasers: Theory and Applications*, Plenum Press, New York.

Yariv A. (1985) *Optical Electronics*, Holt Rinehart and Winston, New York.

Chapters 11–14

Adams, M.J. (1981) *An Introduction to Optical Waveguides*, John Wiley, Chichester.

Alferness, R.C. (1981) Guided wave devices for optical communication, *IEEE J. Quant. Electron.* **QE-17**, 946.

Alferness R.C. and Schmidt R.V. (1978) Tunable optical waveguide directional coupler filter, *Appl. Phys. Letts.* **33**, 161.

Alferness, R.C., Schmidt, R.V. and Turner, E.H. (1979) Characteristics of Ti diffused LiNbO$_3$ optical directional couplers, *Appl. Opt.* **18**, 4012.

Anderson, D.B. (1978) Integrated optical spectrum analyser: an imminent chip, *IEEE spectrum* **15** (12), 22.

Arnaud, J.A. (1976) *Beam and Fiber Optics*, Academic Press, New York.

Blank, L.C., Bickers, L. and Walker, S.D. (1985) Long span optical transmission experiments at 34 and 140 Mbit/s, *J. Lightwave Technol* **LT-3**, 1017.

Cohen, L.G. (1985) Comparison of single mode fiber dispersion measurement techniques, *J. Lightwave Tech.* **LT-3**, 958.

Conwell, E.M. (1973) Modes in optical waveguides formed by diffusion, *Appl. Phys. Lett.* **23**, 328.

Gedeon A. (1974) Comparison between rigorous theory and WKB analysis of modes in graded-index waveguides, *Opt. Commun* **12**, 329.

Ghatak, A.K. (1985) Leaky modes in optical waveguides, *Opt. Quant. Electron.* **17**, 311.

Ghatak, A.K., Khular, E. and Thyagarajan, K. (1978) Modes in optical waveguides formed by silver-sodium ion exchange, *IEEE J. Quant. Electron.* **QE-14**, 389.

Ghatak, A.K. and Lokanathan, S. (1984) *Quantum Mechanics: Theory and Applications*, Macmillan India, New Delhi.

Ghatak, A.K. and Sharma, A. (1986) Single mode fiber characteristics, *J. Inst. Electronics and Telecom. Engrs (India)* **32**, 213.

Ghatak, A.K. and Thyagarajan K. (1975) Ray and energy propagation in graded index media, *J. Opt. Soc. Am.* **65**, 169.

Ghatak, A.K. and Thyagarajan K. (1978) *Contemporary Optics*, Plenum Press New York.

Ghatak, A.K. and Thyagarajan K. (1980) Graded index optical wave guides; a review, *Progress in Optics* (Ed. E. Wolf) North Holland, Amsterdam, Vol XVIII, pp. 1–126.

Ghatak, A.K., Thyagarajan, K. and Shenoy, M.R. (1987) Numerical analysis of planar optical waveguides using matrix approach, *J. Lightwave Technol.* **LT-5**, 660.

Gloge, D. (1971) Weakly guiding fibers, *Appl. Opt.* **10**, 2252

Gloge, D. and Marcatili, E.A.J. (1973) Multimode theory of graded-core fibers, *Bell. Syst. Tech. J.* **52**, 1563.

Goell, J.E. (1969) A circular harmonic computer analysis of rectangular dielectric waveguides, *Bell Syst. Tech. J.* **48**, 2133.

Goyal, I.C. (1986) Dispersion in telecommunication optical fibers. *J. Inst. Electronics Telecom. Engrs* **32**, 196.

Gradshtein, I.S. and Ryzhik, I.M. (1965) *Tables of integrals, series and products*, Academic Press, New York.

Grau, G. (1978) *Quantenelektronix*, Vieweg Verlag, Braunschweig.

Grau, G. (1986) *Optische Nachrichtentechnik*, Springer Verlag, Berlin.

Henry, P.S. (1985) Introduction to lightwave transmission, *IEEE Communications Magazine* **23**, 12.

Hocker, G.B. and Burns W.K. (1975) Modes in diffused optical waveguide of arbitrary index profile *IEEE J. Quant. Electron.* **QE-11**, 270

Irving, J. and Mullineux, I. (1959) *Mathematics in Physics and Engineering*, Academic Press, New York.

Ito, H., Ogawa, Y, and Inaba, H., (1979) Integrated bistable optical device using Mach–Zehnder interferometric optical waveguide, *Electronics Letts.* **15**, 283.

Janta, J. and Ctyroky, J. (1978) On the accuracy of WKB analysis of TE and TM modes in planar graded index waveguides, *Opt. Commun.* **25**, 49.

Jeunhomme, L.B. (1983) *Single mode fiber optics*, Marcel Dekker, New York.

Kaminow, I.P., Stulz, L.W. and Turner, E.H. (1975) Efficient strip–waveguide modulator, *Appl. Phys. Letts.* **27**, 555.

Kimura, T. (1979) Single-mode systems and components for longer wavelengths, *IEEE Trans. Circuits and Systems* **CAS-26**, 987.

Kimura, T. (1980) Single mode digital transmission technology, *Proc. IEEE* **68**, 1263.

King, G.D.H. and Jackson, J.D. (1980) An integrated electro optic analog to digital converter, *Proc. 6th Europ. Conf. Opt. Comm.*, IEE Press, London, p. 256.

Kogelnik, H. and Schmidt, R.V. (1976) Switched directional couplers with alternating $\Delta\beta$, *IEEE J. Quantum Electron* **QE-12**, 396.

Korotky, S.K., Marcatili, E.A.J., Vaselka, J.J. and Bosworth, R.H. (1986) Greatly reduced losses for small radius bends in Ti: LiNbO$_3$ waveguides, *Appl. Phys. Letts.* **48**, 92.

Kumar, A., Thyagarajan, K. and Ghatak, A.K. (1983) Analysis of rectangular core waveguides: An accurate perturbation approach, *Opt. Letts.* **8**. 63.

Leonberger, F.J. (1980) High speed operation of LiNbO$_3$ electro optic interferometric waveguide modulators, *Opt. Lett.* **5**, 312.

Leonberger, F.J. (1980) High speed operation of LiNbO$_3$ electro optic interferometric high speed electro optic A/D converter, *IEEE Trans. Circuits and Systems* **CAS-26**, 1125.

L' Optique Guidee Monomode (1985) by a group of engineers from Thomson-CSF, Masson, Paris.

Love, J. and Ghatak, A. (1979) Exact solutions for TM modes in graded index slab waveguides, *IEEE J. Quantum Electron.* **QE-15**, 14.

Marcatili, E.A.J. (1969) Dielectric rectangular waveguide and directional coupler for integrated optics, *Bell Syst. Tech. J.* **48**, 2071.

Marcuse, D. (1972) *Theory of Dielectric Optical Waveguides*, Academic Press, New York.

Marcuse D. (Ed.) (1973) *Integrated Optics*, IEEE Press, New York.

Marcuse, D. (1977) Loss analysis of single mode fibre splices, *Bell Syst. Tech. J.* **56**, 703.

Marcuse, D. (1978) Gaussian approximation of the fundamental modes of graded index fibers, *J. Opt. Soc. Am.* **68**, 103.

Marcuse, D. (1979) Interdependence of waveguide and material dispersion, *Appl. Opt.* **18**, 2930.

Midwinter, J. (1979) *Optical Fibers for Transmission,* John Wiley, New York.

Miya, T., Terunama, Y., Hosaka, T. and Miyashita, T. (1979) An ultimate low-loss single mode fiber at 1.55 μm, *Electron, Lett.* **15**, 106.

Olshansky, R. and Keck, D.B. (1976) Pulse broadening in graded-index optical fibers, *Appl. Opt.* **15**, 483.

Olsson, N.A., Hegartz, J., Logen, R.A., Johnson, L.F., Walker, K.L., Cohen, L.G., Kasper, B.L. and Campbell, J.C. (1985) 68.3 km transmission with 1.37 Tbit km/s capacity using wavelength division multiplexing of ten single frequency lasers at 1.5 μm, *Electron Letts.* **21** 105.

Paek, U.C., Peterson, G.E. and Carnevale, A. (1981) Dispersionless single mode light guides with α-index profiles, *Bell Syst. Tech. J.* **60**, 583.

Pal, B.P. (1979) Optical communication fiber waveguide fabrication: a review, *Fiber and Integrated Optics* **2**, 195.

Papuchon, M. (1978) *Utilization des coupleurs directionelles pour la commutation en optique integree*, Ph.D. thesis University of Nice.

Papuchon, M., Combemale, Y., Mathieu, X., Ostrowsky, D.B., Reiber, L., Roy, A.M., Sejourne, B. and Werner, M. (1975) Electrically switched optical directional coupler: *COBRA, Appl. Phys. Lett.* **27**, 289.

Papuchon, M. and Puech, C. (1977) L' optique integree: De nouvelles possibilities pur la modulation et la commutation de la lumiere, *Entropie* 13.

Rudolph, H.D. and Neumann, E.G. (1976) Approximations for the eigen values of the fundamental mode of a step index glass fiber waveguide, *Nachrichtentech. Z.* **4**, 328.

Sammut, R.A. (1979) Analysis of approximations for the mode dispersion in monomode fibers, *Electron. Lett.* **15**, 590.

Schmidt, R.V. and Cross, P.S. (1978) Efficient optical waveguide switch/modulator, *Optics Letts.* **2**, 45.

Schnapper, A., Papuchon, M., and Puech, C. (1979) Optical bistability using an integrated two arm interferometer, *Optics Commun.* **29**, 364.

Sharma, A. (1988) On constructing linear combinations of LP modes to obtain zeroeth order vector modes of optical fibers, *Appl. Opt.*, **27**, 2647.

Snitzer, E. (1961) Cylindrical dielectric waveguide modes, *J. Opt. Soc. Am.* **51**, 491.

Snyder, A.W. and Love, J.D. (1983) *Optical Waveguide Theory*, Chapman and Hall, London.

Sodha, M.S. and Ghatak, A.K. (1977) *Inhomogeneous Optical Waveguides*, Plenum Press, New York.

Stewart, G., Millar, C.A., Laybourn, P.J.R., Wilkinson, C.D.W. and De la Rue, R.H. (1977) Planar optical waveguides formed by silver ion migration in glass, *IEEE J. Quantum Electron.* **QE-13**, 192.

Tamir, T. (Ed.) (1979) *Integrated Optics*, Springer Verlag, Berlin.

Taylor, H.F. (1973) Frequency selective coupling in parallel dielectric waveguides, *Optics Commun.* **8**, 421.

Thyagarajan, K., Diggavi, S., and Ghatak A.K. (1987) Analytical investigations of leaky and absorbing planar structures, *Opt. Quantum Electron* **19**, 131.

Thayagarajan, K., Shenoy, M.R., and Ghatak, A.K. (1987) Accurate numerical method for the calculation of bending loss in optical waveguides using a matrix approach, *Opt. Letters.* **12**, 296. Erratum: Optics Letters March, 1989.

Tien P.K. (1977) Integrated optics and new wave phenomena in optical waveguides, *Rev. Mod. Phys.* **49**, 361.

Verdeyen, J.T. (1981) Laser Electronics, Prentice Hall, Englewood Cliffs, New Jersey.

Chapters 15–20

Adler, R. (1967) Interaction between light and sound, *IEEE Spectrum* May, 12.

Anderson, D.B. (1978), Integrated optic spectrum analyser: an imminent chip, *IEEE Spectrum* Dec issue, p. 22.

Akhmanov, S.A., Sukhorukov, A.P. and Khokhlov, R.V. (1968) Self-focussing and diffraction of light in a non-linear medium, *Sov. Fiz. Usp.* **10**, 609.

Arfken, G. (1970) *Mathematical Methods for Physicists*, Academic Press, New York.

Ashkin, A., Boyd, G.D. and Dziedic, J.M. (1963) Observation of continuous optical harmonic generation with gas masers, *Phys. Rev. Letts.* **11**, 14.

Biazzo, M.R. (1971) Fabrication of a lithium tantalate temperature stablized optical modulator, *Appl. Opt.* **10**, 1016.

Bloembergen, N. (1965) *Nonlinear Optics*, W.A. Benjamin Inc., New York.

Casasent, D. (1981) A review of optical signal processing, *IEEE Comm. Magazine* Sept., 40.

Denton, R.T. (1972) Modulation techniques, in *Laser Handbook*, Vol. I (Eds. F.T. Arecchi and E.O. Schulz-Dubois) North Holland, Amsterdam.

Denton, R.T., Chen, F.S. and Ballman, A.A., (1967) Lithium tantalate light modulators, *J. Appl. Phys.* **38**, 1611.

Franken, P.A., Hill, A.E., Peters, C.W. and Weinreich, G. (1961) Generation of optical harmonics, *Phys. Rev. Lett.* **7**, 118.

Garmire, E., Chiao, R.V., and Townes, C.H. (1966) Dynamics of characteristics of self-trapping of intense light beams, *Phys. Rev. Lett.* **16** 347.

Gaylord, T.K. and Moharam, M.G. (1985) Analysis and applications of optical diffraction gratings, *Proc. IEEE* **73**, 894.

Ghatak, A. and Thyagarajan, K. (1978) *Contemporary Optics*, Plenum Press, New York.

Gottlieb, M., Ireland, C.L.M. and Ley, J.M. (1983) *Electro Optic and Acousto Optic Scanning and Deflection*, Marcel Dekker, New York.

Hartfield, E. and Thompson, B.J. (1978) Optical modulators, in *Handbook of Optics* (Eds. W.G. Driscoll and W. Vaughan) McGraw Hill, New York.

Kaminow, I.P. (1963) Splitting of Fabry-Perot rings by microwave modulation of light, *Appl. Phys. Letts.* **2**, 41.

Kaminow, I.P. (1974) *An Introduction to Electro Optic Devices*, Academic Press, New York.

Kaminow, I.P. and Liu, J. (1963) Propagation characteristics of partially loaded two conductor transmission line for broadband light modulators., *Proc. IEEE* **51**, 132.

Korpel, A. (1980) Acousto optics, in *Applied Optics and Optical Engineering Vol VI* (Eds. R. Kingslake and B.J. Thompson) Academic Press, New York.

Korpel, A, Adler, R, Demares, P. and Watson, W. (1966) A television display using acoustic deflection and modulation of coherent light, *Proc. IEEE* **54**, 1429.

Kurtz, S.K. (1972) Nonlinear optical materials, in *Laser Handbook*, Vol. I (Eds: FT Arecchi and E.O. Schulz-Dubois) North Holland, Amsterdam.

Maydan, D. (1970) Acoustooptical pulse modulators, *IEEE J. Quant. Electron.* **QE-6**, 15.

Panofsky, W.K.H. and Phillips, M. (1962) *Classical Electricity and Magnetism*, Addison Wesley, Reading, Massachusetts.

Peters, C.J. (1963) Gigacycle bandwidth coherent light travelling wave phase modulator, *Proc. IEEE* **51**, 147.

Peters, C.J. (1965) Gigacycle bandwidth coherent light travelling wave amplitude modulator, *Proc. IEEE* **53**, 455.

Pinnow, D.A. (1972) Elasto Optic Materials, in *Laser Handbook* Vol. I (Eds: FT Arecchi and E.O. Schulz-Dubois) North Holland, Amsterdam.

Sodha, M.S. (1973), Theory of non-linear refraction: Self focussing of laser beams, *J. Phys. Educ. (India)* **1**, 3.

Sodha, M.S., Ghatak, A.K. and Tripathi, V.K. (1974) *Self-focusing of Laser Beams in Dielectrics, Plasmas and Semiconductors*, Tata McGraw Hill, New Delhi.

Sodha, M.S., Ghatak, A.K., and Tripathi, V.K. (1976) Self-focusing of laser beams in plasmas and semiconductors (a review), in *Progress in Optics*, Vol. XIII (E. Wolf, ed.), p. 169, North Holland, Amsterdam

Solymar, L. and Cooke, D.J. (1981) *Volume Holography and Volume Gratings*, Academic Press, London.

Spencer, E.G., Lenzo, P.V. and Ballman, A.A. (1967) Dielectric materials for electrooptic, elasto optic and ultrsonic device applications, *Proc. IEEE* **55**, 2074.

Suhara, T. and Nishihara, H. (1986) Integrated optics components and devices using periodic structures, *IEEE J. Quantum. Electron* **QE-22**, 845.

Svelto, O. (1974) Self-focusing, self-trapping and self-phase modulation of laser beams, in: *Progress in Optics*, Vol. XVII (E. Wolf, ed.) p. 1 North Holland, Amsterdam.

Takizawa, K. and Okada, M (1979) Analog to digital converter using electro optic light modulators, *Jap. J. Appl. Phys.* **18**, 1417.

Taylor, H.F. (1975) An electro optic analog to digital converter, *Proc. IEEE* **63**, 1524.

Taylor, H.F. (1979) An optical analog-to-digital converter – design and analysis, *IEEE J. Quant. Electron.* **QE-15**, 210.

Taylor, H.F., Taylor, M.J. and Bauer, P.J. (1978) Electrooptic analog to digital conversion using channel waveguide modulators, *Appl. Phys. Letts.* **32**, 559.

Wagner, W.G., Haus, H.A. and Marburger, J.H. (1968) Larger scale self trapping of optical beams in the paraxial ray approximation, *Phys. Rev.* **175**, 256.

Watson, G.N. (1958) *A Treatise on the Theory of Bessel Functions*, 2nd edn., Cambridge Univerisity Press, Cambridge.

Wemple, S.H. (1972) Electro-optic materials, in *Laser Handbook* Vol. I (Eds: FT Arecchi and E.O. Schulz-Dubois), North Holland, Amsterdam.

Yariv A. (1985) *Optical Electronics*, Holt Rinehart and Winston, New York.

Yariv A. and Yeh, P. (1984) *Optical Waves in Crystals* John Wiley, New York.

Young, E.H. and Yao, S. (1981), Design considerations for acousto optic devices, *Proc. IEEE* **69**, 54.

Zernike F. and Midwinter, J.E. (1973) *Applied Nonlinear Optics*, John Wiley, New York.

Appendices

Ghatak, A. (1977) *Optics*, Tata McGraw Hill, New Delhi.

Ghatak, A.K. and Lokanathan S. (1984) *Quantum Mechanics: Theory and Applications*, Macmillan, India, New Delhi.

Gloge, D. and Marcatili, E.A.J. (1973) Multimode theory of graded-core fibers, *Bell. Syst. Tech. J.* **52**, 1563

Olshansky, R. and Keck, D.B. (1976) Pulse broadening in gradex-index optical fibers, *Appl. Opt.* **15**, 483.

Index